12000

THERMODYNAMIC PROPERTIES OF HELIUM

NATIONAL STANDARD REFERENCE DATA SERVICE OF THE USSR: A Series of Property Tables

1. Thermodynamic Properties of Helium
2. Thermodynamic Properties of Nitrogen
3. Thermodynamic Properties of Methane
4. Thermodynamic Properties of Ethane
5. Thermodynamic Properties of Oxygen
6. Thermodynamic Properties of Air
7. Thermodynamic Properties of Ethylene
8. Thermophysical Properties of Freons, Part 1
9. Thermophysical Properties of Freons, Part 2
10. Thermophysical Properties of Neon, Argon, Krypton, and Xenon

In Preparation

Thermodynamic Properties of n-Hexane
Thermodynamic Properties of Propane
Thermodynamic Properties of Hydrogen

THERMODYNAMIC PROPERTIES OF HELIUM

V. V. Sychev
A. A. Vasserman
A. D. Kozlov
G. A. Spiridonov
V. A. Tsymarny

Theodore B. Selover, Jr.
English-Language Edition Editor

◉ HEMISPHERE PUBLISHING CORPORATION
A subsidiary of Harper & Row, Publishers, Inc.
Washington   New York   London

DISTRIBUTION OUTSIDE NORTH AMERICA
SPRINGER–VERLAG
Berlin   Heidelberg   New York   London   Paris   Tokyo

Thermodynamic Properties of Helium

Copyright © 1987 by Hemisphere Publishing Corporation. All rights reserved. Printed in the United States of America. Except as permitted under the United States Copyright Act of 1976, no part of this publication may be reproduced or distributed in any form or by any means, or stored in a data base or retrieval system, without the prior written permission of the publisher.

Originally published by Standards Publishers, Moscow, 1984 as Termodinamicheskiye Svoystva Geliya in the Monograph Series of the National Standard Reference Data Service of the USSR; State Committee on Standards of the Council of Ministers of the USSR.

English translation by G. E. Slark.

1 2 3 4 5 6 7 8 9 0   B C B C   8 9 8 7 6

This book was set in English Times by Hemisphere Publishing Corporation.
The cover designer was Sharon Martin DePass; the production supervisor was Peggy M. Rote; and the typesetter was Sandra F. Watts.
BookCrafters, Inc. was printer and binder.

Library of Congress Cataloging in Publication Data

Termodinamicheskie svoĭstva geliĭa. English.
  Thermodynamic properties of helium.

  (National standard reference data service of the USSR)
  Translation of: Termodinamicheskie svoĭstva geliĭa.
  1. Liquid helium—Thermal properties. I. Sychev,
V. V. (Vîacheslav Vladimirovich), date.
II. Selover, Theodore B., date. III. Series.
QC145.45.H4T4513 1987   546'.751569   86-22858
ISBN 0-89116-613-0 Hemisphere Publishing Corporation

DISTRIBUTION OUTSIDE NORTH AMERICA:
ISBN 3-540-17278-5 Springer-Verlag Berlin

CONTENTS

Preface to the Series   ix
Preface   xi
Foreword   xv

Part I

1 EXPERIMENTAL DATA ON THE THERMODYNAMIC PROPERTIES OF HELIUM-4   3

    1.1 Thermodynamic Properties on Phase Transition Lines   3
    1.2 Thermodynamic Properties in the Single-Phase Region   25

2 METHOD OF CONSTRUCTING A SINGLE EQUATION OF STATE AND COMPILATION OF THERMODYNAMIC PROPERTIES   47

3 EQUATIONS OF STATE AND TABLES OF THERMODYNAMIC FUNCTIONS OF HELIUM-4   53

    3.1 Thermodynamic Functions in the Ideal Gas State   53
    3.2 Equations for Calculating Thermodynamic Properties of He^4 54

| 3.3 | Estimated Reliability of Calculated Thermodynamic Functions | 68 |
| 3.4 | Comparison with Previously Published Data | 86 |

Part II

TABLES OF THERMODYNAMIC PROPERTIES OF HELIUM-4 131

Basic Numerical Values 131
Definitions of Symbols and Units 131

THERMODYNAMIC PROPERTIES OF LIQUID HELIUM-4 ON THE FREEZING CURVE (AT THE INDICATED TEMPERATURE) 133

Table II.1 $(T, p, \Phi, d\pi/d\tau, d^2\pi/d^2\tau)$ 133
Table II.2 $(T, \rho, h, s, c_v, c_p, c_m)$ 133
Table II.3 $(T, w, \mu, k, f, \alpha/\alpha_0, \gamma/\gamma_0)$ 134

THERMODYNAMIC PROPERTIES OF LIQUID HELIUM-4 ON THE FREEZING CURVE (AT THE INDICATED PRESSURE) 135

Table II.4 $(p, T, \Phi, d\pi/d\tau, d^2\pi/d\tau^2)$ 135
Table II.5 $(p, \rho, h, s, c_v, c_p, c_m)$ 136
Table II.6 $(p, w, \mu, k, f, \alpha/\alpha_0, \gamma/\gamma_0)$ 136

THERMODYNAMIC PROPERTIES OF HELIUM-4 ON THE SATURATION CURVE (AT THE INDICATED TEMPERATURE) 137

Table II.7 $(T, p, \Phi, r, d\pi/d\tau, d^2\pi/d\tau^2)$ 137
Table II.8 $(T, \rho', \rho'', h', h'', s', s'')$ 138
Table II.9 $(T, c_v', c_v'', c_p', c_s', c_s'')$ 139
Table II.10 $(T, w', w'', \mu', \mu'', k', k'')$ 139
Table II.11 $(T, f', f'', \alpha'/\alpha_0, \alpha''/\alpha_0, \gamma'/\gamma_0, \gamma''/\gamma_0)$ 140

THERMODYNAMIC PROPERTIES OF HELIUM-4 ON THE SATURATION CURVE (AT THE INDICATED PRESSURE) — 141

Table II.12 $(p, T, \Phi, r, d\pi/d\tau, d^2\pi/d\tau^2)$ — 141
Table II.13 $(p, \rho', \rho'', h', h'', s', s'')$ — 142
Table II.14 $(p, c_v', c_v'', c_p', c_p'', c_s', c_s'')$ — 142
Table II.15 $(p, w', w'', \mu', \mu'', k', k'')$ — 143
Table II.16 $(p, f', f'', \alpha'/\alpha_0, \alpha''/\alpha_0, \gamma'/\gamma_0, \gamma''/\gamma_0)$ — 143

THERMODYNAMIC PROPERTIES OF HELIUM-4 IN THE SINGLE-PHASE REGION — 145

Table II.17 $(p, \rho, z, h, s, c_v, c_p)$ — 145
Table II.18 $(p, w, \mu, k, f, \alpha/\alpha_0, \gamma/\gamma_0)$ — 227

REFERENCES — 309

PREFACE TO THE SERIES

This treatise is part of a continuing series on thermodynamic properties of technologically important fluids. These very important contributions by scientists and engineers working through the aegis of the Soviet National Service for Standard Reference Data have been released to Hemisphere Publishing Corporation for translation to make them available to the English reading technical community. The authors are Soviet experts in the field.

While a team of translators was involved in producing the English versions, the overall series is being published under the technical editorship of T. B. Selover, Jr.

Each volume presents a comprehensive survey of the world's literature up to its publication date. A special effort has been made to give a thorough presentation of Russian as well as other work. Many studies not previously known to Western counterparts are included. The results have been to broaden the range of applicability of data and to improve upon equations of state to provide accurate computation methods for generating smoothed tables of properties.

For some volumes there are no equivalent comprehensive surveys available in English. Thus, a valuable service has been fulfilled to workers in the fields of process design, equipment development, custody transfer, and safety.

Each volume is set up in the same way with Part I dealing with a study of all necessary aspects of experimental data interpretation and analysis. Then in Part II the fundamental constants, symbols with units, and data tables can be found. The use of SI units is consistent throughout.

The section on experimental data is particularly important because it covers, in detail, the key studies. Possible errors in measurement or data analysis that could have led to inaccurate tabular results in publications are pointed out.

Methods of constructing the equation of state and procedures for computing the data tables are covered thoroughly. There is a detailed error analysis of data generated from the equation of state relative to literature values.

Properties of each fluid in phase equilibria are treated at the freezing curve and the saturation curve with tables given for both temperature and pressure dependence. Properties cover the range of triple point to critical point in the condensed phase. Single phase properties in each volume cover a range of temperature and pressure wide enough to include most practical applications. Ideal gas property equations and calculation methods are included. The scope of properties generated and covered in the tables is more comprehensive than is typically found in any one English-language treatise. This gives the advantage of internal consistency.

An extensive bibliographic listing is included with each volume. All Russian citations have been translated. In some cases the availability in English of translated Russian sources is given.

Frequent reference to key English-language references has been made to help in providing correct translations. In some cases the original symbols have been changed to avoid confusion within the text or to avoid misuse of terms commonly found in English. Where mistakes in the original Russian text have been found, they have been corrected and so noted as editor's changes. Added descriptive clarification has been used in some places for tables, figures, and text to clarify meaning.

Although we recognize that key review papers have been published in 1986 for some of the fluids in this set, the series stands on its own merit. It represents a vast accumulation of knowledge never before available to Western countries. Through careful study of these volumes, workers in the field will develop an appreciation for both the scope of studies and where differences exist. Hopefully, this will lead to more dialogue between the authors and their Western counterparts.

Theodore B. Selover, Jr.

PREFACE

Terrestrial helium was discovered by Ramsey in 1895 and immediately became the object of attention. It has continued to be so to this day. Helium has the lowest known critical temperature, which allows the lowest temperature cycle to be carried out. Physicists have been attracted to the study of helium by its anomalous liquid state (helium II) with its exotic properties such as superfluidity and the λ-transition. The molecular simplicity, high symmetry, and, consequently, the weak intermolecular interaction of helium have allowed very interesting quantum-mechanical analyses to be carried out.

Natural helium consists of a mixture of the two isotopes He^4 and He^3. The concentration of He^3 in natural helium is extremely small, i.e., between 2×10^{-8} and 1.2×10^{-5}%, depending on the source of the sample. The unstable isotope He^6 has a half-life of about 0.8 s and is the by-product of certain nuclear reactions (e.g., neutron bombardment of Be^9).

Helium is the second most common substance in the universe (after hydrogen) and accounts for about 23% of the mass of the universe. The concentration of helium in the atmosphere is only 0.0005% by volume (up to 10% in some natural gases).

The technical importance of helium has grown with the intensive development of new technologies. Helium is used for the cooling of superconducting devices, in the creation of high vacuums, in atomic energy, in rocketry, and in various other fields of technology.

Increased demand for helium and its low abundance have created a need for reliable data on its thermodynamic properties under a wide range of conditions. Such data are necessary in the design of equipment used to extract helium from natural gases and in the design of cryogenic apparatus.

A great deal of effort has been devoted to experimental and theoretical investigation of helium in laboratories across the world. Experimental data on the thermal properties of helium are reviewed in a number of monographs and handbooks [8, 11, 24, 25, 32–34, 112, 169].

One of the most detailed publications is the 1949 monograph by W. Keesom [11], who contributed to the early work on helium. His monograph contains extensive data and detailed descriptions of published work. Several chapters are devoted to superfluidity. Keesom's monograph [11] cannot be viewed as a handbook on the thermodynamic properties of helium because it does not contain adequate tables, but it can be considered a kind of history of the discovery of helium and of the early work on it. A vast amount of experimental and theoretical data has accumulated since its publication.

An early Soviet attempt to describe in detail the thermodynamic properties of helium was made in [24, 25]. Unfortunately, a very simple equation of state was used in this description. It includes two virial coefficients and is based on limited experimental data. The tables presented in [24, 25] contain a minimal collection of properties (we are not concerned here with thermal conductivity or viscosity) and do not cover temperatures below 0 °C.

The properties of liquid and solid He^3 and He^4 are reviewed in [8]. However, [8] does not give an equation of state or reference tables.

The most reliable Western source devoted to the thermodynamic properties of helium was published in 1977 [32]. It is intended for scientists and engineers; it contains tables of thermodynamic functions, including specific heats and the speed of sound. It covers the whole range of temperatures and pressures of interest in science and technology, including the liquid phase and phase boundaries. However, three equations of state are used in [32] to calculate tables of functions for three ranges of parameter values, which is a distinct drawback when these equations are applied to processes and thermodynamic cycles. Moreover, the values of thermodynamic functions on the saturation line are calculated by different methods in different tables. Finally, the tables do not take account of the most recent investigations and virtually ignore Soviet work.

All this has stimulated the preparation of the present monograph, which is dedicated to the thermodynamic properties of the most common isotope of helium, He^4, in the temperature range between the λ-point and 1500 K. At temperatures below the λ-point, helium is in the superfluid state (helium II) and exhibits a range of different and unusual properties. These properties are of great interest in physics but, because of the singular behavior of superfluid helium, they will not be discussed in this monograph.

The USSR Academy of Sciences has laid down a program for the investigation of the thermodynamic properties of technologically important gases and liquids, including experimental work on helium and the setting up of tables of data. In this monograph we review data on the thermodynamic properties of helium up to 1982 inclusive. Although there is a large amount of information on helium, these data are not uniformly distributed across the range of parameter values. The thermal properties of helium have not been studied in adequate detail. This hampers the formulation of an equation of state and does not allow the accuracy of the calculated values of thermal properties to be determined by comparison with experiment. As in previous monographs in this series, the tabulated data were calculated from an average equation of state, based mostly on experimental p, v, T data, existing values of virial coefficients, isochoric heat capacities, and the derivatives $(\partial P/\partial T)_v$ and $(\partial P/\partial v)_T$. The probable random errors in the tabulated values of the thermodynamic functions were determined with the help of a set of equations of state that describe the p, v, T data with equivalent precision.

Values of thermodynamic functions calculated from the equation of state of He^4 have been compared with experiment, and this has led to an equation of acceptable precision. We have used this equation to calculate the thermodynamic properties of He^4 between the λ-point and 1500 K for pressures up to 100 MPa. We cover a greater range of parameter values and tabulated functions than other publications and indicate the precision of these results.

This monograph is one of a series of handbooks on the thermal properties of technologically important gases and liquids. It correlates and generalizes investigations of the properties of helium carried out at the Moscow Power Institute, the Odessa Institute of Marine Engineering, and the All-Union Science Research Center of the USSR State Committee for Standards.

We are grateful to the staff of these institutions, including Professor V. A. Rabinovich for his participation in the analysis and evaluation of experimental data on the thermodynamic properties of helium, and to A. Ya. Kreizerov, Yu. I. Kaśyanov, L. R. Malov, and N. A. Kochetov for help with data processing and for carrying out calculations.

We are greatly indebted to Dr. F. I. Amirkhanov and Dr. A. V. Kletzkii for reading the manuscript and for valuable suggestions. Comments from readers will be gratefully received.

V. V. Sychev
A. A. Vasserman
A. D. Kozlov
G. A. Spiridonov
V. A. Tsymarny

FOREWORD

The acceleration of scientific and technological progress is one of the major tasks of the tenth Five-Year Plan, promulgated by the Twenty-fifth Congress of the Communist Party of the Soviet Union. In light of this task, it is important to develop and expand the sphere covered by standardization, since such expansion would allow accumulation of the latest scientific and technological achievements, organically combine pure and applied sciences, and promote rapid and practical implementation of scientific achievements.

The State Service for Standard Reference Data oversees one of the new trends in standardization, that of standardization of the most reliable data available on the physical constants and properties of materials and substances.

The development of standard reference data is a multifaceted scientific and technological task that must be based on the Soviet and worldwide practice of selection and estimation of the reliability of data. An important stage in this work consists of preparing standard reference publications that contain not only data with an estimated reliability but also modern methods of obtaining such estimates. It is precisely these such publications that should serve as a methodological basis for the development of official tables of standard reference data on the properties of materials and substances.

The experience accumulated in working out tables of data on the thermodynamic properties of gases and liquids is of great interest from this point of view. This work has been particularly developed in the USSR. The work of Soviet investigators in thermophysics, which has established the basis for Soviet and international tables of thermophysical properties, has been acclaimed worldwide. Currently, this work is carried out by over 40 research organizations in the USSR under the auspices of the scientific program of the

State Service of Standard Reference Data and the Working Group on Thermodynamic Tables of the Soviet National Committee on Numerical Scientific and Technological Data of the Presidium of the USSR Academy of Sciences.

The publication of this series of books will supply reliable reference information to a wide circle of engineers and scientists and will serve as a basis for the development and practical utilization of pertinent official tables of standard reference data. The use of such data, obtained on the basis of the most up-to-date and accurate methods of study, is one of the conditions necessary for improving the level of scientific and developmental work. It ensures effective monitoring of industrial processes and quality of industrial output and promotes efficient utilization and accounting of the consumption of raw and finished materials, fuel, and energy. It is thus obvious that supplying reliable data on the properties of substances and materials is an important economic activity. The State Service of Standard Reference Data has been established primarily to supply this need.

V. V. Boytsov

PART ONE

CHAPTER
ONE
EXPERIMENTAL DATA ON THE THERMODYNAMIC PROPERTIES OF HELIUM-4

Experimental studies of the thermodynamic properties of helium began in the 19th century and have continued ever since. Extensive data are now available on the density of gaseous and liquid helium, and on the properties of phase equilibria curves. The thermal and acoustic properties of helium (and of other technologically important media), have been studied in less detail.

In this chapter we present a survey of existing experimental work and devote particular attention to work which is later used to establish an equation of state.

1.1 THERMODYNAMIC PROPERTIES ON PHASE TRANSITION LINES

1.1.1 Phase Diagram of He^4

The phase diagram of He^4 (Fig. 1) differs from that of other media by its unusual melting curve, the existence of an anomalous liquid state (helium II), the λ-curve, and several triple points. Atypically, the melting curve does not cross the vaporization curve, As the temperature approaches absolute zero, the curve becomes practically horizontal on the p, T diagram, crossing the pressure axis at 2.53 MPa. This implies that crystalline helium does not exist at any temperature at pressures below 2.53 MPa.

The line AB separates the regions of existence of helium II and helium I and corresponds to parameter values for which there is a second-order phase transition. It is referred to as the λ-line, and its points of intersection with the vaporization (A) and melting (B) curves are called the lower and upper λ-points respectively.

The triple points E, F, and G are points at which liquid helium co-exists with two different crystalline forms of helium. At the point E liquid helium II coexists with a hexagonal crystalline phase (α-phase) and a body centered cubic phase (γ-phase). At point F liquid helium I coexists with both forms (α and γ), and at G liquid helium I coexists with the hexagonal phase (α-phase) and a face-centered cubic phase (β-phase).

Our tables cover the temperature range down to 2.2 K but do not refer to the region of existence of helium II. Figure 2 shows, for completeness, the phase diagram of He^4 on the p,v plane.

1.1.2 Saturated Vapour Pressure of He^4 and Temperature Scales

Traditional assumptions about the precision of determinations of the temperature dependence of saturated vapor pressure are invalid in the case of helium. In

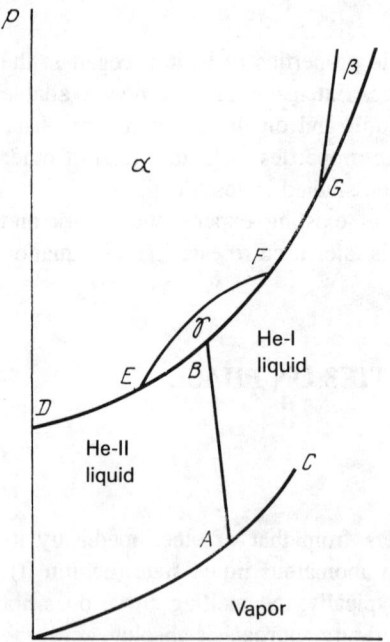

Figure 1 The phase diagram of He^4.

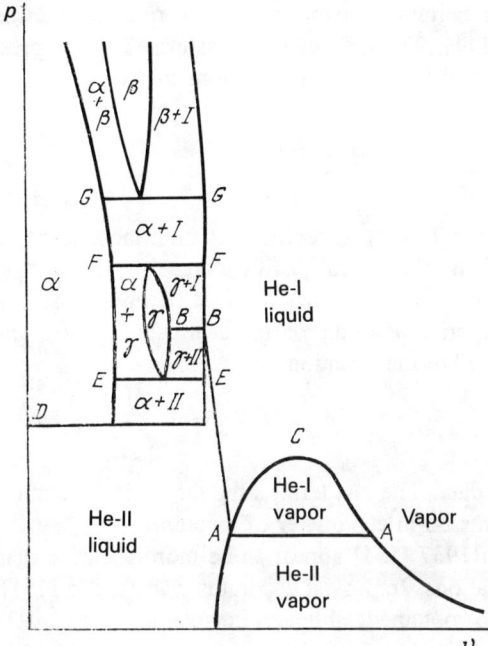

Figure 2 The thermodynamic diagram of He^4 on the p, v plane.

principle, there are three standard ways of determining this dependence; namely, (a) direct measurement with a gas thermometer, (b) calculation, using a limited number of measurements carried out with a gas thermometer, and (c) measurement with a secondary thermometer, calibrated against a gas thermometer.

Gas thermometers are difficult to use at moderate temperatures, and these difficulties are exacerbated at low temperatures. This has forced metrologists to look for alternative ways of measuring temperature in the "helium range," i.e., below 5.2 K.

In early experiments the thermodynamic temperature in the helium region was measured with a helium-filled constant-volume gas thermometer, and the p, V isothermals measured in this way were described by the equation

$$\frac{pV}{N} = RT \left[1 + B\left(\frac{N}{V}\right) + C\left(\frac{N}{V}\right)^2 + \ldots \right] \qquad (1.1)$$

Extrapolation of these isothermals to zero density (N/V → 0) then yielded the required temperatures. As a rule, the saturated vapor pressure of the helium in the thermostat bath was measured at the same time. This proved to be valuable because it meant that a temperature scale based on the relatively easily reproducible values of the saturated vapor pressure could be developed, and the gas-thermometer measurements could be replaced with the simpler measurement of saturated vapor pressure.

The first attempts to create a helium temperature scale were undertaken in the Leiden laboratory in 1929 [111]. The then available saturated vapor pressures for $T > 0.9$ K were described [111] by two equations of the form

$$\lg p_s = -\frac{A}{T} + B \lg T + C \tag{1.2}$$

in which the empirical constants for $T > T_\lambda$ were different from those for $T < T_\lambda$. These constants did not have any theoretical justification.

In 1932, Keesom [103] described the p_s values for $T < T_\lambda$ on the basis of theoretical considerations and experimental data on the heat capacity of liquid helium [104]. The coefficient of T^4 in the equation

$$\lg_{10} p_s = -3.018/T + 2.484 \lg_{10} T - 0.000297 T^4 + 1.197 \tag{1.3}$$

was deduced from heat capacity data. The last term is the theoretical chemical constant and the pressure is expressed in centimeter of mercury.

The values of p_s obtained in 1937 [151] appear to be more accurate than those obtained in earlier investigations. Improved apparatus was used in [151]; but an equation for $p_s(T)$ was not obtained, although corrections to the 1932 scale were introduced in a graph. It is clear from this graph that the discrepancy between the 1932 and 1937 scales is significant. At any rate, the 1937 scale is a few hundredths of a degree lower than the 1932 scale for the $T < 1$ K.

Further improvements in the temperature scale called for careful measurements and calculations, and it was not until 1948 that the first official scale was adopted on the basis of extensive measurements of low temperatures. It is known as the "1948 scale" (T48) and, sometimes, as the 1949 scale. A detailed table of $p_s(T)$ on this scale is presented in [60].

In the range 1–1.6 K, the 1948 scale is based on the calculations reported in [46]. For temperatures in the range 1.6–5.2 K, it is based on the measurements of p_s reported in [151] and the corresponding gas thermometer measurements of temperature.

The 1948 scale was criticized from the moment of its adoption. In particular, deviations of the T48 scale from the thermodynamic scale were noted in [124, 125]. Nevertheless, the 1948 scale was adopted by the International Conference on Weights and Measures because of the acute need for the standardization of low temperature measurements. The reasons for these deviations, and their precise values, were established, only after difficult and extensive measurements.

Two new scales were examined at the Paris conference in 1955; (1) the L55 scale developed in [61] and based on thermodynamic calculations of p_s, using the assumption of equal Gibbs potentials of coexisting liquid and vapor, and (2) the 55E scale developed [56] at the Australian National Research Laboratory and based on a large number of direct measurements of the temperature dependence of p_s [29, 30, 42, 57, 77, 114].

The L55 scale defines p_s as a function of T in the range 0.9–4.25 K. An attempt was made subsequently to use the gas-thermometer measurements [57] to extend this scale to 5.2 K.

The L55 and 55E scales are not identical. They deviate from the thermodynamic scale (T) by differing amounts. The maximum T-55E difference occurs at 2.6 K (+1.3 mK) and at 3.8 K (−1.3 mK), and the maximum T-L55 difference occurs at 2 K (+4 mK) and at 3.85 K (−4.5 mK). However, the source of these discrepancies between the scales had not been established by the beginning of the 1955 conference. It was therefore not possible to choose between them, and the possible use of both scales was considered. However, the 55E scale seemed to be better founded from the thermodynamic viewpoint and also from the point of view of direct p, T measurements.

Improved measurements at low temperature [29, 42, 56, 57, 60, 77, 113, 114, 171] created the basis for the 1958 scale (T58).

Magnetic and carbon thermometers have played an important part in the investigation of $p_s(T)$ [29, 60, 77]. They have proved to be particularly useful in interpolations between reference points established with the gas thermometer.

The T58 scale is not without its deficiencies, so that a series of temperature scales for low temperatures had to be investigated. Each of them resulted from improvements in the precision of existing scales and their approach to the thermodynamic scale. A large number of measurements were carried out in the temperature range 5.2–13.8 K with gas and acoustic thermometers, and with germanium resistance thermometers and magnetic interpolation thermometers calibrated against them.

The 1962 helium scale (T62) [153] is based on a combination of thermodynamic calculations, isothermal comparisons of the saturated vapor pressures of He^3 and He^4, and magnetic measurements. It is recommended for the temperature range 0.2–3.2 K. The temperatures measured on the T58 and T62 scales agree to within 0.3 mK, and the discrepancies lie within the range of random error.

Careful measurements in the range 2–20 K and a new temperature scale based on acoustic measurements were reported in [143]. It is well-known that there is a simple relationship between the speed of sound and thermodynamic temperature. Detailed knowledge of reference points is not required in the measurement of temperature by acoustic methods, and it is also possible to avoid the use of virial coefficients which are not entirely mutually consistent. The acoustic thermometer allows temperatures to be determined to within ±2.5 mK at about 4.2 K and ±5 mK at 20 K. These uncertainties exceed somewhat the errors introduced when the equation given in [62] is used for p_s. However, the more significant point is that deviations of the T58 scale from the thermodynamic (acoustic) scale were reported in [143]. These deviations amount to 5–11 mK and somewhat exceed earlier estimates of uncertainties in T58 and the uncertainties in acoustic measurements. Deviations of the T58 scale from the thermodynamic scale by amounts of the order of 3–7 mK in the range 2.1–3.3 K were confirmed by the acoustic measurements reported in [87]. The scale presented in [143] will henceforth be designated T65.

Investigations of the properties of paramagnetic salts at low temperatures have led to the development of a device that can be used to construct a temperature scale. The paramagnetic temperature scale (T_x) for the range 0.9–34 K was developed at Iowa State University and is reported in [54, 162]. This scale is thermodynamically smooth, internally consistent, and identical for both salts used in these experiments. It corroborates the internal consistency of the scale NBS-55 in the range 18–34 K and its unreliability at lower temperatures. IPTS-68 is internally consistent in the range 15–27 K, but deviates from the temperature dependence of magnetic susceptibility extrapolated to 34 K. The NBS-55 and IPTS-68 scales differ by 8.5 mK in the range 18–27 K, but the scale T_x does not allow a judgment to be made on the correctness of either of these scales in an absolute sense.

The deviation of T_x from the national temperature scales is analyzed in [54, 162]. It is concluded that the resultant smoothed scale must lie between NBS-55 and IPTS-68. This scale demands a lowering of the presently assumed boiling point of helium (T_B) by 8.5 ± 1.5 mK, and agrees with the boiling point of hydrogen assumed in the IPTS-68 scale. The precision of susceptibility data can be fully exploited only if a thermodynamic scale is established in the range 15–34 K with an uncertainty of 1–2 mK.

The scale developed in 1975 at the British National Physical Laboratory (NPL-75) is based on gas-thermometer measurements in the range 2.6–27.1 K [31]. The origin of this scale is the boiling point of equilibrium hydrogen, set equal to 20.2714 K. The reference points determined for NPL-75 differ somewhat from the reference point of IPTS-68 (see Table 1.1).

The Practical Temperature Scale of 1976 (T76) is based on a comparison of national low temperature scales. It was also developed at Iowa State University

Table 1.1 Temperature reference points on the IPTS-68 and NPL-75 scales

Reference point	IPTS-68	NPL-75	Δ(T68–T75), mK
Triple point of equilibrium hydrogen	13.8100	13.8039	6.1
The point at 17.042 K	17.042	17.0357	6.3
Boiling point of equilibrium hydrogen	20.2800	20.2714	8.6
Boiling point of neon	27.1020	27.0979	4.1

and appears to agree most closely with the thermodynamic scale in the range 1.1–30 K. Unfortunately, this scale has not been accepted internationally, so far.

Magnetic and gas temperature scales for the range 4.2–27 K have been developed in the USSR [1].

The T58* scale was proposed in [32] for the range 2.2–5.2 K on the basis of results reported in [87, 143, 149]. This scale can be related to T58 by the formula (T58*) = [1.002(T58) + 0.001]K.

Yet another scale is proposed in [150] and is based on measurements with the gas thermometer and the paramagnetic salt thermometer. Its deviations from the T58 and T62 scales are determined in [150]. Measurements of temperature made on this scale agree with the NPL-75 scale to within 0.2 mK above 2.6 K. In the range 1.3–2.6 K, they agree with the magnetic scale T_x to the same precision.

The problem of the temperature scale at helium temperatures is definitely not settled, and some mistakes have been made. An example is provided by the very common scale T58 which has been shown by recent investigations to deviate from the thermodynamic scale to a greater extent than the 1948 scale over a significant part of the temperature interval. Further improvements of the temperature scale will, of course, be made in the future.

We have used T58 in the range 2.2–5.2 K. It is by no means the most perfect scale, but it is very widely used in experimental investigations. Table 1.2 shows the deviations of different scales from T58 and will be used to convert values of T from one scale to another. The table does not claim to be exhaustive and contains information only on the most frequently met scales.

The closest possible agreement between the temperature and the thermodynamic scale is particularly important in phase equilibria and in determinations of thermal and acoustic properties. In most measurements of the density of He^4 at low temperatures, the scatter of data significantly exceeds uncertainties due to inaccuracies of the temperature scale.

Table 1.3 gives the saturated vapor pressure and its temperature dependence on the T58 and T58* scales, showing the obvious effect of the choice of temperature scale on p_s.

1.1.3 Characteristic Points of the Phase Diagram

Lower λ-point At first sight, the evaporation curve of He^4 does not seem to have singularities even at the lowest temperatures. At any rate, such singularities were not noticed in the range 4.2–1.47 K in an investigation carried out in 1915 [102]. There was no obvious discontinuity when log p_s was plotted against $1/T$. However, the fact that all the evaporation curves could not be described by a single equation, testified to the existence of singularity. Subsequent investigations, carried out with greater precision, did not eliminate this difficulty.

An anomaly on the evaporation curve at the so-called λ-point was discov-

Table 1.2 Deviations of different temperature scales from the T58 scale

Values of Δ = T58 − T_i (mK) for scale T_i

T, K	T24	T29	T37	T48	T55E	T55L	T58*	T62	T76
2.2	−111.2	−8.22	−9.32	−9.42	−0.82	2.18	−5.40	−0.18	—
2.3	−111.2	−6.79	−8.13	−8.18	−0.44	2.06	−5.60	−0.20	—
2.4	−111.2	−5.35	−6.95	−6.95	−0.05	1.95	−5.80	−0.22	—
2.5	−106.8	−3.78	−5.48	−5.58	0.42	1.92	−6.00	−0.23	−5.08
2.6	−100.7	−2.12	−4.42	−5.02	0.68	1.78	−6.20	−0.25	—
2.7	−92.9	−0.62	−3.91	−4.31	0.89	1.54	−6.40	−0.27	—
2.8	−85.2	0.89	−3.41	−3.61	1.09	1.29	−6.60	−0.29	—
2.9	−76.0	2.18	−3.82	−4.62	1.08	0.92	−6.80	−0.28	—
3.0	−66.8	3.47	−4.23	−5.63	1.07	0.57	−7.00	−0.27	−6.13
3.1	−57.2	4.32	−5.48	−6.84	0.96	0.16	−7.20	−0.26	—
3.2	−47.7	5.16	−6.74	−8.04	0.86	−0.24	−7.40	−0.24	—
3.3	−38.6	5.55	−7.70	−8.35	0.70	−0.75	−7.60	—	—
3.4	−29.4	5.94	−8.66	−8.66	0.54	−1.26	−7.80	—	—
3.5	−21.0	5.76	−8.94	−8.84	0.26	−1.94	−8.00	—	−6.54
3.6	−13.2	5.50	−9.00	−8.70	0.10	−2.50	−8.20	—	—
3.7	−6.6	4.72	−8.24	−8.18	0.07	−2.84	−8.40	—	—
3.8	0.1	3.93	−7.47	−7.67	0.03	−3.17	−8.60	—	—
3.9	4.8	2.42	−6.08	−6.33	0.12	−3.09	−8.80	—	—
4.0	9.6	0.90	−4.70	−5.00	0.20	−3.00	−9.00	—	−6.80
4.1	12.1	−1.40	−2.90	−2.75	0.40	−2.10	−9.20	—	—
4.2	14.6	−3.70	−1.10	−0.5	0.60	−1.20	−9.40	—	−6.90
4.3	14.6	−6.85	—	−0.7	1.00	−0.15	−9.60	—	—
4.4	14.6	−10.0	—	−0.9	1.40	0.90	−9.80	—	—
4.5	12.4	−14.0	—	−3.83	1.88	1.48	−10.0	—	−7.12
4.6	8.8	−18.5	—	−7.67	2.33	1.73	−10.2	—	—
4.7	2.9	−23.9	—	−11.73	2.77	1.32	−10.4	—	—
4.8	−3.0	−29.3	—	−15.79	3.21	0.91	−10.6	—	—
4.9	−11.9	−36.0	—	−17.40	3.80	−0.45	−10.8	—	—
5.0	−20.8	−42.6	—	−19.02	4.38	−1.72	−11.0	—	−7.32
5.1	—	—	—	−12.76	4.69	−4.01	−11.2	—	—
5.2	—	—	—	−6.5	5.00	−6.30	−11.4	—	—

ered in an experimental study of the temperature dependence of the heat of evaporation of helium [59]. This singularity was fully consistent with the presence of a discontinuity in the heat capacity of liquid He^4 at the λ-point. Two branches of the evaporation curve meet at this point with different curvatures, and the combined curve appears to be smooth because deviations due to curvature difference are of the order of a few thousandths of a degree and are close to the experimental uncertainty.

Table 1.3 Saturated vapor pressure of He^4

T58, K	T58*, K	p_s, kPa	T58*, K	T58, K	p_s, kPa
2.2	2.2054	5.395	2.2	2.1946	5.326
2.4	2.4058	8.440	2.4	2.3942	8.337
2.6	2.6062	12.50	2.6	2.5938	12.35
2.8	2.8066	17.72	2.8	2.7934	17.53
3.0	3.0070	24.27	3.0	2.9930	24.02
3.2	3.2074	32.30	3.2	3.1926	31.91
3.4	3.4078	41.96	3.4	3.3922	41.55
3.6	3.6082	53.39	3.6	3.5918	52.89
3.8	3.8086	66.75	3.8	3.7914	66.14
4.0	4.0090	82.20	4.0	3.9910	81.47
4.2	4.2094	90.90	4.2	4.1906	99.02
4.4	4.4098	120.0	4.4	4.3902	119.0
4.6	4.6102	142.8	4.6	4.5898	141.6
4.8	4.8106	168.4	4.8	4.7894	167.0
5.0	5.0110	197.1	5.0	4.9890	195.4
5.19	5.2014	—	5.2	5.1886	227.5

Table 1.4 lists the sources from which the parameters at the lower λ-point were taken. The original values of $T_\lambda^{(L)}$ are presented and the temperature scale used is indicated.

Edwards [68] performed special measurements in a very narrow range of $T - T_\lambda$ (±100 mK) and obtained 24 values of the refractive index. These were then used to calculate the density of the liquid and the coefficient of expansion. A pycnometer was used in [115, 117] to measure the density ρ. This improved the reliability of the density values. The permittivity was measured in [76] and was used to calculate the density from the Clausius-Mossotti formula.

Table 1.4 Parameters of the lower λ-point

Year	Author	Scale	$T_\lambda^{(L)}$, K	$p_\lambda^{(L)}$, Pa	$\rho_\lambda^{(L)}$, g/cm^3
1957	Kerr [115]	T55E	2.172	—	0.14657†
1958	Edwards [68]	T55E	2.1728	—	0.14596†
1964	Kerr and Taylor [117]	T58	2.1720	—	0.14615†
1967	Elwell and Meyer [76]	T58	2.17312	5040 ± 10‡	0.14622
1967	Kierstead [120]	T58	2.1720	5040 ± 1	0.14615

†Estimated.
‡Calculated value.

12 THERMODYNAMIC PROPERTIES OF HELIUM

The pressure p_λ and the derivatives $(\partial p/\partial T)_\lambda$ and $(\partial \rho/\partial T)_\lambda$ were measured directly in a thorough study of the λ-curve reported in [120]. The parameters at the lower λ-point obtained in these experiments appear to be the most reliable. The uncertainty in temperature $T_\lambda^{(L)}$ is estimated in [120] as being of the order of 0.1 mK, and $T_\lambda^{(L)}$ agrees well with the results of other experiments. The estimated precision of the measured $p_\lambda^{(L)}$ is 1 Pa, i.e., less than 0.02%. The values of $(\partial \rho/\partial T)_\lambda$ are used in [76] to calculate the density $\rho_\lambda^{(L)}$, and the results agree with [117, 120].

Thus, having the basis for adoption, the significant parameters for the lower λ-point from [120] are: $T_\lambda^{(L)} = (2.1720 \pm 0.0001)$ K on the T58 scale, $p_\lambda^{(L)} = (5040 \pm 1)$ Pa, $\rho_\lambda^{(L)} = 0.14615$ gm/cm^3.

Upper λ-point The melting curve has a singularity analogous to that of the lower λ-point. According to [107], this singularity is the intersection of the λ-line with the melting curve, and is called the upper λ-point. Table 1.5 lists the measured parameters of the upper λ-point together with the sources of these data. The results reported in [107] and [160] deviate most from the average value of $T_\lambda^{(U)}$. Naturally, these data also show the largest deviation of $p_\lambda^{(U)}$. Without dwelling too long on old results such as those in [107], we noted that there is a mistake in [160] which is pointed out in a subsequent publication [161]. The data from [107] and [160] must not be taken into account when the parameters of the upper λ-point are determined.

As for the lower λ-point, we have used the results from [120] to determine the parameters of the upper λ-point as follows: $T_\lambda^{(U)} = 1.7633$ K, $p_\lambda^{(U)} = 3.013$ MPa, and $\rho_\lambda^{(U)} = 0.18044$ gm/cm^3.

Table 1.5 Parameters of the upper λ-point

Year	Author	Scale	$T_\lambda^{(U)}$, K	$p_\lambda^{(U)}$, MPa	$\rho_\lambda^{(U)}$, g/cm^3
1933	Keesom and Keesom [107]	T32	1.753	3.031	—
1952	Swenson [160]	T48	1.743	2.951	—
1953	Swenson [161]	T48	1.764	3.003	—
1960	Lounasmaa and Kaunisto [128]	T58	1.765	3.010	0.1804
1962	Grilly and Mills [86]	T58	1.760	3.006	0.17989
1964	Ahlers [28]	T58	1.7628	—	—
1965	Kierstead [118]	T58	1.7633	3.024	—
1966	Kierstead [119]	T58	1.7732	3.022	—
1966	Edwards and Pandorf [71]	T58	1.763	—	—
1966	Vignos and Fairbank [165]	T58	1.765	3.030	—
1967	Kierstead [120]	T58	1.7633	3.013	0.18044

Table 1.6 Parameters of the α-γ-I triple point

Year	Author	Scale	T, K	p_{tr}^{γ}, MPa
1962	Grilly and Mills [86]	T58	1.760	3.006
1964	Ahlers [28]	T58	1.7730	—
1966	Vignos and Fairbank [65]	T58	1.778	3.068
1966	Edwards and Pandorf [71]	T58	1.7715	—
1966	Kiersted [119]	—	—	3.044
1967	Kiersted [120]	T58	1.7733	3.042
1973	Grilly [84]	—	1.772	3.040

Triple points We have already mentioned that the solidification curve of He^4 has three triple points which correspond to the coexistence of two different crystal structures and liquid helium I or II. The triple point of the coexistent α and γ crystal phases and liquid helium II lies beyond the range of parameters presented here. The other triple points are points of equilibrium with liquid helium I. They are significant for the correct representation of the melting curve.

The parameters of the α-γ-I triple point (the γ-point, for short) are very close to the parameters of the upper λ-point. They have been investigated in a number of papers (Table 1.6). The values reported in [86] appear to be wrong because the parameters of the triple point seem to coincide with the parameters of the upper λ-point given in [86]. The pressure given in [165] is also erroneous because of a fault in the manometer system.

The high precision of the measurements reported in [119, 120] allows us to adopt these values of $T_{tr}(\gamma)$ and $P_{tr}(\gamma)$, which also appears to be logical from the point of view of their correlation with the previously adopted values of P and T at the λ-point. Our values of the γ-point parameters based on [119, 120] are: $T_{tr}^{\gamma} = (1.7733 \pm 0.0001)$ K (T58) and $p_{tr}^{\gamma} = (3.043 \pm 0.005)$ MPa.

The temperature at the α-β-I triple point (the β-point) was measured only in [64, 65]. The results were: $T_{tr}^{(\beta)} = 14.9$ K [65] and $T_{tr}^{(\beta)} = 15.01$ K [64]. When the latter result is taken in conjunction with the melting curve [134], the pressure turns out to be 128 MPa.

Critical point The usual difficulties associated with the realization of the critical state and with the determination of the critical-point parameters (in the first instance, the determination of critical temperature), are exacerbated in the case of helium by uncertainties in the temperature scales at this point. Most of these scales are based on measurements of the saturated vapor pressure and have the lowest precision precisely at the critical temperature.

Table 1.7 lists some (mostly new) investigations of the parameters of the critical point. Of course, they are not equivalent from the point of view of precision.

One of the first investigations of the critical parameters of helium was carried out in 1911 [100]. The critical pressure was determined at the onset of critical opalescence and the result was used to find the critical temperature. The method used to determine p_{crit} and the uncertainties in the temperature scale prevent us from regarding these results as sufficiently trustworthy.

The critical temperature and density were determined in [157] by the rectilinear diameter method which has been repeatedly criticized. This method was also used in [40, 73]. In [115], the critical density alone was determined as an average of values obtained by four different methods.

Measurements of the critical parameters were subsequently repeated a number of times in [145–147]. The results reported in [145, 147] are of special interest insofar as three different methods were used in [145] (analysis of p, v,

Table 1.7 Parameters of the critical point

Year	Author	Scale	T_{crit}, K	p_{crit}, kPa	ρ_{crit}, kg/m^3
1911	Kamerlingh Onnes [100]	—	5.1994	229.0	—
1925	Mathias et al. [131]	T24	5.19	—	69.30
1957	Kerr [115]	T55E	—	—	67.5
1958	Berman and Mate [40]	T55E	5.194	229.0	68.0
1960	Van Dijk et al. [62]	T58	5.1994	229.0	—
1963	Edwards and Woodbury [73]	T58	—	—	69.48
1966[†]	Roach and Douglas [146]	T58	5.1890	227.3	70.1
1967[†]	Roach and Douglas [147]	T58	5.191	227.7	69.0
1968	Roach [145][‡]	T58	5.198	227.31	69.0
			5.193	227.98	69.0
			5.191	227.71	68.5
1968	Edwards [69]	T58	5.1897	227.43	—
			5.18988	—	69.323
1969	Moldover [137]	T58	5.1891	227.32	69.58
1969	El Hadi and Durieux [74]	—	5.189	—	69.77
1971	Kiersted [121]	T58	5.18992	227.464	69.64
1973	Kiersted [122]	T58	5.19045	—	69.58
			5.19052	—	—

[†]Changed in proof T.B.S.
[‡]Three values of the critical-point parameters were obtained as a result of an analysis of the evaporation curve, isotherms, and isobars.

T data and isothermal and isobaric measurements). The dielectric constant was measured in [147]. Results differing by 7 mK were obtained.

The value of T_{crit} was determined by two methods in [69], namely, by noting the disappearance of the meniscus and by measuring the refractive index. The values obtained differ by 0.18 mK. In [137], the values of T_{crit}, P_{crit}, ρ_{crit} were deduced from discontinuities in the isochoric specific heat.

The result $p_{crit} = (227.464 \pm 0.013)$ kPa obtained in [121, 122] agrees well with most recent investigations. Furthermore, the high metrological standards of this work, verified by dependable measurements of the λ-point parameters, suggests that the value of p_{crit} given in [32] is the most reliable. This value of p_{crit} is consistent with the result T_{crit} (58) = 5.19 K. The critical density ρ_{crit} was measured directly by the Cailletet method in [122] and the result was ρ_{crit} = 69.580 ± 0.020 mg/cm³. However, if we take for p_{crit} and T_{crit} the values given in [121], we can then take ρ_{crit} to be 69.64 ± 0.07 kg/m³, which is consistent with [121].

The choice of 5.19 K for T_{crit}, based on an examination of the results presented in Table 1.7, does not seem unique, especially since somewhat higher values were obtained in recent work [122]. However, we must remember that the equation in virial form does not pretend to be an exact description of the properties of helium near the critical point, so that the reliability of the value of T_{crit} is not fundamental to our calculations.

1.1.4 The λ-Transition Curve

The λ-transition temperature is a function of pressure. The set of p, T values corresponding to this transition generates the λ-line which lies on the phase diagram between the lower and upper λ-points and was investigated even in [104]. Repeated investigations at the Leiden Laboratory have subsequently made the λ-line data more precise. Work done in the following 30 years is summarized in Table 1.8.

Most of the data are in relatively good agreement. The exceptions are the values of [160, 161], which are too low by 0.3–0.4 bar. This discrepancy may be explained by uncertainties on the temperature scale used by Swenson. One of the most recent and most thorough investigations was made by Kierstead [120] in the range between the lower and upper λ-points. The high resolution of his apparatus (1 mK, 10^{-5} bar, 10^{-8} g/cm³) allowed measurements to be made in very small temperature steps, and values of $(dp/dT)_\lambda$ and $(d\rho/dT)_\lambda$ were obtained. The abrupt change in thermal conductivity was used to identify the moment at which the phase transition occurred.

The experimental values of $(dp/dT)_\lambda$ were fitted in [120] with an equation which was integrated to obtain $p(T_\lambda)$ in the form

$$\frac{p}{p_\lambda^L} - 1 = \sum_{i=0}^{3} b_i \Theta^{i+1} + b_4(1 - e^{b_5 \Theta})$$

Table 1.8 Experimental investigations of the properties of He^4 on the λ-line

Year	Author	Scale	Parameter range		Number of points	Measured quantities
			ΔT, K	Δp, MPa		
1952	Swenson [160]	T48	2.15–1.743	0.3–2.951	6	p–T
1953	Swenson [161]	T48	2.186–1.764	0.002–3.003	10	p–T
1959	Lounasmaa and Kojo [129]	T58	2.172–1.771	—	9	p–T
1960	Lounasmaa and Kaunisto [128]	T58	2.0868–1.7622	0.79–3.011	6	p–T, ρ
1966	Vignos and Fairbank [165]	T58	2.146–1.778	0.278–2.951	13	p–T
1966	Grilly [83]	T58	2.050–1.799	1.106–2.811	7	p–T
1967	Elwell and Meyer [76]	T58	2.173–1.7935	0.005–2.8	23	p–T, ρ dp/dT, $d\rho/dT$,
1967	Kiersted [120]	T58	2.1720–1.7633	0.005–3.013	Equation	p–T

where $\Theta = (T/T_\lambda^{(L)}) - 1$ and $p_\lambda^{(L)}$, $T_\lambda^{(L)}$ are the pressure and temperature at the lower λ-point, respectively.

This equation agrees with most of the measured pressures to within ± 0.2 bar. The data reported in [161] are the exceptions.

The function $\rho(T_\lambda)$ was measured in [129] and also in the series of experiments reported in [76, 120, 128] and dedicated, basically, to an investigation of the p-T dependence.

The specific heat c_v was measured in [129] on nine isochores. The function $c_v(\rho)$ was then used to obtain the density along the λ-line, and the results were presented in a table of values of T_λ for nine smoothed values of ρ. The error in T_λ was estimated to be about 5 mK.

The dielectric constant was measured in [76] and the density was calculated from the Clausius-Mossotti equation. A carbon resistance thermometer was used [76], and the experimental error in the measured pressure was 2.6×10^{-4} bar for $p < 1$ bar and 0.05 bar for $p > 1$ bar.

The values of $(d\rho/dT)_\lambda$ were determined in [120], and an equation for the temperature dependence was established. Integration of this equation gives the function $\rho(T_\lambda)$ in the form

$$\frac{\rho}{\rho_\lambda^L} - 1 = \sum_{i=0}^{3} c_i \Theta^{i+1} + c_4(1 - e^{c_5 \Theta})$$

where $\rho_\lambda^{(L)}$ is the density of liquid helium I at the lower λ-point. Densities calculated from this equation agree with all experimental values of ρ_λ to within ± 0.0003 g/cm^3.

1.1.5 The Melting Curve

Kamerlingh Onnes carried out experiments at temperatures below 1 K but did not manage to obtain solid helium at the saturated vapor pressure so that further experiments at higher pressure became necessary. Subsequent experiments by Keesom yielded the p, T dependence along the melting curve between 1.22 and 4.2 K, i.e., between 2.6 and 14.2 MPa.

The shape of the melting curve of He^4, the presence of more than one crystal phase, and the corresponding triple points were all referred to above. The parameter range covered by our tables contains only part of the melting curve. From the upper λ-point, the melting curve extends to lower temperatures, beyond the limits of these tables, down to $T = 0$, and separates the regions of existence of crystalline helium and the anomalous liquid helium II. We note that, in the first approximation, the slope of the melting curve tends to zero at low temperatures. This implies that, at $T = 0$, the entropy of the crystal is equal to that of the liquid, in agreement with the Nernst theorem as applied to the liquid-crystal transition in helium. However, the above feature ($dp/dT \to 0$

18 THERMODYNAMIC PROPERTIES OF HELIUM

as $T \to 0$) was not confirmed experimentally for a long time. On the other hand, there were no data to refute its presence. In 1927, Keesom suggested that the pressure might increase on the melting curve at $T \to 0$, after passing through a minimum. This hypothesis was supported by the results reported in [156] which disclosed a minimum on the melting curve at 0.775 K. As the temperature was reduced the pressure was found to increase slightly, and then became constant, i.e., the derivative (dp/dT) tended to approach zero.

For $p > 128$ MPa, i.e., above the α-β-I point, the melting curve extends to high pressures without any apparent special points. At any rate, the melting pressures at temperatures of 60, 77, and 97 K [58, 136] lie on a smooth continuation of the curve.

The melting curve consists of two parts in the parameter range of interest to us. Between the upper λ-point and the α-γ-I triple point, it separates the crystalline γ-phase from liquid helium I. Both these points have been established quite reliably. This region is very short and encompasses a temperature interval of only 10 mK. The second part of the melting curve corresponds to the equilibrium between the crystalline α-phase and liquid helium I. The α-β-I triple point lies beyond our parameter range, and is determined less accurately.

Analysis of the experimental melting curve is made difficult by the absence of a generally accepted temperature scale in the range 5.2–13.81 K and by our inadequate knowledge of the relationships between existing scales and the thermodynamic scale. The most reliable points lie at the ends of the range. There is very little published experimental information, so that, as a rule, it is impossible to determine the corrections needed to reduce a given set of experimental data to a unified temperature scale.

All this inevitably leads to a relatively inadequate experimental melting curve. In the region of interest to us, the amount of experimental data is not large. Actually, it is clear from Table 1.9 that much of the information that *is* available for the melting curve lies beyond our parameter range. Some data have been published for $T < T_\lambda^{(U)}$, but most other data refer to pressures $p > 100$ MPa and lie outside the scope of this monograph.

From the point of view of the equation of state, the most interesting results come from those investigations on which the values of p, T and the corresponding densities (or specific volumes) are measured on the melting curve. Such data are quite rare [67, 84, 85, 135] and, unfortunately, the p, T, ρ data that *are* available are distributed very unevenly along the melting curve. The temperature region 4.7–14.7 K and the corresponding pressure region 17–110 MPa were not investigated in these publications.

The data reported in [81] are a further significant contribution. They were obtained as a result of measurements on isochores, some of which passed through the two-phase region. Such isochores have a discontinuity and coincide with the melting curve in the two-phase region. It is possible to use the data given in [81] to establish directly an equation of state, the data being close to

Table 1.9 Measurement of the properties of He^4 on the melting curve

Year	Author	Parameter range ΔT, K	Parameter range Δp, MPa	Number of points
1929	Simon et al. [154][†]	12.15–20.1	81–179	9
1929	Simon et al. [155][†]	12.2–42.0	81–552	26
1950	Holland et al. [97]	5–50	19–737	10
1950	Swenson [159]	0–4	2.5–13	16
1952	Swenson [160]	1.6–4.0	2.7–13	11
1953	Dugdale and Simon [65][‡]	4.2–12.1	14.3–78.9	14
1953	Swenson [161]	1.4–4.0	2.6–13	23
1955	Mills and Grilly [134][†]	2.1–30.5	3.7–355	—
1955	Mills and Grilly [135][§]	—	8–304	9
1958	Edeskuty and Sherman [67][§]	2–4	—	5
1959	Grilly and Mills [85][§]	1.35–30.77	2.6–349	18
1962	Grilly and Mills [86]	1.76–2.00	3.0–3.78	7
1964	Bogoyavlenskii et al. [2]	1.63–4.14	2.8–14.4	34
1965	Edwards and Pandorf [70]	1.86–4.27	3.2–14.3	8
1966	Glassford and Smith [81]	4.55–16.4	16.1–132	125
1966	Vignos and Fairbank [165]	1.02–2.27	2.6–4.7	14
1971	Crawford and Daniels [58]	13.4–60.9	94–1002	28
1973	Grilly [84][§]	0.3–2.0	2.5–3.78	26
1980	Mills et al. [136]	—	—	—

[†] An equation is presented for the melting curve.
[‡] Calculated specific volumes of the liquid on the solidification curve are presented.
[§] Data on the density of the liquid on the solidification curve are presented.

the melting curve. Moreover, simple analysis of the isochores can be used to determine the p, T coordinates of the discontinuity, and from a knowledge of the density on a given isochore, the p, T, ρ values on the melting curve as well. The precision of this method may not be very high.

A separate equation for the melting curve must be available before the thermodynamic properties of a liquid in the state of liquid-crystal equilibrium can be calculated form the equation of state. One such equation is obtained in

[32] from nine points reported in [85] and 106 points given in [81]. The melting curve in [32] is described by two equations: one for the range 1.7633–1.7733 K (between the upper λ-point and the α, γ, I triple point) and the other for the region between the triple point and 100 MPa.

1.1.6 Density of Helium in the State of Liquid-Vapor Equilibrium

The densities of coexisting liquid and vapor have been measured many times. Some measurements yielded data on both phases, while other investigations were confined to only one of the phase (Table 1.10).

The experiments may be divided into three groups: namely, temperatures below the normal boiling point, temperatures in the range between T_{boil} and T_{crit}, and temperatures close to the critical point. In principle, direct measurements

Table 1.10 Measurements of the density of liquid and gaseous helium in equilibrium

Year	Author	ρ' (Liquid)		ρ'' (Gas)	
		ΔT, K	Number of points	ΔT, K	Number of points
1924	Kamerlingh Onnes and Box [101]	1.20–4.22	20	—	—
1925	Mathias et al. [131]	4.23–4.71	3	4.23–4.71	3
1957	Kerr [115]	1.2–4.4	38	—	—
1958	Edwards [68]	1.6–4.2	29	—	—
1961	Edwards and Woodbury [72]	3.0–5.0	5	—	—
1961	Chase et al. [55]	1.4–4.2	Graph	—	—
1963	Edwards and Woodbury [73]	4.2–5.16	13	4.2–5.15	13
		$\|T - T_{crit}\| \approx 0.25$	44	$\|T - T_{crit}\| \approx 0.25$	32
1964	Kerr and Taylor [117]	0–4.40	151	—	—
1966	Roach and Douglas [146]	4.93–5.19	14	5.00–5.19	25
1968	Roach [145]	4.93–5.19	14	5.00–5.19	24
1969	El Hadi and Durieux [74]	—	—	2.02–5.10	23
1969	El Hadi et al. [75]	1.23–5.11	21	—	—
1973	Kierstead [122]	$T \approx T_{crit}$	28	$T \approx T_{crit}$	28

carried out with a pycnometer are the most reliable because simpler methods are less likely to be subject to error. Experimental data obtained this way are subject to errors that depend only on the accuracy of the measuring apparatus, the purity of the sample under investigation, and the care taken by the experimenter. The deviation of the practical temperature scale employed from the thermodynamic scale is not considered in these experiments.

It is important to note the good agreement between the pycnometer values of ρ' [101] and the results of recent investigations. The random spread in the data reported in [115] is $\pm 0.3\%$.

Practically the entire range of temperatures that is of interest to us is covered in [74, 75]. These data extend to within only 0.1 K of the critical temperature.

In a series of investigations the density was determined from measurements of the dielectric constant and refractive index. The latter was measured in [68, 72, 73] and the density was calculated from the Lorentz-Lorenz formula

$$\frac{n^2-1}{n^2+2} = \frac{4\pi}{3M}(N_0\alpha)\rho \qquad (1.1.6a)$$

where n is the refractive index, M is the molar mass of helium, $N_0\alpha$ is the molar polarizability, and ρ is the density.

It was assumed in these calculations that the molar polarizability remained constant. The adopted value was deduced from the density data [115] at $T = 3.7$ K. This temperature was chosen for the determination of $N_0\alpha$ because, in the authors' opinion, it corresponded to the most accurate refractive index. It was found that $N_0\alpha = 0.12454$ cm^3/mole, and this was subsequently used in density calculations.

The density was calculated from dielectric constant data [55, 82, 145, 146], using the Clausius-Mossotti equation

$$\frac{\epsilon-1}{\epsilon+2} = \frac{4\pi}{3M}(N_0\alpha)\rho \qquad (1.1.6b)^\dagger$$

where ϵ is the dielectric constant and the other symbols are the same as in the previous formula.

As in the case of density calculations based on the refractive index, the molar polarizability is assumed constant. Roach adopted the value $N_0\alpha = 0.1230$ cm^3/mole. However, the constancy of this product has been an open question. On the other hand, investigations of He^3 [116] suggest that $N_0\alpha$ is a function of density. Naturally, this may also be true for He^4.

On the strength of what has been said, the density data deduced from the refractive index and dielectric constant measurements must be regarded as indirect.

†Changed in proof.—T.B.S.

22 THERMODYNAMIC PROPERTIES OF HELIUM

1.1.7 Thermal and Acoustic Properties on Phase Equilibrium Lines

The thermal and acoustic properties of He^4 have been extensively investigated (Table 1.11). Much of this work has been concerned with the properties of helium in both the state of phase equilibrium and in the single-phase state.

The isochoric specific heat $c_v^{(m)}$ of liquid He I in equilibrium with the crystal is studied in [105, 108, 129], the first two of which do not claim high precision. The isochoric specific heat of liquid helium in equilibrium with the vapor is measured in [105]. Some of the values of c_v' in this work refer to region of existence of He II. In [108], the isochoric specific heat is presented together with the calculated latent heat of melting r_m. Unfortunately, about half of the values of $c_v^{(m)}$ in [108] refer to temperatures $T < T_\lambda^{(U)}$.

Relatively detailed investigations of the specific heat, c_s', of saturated liquid are reported in [90]. The copper calorimeter used in [90] was filled with a thin copper wire to improve its heat transfer properties and to reduce the time necessary to reach thermal equilibrium. A stable temperature is reached in approximately 10 s. This time increases somewhat as the temperature approaches T_{crit}, but does not exceed 1 min. The calorimeter is only partially filled with liquid helium, so that the state of liquid-vapor phase equilibrium is established. Measurements of the heat supplied to the substance in the calorimeter must be corrected for the heat used to evaporate the liquid and to heat up the vapor. This correction is determined from thermodynamic relationships and is important. The temperature was measured on the T55E scale with a carbon resistance thermometer which was calibrated against the saturated vapor pressure of helium before each set of experiments. The uncertainty in the values of c_s' obtained in [90] for $T > T_\lambda$ did not exceed 1%.

Some of the experimental values of c_s' (9 out of 46) refer to temperatures $T < T_\lambda^{(L)}$. The specific heat capacity c_s' in the vicinity of the λ-line was obtained in a very narrow range of temperatures, using a method similar to that in [129]. Most of the values of c_s' (21 out of 31) were for $T < T_\lambda^{(L)}$.

Most thermal measurements have been concerned with the latent heat of vaporization r which appears in the formula used to calculate the saturated vapor pressure and therefore influences the temperature scales at helium temperatures. This explains the interest experimentalists have in this subject. The first thorough measurements of the latent heat of vaporization were carried out in [59] at atmospheric and lower pressures. Because of the special difficulties of thermal measurements at ultralow temperatures, the authors of [59] strived to devise a method of utmost simplicity. This eventually came down to the determination of the level of liquid helium in the calorimeter as heat was supplied to it. The largest error (up to 0.5%) was associated with the measurement of the height of the liquid column in the calorimeter capillary. Errors in the measured electric power and in the heat-supply time do not exceed 0.1% each.

In contrast to [59], the authors of [41] measured the quantity of vapor formed

Table 1.11 Thermal and acoustic properties of helium in crystal-liquid and liquid-vapor equilibria

Property	Year	Author	Temperature interval ΔT, K	Number of points
c_v^m	1922	Keesom and Clusius [105]	2.55–4.01	7
	1936	Keesom and Keesom [108]	3.18–4.36	28
	1959	Lounasmaa and Kojo [129]	1.49–1.89	25
c_v'	1922	Keesom and Clusius [105]	1.4–4.1	31
c_s'	1957	Hill and Lounasmaa [90]	1.8–5.05	46
	1959	Lounasmaa and Kojo [129]	1.47–2.79	31
r_m	1936	Keesom and Keesom [108][†]	2.5–3.5	5
r	1926	Dana and Kamerlingh Onnes [59]	1.74–4.21	16
	1952	Berman and Poulter [41]	2.87–4.20	7
	1954	Berman and Swenson [42]	4.2–4.5	7
	1958	Berman and Mate [40]	2.2–5.2	18
	1967	Ter Harmsel et al. [163]	2.2–5.0	29
w'	1938	Findley et al. [78]	1.76–4.22	8
	1947	Pellam and Square [141][‡]	$T \approx T_\lambda$	—
	1951	Atkins and Chase [35][‡]	1.2–4.2	—
	1953	Atkins and Stasior [36]	1.25–4.20	14
	1966	Vignos and Fairbank [165]	1–4	7
w''	1958	Itterbeck and de Laet [99]	2.2–4.2	8

[†]Calculated.
[‡]Graphical data.

24 THERMODYNAMIC PROPERTIES OF HELIUM

when a given amount of heat was supplied to liquid helium. This yields the "apparent" latent heat of evaporation r_a. The "true" value is calculated from the formula $r = r_a (1 - \rho_v/\rho_L)$ where ρ_v is the density of the vapor and ρ_l the density of the liquid. The quantity r_a was measured in [41] in the range 2.867–4.512 K, and the final values of r were presented for 2.867–4.198 K. Only the directly measured quantities were reported for temperatures above 4.198 K, evidently because there were doubts about the reliability of the data at these temperatures. This is supported by the fact that the estimated error in the values of r_a for $T > 4.198$ K is greater by a factor of 5 than the error for $T < 4.198$ K.

Having striven to make the temperature scale more precise, the authors of [42] measured r_a in a very narrow range of temperatures, which extended the earlier investigations [41]. We note that the value $T_{\text{crit}} = 5.2060$ K reported in [42] is higher than the values obtained in most other experiments.

Berman and Mate [40] measured the heat of evaporation of He I in the widest temperature range, practically from $T_\lambda^{(L)}$ to T_{crit}. These results exceed by almost 1% the earlier data [42], and the discrepancy increases with increasing temperature. It was partially (up to 0.4%) accounted for by the temperature dependence of the density of mercury which was ignored earlier. However, a residual discrepancy remains. The scatter of the data reported in [40] is approximately 0.5% at almost all temperatures. The error in the smoothed values of r [40] is estimated at a few tenths of a percent. At the same time, the values of r differ from the results in [42, 59] and from the smoothed values of r in [11]. The new results are higher by about 0.5% at $T = 2.4$ K and by 1.7% at $T = 4.2$ K.

In their determination of the heat of evaporation, Ter Harmsel et al. [163] also measured the quantity of heat supplied to the liquid and the volume of gas formed. All the quantities needed for the determination of r are presented in [163] for 51 points, including the power generated by the heater, the duration of heating, the number of moles of gas formed, and the apparent heat of evaporation r_a. The error δr_a is estimated as 0.3%. A table of smoothed values of r is presented in [163] for rounded values of temperatures measured on the T58 scale in the region 2.2–5.0 K. The values of the derivative (dp_s/dT) are also presented in this table.

Among measurements of the speed of sound, the data given in [35, 141] are of minor interest because they are presented in small-format graphs, and only a qualitative picture of the function $w'(T)$ is given. The results obtained earlier in [78] are corroborated in [35, 141] in the region of the λ-line.

The speed of sound w' in liquid helium was measured in [78] with an interferometer. The mirror displacement was in the range 25–100 half-wavelengths at 1.3 MHz.

A previously undiscovered maximum was noted in the value of w' in the vicinity of 2.5 K. Far from the λ-point, the data in [78] agree well with the classical thermodynamic values. Three of the eight values of w' lie outside the region of interest to us, i.e., $T < T_\lambda^{(L)}$. Unfortunately, not much information is given in [78] on the experimental technique employed, or on temperature measurements, so it is impossible to make a judgment as to the quality of the data.

The pulse method was used in [36] to measure w'. The speed of sound in the single-phase region was also measured under pressure. Seven of the fourteen values of w' were obtained for $T < T_\lambda^{(L)}$. For 2.2–4.2 K, the error $\delta w'$ was reported to be about 1%. The single-phase region and part of the melting and evaporation curves were investigated in [165], where most of the data on the melting curve refer to $T < T_\lambda^{(L)}$, i.e., to He II. Seven points were obtained on the evaporation curve, three of which were in the region $T < T_\lambda^{(L)}$.

The speed of sound in gaseous He under pressure was measured in [99], and data on the evaporation curve were obtained at the same time. The vicinity of the λ-line was investigated, and measurements of w' were made on eight isotherms for $T < T_\lambda$. One value was obtained at T_λ.

The above data on the thermal and acoustic properties of He^4 in phase equilibrium will be used later when we examine the accuracy of the low-temperature equation of state.

1.2 THERMODYNAMIC PROPERTIES IN THE SINGLE-PHASE REGION

1.2.1 Density

The thermodynamic functions, of helium, especially its density (compressibility), in a wide range of parameters of state are of great interest to theorists and experimentalists alike, and have been under investigation for a long time. This was initially stimulated by the demands of gas thermometry, but, in recent years, helium has become the subject of basic investigations at low and ultralow temperatures at which quantum mechanical effects and the complexity of phase transitions become apparent. The growth of interest in accurate data is due to the increasing application of helium, e.g., in superconducting systems and in nuclear power engineering.

There is, however, a lack of data on p, v, T relationships that are the basis for an equation of state. Basic investigations carried out in the last 50–60 years are summarized in Table 1.12. The information presented in [49, 100-102, 106, 107] has been made significantly more precise by more recent work. At low temperatures, the data presented in these investigations must be corrected for subsequent changes in the temperature scale. However, such corrections are not easy to make because of uncertainties in the temperature scale used and the experimental techniques employed. Of course, all this lessens the value of helium data published in the 1920s. Quantum mechanical effects, solid state properties, and properties in immediate vicinity of absolute zero are investigated in detail in specialist publications.

The data summarized in Table 1.12 can be conveniently divided into high

Table 1.12 Measurements of the density of helium in the single-phase region

Year	Author	Parameter range ΔT, K	Δp, MPa	Number of points
1915	Holborn and Schultze [96]	273–373	2.0–5.1	12
1922	Palacios and Kamerlingh Onnes [140]	20.5	0.01–0.09	4
1922	Holborn and Otto [92][†]	15.15–20.35	0.13–10.7	9
1924	Boks and Kamerlingh Onnes [47]	14–293	1.6–6.4	107
1924	Bridgeman [49]	338	294–1471	13
1924	Holborn and Otto [94]	493–673	2.3–10	19
1924	Holborn and Otto [93]	90–223	1.9–10	27
1924	Kamerlingh Onnes and Boks [101]	2.57–4.21	$1.5 \cdot 10^{-2}$–0.1	18
1924	Keesom and Keesom [106]	1.11–4.24	0.2–2.8	89
1924	Keesom and Keesom [107]	1.64–4.23	2.23–3.82	54
1925	Van Agt and Kamerlingh Onnes [27]	16.6–69.9	$4.6 \cdot 10^{-3}$–0.17	30
1925	Holborn and Otto [95][†]	90–673	0–10.7	98
1927	Nijhoff and Keesom [138]	72; 90	0.3–0.8	9
1927	Nijhoff et al. [139]	14–170	0.1–1.4	24
1929	Gibby et al. [80][‡]	298–448	<12.7	—
1931	Wiebe et al. [168]	203–473	10–101	36
1936	Barnett [52]	273–323	0.2–13	—
1941	Michels and Wouters [133]	273–423	0.9–29.6	119
1946	Kistemaker and Keesom [125]	1.59–2.72	$5 \cdot 10^{-4}$–$2 \cdot 10^{-3}$	67
1947	Keesom and Walstra [110]	9.6–20.5	—	47
1955	Keller [113]	2.15–3.96	$1.4 \cdot 10^{-3}$–0.03	64
1958	Edeskuty and Sherman [67]	2.2–4.2	0.1–14.2	88
1959	Beenakker et al. [39]	20.4	0.045–0.1	5
1960	White et al. [167]	20.6–300	0.1–3.4	152
1960	Hill and Lounasmaa [91][§]	2.9–21.4	0.1–10.3	262

Table 1.12 Measurements of the density of helium in the single-phase region (*Continued*)

Year	Author	Parameter range ΔT, K	Parameter range Δp, MPa	Number of points
1960	Stroud et al. [157]*	250–311	0.6–28.1	68
1961	Edwards and Woodbury [72]	3.0–5.0	p_s–0.46	38
1965	Dobrovol'ski and Golubev [7][†]	20.3–164.2	5.0–49.1	48
1965	Canfield et al. [53]*	133–273	0.2–54	149
1966	Glassford and Smith [81]	4.3–22.5	6.8–136	254
1967	Sullivan and Sonntag [158]*	70–120	1.3–70.3	99
1967	Elwell and Meyer [76]•	1.25–4.2	0.05–2.8	40
1968	Roach [145]•	4.85–5.91	0.22–0.23	350
1969	Briggs et al. [51]*	273	0.1–101	22
1969	Weems and Miller [166]	273–308	0.26–6.9	43
1970	Blancett et al. [45]*	223, 273, 323	0.3–69.3	72
1970	Briggs [50]*	268–353	0.35–82	175
1970	Hall and Canfield [89]*	83–113	0.2–72	71
1971	Provine and Canfield [144]*	143–183	0.26–70	72
1972	Karnus and Rudenko [10][§]	14–53	0.54–11.2	108
1972	Petrov [15][§]	123–293	1.8–40	106
1972	Tsederberg et al. [26]	123–293	1.9–40	108
1976	Bogoyavlenskii and Yurchenko [4]•	1.5–4.2	0.1–10	262
1976	Kalenkov [9][§]	14–273	5.1–100	132
1978	Bogoyavlenskii et al. [3]•	4.2–20	0.01–3.8	406
1978	Dillard et al. [63]*	223, 273, 323	0.7–15	83
1979	Popov et al. [17][§]	6.91–20.8	0.08–13.1	28
1980	Kukarin et al. [13]•	5.06–5.38	0.2–0.27	620
1981	Popov [16]	6.91–31.1	0.08–14	86

[†]Smoothed data.
[‡]Data presented in the form of equations of isotherms.
[§]Data presented on isochores.
*Measurements carried out by Burnett's method.
•Measurements of dielectric constant.

and low temperature groups. The dividing line lies somewhere near the 20 K isotherm.

This is a somewhat arbitrary subdivision although it has become clear that it is precisely in the neighborhood of this isotherm that the equations describing the thermodynamic surface of helium could be matched. In addition, there are no difficulties with temperature measurements for $T > 20$ K, and all the difficult parts of the thermodynamic surface, including the phase boundary lines, lie in the region $T < 20$ K.

Unfortunately, it is not now possible to make an *a priori* appraisal of the experimental methods used in these investigations. It is still more difficult to estimate the mutual agreement of different series of data, particularly if the data obtained do not overlap. As a rule, such an estimation can only be made after an equation of state has been established and the deviation of the experimental data from it have been calculated. Statistics therefore plays an important part, i.e., it is desirable to have several independent sets of data in each range of parameter values, regardless of the labor and cost involved in this approach. From this point of view, existing measurements of the density of helium do not look too good.

In the high-temperature range, researchers have most often been attracted to temperatures between 200 and 350 K for which the results of 14 investigations overlap (Fig. 3). All of them report data at 273 K.

There are no data on the density of helium for $T > 673$ K. However, this is not of any fundamental significance because the second virial coefficient suffices in this region.

Currently available data are not uniformly distributed along the pressure axis. Practically all the work was done for $p < 30$ MPa. As the pressure increases, the number of publications systematically decreases. In the range 80–100 MPa there are four sets of measurements, and for $p > 130$ MPa the only results are those in [85].

The high-temperature data cover different temperature ranges and vary in quality and "consistency." For example, narrow ranges of temperature and pressure are investigated in [133] with great accuracy; while a broad range of temperature and pressure is examined in [9], but with a lower degree of accuracy and self-consistency.

A method typical of the work of Michels was employed in [133]: measurements were made of pv on isotherms in Amagat units. In the variable-volume piezometer used in these experiments, an adjustable mercury column confined the gas to spherical containers separated by short capillary sections. Particular attention was devoted to the determination of corrections. The compressibility was obtained for seven isotherms, and the virial coefficients were calculated for each isotherm from second and third order expansions. The results were then compared with earlier work [80, 168].

Although the purity of the helium used was not indicated in [133] we may assume that the measurements were of the highest quality and that the experi-

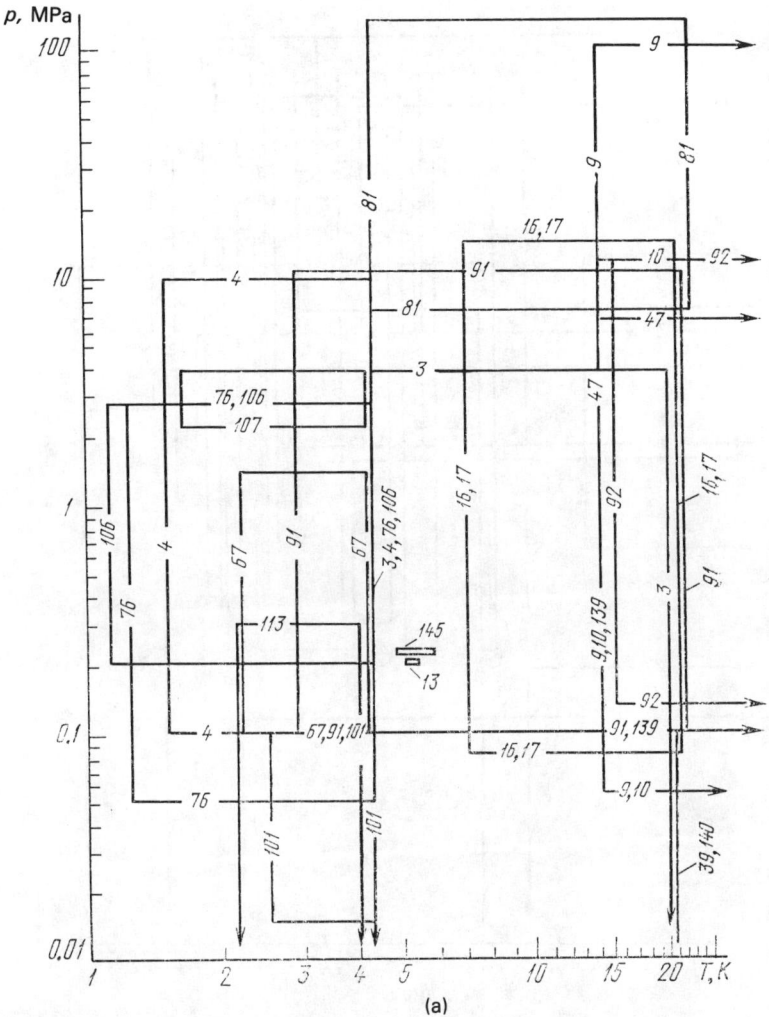

Figure 3 Ranges of parameters of state in which there are experimental data on the density of helium: (a) $T < 25$ K.

mental uncertainty in the great majority of the data is typically no more than 0.1%.

The work of Michels et al. [133] was preceeded by the measurements of Wiebe et al. [168] who covered a relatively wide range of temperature and pressure. The experimental methods employed were described in detail in [38]. Measurements were carried out on the isotherms. A cylinder of known volume was filled at a fixed temperature with the gas at the desired pressure. The quantity of gas was measured later. The dependence of the volume of the cylin-

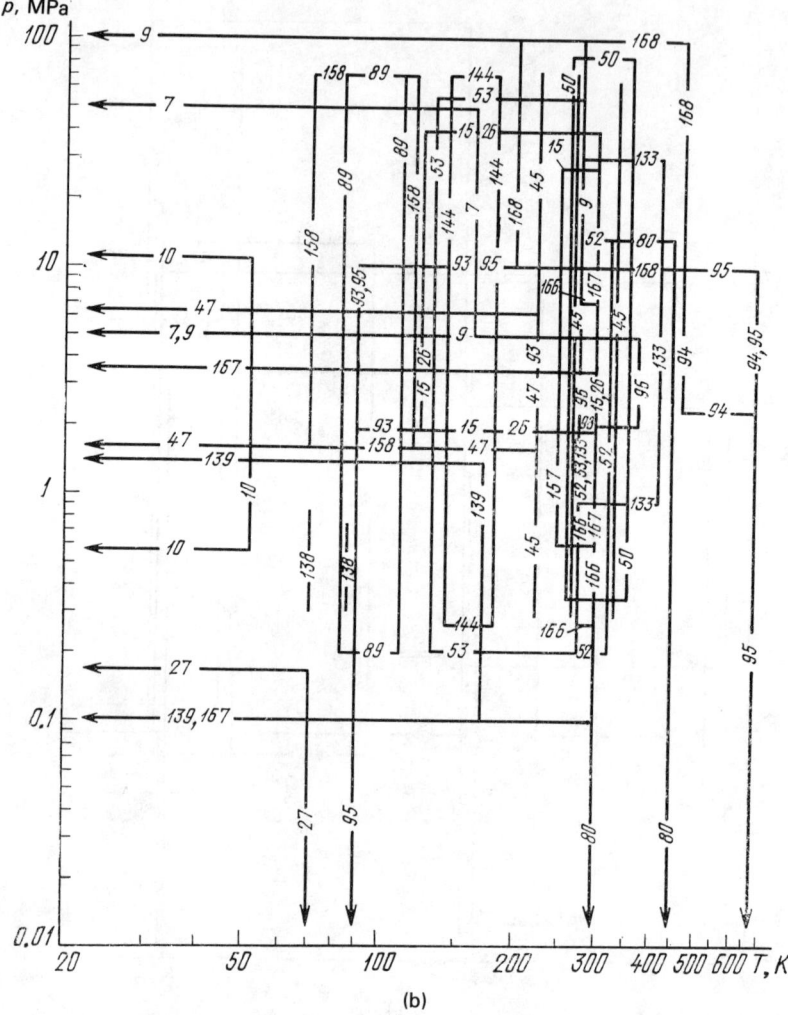

Figure 3 Ranges of parameters of state in which there are experimental data on the density of helium: (*Continued*) (*b*) $T > 20$ K.

der on temperature and pressure was taken into account by calculation. Commercial helium containing more than 2% of contaminants was subjected to a two-stage purification process. Subsequent analysis of the helium gas by the refractometer method at the U.S. National Bureau of Standards revealed that the final impurity concentration did not exceed 0.05%. The precision of the thermostat was 0.05 K, and the precision of the piston manometer (nominal maximum pressure rating of 100 MPa) was 2×10^{-4} at 20 MPa. The temperature was measured with a triple-junction Chromel-Copel thermocouple for

$T > 273$ K and with two five-junction Chromel-Copel thermocouples for $T < 273$ K. The error was estimated from the deviation of the experimental points from a smooth curve, and did not exceed 1%. Two points are too low and one too high by 1%. It seems that the average uncertainty in [168] is much less than this, i.e., 0.2–0.3%.

The melting curves of He^3 and He^4 were investigated in [85] in a wide range of pressure. Molar volumes of the liquid in equilibrium with the crystal were determined. They correspond to the limiting densities for the equation of state constructed in the present monograph. The investigation described in [85] constitutes the first direct measurement of several variables on the melting curve at $T > 4$ K, including the molar volume of the liquid. Moreover, the temperature range was sufficient to ensure that high pressures could be attained.

The helium under investigation was held in a piezometer joined to a U-shaped high-pressure differential manometer by a heated capillary. The mercury level in the tube was maintained in a fixed position to within ± 0.5 mm and was monitored with a floating magnetic indicator. The pressure was measured by two piston manometers with working ranges of 0.5–12 and 5–343 MPa. The former was calibrated against the saturated vapor pressure of CO_2 at 0 °C, and the latter against a standard piston manometer. Temperature stability in the liquid thermostat was secured by establishing constant saturated vapor pressures of the thermostatic fluid (helium, hydrogen, and nitrogen) in the temperature ranges 1.2–5, 14–24.5, and 24.5–31 K respectively. The piezometer volume was determined from the measured mass of mercury necessary to fill it (with corrections for temperature, pressure, and dead volume). The purity of helium, as determined by a mass spectrograph, was at least 99.95%. The concentrations of hydrogen, nitrogen, and oxygen did not exceed 0.01, 0.03, and 0.002%, respectively.

Each molar volume presented in [85] is the average of 2–5 measurements. The results agree with those in [65] to within 1%. This uncertainty is apparently also typical of the data reported in [85], owing to the difficulty of working under high pressures in the state of liquid-crystal equilibrium, and the somewhat controversial nature of the apparatus employed. Nevertheless, these data are very important because they lie at limit of the range of state variables that we are investigating here.

In [167] the compressibility of helium was measured in a wide range of temperature and a narrow range of pressure. This was dictated by the basic aim of the experiment, which was to determine the second virial coefficient and the departure from the ideal gas law. Particular attention was paid to the purity of helium and to the determination of errors. The helium was distilled three times and was purified with activated charcoal at the boiling point of N_2. The final purity was 99.99% (the remaining contaminants were traces of N_2 and O_2). The measured pressure was corrected for the temperature dependence of the density of mercury, the capillary depression, the true value of the acceleration due to gravity, and so on.

The compressibility of helium was measured in [157] by the Burnett method within a comparatively narrow range of parameter values. It will not be necessary to describe this method in detail because it is widely used in thermophysical research. Errors were estimated in [157] as 0.03 K in temperature and 7×10^{-5} MPa in pressure. The resultant uncertainty was put at 0.1%. It is difficult to appraise this work because the apparatus is not fully described. However, these results overlap in temperature and pressure the results reported by other workers, and this allows a comparison to be made between them.

The data presented in [45, 53, 89, 144] are mutually complementary. All of them use the Burnett method, but they cover different ranges of parameter values. Temperatures were measured on the 1948 scale, using a platinum resistance thermometer and a d.c. bridge. Pressures were measured to within 0.01% with two piston manometers (3.5×10^{-2}–16.8 and 0.2–70 MPa). The uncertainty in temperature, the precision of the thermostat, and the temperature gradients were somewhat different in different reported experiments. The aim was to examine the properties of mixtures, and data on pure materials were used only to check the equipment and experimental procedure. Nevertheless, significant data were obtained for helium.

The uncertainty in the measured temperature was estimated in [53] as 0.01 K. There is an indication that a mass-spectrometer was used to monitor the purity of helium, but there is no information about the amount of contamination. Comparison of the coefficients of compressibility found in [53] with the values obtained in [152, 167] shows that the data in [53] are subject to an uncertainty of $\delta z = 0.03$–0.05%.

A differential copper-constantan thermocouple was used in [45] to measure the temperature gradient in two parts of the apparatus. The precision of the thermostat was 0.03 K, and a comparable maximum temperature gradient occurred at 50 °C. The system contained two dividing diaphragms, one of which was situated in the cryostat, the other outside it. Corrections were made for the hydrostatic pressure of the oil and for the gas column. The estimated total uncertainty varied from 0.001% at the lowest pressure to 0.006% at the highest pressure. The range of parameter values investigated in [144] does not go beyond that of the earlier investigations in the same laboratory [45, 53], while improvements introduced in the more recent work are not of fundamental significance.

The work reported in [89] differs from earlier studies by the design of the cryostat, which significantly extends range of the investigation to lower temperatures. This improved cryostat produced greater temperature uniformity: the temperature gradient did not exceed 0.005 K.

The measurements reported in [158] cover a relatively wide pressure range and encompass temperature intervals that were investigated at only one or two other laboratories. Once again, the Burnett method was employed. The thermostat maintained the temperature to within ±0.005 K and the temperature gradient did not exceed 0.005 K. The uncertainty in the measured temperature was

0.015 K. Although the temperature scale was not explicitly identified, we may assume that it was the 1948 scale. The pressure was measured with a differential divider and a piston manometer. The helium used was carefully purified and contained traces of hydrogen, water vapor, oxygen, and carbon dioxide with a total concentration of about 2×10^{-6}. The concentrations of nitrogen and neon were 4×10^{-6} and 14×10^{-6}, respectively, and had little effect on the accuracy of the data. It was noted that the data obtained in the range 0–10 MPa agreed with the data in [92] to within 0.3%, and the maximum uncertainty on the 80 K isotherm ranged from 0.05 to 0.13% at pressures of between 2 and 60 MPa, respectively. This is an indication of the satisfactory quality of the data reported in [158].

The authors of [51] reformulated the problem after examining the virtues and deficiencies of the Burnett method. The method does not actually require a knowledge of the exact volume of the piezometer, or the number of moles of the gas. It ensures high accuracy, but does not allow a direct determination of z without a knowledge of the function $z(\rho)$ or a graphical representation of it. The work reported in [51] had two aims, namely, to choose this function and to determine the effect of the weighting factor on its description. A total of 22 sets of measurements by the Burnett method were carried out at a temperature of 0 °C, practically, and pressures up to 101 MPa [51]. The helium used contained traces of Ne and N_2 with concentrations of less than 45×10^{-6}. The thermostat maintained the temperature to within ± 0.0015 K with a temperature gradient of 0.0025 K. The temperature was measured to within ± 0.001 K. In different sets of measurements, the temperature ranged from -0.006 to $+0.004$ °C on the 1948 scale. The authors of [51] estimated the uncertainty in z as 0.01%, which does not appear to be correct. The comparison made in [51] with six other investigations by different authors shows that this figure is indeed unsatisfactory because most points have a spread that is several times greater than the above figure, and for a few points [53, 133, 168] the spread reaches 0.13–0.3%.

A similar method was used by Briggs [50], except that the range of temperatures was wider and the uncertainty in z was estimated as 0.05%.

A very narrow range of temperature and pressure is covered in [166]. The Burnett method and a very sensitive manometer (7×10^{-4} bar) were employed. Corrections were introduced for the temperature dependence of the piezometer volume and for the dead volume. Very pure (99.9991 mol %) helium was used. However, the use of a mercury thermometer reduced the precision of temperature measurement (0.2 °C on the 0°, 10°, and 20 °C isotherms; 0.3 °C on the 23° and 35 °C isotherms). The error in z was estimated as 0.1%. In the context of the extensive measurements by different authors, the work reported in [166] is hardly an important contribution to our knowledge of the density of helium.

Comparatively recently, the Burnett method was used with reasonable accuracy [63] to investigate helium containing no more than 5×10^{-3} mol % of N_2, O_2, Ar, Ne, H_2O, and CO_2. These results agree with other published data.

However, the measurements were carried out by traditional methods in a comparatively narrow and well investigated range of the parameters of state.

A series of investigations of the thermodynamic properties of helium was carried out in recent years in the U.S.S.R. It includes [9, 10, 15, 16] and also a few original investigations of $T < 20$ K.

A modified Burnett method was used in [15] to determine the density of helium. Apart from the two piezometers used in the traditional Burnett method, the apparatus [15] included a volume manometer and a room temperature thermostat. The gas was removed from the second piezometer into volume manometer after every expansion, so that the constant of the apparatus could be determined with great precision. This constant was then used in the analysis of results. The gas under investigation was isolated from the pressure-measuring system by a bellows-type null indicator with an inductive displacement transducer and a membrane indicator incorporating a capacitive transducer. Pressure was measured with piston manometers (type MP-60 and MP-600). The overall uncertainty δp was 0.05% for $p \geq 6$ MPa and 0.08% for $p \approx 2$ MPa. The gas pressure in the volume manometer was measured with an absolute pressure manometer (type MAD 2500) with an estimated uncertainty of 0.03%. The temperature in the thermostat was measured with a TSPN-1 platinum resistance thermometer, and in the volume manometer with a PTS-10 thermometer working with a R308 potentiometer and the R321 standard resistance coil (precision class 0.01). Total uncertainty in the measured temperature was 0.02 K. The helium used in [15] contained not more than 0.01% of contaminants. The experimental uncertainty in z was estimated to be 0.001%. Regardless of the care with which the individual quantities were measured, this degree of accuracy seems to be somewhat optimistic.

Experimental data in the relatively unexplored temperature range 14–54 K was obtained in [10]. A constant-volume piezometer with a sealed capillary was used. The temperature was measured with a platinum resistance thermometer to within ± 0.02 K. The pressure was determined with a standard manometer, but the accuracy attained ($\pm 0.4\%$) falls short of modern standards of measurement. Analysis of the sources of error showed that the experimental uncertainty in density was $\delta \rho < 0.4\%$. No quantitative information is given on the presence of contaminants. All we know is that, before the helium was allowed into the piezometer, it was passed slowly through a layer of active carbon at a temperature of boiling hydrogen. Circumstantial evidence suggests that the 1968 scale was employed in [10]. The authors of [10] state that their data agreed well with the results reported in [91], and satisfactorily with those in [92] and [167]. They note a significant deviation from the results in [47].

The author of [9] points out that most of the earlier work covered pressures below 10 MPa, and there were discrepancies of up to 1% at temperatures of 80–273 K. Although the first of these statements is not quite accurate, the investigation reported in [9] is undoubtedly of interest because it was the first time that measurements of compressibility covered such a wide range of temperature and pressure (14–273 K, $p < 100$ MPa). A constant-volume piezometer was used in [9]. The spherical shape of the loaded piezometer allowed the necessary corrections to be determined more accurately. The thermostat was accurate to within ± 0.005 K. The temperature was measured with a platinum resistance thermometer on the 1968 scale, using a R348 potentiometer (precision class 0.02) and a standard resistance coil (precision class 0.01). Helium in the piezometer was separated from the pressure-measuring system by bellows-type differential manometers with sensitivity of 1×10^{-4} MPa (MP-60, MP-600, or MP-2500; precision class 0.05). A volume manometer with a working volume of 26 dm^3 was used to determine the quantity of gas, and was held at 20 °C by a thermostat. Corrections for the dead volume were determined by the sealed-capillary method. The total combined concentration of hydrogen, nitrogen, oxygen, neon and carbon dioxide CO_2 did not exceed 0.015% by volume. The description of the method and procedure used in [9] suggest that this work conforms to modern standards of measurement. The author of [9] reports that the uncertainty δz varied with temperature from 0.25% at 14 K to 0.18% at 40 K, and was 0.15% in the range 50–273 K. Comparison with earlier results showed that the values obtained agree well with the data in [45, 90, 95, 144, 157, 158, 167]. Data of [168] at the greatest pressure deviate by 0.2%. A 1% discrepancy with [7] was found at 5 MPa, but tends to fall with increasing pressure. The disagreement with [81] in the range 14–20 K is up to 3%. This places the estimate of uncertainty given in [9] under some doubt.

The most important thermodynamic properties of helium are concentrated in the low temperature region, which we have arbitrarily defined as lying below 20 K, but very little data is available in this range. Most investigations have been confined to pressures below 100 MPa, and only in three cases does the pressure reach 100 MPa. A series of investigations [101, 106, 107, 140] was carried out before 1930 in the laboratory of Kamerlingh Onnes. Some of the data in [27, 47, 139] refer to $T < 20$ K, but density data at higher temperatures are also reported. Most of this work was concerned with measurements at low pressures and is now largely obsolete in the light of modern standards of measurement, and is of historical interest only. Data obtained at atmospheric and very low pressures are reported in [72, 110, 113, 125], while a narrow, near-critical, range of parameters was investigated in [13, 145].

The two investigations performed in 1946–1947 [110, 125] are methodologically similar and cover a similar range of pressure, i.e., well below atmospheric pressure. Unfortunately, the temperature ranges investigated in [110] and [125] do not overlap. Measurements were carried out on isotherms, and the results are presented as values of pv in Amagat units. The greatest complications that had to be overcome in [125] were related to measurements at $T < 2.6$ K where the isotherms cover a very narrow density range in the single-phase region, so that the measurement of pressure and the maintainance of constant temperature for long periods of time are difficult. A special method was employed in [125] to measure pressures to within ± 3 μm Hg. The thermomolecular effect was found to be significant and had to be taken into account separately. A special method was used to increase the precision of temperature measurements because two accurate values of pv at constant temperature were needed to determine the virial coefficient. This method relied on the use of two reservoirs under identical temperature conditions, so that two measurements could be made simultaneously. The temperature was measured on the 1937 scale with an uncertainty of 2×10^{-3} K, using a platinum resistance thermometer. The temperature was also determined from the vapor pressure in the cryostat and from the coefficient A_a in the virial expansion, extrapolated to zero density. The mean difference between temperatures measured by these different methods was 0.01 K, i.e., substantially greater than the thermometer uncertainty. The authors of [125] attribute this to possible errors on the 1937 scale.

The basic task in [113] was to improve the accuracy of data required for gas thermometry. Results at pressures up to 230 mm Hg were obtained, and the second virial coefficients were determined from them for two- and three-parameter isotherm equations. Keller developed a new method of obtaining pv isotherms at low temperatures and set up an apparatus for these measurements. The dead volume was taken into account experimentally using a volume manometer. The temperature was measured with a condensation helium thermometer, and the pressure with mercury and oil manometers. The results were presented on the 1948 temperature scale but led to doubts about the correctness of this scale, and possible corrections were considered.

The first measurements of the density of liquid helium at pressures up to 14 MPa were reported in [67]. The method used involved the successive discharge of small quantities of helium from a piezometer into a volume manometer. The temperature was measured by a helium condensation thermometer on the T55E scale proposed by Clement. Pressures in the ranges <0.1 MPa, 0.1–0.35 MPa, and >0.35 MPa were measured by a mercury manometer, a Bourdon gauge, and a piston manometer respectively. For $p > 0.35$ MPa, a magnetic float was used to monitor the level of mercury in the U-shaped divider. The uncertainty in the determination of the mercury meniscus corresponded to an uncertainty of 0.01% in density. Only a limited comparison of the results in [67] with the data of different authors is possible. At low pressures, the results may be compared with those in [107] where the density is too low by 0.3%. However, at $T = 2.2$ K, the values calculated in [115] are higher by about the same amount (0.34%).

The data in [39] were obtained at the single temperature of 20.4 K in a narrow range of pressure. No information is given on the helium used, but experiments on He-H mixtures used samples of mediocre purity, which casts doubt on the purity of helium. The various methods used in [39] are of basic interest for the determination of virial coefficients of mixtures.

In the complex investigation reported in [91], the p, v, T relations were measured on 25 isochores in the density range 0.488–2.923 g/cm^3. The isochoric heat capacity, the derivatives $\partial p/\partial v$, $\partial p/\partial T$, and their temperature dependence were also measured (this part of the experiments will be discussed later).

One working vessel, which serves as a piezometer and a calorimeter, was used in all these measurements. Its ingenious design had to satisfy contradictory demands. It had to be stable and have low heat capacity. Apart from the calorimeter (the name given to the working vessel), there is a volume manometer consisting of two vessels of approximately 200 cm^3 each, two mercury-in-glass manometers, a Bourdon gauge, and a piston manometer. Valves which allow the necessary experimental operations to be carried out are provided. The apparatus is thus a constant-volume piezometer with a dead volume and a volume manometer.

The temperature in the ranges 10–20 K and 3–11 K was measured with constantan and carbon thermometers, respectively. Each of these thermometers has its merits and limitations. The constantan thermometer is less sensitive, but is more stable. The carbon thermometer has a very simple temperature dependence, but its resistance becomes very low with increasing temperature. Before each experiment, both thermometers were calibrated in the range 12–20 K against the hydrogen vapor pressure, using the equation for p_s proposed by Woolley et al. [175].[†] At lower temperatures the thermometers were calibrated against helium vapor pressure on the T55E scale which, for our purposes, may be considered to be identical with the T55 and T58 scales. The temperature was measured to within ±0.01 K for $T > 4$ K and approximately ±0.003 K for $T < 4$ K.

For the sake of convenience the experiments reported in [91] were carried out in two stages. Initially, for temperatures of 3–11 K, the carbon thermometer was used and helium was the thermostatic medium; later, for temperatures of 10–20 K, the constantan thermometer and liquid hydrogen were employed. Pressures below 4.5 MPa were measured with a piston manometer, and the Bourdon gauge was used at higher pressures. At low-pressures, the error δp depends, basically, on the precision of the piston manometer. In [91], this was 0.01%, but for $p < 0.4$ MPa and $p > 4.5$ MPa this figure rose to 0.2%. The volume of the piezometer and the number of moles of gas were measured to within 0.1% and 0.2%, respectively, which means that δv was 0.3%.

Commercial helium was purified with activated carbon at the temperature of boiling oxygen and hydrogen. The total uncertainty in the measured pressure, taking into account the errors δv and δT, was estimated as ~1%, and the probable uncertainty was 0.6%.

The authors of [91] compared their values with the data in [107] and [93]. In [107], the temperature range was 1.15 to 4.2 K and the pressure was up to 35 MPa. The densities reported in [107] were lower by about 0.3%, but the devia-

---

[†]Changed in proof.—T.B.S.

tion decreases with increasing density. The densities reported in [93] are higher than those in [91] by about 0.35% in the range 15–20 K.

Measurements of the density and internal energy of helium were reported in [81] and extend the data in [91] to higher pressures. The method used and certain parts of the apparatus were similar in these two experiments. The essential difference was the use of the helium-filled gas thermometer in [81]. The pressure of the thermometric gas was measured with an oil manometer and a cathetometer, and corrections were introduced. Measurements on each isochore were preceeded by calibration of the thermometer against helium vapor pressure. The temperature of the helium bath was measured to better than $\pm 10^{-4}$ K. The calibration was carried out on the 1958 scale. The total experimental uncertainty in temperature was estimated in [81] as less than 0.005 K.

In the range 3.5–140 MPa, the pressure was measured in [81] with a piston manometer calibrated at NBS. The total uncertainty, including the uncertainty introduced by the differential divider, was 0.1% at 7 MPa and 0.005% at 140 MPa. The mass of gas was measured with a volume manometer at atmospheric pressure and room temperature, and corrections were introduced for the mass of gas in the feedline.

Experimental values of $p(T)$ were obtained for 12 isochores in the density range 3.08–5.21 g/cm^3. The measured pressures on the melting curve lay on a smooth curve to within 0.1%. Measured p, v, and T at temperatures of 7, 12, and 17 K were also compared with extrapolated values deduced from the Lounasmaa equation, using data for $p < 10$ MPa. Good agreement was observed for $T < 17$ K. At lower temperatures and, particularly, at high pressures, extrapolation toward the melting curve becomes unreliable. This substantially enhances the value of the experimental data in [81].

The investigations reported in [145] and [13] are very similar. The dielectric constant was measured in both experiments, and the density was calculated from the Clausius-Mossotti formula. Both investigations were confined to the region close to the critical point. The virial form of the equation of state adopted in our monograph does not claim to describe accurately the properties of helium near the critical point. The information contained in [13, 145] that is of use to us here relates only to the critical parameters, and will not be analyzed in greater detail.

The molar volume of liquid He^4 was determined on isobars in [76], again by measuring the dielectric constant and applying the Clausius-Mossotti equation with molar polarizability equal to 0.1230. A carbon resistance thermometer was used in these experiments to record the temperature on the 1958 scale. The pressure was measured to within 3×10^{-4} MPa for $p < 0.1$ MPa and 5×10^{-3} MPa for $p > 0.1$ MPa. No estimates were given of the accuracy of the results obtained.

The density of He^4 was measured between the years 1977 and 1979 across a wide range of state parameters [16, 17]. In effect, these investigations provide a a single, more or less complete, set of results. The experimental methods and the apparatus are described in great detail in [16]. A constant-volume piezometer was used and was connected by a capillary to a separating set of piston manometers. A valve separated the piezometer from the volume manometer. The divider was in the form of a membrane, sealed with mercury on one side, and provided with a viewing device. The piezometer was mounted in a cryostat, the temperature of which was monitored by a germanium resistance thermometer. The temperature gradient was determined with a gold-chromel thermocouple, the junctions of which were held in the upper and lower parts of the piezometer. The volume manometer was located in a special thermostat. The temperatures of the piezometer and the volume manometer were held constant to within ± 0.001 K and ± 0.005 K, respectively. Measurements consisted of successive admissions of helium from the piezometer into the volume manometer which was carefully evacuated before each such admission. The volumes of the piezometer and volume manometer were carefully measured (by determining the mass of water necessary to fill the volumes), and the dead volume was determined. Corrections were made for the zero shift of the divider during pressure measurements, and for the hydrostatic pressure in the mercury and oil between the membrane and the piston manometer. The temperature of the piezometer was determined with platinum and germanium resistance thermometers, and the temperature of the volume manometer was measured with a platinum thermometer. Pressure was measured with MP-600 manometers (precision class 0.05) or MP-60 and MP-6 manometers (precision class 0.005). The pressure in the volume manometer was determined with a constant-volume mercury manometer and a cathetometer. Corrections were made for the temperature of mercury in the manometer.

The helium used in [16, 17] was 99.985% pure. Estimates of uncertainties in the measured mass and volume showed [16] that the coefficient of compressibility z was obtained to within 0.4–0.2% at temperatures of 6.9–20.8 K and pressure of 0.2 MPa. This uncertainty fell to 0.13% when the temperature and pressure rose beyond these levels.

The data obtained were found to agree with [167] to within the overall experimental uncertainty (0.1% at 290 K and 0.15% at 124.6 K). An analysis was made in [16] of the influence of the choice of the temperature scale in different experiments on the consistency of the resulting compressibilities. Of course, the discrepancies increase with decreasing temperature and can reach 0.2% as $T = 13$ K is approached.

The results obtained in [16, 17] are of considerable interest because they were carried out over a wide range of parameter values, using the same apparatus and methods. However, there is some doubt about the uncertainty estimated in [16, 17].

The experiments reported in [9, 10] were described above in our discussion of the high-temperature region. The dielectric constant was measured in [3, 4] as in [13, 76, 145]. The investigations carried out in [3] and [4] are complementary to a considerable extent because they cover adjacent temperature regions ($T < 4.2$ and $T > 4.2$). The dielectric constant was determined from the measured resonance frequency of the empty and filled resonator. In the ranges 2–5 K and 10–20 K, the temperature was measured on the 1958 scale and the standard hydrogen scale. The accuracy of the thermostat was 0.01 K for 4.2–14 K and 0.001 K for $T < 4.2$ K. The precision of the thermometer calibration was 0.01 K at 2–5 K and 10–20 K, and 0.05 K at 7–8 K. The pressure was measured with a Boudon gauge to within $<0.2\%$. It was noted that the uncertainty in the specific volume was found to increase from 0.2% to 3% as the specific volume increased from 25 to 1000 cm^3/mole. The results in [4] agree with [67] and [91] to within the experimental uncertainty. The data in [3] are said to be in "good" agreement with [91] and [10] with the exception of a "small" (up to 5%) discrepancy at $T = 6-8$ K, while for $T < 4.2$ K the data agree with [4]. It is difficult to accept this characterization of the quality of the data.

Since the data in [3, 4, 76] were obtained by an indirect method and the experimental uncertainties depend on factors that are difficult to determine, and if we take into account the estimated uncertainties in [3, 4], we are led to the conclusion these data must be given less weight in our derivation of the equation of state.

1.2.2 The Second Virial Coefficient

The widespread use of helium gas thermometers has necessitated a careful determination of the second virial coefficient. This had led to a large number of investigations, in the majority of which either the absolute pv-isotherm method or the Burnett method were used (Table 1.13).

The data on the second virial coefficient in the ranges $T < 20$ K and $T > 673$ K are of particular interest. In the first of these regions the virial coefficient varies with temperature in a simple way, and the precision with which it is determined has a considerable influence on the precision of the equation of state. In the second of these regions, there is a lack of experimental data on density, but the pv-isotherms are described accurately by a linear function of the form $pv = A + Bp$. The pattern of the data obtained by different workers along with temperature range can be seen in Table 1.13.

The work reported in [172] should, perhaps, be excluded from these data because it was performed at very high temperatures, using a different method.

A modified Burnett method was used in [172] in which one part of the apparatus was placed at a high temperature and the other at 0 °C. This substantially simplified the design of the apparatus and, in particular, the form of the high-temperature vessel. The thermostat of this particular part of the apparatus proved to be simpler and guaranteed better temperature stability (± 0.03 K). The temperature was measured with a platinum-platinum/rhodium thermocouple, i.e., essentially at a point so that the uniformity of temperature could be

Table 1.13 Measurements of the second virial coefficient of He^4

Year	Author	Temperature interval ΔT, K	Number of points
1923	Penning and Kamerlingh Onnes [142]	23–68	8
1924	Boks and Kamerlingh Onnes [47]	14–293	14
1924	Holborn and Otto [94][†]	473–673	3
1925	Van Agt and Kamerlingh Onnes [27][†]	16.6–69.9	6
1925	Holborn and Otto [95][†]	90–673	4
1927	Nijhoff and Keesom [138]	71–90	2
1927	Nijhoff et al. [139]	13–170	8
1929	Gibby et al. [80]	298–448	7
1931	Wiebe et al. [168][†]	203–473	6
1935	Keesom and Kraak [109]	2.6–4.2	4
1941	Michels and Wouters [133][†]	273–423	7
1946	Kistemaker and Keesom [125]	1.6–2.7	7
1947	Keesom and Walstra [110]	9.6–20.5	6
1949	Schneider and Duffie [152]	273–873	7
1950	Yntema and Schneider [172][‡]	873–1473	4
1954	Kilpatrick et al. [123]	0.3–60	96
1955	Keller [113][†]	2.15–4	5
1959	Beenakker et al. [39]	20.4	1
1959	Varekamp and Beenakker [164]	14–21	8
1960	White et al. [167][†]	20.6–300	22
1960	Stroud et al. [157]	250–328	7
1965	Canfield et al. [53]	133–273	6
1967	Sullivan and Sonntag [158][‡]	70–120	4
1968	Boyd et al. [48][§]	2–20	—
1969	Weems and Miller [166]	273–308	5
1969	Briggs et al. [51]	273	1
1970	Blanset et al. [45][†]	223–323	3
1970	Hall and Canfield [89][†]	83–113	3
1971	Provine and Canfield [144][†]	143–183	3
1972	Berry [43]	4.2–20.4	3
1978	Dillard et al. [63][†]	223–323	3
1979	Berry [44]*	2.6–27.1	8
1980	Gugan and Michel [88][†]	4.2–27.2	8

[†]The value of the third virial coefficient is presented in the same temperature interval.
[‡]The data are presented in graphical form.
[§]Data are presented in the form of an equation.
*Three values of the third virial coefficient are presented in the temperature range 3.3–7.2 K.

monitored. The temperature gradient was determined using a stainless-steel model of the high-temperature vessel. Three thermocouples were attached to this model. The temperature difference between the ends of the apparatus and the center did not exceed 0.5 K. The pressure was measured with a piston manometer calibrated against the saturated vapour pressure of carbon dioxide.

New values of the second virial coefficient at 600, 800, 1000, and 1200 °C are presented in [172] together with the earlier values [152] obtained in the range 0–600 K. The two sets of data are in agreement.

It was noted in [18, 19, 32] that the values of the second virial coefficient obtained in different experiments were, on the whole, in good agreement. At low temperatures, the spread did not, as a rule, exceed 10 cm^3/mol. At temperatures above 20 K the spread falls to 3.1 cm^3/mol. The data reported in [47] show a considerable deviation and are the exception.

1.2.3 Thermal and Acoustic Properties

Most work on acoustic and thermal properties of helium have been devoted to the determination of the isochoric specific heat c_v (Table 1.14).

The data reported in [91, 129] constitute a single set of measurements because they were obtained in the same laboratory, using similar methods and identical apparatus. Isochoric data were reported in both investigations. The range of parameters in [91] is significantly greater than in [129]. It is important to note that the data in the former paper were obtained for both liquid and gas phases.

The method employed in [91] yielded values of $(\partial p/\partial T)_v$ which were used with the measured T, c_v and the calculated $(\partial v/\partial p)_T$ to determine c_p with sufficient accuracy. Smoothed values of c_p are presented in [91] for temperatures in the range 3–20 K and pressures of 0–10.1 MPa.

The temperature scales used in [91, 129] were T55E and T58. However,

Table 1.14 Measurements of the isochoric specific heat of helium in the single-phase region

Year	Author	ΔT, K	$\Delta \rho$, g/cm^3	Number of points
1959	Lounasmaa and Kojo [129]	1.47–2.96	0.1439–0.1834	209
1960	Hill and Lounasmaa [91]	2.0–19.85	0.0220–0.1995	388
1964	Dugdale and Franck [64]	7–29	0.24–0.34	45
1969	Moldover [137]	4.83–5.26	0.069–0.080	236

these scales were deemed coincident in the region investigated in [91]. The values of c_v were determined with an estimated uncertainty of approximately 0.5%.

The isochoric specific heat c_v of crystalline and liquid helium was measured in [64], and smoothed values of c_v were presented in a table for four values of the density at temperatures in the range 7–29 K. The smallest number of points are given for the isochore with the highest density, and correspond to the highest temperature.

Values of c_v on seven isochores are presented in [137] for a region very close to critical density in a very narrow temperature range.

The isobaric specific heat has been measured in only one contemporary investigation. In [14], 99.992% pure helium was used, and the continuous-flow method was employed. The temperature was measured with a platinum resistance thermometer and the pressure with a piston manometer (precision class 0.02). Specific methods were used to secure a constant flow rate. In seven series of experiments, 44 values of c_p were obtained at 16.6 to 40.9 K and pressures in the range 0.5–4 MPa. The results reported in [14] agree with those in [91] to within an average of 0.8% although the uncertainty in the experimental values of c_p was estimated to be 1.1%. We note that the values of c_p reported in [91] cannot, strictly speaking, be regarded as experimental.

Calculations of c_p, c_v, and c_p/c_v from speed of sound data are presented in [99] for temperatures of 2.08–4.23 K at sub-atmospheric pressures.

The integral Joule-Thompson effect was investigated in [148] by throttling helium through a porous membrane in a closed circuit. The pressure and temperature were measured in front of (p_1, T_1) and behind (p_2, T_2) the membrane. Contaminating N_2 was removed by the condensation method and by absorption in active carbon at the boiling point of liquid air. Nevertheless, the final concentration of N_2 was 0.1%. It was considered that this level of contamination would not substantially influence the final results obtained from the Joule-Thompson effect. Values of the J-T coefficient μ were obtained in [148] for temperatures in the range 81–576 K. It was noted that they were negative and not strongly dependent on pressure. The numerical value of μ did not vary much, except for data obtained on two isotherms at the lowest temperatures.

The enthalpy of helium-nitrogen mixtures was investigated in [130]. The conclusion reported in [148] that μ was independent of pressure has been the subject of some criticism. The authors of [130] note that the detailed description given in [148] can be used to determine corrections to these data.

The enthalpy and the integral Joule-Thompson effect were investigated in the two methodologically similar experiments described in [173, 174]. In [173], the Joule-Thompson effect was measured by the continuous-flow method. A total of 31 values of μ were obtained at initial temperatures of 6, 8, 10, 15.5 and 17.5 K and initial pressures of 0.4–5.6 MPa. The enthalpy of helium at various temperatures from 4.3 to 23.26 K at atmospheric pressure was measured by a calorimetric method in [174]. A total of 49 experimental values of the enthalpy was obtained, and the smoothed values of the enthalpy were presented in a table. An equation for c_p was also presented.

Table 1.15 lists measurements of the speed of sound in both gas and liquid phases.

Findlay et al. carried out two sets of measurements of the speed of sound in helium. The first [78] was made on the liquid-vapor equilibrium curve. Subsequent investigation [79] improved the data in the λ-point region $T \simeq T_\lambda$ on four isobars. Unfortunately, only a graph is reproduced. The speed of sound was measured in [36, 165] by the pulse method in the liquid phase. No dispersion was observed in [165] in the working frequency range of 0.2–15 MHz. The frequency used in [36] (11.8 MHz) was within this range and therefore no dispersion occurred. The speed of sound w on the λ-line was obtained in both these investigations, but in [36] the corresponding data were excluded from the table, apparently because of doubts about reliability. The authors of [165] estimate the typical uncertainty as $\delta w = 0.1$–0.3%. Only a graph is given in [36], but the smoothed values of w are presented in a table with an estimated uncertainty of 1%.

The speed of sound in the gas phase was measured in [87, 99] at "helium temperatures." The function $w(p)$ was extrapolated to zero pressure, and the limiting value w_0 obtained. The pressure dependence of the speed of sound was also measured in [143] for 21 isotherms, each covering a narrow range of

Table 1.15 Experimental investigations of the speed of sound in He^4

Year	Author	Parameter range		Number of points
		ΔT, K	Δp, MPa	
1939	Findlay et al. [78, 79]	$T \approx T_\lambda^{(L)}$	p_s–0.56	Graphic
1953	Atkins and Stasior [36]	1.25–4.20	p_s–7.1	114
1958	van Itterbeek and de Laet [99]	2.08–4.23	0–0.09	60+8[†]
1966	Vignos and Fairbank [165]	1.0–4.0	0.25–5.07	55
1966	Plumb and Cataland [143]	2.3–20	$2 \cdot 10^{-3}$–0.14	150
1967	Grimsrud and Werntz [87]	2.1–3.8	$(1.6$–$38) \cdot 10^{-3}$	70+8[†]
1968	Barmas and Rudnick [37]	$T \approx T_\lambda$	—	Graphic

[†]Meaning w obtained by extrapolating to $p = 0$.

pressure. The principal aim of this experiment was to measure the temperature by an acoustic method. The authors of [143] obtained the value of w_0 for each isotherm by extrapolation, as in [87, 99]. For a monatomic gas, this value is related to the corresponding thermodynamic temperature by the equation $w_0 = \sqrt{5RT/3M}$ where M is the atomic mass. The detailed character of this investigation and its temperature range were dictated by metrological considerations. The problem was to set up an acoustic temperature scale for the range 2–20 K. This scale is known as the Provisional Temperature Scale (NBS 1965).

The isochoric specific heats obtained in [91] can be used when the equation of state is established by the linear procedure of determining the coefficients from specific heat data. Speed of sound data are used subsequently as a check on the resulting equation of state.

CHAPTER
TWO

METHOD OF CONSTRUCTING A SINGLE EQUATION OF STATE AND COMPILATION OF THERMODYNAMIC PROPERTIES

A method by which a single equation of state for the gas and liquid phases is deduced from experimental data for thermodynamic and thermal properties is reported in [21, 22]. Here, we present only the fundamentals of the method and the changes that are necessary for applications to helium.

The data on the thermodynamic properties of helium may be described by an equation of state of the form

$$z = 1 + \sum_{i=1}^{r} \sum_{j=0}^{S_i} b_{ij} \frac{\omega^i}{\tau^j} \qquad (2.1)$$

where $z = pV/RT$ is the coefficient of compressibility, $\omega = \rho/\rho_{\text{crit}}$ is the reduced density, and $\tau = T/T_{\text{crit}}$ is the reduced temperature.

Experimental data for the properties of helium cover a very wide range of reduced-parameter values, namely, $\tau = 0.34$–130 and $\omega = 0.001$–517. This means that acceptable precision can only be achieved by using two equations of state: one for the temperature range from the λ-line to 24 K and the other from 15 K to 1500 K. The temperature range from 15 to 25 K is common to both equations of state because, in this range, both equations provide an acceptable fit to the experimental data.

The coefficients b_{ij} were determined by the method of least squares. The function to be minimized contains the p, v, T data used to formulate the equation of state, as well as the values of the coefficients B_1, B_2 for $T > 15$ K and the isochoric specific heat. For $T < 25$ K, it contains the derivatives $(\partial p/\partial v)_T$, $(\partial p/\partial T)_v$ and a term used to satisfy Maxwell's rule.

Our formula for the weights of the measured coefficients of compressibility is more accurate than that given in [21, 22] and allows us to take into account the effect of the correlation between measured and calculated values of z on the weights [5].

When the experimental data are fitted by the method of least squares, and both the function y and the independent variables $x_1 \ldots, x_N$ are subject to uncertainty, the weights w are given by

$$w = 1/\sigma_y^2 \tag{2.2}$$

where σ_y^2 is the variance of the uncertainty and $\epsilon_y = y_e - y_a^\dagger (x_1 \ldots, x_n)$ [23]. The subscripts e and a label experimental and calculated values respectively.

It follows that the determination of the weight of an experimental value z_e reduces to estimating the variance σ_z^2 of the quantity $\epsilon_z = z_e - z_a$ from the variances of experimental p, v, T data. These variances are unknown, but can be estimated within a definite confidence interval from the limiting values of experimental uncertainties.

The general formula for the variance of the difference between two random values [12] gives

$$\sigma_z^2 = \sigma_{ze}^2 + \sigma_{za}^2 - 2 \text{ cov } \{z_e, z_a\} \tag{2.3}^\dagger$$

where $$\text{cov } \{z_e, z_a\} = M(z_e - Mz_e)(z_a - Mz_a)^\dagger$$

and M is the expectation value.

The quantities z_e and z_a are not independent since the same values of T and ρ are used to determine them, so that $\{\text{cov } z_e, z_a\} \neq 0$.

The differences $z_e - Mz_e$ and $z_a - Mz_a$ are the uncertainties in z_e (p, T, ρ) and z_a (T, ρ), and we can use this to transform the expression for cov $\{z_e, z_a\}$ and express it in terms of the absolute uncertainties Δp, ΔT, $\Delta \rho$ in the experimental data:

$$\text{cov } (z_e, z_a) = M\left[z_e\left(\frac{\Delta p}{p} - \frac{\Delta T}{T} - \frac{\Delta \rho}{\rho}\right) \cdot \left(\frac{\partial z_a}{\partial T}\Delta T + \frac{\partial z_a}{\partial \rho}\Delta \rho\right)\right] \tag{2.4}^\dagger$$

---

†Changed in proof.—T.B.S.

The quantities z_e, the parameters of state, and the derivatives may be regarded as constants, but the uncertainties Δp, ΔT, and $\Delta \rho$ are independent and symmetrically distributed about zero. Thus, using the well-known properties of expectation values, we obtain

$$\text{cov}\{z_e, z_a\} = -\frac{z_e}{T} \cdot \frac{\partial z_a}{\partial T} \sigma_T^2 - \frac{z_e}{\rho} \cdot \frac{\partial z_a}{\partial \rho} \sigma_\rho^2 \qquad (2.5)^\dagger$$

The variances σ_{ze}^2 and σ_{za}^2 may be determined by using the rule for the combination of errors [23]:

$$\sigma_{ze}^2 = z_e^2 \left(\frac{\sigma_p^2}{p^2} + \frac{\sigma_T^2}{T^2} + \frac{\sigma_\rho^2}{\rho^2} \right) \qquad (2.6)^\dagger$$

$$\sigma_{ze}^2 = \left(\frac{\partial z_a}{\partial T} \right)^2 \sigma_T^2 + \left(\frac{\partial z_a}{\partial \rho} \right)^2 \sigma_\rho^2 \qquad (2.7)^\dagger$$

Substituting (2.5)–(2.7) in (2.3), we obtain the following expression for the variances σ_z^2:

$$\sigma_z^2 = \frac{z_e^2}{p^2} \sigma_p^2 + \left(\frac{z_e}{T} + \frac{\partial z}{\partial T} \right)^2 \sigma_T^2 + \left(\frac{z_e}{\rho} + \frac{\partial z}{\partial \rho} \right) \sigma_\rho^2 \qquad (2.8)^\dagger$$

For normally distributed random uncertainties (such as those encountered in thermophysical experiments), the variances of the experimental values of p, T, and ρ that appear in (2.8) are given by the following expressions with a probability of 0.954:

$$\sigma_p^2 = \left(\frac{p\delta p}{2} \right)^2; \quad \sigma_T^2 = \left(\frac{T\delta T}{2} \right)^2; \text{ and } \sigma_\rho^2 = \left(\frac{\rho\delta\rho}{2} \right)^2 \qquad (2.9)^\dagger$$

where δp, δT and $\delta \rho$ are the relative uncertainties.

The same result is obtained when the variance σ_z^2 of the function $\epsilon_z = z_e(p, T, \rho) - z_a(T, \rho)$ is estimated using the rule for combination errors and taking account of the fact that the same values of T and ρ were used in the determination of z_e and z_a.

We may assume that the values of T and ρ are accurate, and can be used to analyze experimental data, because the effect of the uncertainties in the independent variables on the variance of the uncertainty ϵ_z has been taken into account.

When the weights of experimental data were calculated in [21, 22], the quantities δp and δT were assumed to be zero. The replacement of variance of the function z_e with the square of the maximum uncertainty, as in [21, 26], has

---

†Changed in proof.—T.B.S.

no effect on the calculated result. Moreover, it is clear from (2.5)–(2.7) that $\sigma_{z_e}^2 + \sigma_{z_a}^2 \geq 2 |\text{cov}\{z_e, z_a\}|$, and additional calculations show that, for most of the experimental points that were used, the above sum is a few times larger than $2|\text{cov}\{z_e, z_a\}|$. Hence, the absence of the latter quantity from the formula for the weight in [21, 22] had no effect on the final results.

When the equation of state is determined from thermal data, the weights are chosen in the light of the precision of the particular data, the number of points available, and the type of approximation employed for each category of data.

The program used produces a series of equations of state that are equivalent from the point of view of the precision with which they approximate the experimental data [6, 22]. This is achieved by eliminating from each successive equation the least significant coefficient for which the ratio of the absolute value of the coefficient to the value of the uncertainty is a minimum. Because the remaining coefficients are correspondingly corrected, the accuracy of the next equation in the series is found to be the same as that of the previous equation, and in a few cases may even be higher because of the reduced effect of computational instabilities. The precision of the fit to the experimental data falls significantly only when the number of coefficients is substantially reduced (down to 20 or less).

At temperatures between the λ-point and 20 K, the same equation of state was used for both gas and liquid to calculate the tables of thermodynamic functions of He^4. Above 20 K, the equation of state for the gas was employed. The coefficients of these equations were obtained by averaging the corresponding coefficients in the series of equations of state of equal precision in the respective temperature intervals.

It is convenient to introduce the following notation:

$$\left. \begin{aligned} A_0 &= \sum_{i=1}^{r} \sum_{j=0}^{S_i} b_{ij} \frac{\omega^i}{\tau^j}; \\ A_1 &= \sum_{i=1}^{r} \sum_{j=0}^{S_i} (i+1) b_{ij} \frac{\omega^i}{\tau^j}; \\ A_2 &= -\sum_{i=1}^{r} \sum_{j=0}^{S_i} (j-1) b_{ij} \frac{\omega^i}{\tau^j}; \\ A_3 &= \sum_{i=1}^{r} \sum_{j=0}^{S_i} \frac{i+j}{i} b_{ij} \frac{\omega^i}{\tau^j}; \\ A_4 &= \sum_{i=1}^{r} \sum_{j=0}^{S_i} \frac{j-1}{i} b_{ij} \frac{\omega^i}{\tau^j}; \\ A_5 &= -\sum_{i=1}^{r} \sum_{j=0}^{S_i} \frac{j(j-1)}{i} b_{ij} \frac{\omega^i}{\tau^j}. \end{aligned} \right\} \quad (2.10)$$

The thermodynamic functions in the single-phase region were calculated from the following formula, obtained from well known thermodynamic relationships:

$$\begin{aligned}
&z = 1 + A_0 ; \\
&h/RT = h_0/RT + A_3 ; \\
&s/R = s_0/R - \ln(\omega/\omega_0) + A_4 ; \\
&u/RT = h/RT - z ; \\
&F/RT = u/RT - s/R ; \\
&\Phi/RT = h/RT - s/R ; \\
&c_v/R = c_{v_0}/R + A_5 ; \\
&c_p/R = c_v/R + (1+A_2)^2/(1+A_1) ; \\
&w/w_0 = \sqrt{1+A_1} ; \\
&\delta/\delta_0 = (A_2 - A_1)/(1+A_1) ; \\
&\mu/\mu_0 = (A_2 - A_1)/(1+A_1) ; \\
&\alpha/\alpha_0 = (1+A_2)/(1+A_1) ; \\
&\beta/\beta_0 = (1+A_0)/(1+A_1) ; \\
&\gamma/\gamma_0 = (1+A_2)/(1+A_0) ; \\
&k/k_0 = (1+A_1)/(1+A_0) ; \\
&f/f_0 = \exp(A_3 - A_4) .
\end{aligned} \quad (2.11)$$

where h_0/RT, s_0/R and c_{v_0}/R are, respectively reduced enthalpy, entropy, and isochoric specific heat in the ideal-gas state, and ω_0, w_0, δ_0, μ_0, α_0, β_0, γ_0, k_0, f_0 are thermodynamic normalizing functions given by

$$\begin{aligned}
&\omega_0 = p_{ST}/\rho_{crit} RT; \\
&(p_{ST} = 0.101325 \text{ MPa}); \\
&w_0 = \sqrt{RT c_p/c_v}; \\
&\delta_0 = 1/\rho; \\
&\mu_0 = 1/\rho c_p; \\
&\alpha_0 = 1/T; \\
&\beta_0 = 1/p; \\
&\gamma_0 = 1/T; \\
&k_0 = c_p/c_v ; \\
&f_0 = \rho RT .
\end{aligned} \quad (2.12)$$

52 THERMODYNAMIC PROPERTIES OF HELIUM

Functions on the melting curve are calculated from (2.11) and also from the equations of this curve. The specific heat c_m of the liquid on the melting curve was calculated from the formula

$$\frac{c_m}{R} = \frac{c_p}{R} - \frac{\alpha}{\alpha_0} \cdot \frac{Z_{\text{crit}}}{\omega_m} \cdot \frac{d\pi_m}{d\tau_m} \tag{2.13}$$

where c_p/R, α/α_0, and $\omega_m = \rho_m/\rho_{\text{crit}}$ were determined on this curve on the liquid side.

Functions on the saturation line were calculated from the common equation of state, using (2.11) and assuming Maxwell's rule. The specific heat capacities of the liquid (c_s') and vapor (c_s'') were calculated along the saturation line from

$$\frac{c_s'}{R} = \frac{c_p'}{R} - \frac{\alpha}{\alpha_0} \cdot \frac{Z_{\text{crit}}}{\omega'} \cdot \frac{d\pi_s}{d\tau_s} \tag{2.14}$$

$$\frac{c_s''}{R} = \frac{c_p''}{R} - \frac{\alpha}{\alpha_0} \cdot \frac{Z_{\text{crit}}}{\omega''} \cdot \frac{d\pi_s}{d\tau_s} \tag{2.15}$$

where

$$\frac{d\pi_s}{d\tau_s} = - \frac{\omega' \cdot \omega''}{Z_{\text{crit}}} \cdot \frac{A_3'' - A_3'}{\omega'' - \omega'}$$

The latent heats of evaporation were calculated from

$$\frac{r}{RT} = A_3'' - A_3'. \tag{2.16}$$

The values of the derivatives $d\pi/d\tau$ and $d^2\pi/d\tau^2$ on the saturation and melting curves were obtained differentiating the corresponding relationships.

CHAPTER
THREE

EQUATIONS OF STATE AND TABLES OF THERMODYNAMIC FUNCTIONS OF HELIUM-4

Most of the equations of state that have been proposed for helium are intended to describe the properties of the gas. In some investigations, the equation of state is derived for the liquid, but there are few investigations in which a single equation of state is found for both the gas and the liquid [20, 32, 132].

As mentioned in Chapter 2, we have constructed two equations of state for He^4 for two temperature intervals, and then used them to obtain two average equations of state for temperatures below 25 K and above 15 K. The use of sets of equations of state of equal precision allowed us to calculate the precision of the thermodynamic functions calculated from these average equations. In this chapter, we compare experimental values of thermodynamic functions obtained by different authors with our calculations. We also compare our calculations with the most extensive range of tabulated data [20, 32] in order to estimate the reliability of earlier tabulations.

3.1 THERMODYNAMIC FUNCTIONS IN THE IDEAL GAS STATE

The thermodynamic functions of He^4 in the ideal gas state are very simply calculated because, as for other monatomic gases, the ideal-gas specific heat c_p^0 of helium is not a function of temperature and is given by

$$c_p^0 = \frac{5}{2}R \qquad (3.1)$$

The enthalpy and entropy of the ideal gas were obtained respectively from:

$$h_0 = \int_{T_0}^{T} c_p^0 \, dT + h_{00} + h_0^0 \qquad (3.2)$$

where h_{00} is the enthalpy at temperature T_0, h_0^0 is the heat of phase transition at $T = 0$ K,

$$s_0 = \int_{T_0}^{T} \frac{c_p^0}{T} \, dT + s_{00} + s_0^0 \qquad (3.3)$$

s_{00} is the entropy at temperature T_0, and s_0^0 is the constant of integration (we let $s_0^0 = 0$).

The heat of phase transition was determined from the data in [127]: $h_0^0 = 59.00$ kJ/kmol = 14.7404 kJ/kg. The initial temperature was $T_0 = 100$ K. The reduced enthalpy and entropy at this temperature are $h_{00}/RT_0 = 2.5$ and $s_{00}/R = 12.4284$, respectively.

3.2 EQUATIONS FOR CALCULATING THERMODYNAMIC PROPERTIES OF He^4

The two sets of equations of state for helium, which are presented in this monograph, shall henceforth be referred to as the low-temperature (LT for $T < 25$ K) and the high-temperature (HT for $T > 15$ K) equations of state, respectively.

The first of these sets consists of nine series of LT equations that describe both the gas and the liquid phases of helium. These equations were determined from the p, v, T data in Table 3.1. We discard 13 experimental values of the density from [91], three from [81], and one from [16]. In the first stage of the calculations, these are all given zero weight. We note that most of the published data have been corrected in order to allow them to be presented on the T58 scale for $T < 5.2$ K and the International Practical Scale of 1968 for $T > 13.8$ K.

In setting up the equations we also used values of the specific heat c_v (312 points in the range 2.5–20 K) and, in some series, the values of $\partial p/\partial v$ and $\partial p/\partial T$ from [91] (305 and 317 points respectively in the range 3–20 K). These data lie in the pressure range 0.1–10.4 MPa. In all these calculations, up to 30 values of the second and third virial coefficients in the temperature range 2.04–25 K were used. They were obtained by analyzing published results. Maxwell's

Table 3.1 The p, v, T data used to construct an equation of state for helium for $T < 25$ K and the uncertainty $\delta\rho_{av}$ in the average LT equation of state

Year	Author	ΔT, K	ΔP, MPa	Number of points	$\delta\rho$ %	$\delta\rho_{av}$ %
1946	Kistemaker and Keesom [125]	1.6–2.7	0.0003–0.002	65	0.1	0.12
1955	Keller [113]	2.2–4.0	0.002–0.03	64	0.1	0.22
1958	Edeskuty and Sherman [67]	2.2–4.2	0.1–14	88	0.1	0.14
1960	White et al. [167]	21–25	0.1–1.0	17	0.05	0.13
1960	Hill and Lounasmaa [91]	2.9–21	0.1–10	249	0.2–0.3	0.46
1966	Glassford and Smith [81]	4.3–23	6.8–136	251	0.1–0.3	0.45
1976	Bogoyavlenskii and Yurchenko [4]	2.0–4.2	0.1–10	234	0.1–0.3	0.23
1981	Popov [16]	6.9–25	0.3–14	65	0.2–0.3	0.45
	Data on saturation curve					
1974	McCarty (ρ'') [132]	2.2–5.0	0.005–0.2	29	0.2	0.41
1974	McCarty (ρ') [132]	2.2–5.0	0.005–0.2	29	0.05	0.08
	Data on λ-curve					
1977	Angus et al. [32]	1.8–2.2	0.005–3.0	10	0.1–0.2	0.12
	Data on melting curve					
1958	Edeskuty and Sherman [67]	2.5–4.0	5.7–13	4	0.1	0.24
1959	Grilly and Mills [85]	2.0–24	3.9–237	14	0.1–0.2	0.27
1966	Glassford and Smith [81]	5.3–17	21–133	8[†]	0.1–0.3	0.44

[†]Data obtained by the present authors by a graphical analysis of the data in [81].

relations were found to be satisfied by using the same p, v, T data [132] for the saturation curve that was introduced into the main p, v, T data file.

The values of $\delta\rho$ adopted in the calculation of the weights of experimental p, v, T data are given in the second column from the right of Table 3.1. The choice of these values was determined by two considerations: the precision estimated by the original experimentalists themselves and by the consistency of the data from different investigations as estimated by the present authors.

The initial equations in the different series were based on different choices of weights for a number of groups of p, v, T data. The weights of the third virial coefficient and the derivatives $\partial p/\partial v$ and $\partial p/\partial T$ were also varied by 2–5% and the weights of the specific heat c_v by 0.5–1%. The number of equations in the different series varied from 7 to 22, and the total number of equations was 133.

The second set consists of 17 series of LT equations of state constructed from the p, v, T data in Table 3.2. The table does not take into account three experimental points from [168] which were given zero weight. In the temperature range 15–1200 K, 30 values of the second and third virial coefficients were used in addition to the p, v, T data. As before, corrections were made to the data published prior to 1968 in order to make them consistent with the IPTS-68 temperature scale. Table 3.2 lists the values of $\delta\rho$ adopted in the calculations of the weights of the p, v, T data.

Table 3.2 Summary of p, v, T data used to set up the equation of state for helium for $T > 15$ K and the deviations $\delta\rho_{av}$ for the average HT equation of state

Year	Author	Parameter range		Number of points	$\delta\rho$ %	$\delta\rho_{av}$ %
		ΔT, K	ΔP, MPa			
1931	Wiebe et al. [168]	203–473	10–101	33	0.2	0.38
1941	Michels and Wouters [133]	273–423	0.9–30	119	0.05	0.08
1960	Stroud et al. [157]	250–328	0.6–28	68	0.05	0.11
1960	White et al. [167]	21–300	0.1–3.4	153	0.05	0.10
1966	Hill and Lounasmaa [91]	19–21	0.9–10	8	0.2	0.32
1967	Sullivan and Sonntag [158]	70–120	1.3–70	99	0.2	0.27
1967	Briggs et al. [51]	273	0.1–81	20	0.1	0.16
1970	Briggs [50]	268–353	0.1–81	200	0.1	0.25
1970	Blancett et al. [45]	223–323	0.3–70	72	0.05	0.13
1970	Hall and Canfield [89]	83–113	0.2–72	71	0.1	0.10
1971	Provine and Canfield [144]	143–183	0.3–71	72	0.1	0.20
1972	Petrov [15]	123–293	1.8–40	108	0.1	0.21
1981	Popov [16]	21–291	0.3–21	188	0.2	0.45
	Calculated value of ρ	15–25	0.5–100	308	0.1–0.3	0.13
	Data on melting curve					
1981	Grilly and Mills [85]	20–31	174–349	6	0.2	0.27

Matching of the HT and LT equations of state presents a serious problem. The virial equations suffer from a well-known deficiency: the calculated thermal parameters at the lower end of the temperature range have lower precision, so that we cannot expect the HT equations determined from p, v, T data above 20 K to match well to the LT equations at this temperature. The densities, the second and third virial coefficients, and the specific heats c_v were therefore calculated for the range 15-25 K from the average LT equation of state, and were used together with empirical data to determine the coefficients of the HT equations. In addition, to improve the relative consistency of the HT and LT equations, the p, v, T data for temperatures below 25 K were excluded beginning with the 12th series.

The initial HT equations of each series were given different weights for B_1 (0.3-1.0%), B_2 (1.5-4%) and c_v (1-4%), and different calculated p, v, T values in the temperature range 15-25 K (0.1-0.3%). The number of equations in the individual series was 6-12, the total number being 172. The large number of HT equations was necessary to secure better agreement between the average HT and LT equations in the common temperature range 15-25 K.

For all the LT and HT equations formulated in this way, the average root mean square deviations of experimental data due to different authors from the calculated results were sufficiently stable to allow us to consider these equations as statistically equivalent.

The two sets of equations of state were used to obtain the average equations in the form

$$z = 1 + \sum_{i=1}^{r} \sum_{j=0}^{s_i} b_{ij} \frac{\omega^i}{\tau^j} \qquad (3.4)$$

where $\omega = \rho/\rho_{\text{crit}}$ and $\tau = T/T_{\text{crit}}$.

The coefficients of the average equations are presented below.

Coefficients of the average equation of state for $T \leq 20$ K

$b_{10} = 0.2819155 \cdot 10^0$
$b_{11} = -0.1292457 \cdot 10^1$
$b_{12} = -0.2129594 \cdot 10^0$
$b_{13} = 0.6437906 \cdot 10^0$
$b_{14} = -0.8326190 \cdot 10^0$
$b_{15} = 0.5006948 \cdot 10^0$
$b_{16} = -0.1412233 \cdot 10^0$
$b_{17} = 0.1488343 \cdot 10^{-1}$

$b_{20} = 0.8462366 \cdot 10^{-1}$
$b_{21} = 0.3001846 \cdot 10^0$
$b_{22} = -0.5251701 \cdot 10^0$
$b_{23} = 0.5410069 \cdot 10^0$
$b_{24} = -0.1832495 \cdot 10^0$
$b_{25} = -0.1714369 \cdot 10^0$
$b_{26} = 0.7279349 \cdot 10^{-1}$

$b_{27} = -0.8634785 \cdot 10^{-2}$
$b_{30} = 0.4704854 \cdot 10^{-1}$
$b_{31} = -0.5334322 \cdot 10^0$
$b_{32} = 0.3341696 \cdot 10^0$
$b_{33} = 0.8362204 \cdot 10^{-1}$
$b_{34} = 0.4843829 \cdot 10^0$
$b_{35} = -0.3192986 \cdot 10^{-1}$
$b_{36} = 0.3671224 \cdot 10^{-2}$

$b_{40} = -0.1260754 \cdot 10^0$
$b_{41} = 0.1101237 \cdot 10^1$
$b_{42} = -0.6353332 \cdot 10^{-1}$
$b_{43} = -0.6627022 \cdot 10^0$
$b_{44} = -0.1973173 \cdot 10^0$
$b_{45} = -0.1678553 \cdot 10^{-1}$

$b_{50} = 0.5636224 \cdot 10^{-1}$
$b_{51} = -0.7136530 \cdot 10^0$
$b_{52} = -0.1632410 \cdot 10^0$
$b_{53} = 0.3670216 \cdot 10^0$
$b_{54} = 0.7594056 \cdot 10^{-1}$
$b_{55} = 0.3843195 \cdot 10^{-2}$

$b_{60} = 0.1531109 \cdot 10^{-1}$
$b_{61} = 0.1095645 \cdot 10^0$
$b_{62} = 0.1901053 \cdot 10^0$
$b_{63} = -0.1029899 \cdot 10^0$
$b_{64} = -0.1194448 \cdot 10^{-1}$

$b_{70} = -0.1170722 \cdot 10^{-1}$
$b_{71} = 0.4836444 \cdot 10^{-1}$
$b_{72} = -0.5169830 \cdot 10^{-1}$
$b_{73} = 0.1255776 \cdot 10^{-1}$

$b_{80} = 0.8140690 \cdot 10^{-3}$
$b_{81} = -0.1915444 \cdot 10^{-1}$
$b_{82} = -0.1581469 \cdot 10^{-3}$

$b_{90} = 0.3304047 \cdot 10^{-3}$
$b_{91} = 0.2581156 \cdot 10^{-2}$
$b_{92} = 0.9684371 \cdot 10^{-3}$

$b_{10,0} = -0.3739834 \cdot 10^{-4}$
$b_{10,1} = -0.1619576 \cdot 10^{-3}$

Coefficients of the average equation of state for $T > 20$ K

$b_{10} = 0.1803041 \cdot 10^0$
$b_{11} = 0.1285745 \cdot 10^1$
$b_{12} = -0.2378314 \cdot 10^2$
$b_{13} = 0.9971745 \cdot 10^2$
$b_{14} = -0.1938884 \cdot 10^3$
$b_{15} = 0.1406779 \cdot 10^3$

$b_{20} = 0.1611295 \cdot 10^{-1}$
$b_{21} = 0.8707625 \cdot 10^0$
$b_{22} = -0.1357183 \cdot 10^1$
$b_{23} = -0.5198535 \cdot 10^1$
$b_{24} = 0.1429547 \cdot 10^2$

$b_{30} = 0.1042847 \cdot 10^0$
$b_{31} = -0.8700183 \cdot 10^0$
$b_{32} = 0.2815541 \cdot 10^1$
$b_{33} = -0.9708081 \cdot 10^0$
$b_{34} = -0.5541532 \cdot 10^1$

$b_{40} = -0.1551514 \cdot 10^0$
$b_{41} = 0.7052546 \cdot 10^0$
$b_{42} = -0.1921619 \cdot 10^1$
$b_{43} = 0.2201513 \cdot 10^1$

$b_{50} = 0.1100556 \cdot 10^0$
$b_{51} = -0.3090680 \cdot 10^0$
$b_{52} = 0.3587898 \cdot 10^0$
$b_{53} = -0.2586436 \cdot 10^0$

$b_{60} = -0.3927200 \cdot 10^{-1}$
$b_{61} = 0.1145860 \cdot 10^0$
$b_{62} = -0.4085258 \cdot 10^{-1}$

$b_{70} = 0.6593721 \cdot 10^{-2}$
$b_{71} = -0.2201105 \cdot 10^{-1}$
$b_{72} = 0.4600854 \cdot 10^{-2}$

$b_{80} = -0.4079607 \cdot 10^{-3}$
$b_{81} = 0.1466608 \cdot 10^{-2}$

The optimum match of the thermodynamic functions calculated from the two equations is obtained at the midpoint of the range 15–25 K, so that the average LT equation was used for $T \leq 20$ K and the average HT equation for $T > 20$ K.

The following values of the critical parameters and gas constant were adopted in calculations: $T_{crit} = 5.19$ K (T58 scale), $p_{crit} = 0.22746$ MPa, $\rho_{crit} = 69.64$ kg/m^3, and $R = 2.077252$ kJ/(kg·K).

The average equations of state describe very satisfactorily most of the p, v, T data used to construct these equations. This is shown by the root mean square deviations of experimental from calculated values (see last column in Tables 3.1 and 3.2). Larger values of $\delta\rho_{av}$ are observed at lower temperatures.

This is due to the considerable spread of experimental points and poor agreement between results reported by different authors for $T < 20$ K (in particular the data in [16, 81, 91]).

The quality of the analytic description of the p, v, T data is illustrated by the histograms of Fig. 4. The histogram for the average LT equation does not include values of $\delta\rho$ for 156 points for which it exceeds 0.5% (71 from [91], 51 from [81] and 19 from [16]). The average deviation $\delta\rho_{av} = 0.36\%$ calculated for all the 1127 points enummerated in Table 3.1 is indicated in the histogram. The histogram for the average HT equation was constructed from the values of $\delta\rho$ for most of the data referred to in Table 3.2 with the exception of the calculated values of $\delta\rho$ for temperatures in the range 15–25 K. It does not include the 71 points for which $\delta\rho > 0.5\%$, but the deviations at these points were taken into account in the determination of $\delta\rho_{av} = 0.25\%$ for the complete file of 1217 points.

Graphs of the deviations $\delta\rho$ presented in Sec. 3.3 allow a detailed analysis to be made of the accuracy with which the experimental values are described analytically. These graphs also allow an estimate to be made of the scatter of experimental points and the mutual consistency of data reported by different authors in identical or comparable parameter ranges.

Table 3.3 presents the standard deviations of $\delta\rho_{av}$ for the experimental data that were not used in constructing the equation of state because their inclusions in the data file during the first stage of the calculations would have lowered the accuracy of the description of more reliable data. It is clear from the table that the average HT equation of state describes quite well the data in [53, 63, 96]. The considerable discrepancies as compared with the data in [3, 9, 10] are due to the large errors in the latter.

It should be noted that the average LT equation of state satisfies Maxwell's relations to a high degree of precision. The value of $\delta p_{s_{av}}$ for this LT equation is 0.04%. This means that the equation can be used to calculate the properties of the liquid by the so-called continuous scheme in which the values of p_s are determined from the equation of state. This secures the consistency of the calculated thermodynamic functions in single-phase regions and on the saturation curve.

The average LT equation of state describes accurately the specific heat c_v and the derivatives $\partial p/\partial T$ and $\partial p/\partial v$ from [91], which were used to construct it. The root mean square deviations are 1.2%, 3.2%, and 4.7%, respectively. The last value was calculated after discarding 10 points lying close to the critical point at which the virial form of the equation is incorrect and where the small value of $\partial p/\partial v$ causes a large relative deviation.

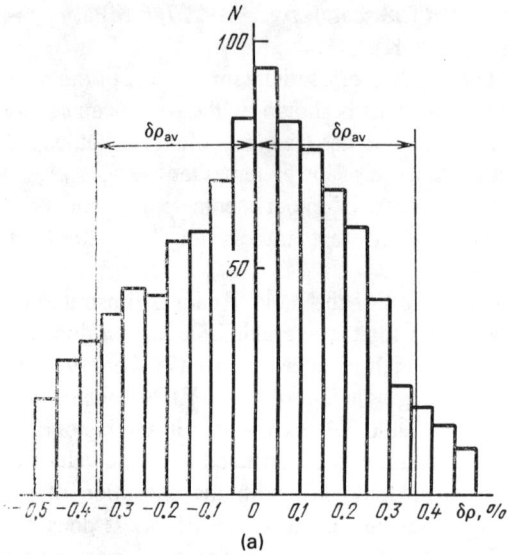

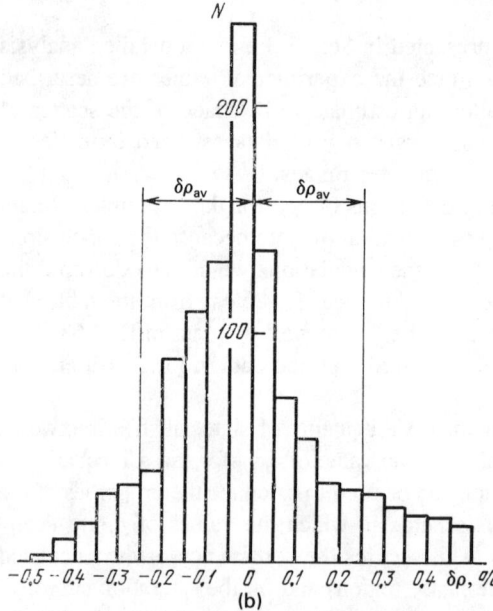

Figure 4 Histogram of deviations of measured density of helium from values calculated from the average equation of state: (a) low temperature (LT), (b) high temperature (HT).

The values of the second and third virial coefficients calculated from the average LT and HT equations of state agree well with the experimental values reported by different authors (Figs. 5 and 6). Figure 7 shows the variation of all the temperature functions of the average equations of state.

Data on the saturated vapor pressure of helium are of independent value, e.g., for low-temperature thermometry. On the other hand, the determination of the values of p_s with the aid of a single equation of state is difficult and demands the use of a large computer. We have therefore constructed a separate interpolation equation for the temperature dependence of the saturated vapor pressure of helium:

$$\pi = \sum_{i=0}^{n} a_i \tau^i \qquad (3.5)$$

where $\pi = p/p_{crit}$ and $\tau = T/T_{crit}$.

The values of the critical parameters are given above. The equation describes p_s in the temperature range between the lower λ point and the critical point on the T58 scale with a mean square deviation $\delta p_{s_{av}} = 0.015\%$.

The smoothed data from [32] on the T58 scale have been used to set up equation (3.5) and thus describe the λ-curve between the lower and upper λ-points with average deviation $\delta p_{\lambda_{av}} = 0.04\%$. The pressures on the melting curve were calculated from (3.5) constructed on the basis of data from [85].

Table 3.3 Standard deviations $\delta \rho_{av}$ for data used in comparisons with calculated densities

Year	Author	Parameter range		Number of points	$\delta\rho_{av}$, %
		ΔT, K	Δp, MPa		
		LT equation			
1922	Holborn and Otto [92]	15–20	0.1–11	18	0.52
1972	Karnus and Rudenko [10]	14–24	0.6–11	43	1.12
1976	Kalenkov [9]	14–21	7.0–58	16	1.19
1978	Bogoyavlenskii et al. [3]	4.2–21	0.01–3.9	406	2.20
		HT equation			
1915	Holborn and Schultze [96]	273–373	0.1–5.3	27	0.11
1922	Holborn and Otto [92]	20–90	0.1–11	27	0.46
1965	Canfield et al. [53]	133–273	0.2–54	149	0.18
1972	Karnus and Rudenko [10]	20–53	0.5–11	87	0.69
1976	Kalenkov [9]	20–273	5.1–100	123	1.05
1978	Dillard et al. [63]	223–323	0.7–15	83	0.11

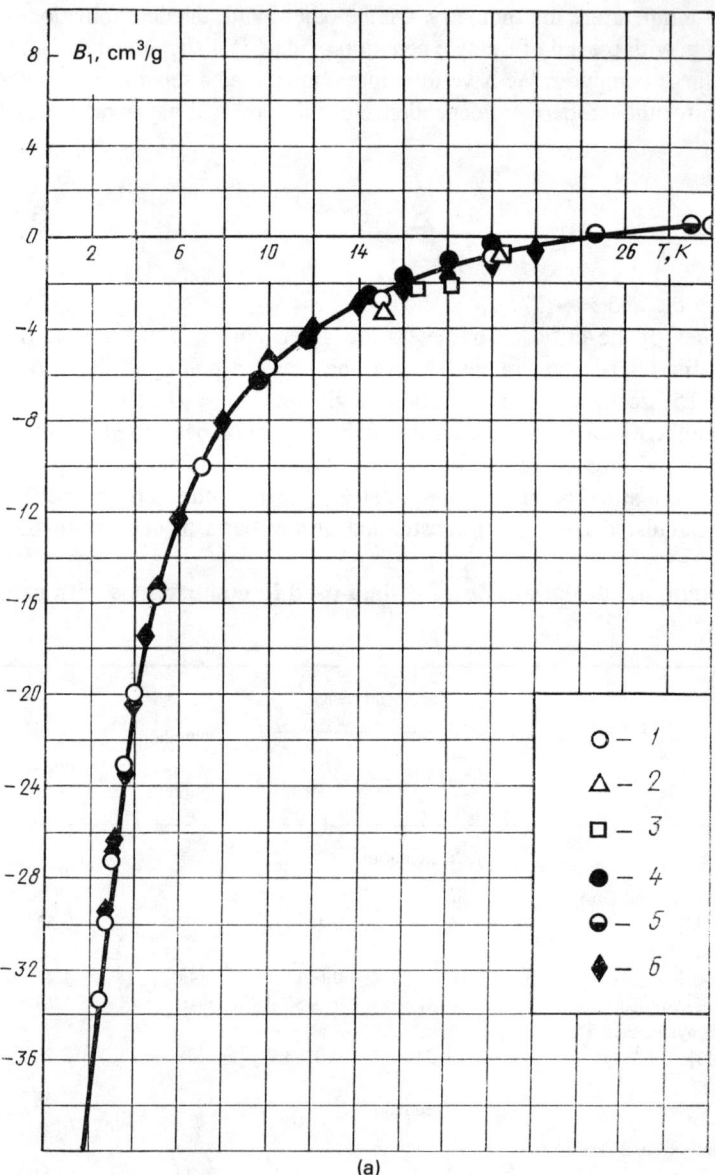

Figure 5 Calculated and experimental values of the second virial coefficient: (a) low temperature region. *(Data from 1-[66]; 2-[95]; 3- [27]; 4-[110]; 5-[167]; 6-[11].)*

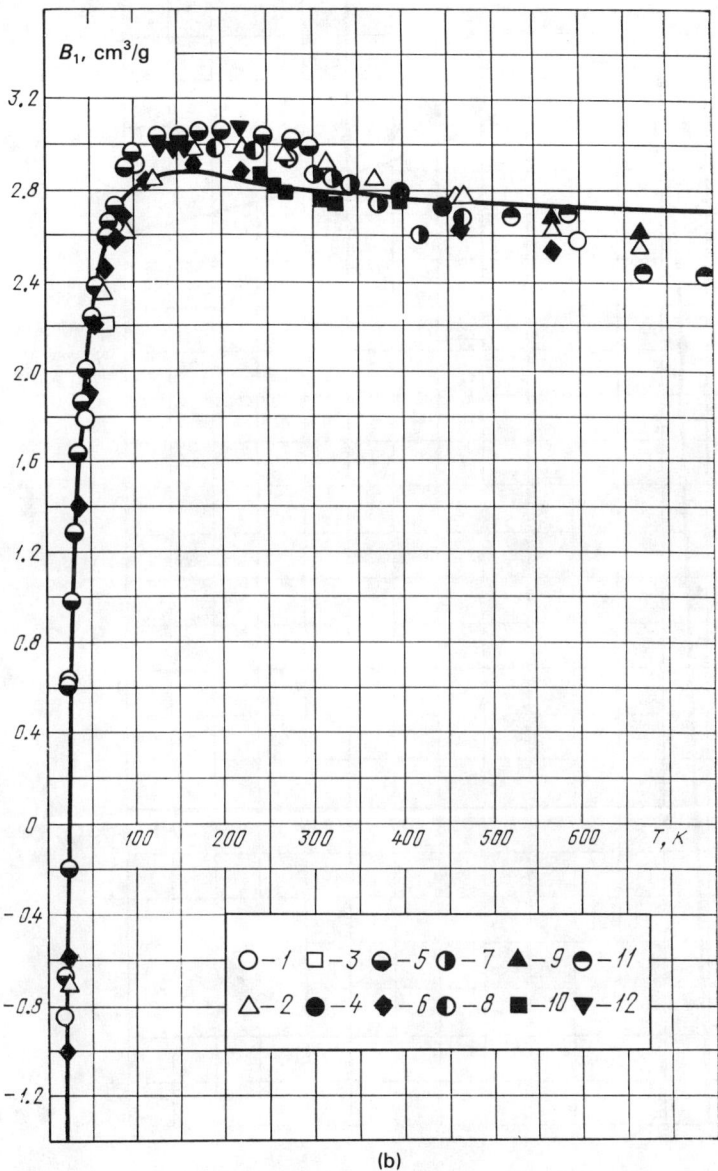

Figure 5 Calculated and experimental values of the second virial coefficient: (*Continued*) (b) high temperature region. *(Data from 1-[66]; 2-[95]; 3-[27]; 4-[110]; 5-[167]; 6-[11]; 7-[80]; 8-[168]; 9-[152]; 10-[157]; 11-[170]; 12-[53].)*

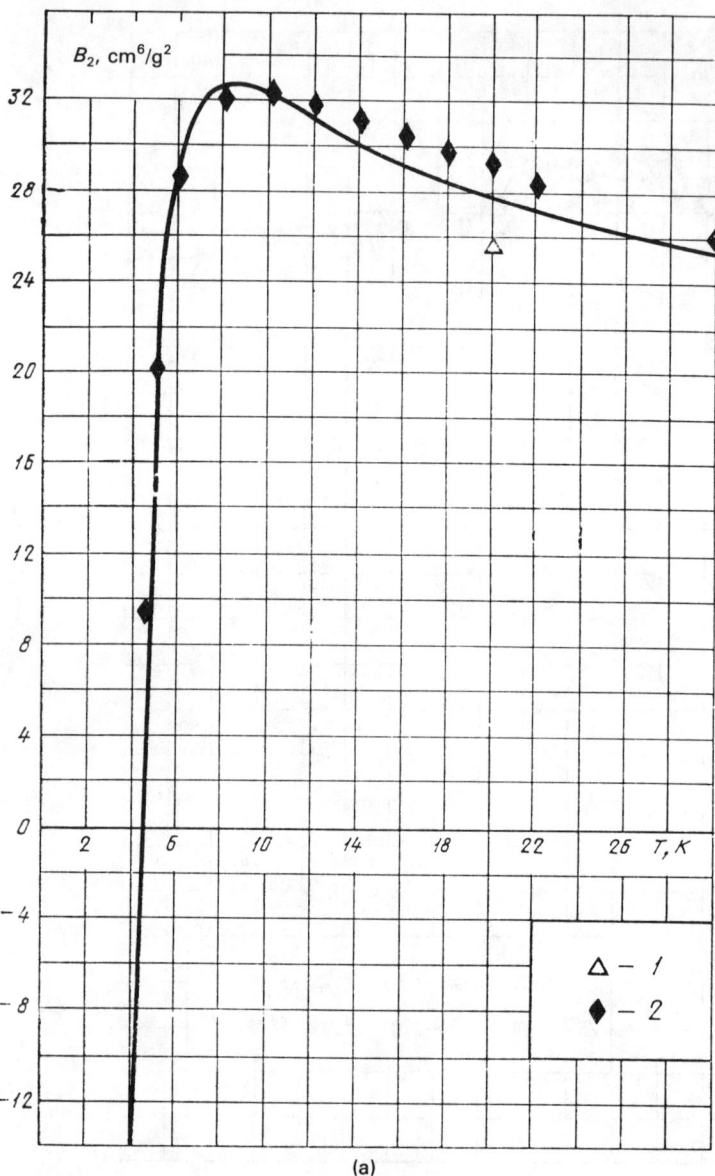

Figure 6 Calculated and experimental values of the third virial coefficient: (a) low temperature region. *(Data from 1-[95]; 2-[11].)*

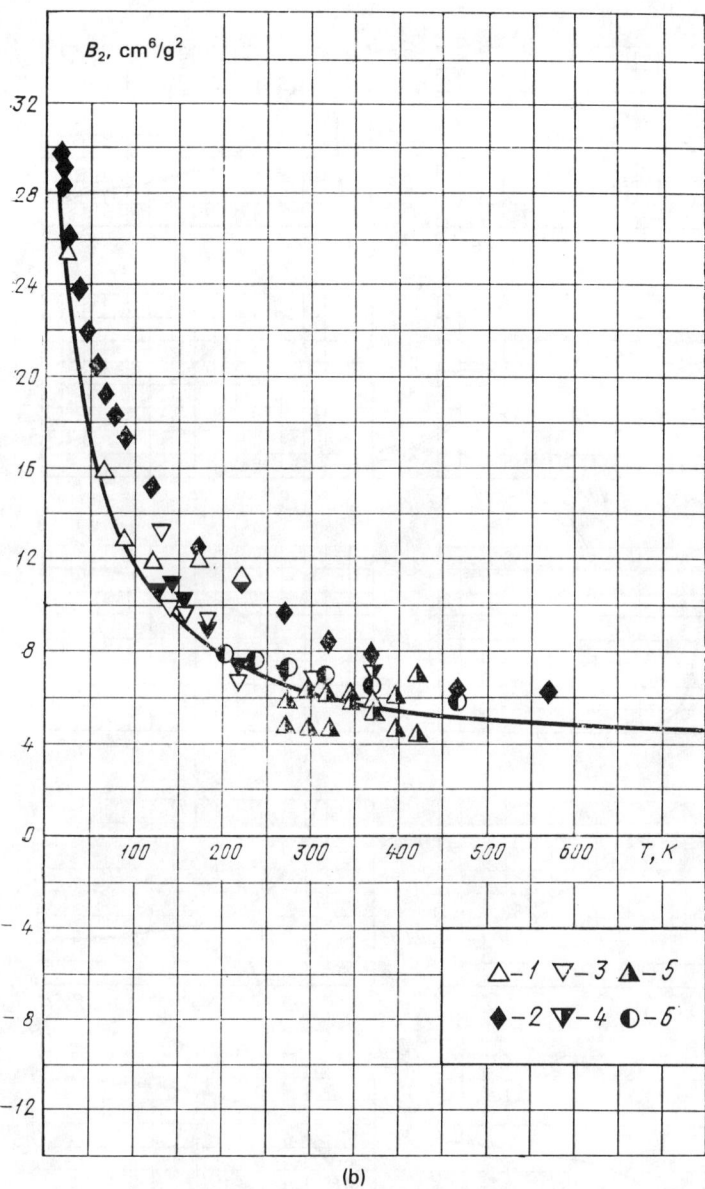

Figure 6 Calculated and experimental values of the third virial coefficient: (*Continued*) (b) high temperature region. *(Data from 1-[95]; 2-[11]; 3-[53]; 4-[126]; 5-[133], 6-[168].)*

66 THERMODYNAMIC PROPERTIES OF HELIUM

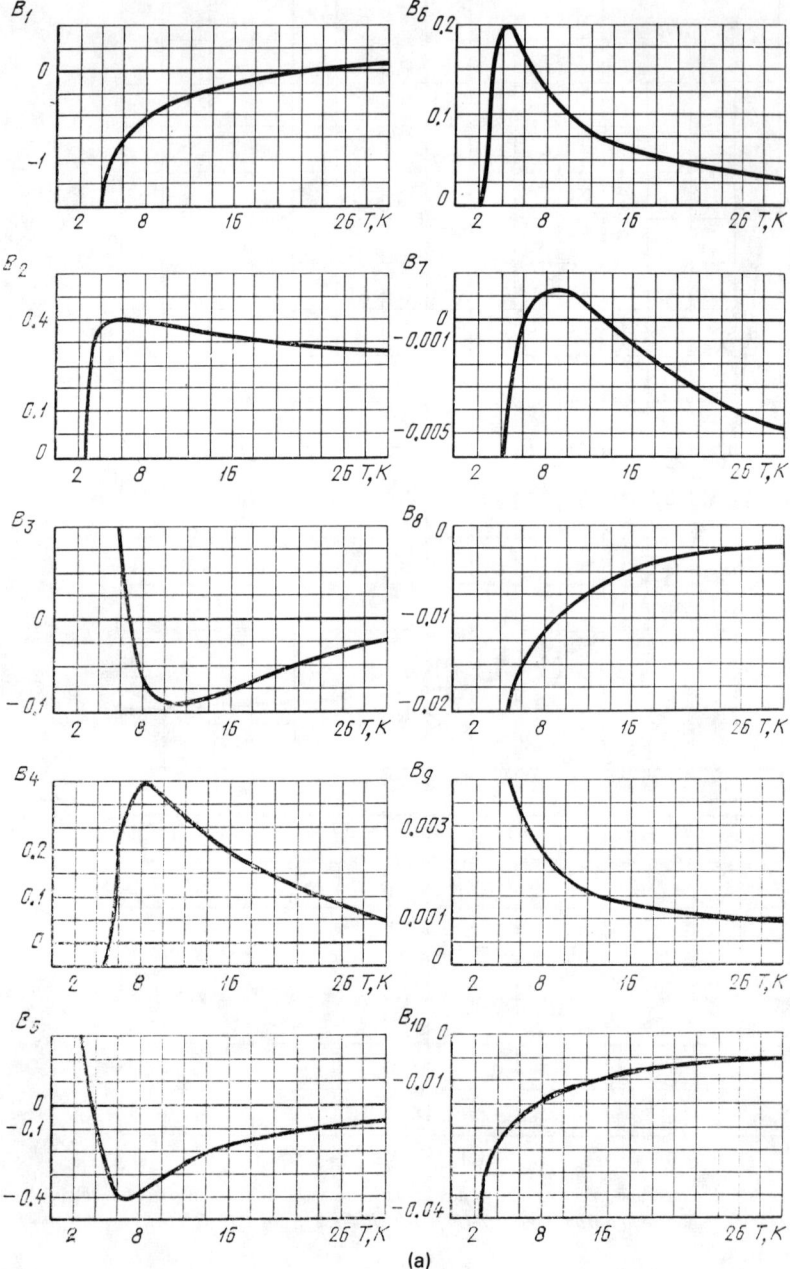

Figure 7 Temperature functions for: (a) average LT equation.

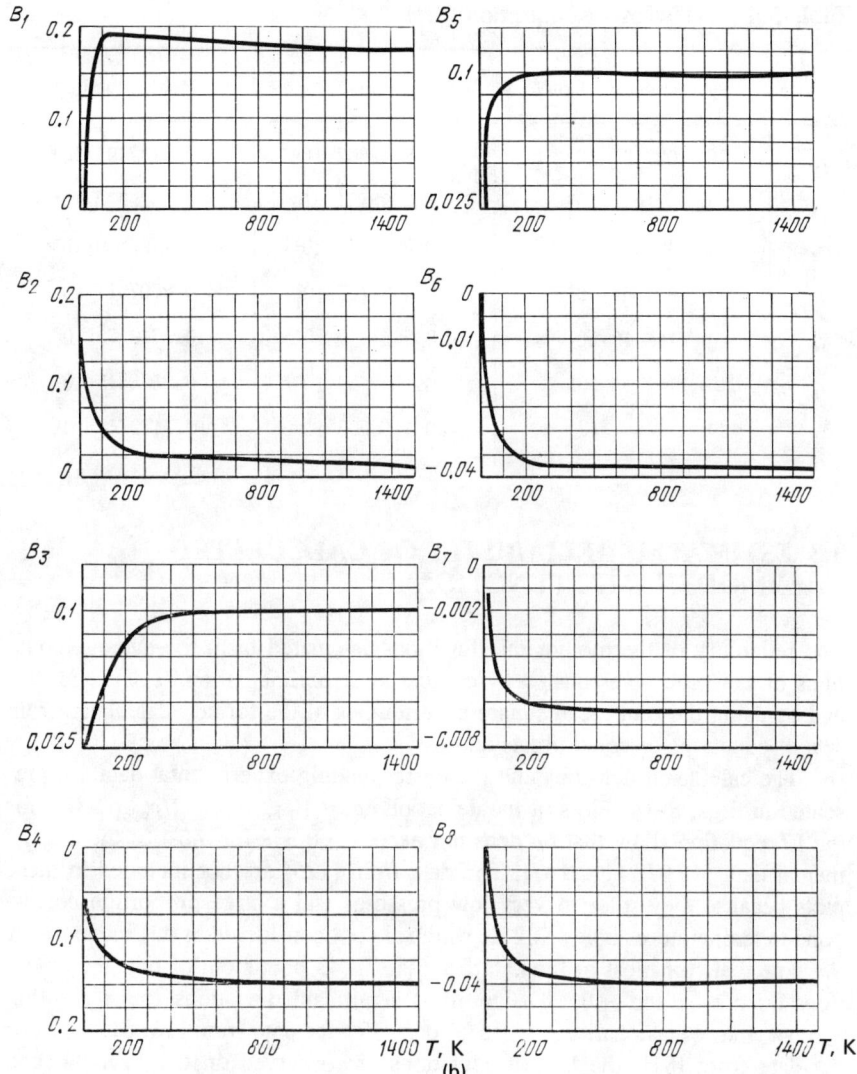

Figure 7 Temperature functions for: (*Continued*) (b) average HT equation.

Equation (3.5) describes these data in the temperature range 1.35–4.67 K (2.6–17.2 MPa) with $\delta p_{m_{av}} = 0.03\%$.

The coefficients of the equations for the above curves are given in Table 3.4. The equations of the melting curve and the λ-curve were used together with the equation of state to calculate the properties of liquid He-I on these curves.

Table 3.4 Coefficients of equation (3.5)

	Liquid evaporation curve	λ-Curve	Melting curve
a_0	$-0.1007457 \cdot 10^0$	$-0.2422802 \cdot 10^3$	$0.2177196 \cdot 10^3$
a_1	$0.9671024 \cdot 10^0$	$0.199847 0 \cdot 10^4$	$-0.2426663 \cdot 10^4$
a_2	$0.3629551 \cdot 10^1$	$-0.3632741 \cdot 10^4$	$0.1127859 \cdot 10^5$
a_3	$0.6071344 \cdot 10^1$	$0.4091732 \cdot 10^3$	$-0.2707971 \cdot 10^5$
a_4	$-0.3448743 \cdot 10^1$	$-0.2551179 \cdot 10^5$	$0.3643173 \cdot 10^5$
a_5	$0.1142905 \cdot 10^1$	$0.1082426 \cdot 10^6$	$-0.2578815 \cdot 10^5$
a_6	—	$-0.1106993 \cdot 10^6$	$0.7472841 \cdot 10^4$

3.3 ESTIMATED RELIABILITY OF CALCULATED THERMODYNAMIC FUNCTIONS

The reliability of thermodynamic functions calculated from the average equations of state can be estimated by comparing experimental and calculated data and by using the root mean square uncertainties of the former, calculated from sets of equations of equal precision.

The calculated densities and the corresponding experimental data are presented in Figs. 8–14. Plots of the deviation $\delta\rho = [(\rho_{exp} - \rho_{calc})/\rho_{calc}]\,100\%$ for the LT equation show that $\delta\rho$ does not exceed $\pm 0.3\%$ for most of the experimental data [4, 67, 113, 167]. The data from [125] are not included in these plots because they refer to very low pressures and a large proportion correspond to temperatures below 2.2 K, which are not considered in this monograph. We note that, for most of the data from [125] (43 points out of 65), the deviations do not exceed $\pm 0.1\%$. The most significant deviations (mostly within $\pm 0.6\%$ and, occasionally, ± 1–1.5%) occur for the data from [16, 81, 91]. For the data from [81], the largest deviations are observed on the 191, 94, and 201.61 kg/m³ isochores, but in the range 4–10 K, their signs are opposite to those of the deviations for the data from [91] on comparable isochores, which shows that there is a lack of consistency here.

It is clear from Figs. 15–18 that the average LT equation provides a satisfactory description of the experimental densities of helium on the saturation, melting and λ-curves. On the melting curve, the deviations of the calculated density from the values quoted in [81] reach 0.6% at some points. However, the latter are not experimental densities because the values of p_m and T_m corresponding to the measured densities were obtained as the coordinates of points of discontinuity on isochores during a transition to the two-phase region.

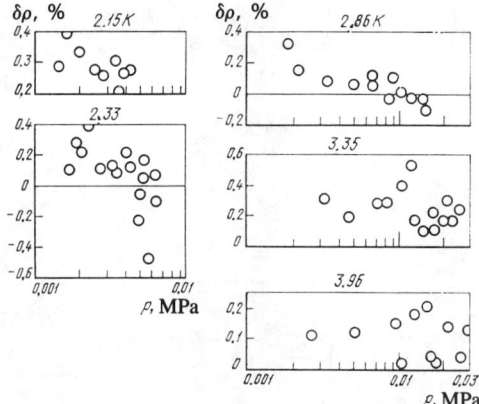

Figure 8 Deviations of experimental data [113] from calculated densities.

The region $T > 20$ K is generally characterized by smaller deviations (Figs. 19–27) than the low temperature region. For most of the experimental data [45, 89, 133, 157, 167] the deviations are in the range $\pm 0.1\%$. The deviations increase with increasing pressure, reaching 0.3–0.5% at the endpoints of some isotherms. The data from [50, 51, 158] are characterized by

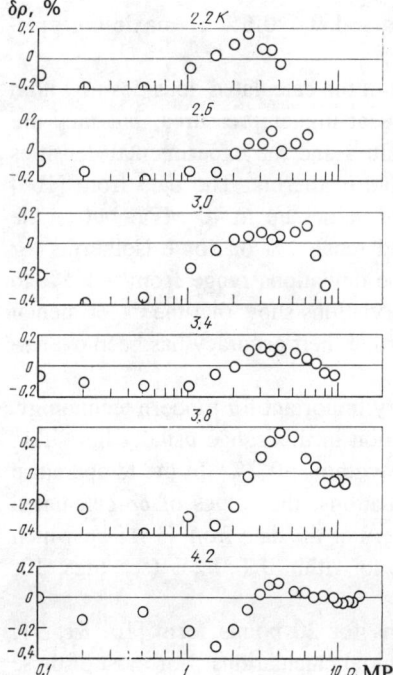

Figure 9 Deviations of the experimental data of Edeskuty and Sherman [67] from calculated densities.

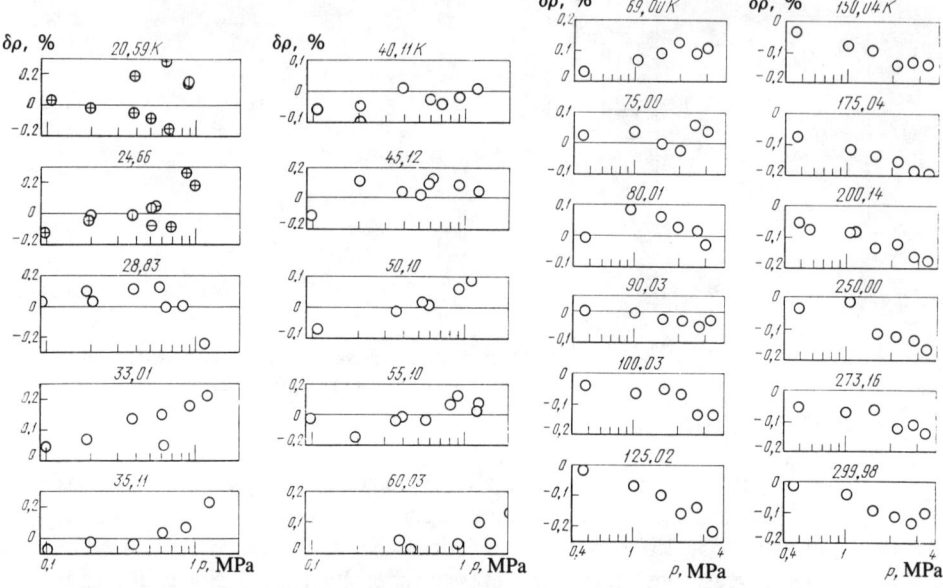

Figure 10 Deviations of experimental data [167] from calculated densities.

higher deviations (±0.2% for most points and 0.4–0.6% at maximum pressure).

The data reported in [15, 144] agree with the calculated densities to within 0.1–0.2% for $p < 10$ MPa and 0.2–0.4% at higher pressures, but they are systematically lower than the calculated values and the experimental densities reported by different authors on comparable isotherms. The data from [168] agree to within ±0.2% with the calculated values up to 40 MPa, but as the pressure increases, the deviations grow and reach 1% on some isotherms for $p = 100$ MPa. For the data from [16], the deviations range from -1.5% to $+1.5\%$, and the random character of the deviations show that the experimental data are subject to considerable scatter, and that their accuracy has been overestimated by the authors.

The pressure range up to 10 MPa is very important for modern technology. The values of $\delta\rho$ for the average HT equation in this range usually lie within ±0.1%, while for the LT equation they lie within ±0.2%. In the temperature range common to both the LT and HT equations, the values of $\delta\rho$ calculated from both equations are essentially the same. For the data from [16, 91], which extend to the highest pressures, they agree to within 0.1–0.2% (see Figs. 10, 11, and 14).

Our graphs do not show the deviations for 20 points form [16, 81, 91, 168]. They were given zero weight in the initial calculations. For most of these

EQUATIONS OF STATE 71

points, the deviations lie in the range 1.5–3%, but for two points from [91] they reach 27 and 32%. We note that significant deviations of some experimental points from the smooth thermodynamic surface were noted in [20].

A comparison of calculated acoustic and thermal data for helium with a number of experimental investigations has also been carried out. The isochoric specific heats were calculated from the LT equation and compared with the smoothed data from [91] and with the experimental values obtained in that investigation. The corresponding deviations are presented in Figs. 28 and 29.

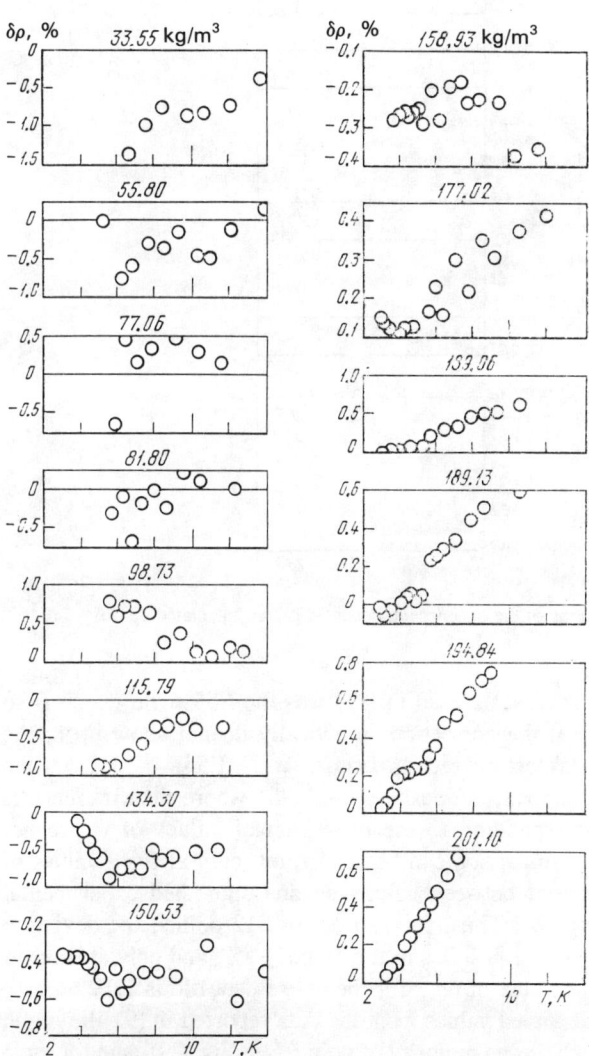

Figure 11 Deviations of the experimental data [91] from calculated densities.

72 THERMODYNAMIC PROPERTIES OF HELIUM

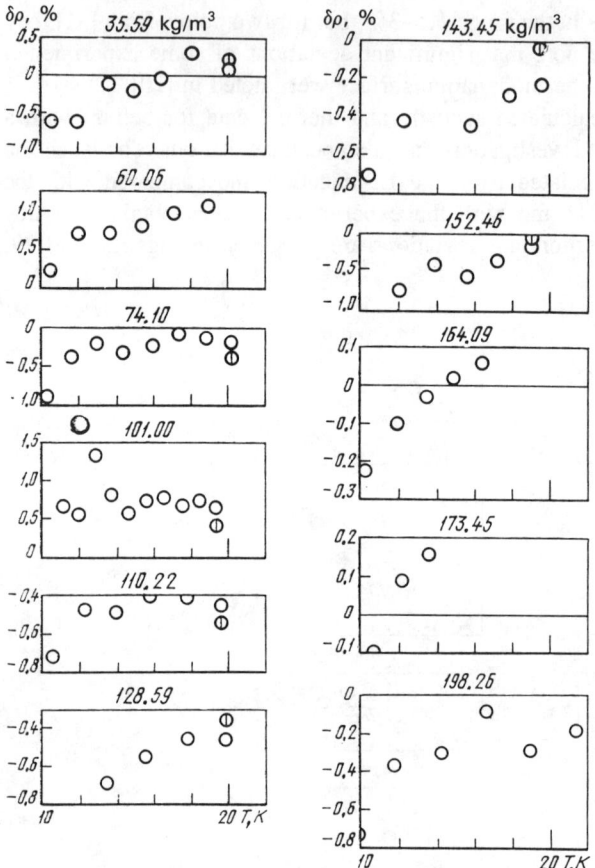

Figure 11 (*Continued*) Deviations of the experimental data [91] from calculated densities.

We note that the temperature scale used in [91] was the E55 scale which is so close to the T58 scale that the corrections practically do not show up in the calculations of c_v. The largest correction occurs at T (E55) = 5 K, when $T(58) = 5.0044$ K, and at reduced density $\omega = 2.964$ where the difference is 0.11. The pressures corresponding to the experimental values of c_v are not quoted in [91], and so a comparison can be carried out only at fixed values of the density. Good agreement between calculated and smoothed experimental values is indicated by Fig. 28. Thus, for 196 out of 312 points, the deviation $\delta c_v = [(c_{v_{\exp}} - c_{v_{\text{calc}}})/c_{v_{\text{calc}}}] \times 100\%$ lies in the range $\pm 1\%$, and only at isolated points reaches 3–4%. The value of $\delta c_{v_{\text{av}}}$ for the entire data file is 1.19%.

The deviations of calculated values from the data reported in [91] lie in the range $\pm 2\%$ for the overwhelming majority of points (see Fig. 29), and for most of the points the deviation does not exceed $\pm 1\%$. Only for 14 points lying close to the saturation curve do the deviations reach 5–30%, and these are therefore not included in the graphs.

When the calculated isochoric specific heats were compared with the experimental data from [129] in the density range 150.5–183.4 kg/m^3, satisfactory agreement was established (deviations from −1.3 to +2.2%) at temperatures above 2.5 K. At lower temperatures, the deviations increased sharply because the polynomial form of the equation of state does not represent the behavior of the specific heat c_v near the λ-curve.

The experimental data from [64] for $T < 29$ K on the 246.3 and 275.1 kg/m^3 isochores deviate from the calculated specific heats by ±8%. On the 327.5

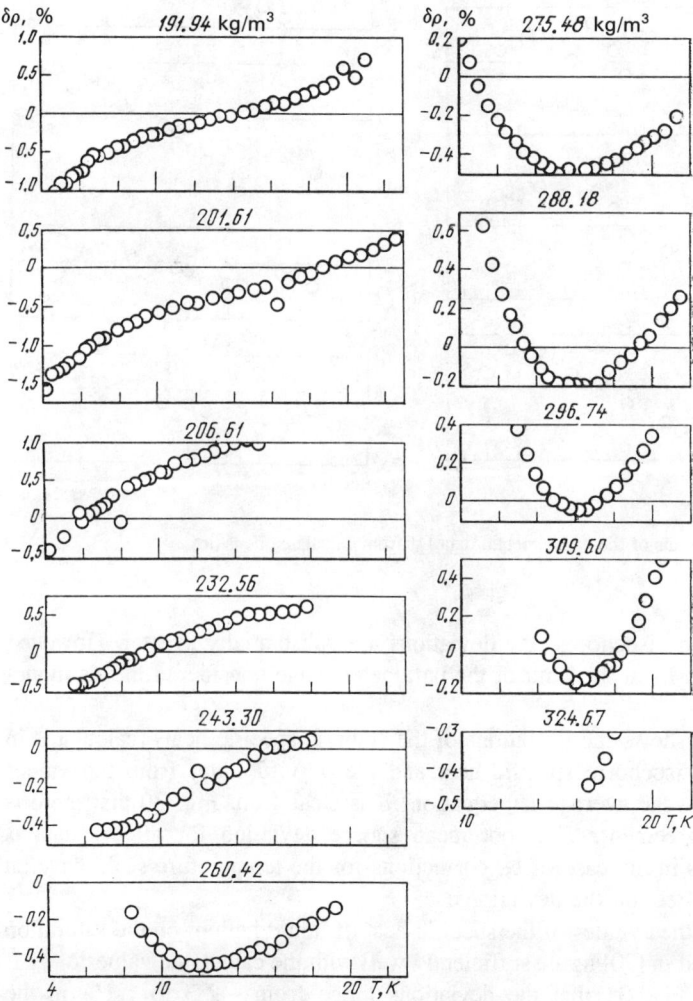

Figure 12 Deviations of experimental data [81] from calculated densities.

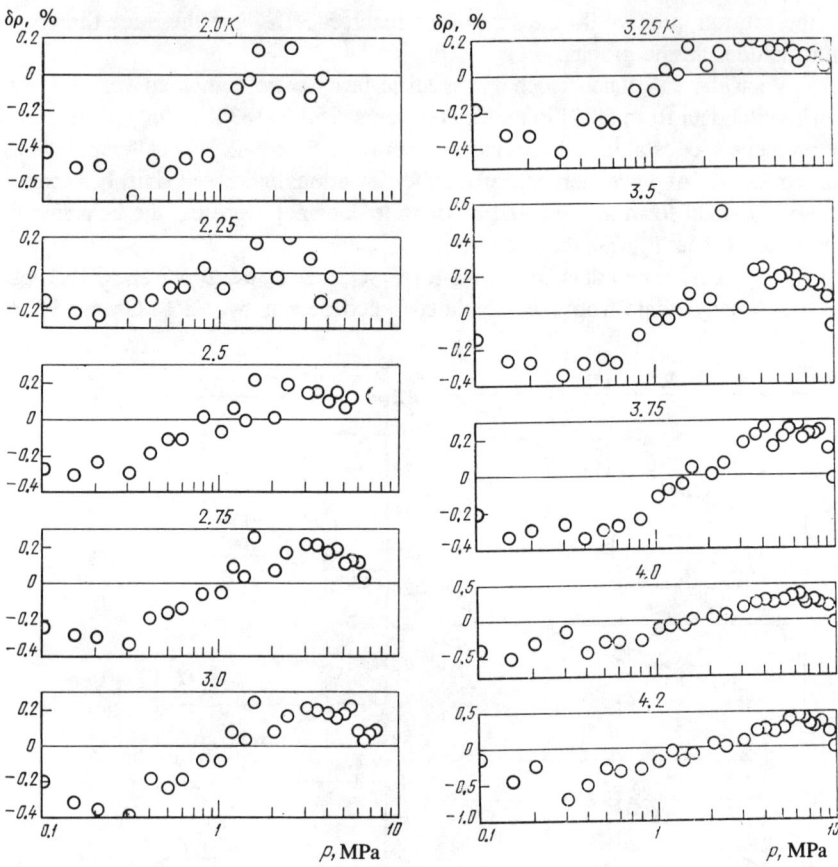

Figure 13 Deviations of the experimental data [4] from calculated densities.

and 340.1 kg/m³ isochores, the deviations are substantially greater. However, these isochores lie at the limit of the parameter range considered in this monograph.

Figure 30 shows the deviations of the isobaric specific heats (calculated in [91] from the isochoric specific heat and the p, v, T data) from the values calculated from the average LT equation. It is clear from Fig. 30 that there is satisfactory agreement. The root mean square deviation for all the data is 2.02% and, as in the case of c_v, corrections for the temperature scale have an insignificant effect on the deviations.

The smoothed values of the specific heat of liquid helium on the saturation curve presented in [90] agree sufficiently well with the calculated values of c_s'. It is clear from Fig. 31 that the deviations range from -2.3 to 1.1% in the temperature range 2.5–4.8 K. Thereafter the deviations grow because the poly-

EQUATIONS OF STATE 75

nomial form of the equation of state does not provide an accurate description of the specific heat c_s near the lower λ point, or close to the critical point.

The authors of [91] measured the derivative $(\partial p/\partial T)_v$ and reported smoothed values of this derivative on isochores. We have used them to construct the LT equations of state. Figure 32 shows the deviations of the smoothed values of $(\partial p/\partial T)_V$ [91] from our calculated values. The deviations lie in the range $\pm 3\%$ for the great majority of points when $\rho \leq 192.6$ kg/m^3 ($\omega \leq 2.76$),

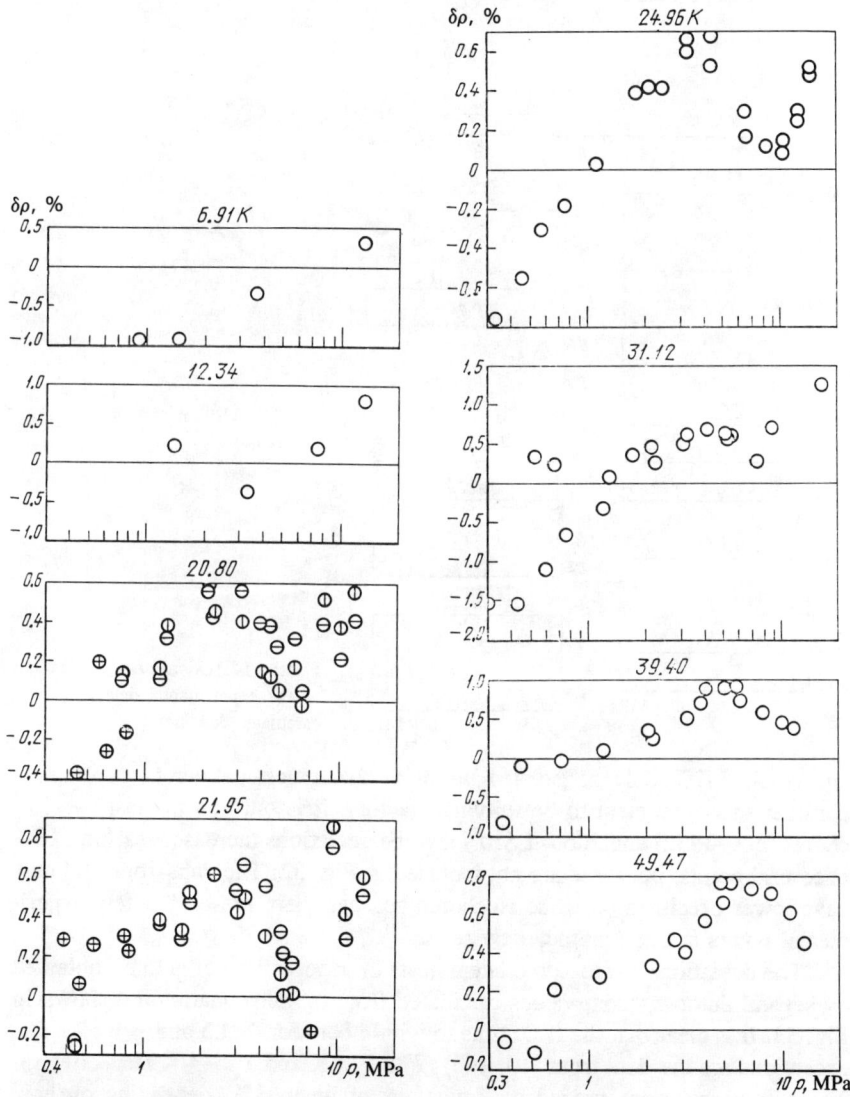

Figure 14 Deviations of the experimental data [16] from calculated densities.

76 THERMODYNAMIC PROPERTIES OF HELIUM

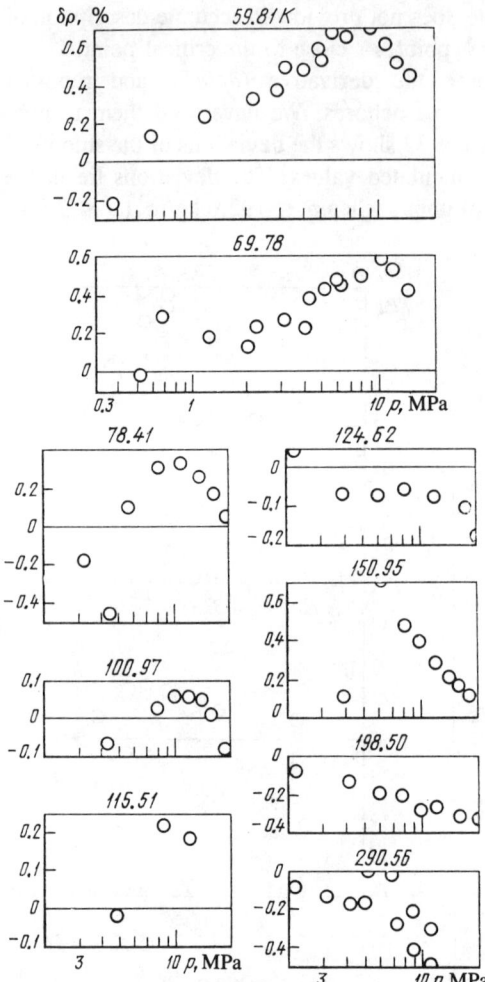

Figure 14 (*Continued*) Deviations of the experimental data [16] from calculated densities.

and reach 5–7% only at isolated points at the lowest temperature. We may consider this agreement to be fully satisfactory. It is only on the last two isochores (ρ = 195.5 and 206.4 kg/m^3) that the deviations increase, reaching 10% at several points, but these are not included in Fig. 32. The data from [91] may have lower precision on these isochores because there are only a few experimental points in this parameter range.

The deviations of measured latent heats of vaporization of helium, obtained by several authors, from values calculated from the LT equation are shown in Fig. 33. It is clear that the two agree to within between −0.5 and +1.5%. The exceptions are the data from [40, 163] (T > 3.6 K and T > 4 K respectively). These data are characterized by deviations of up to 3%, depending on how close they are to the critical temperature.

EQUATIONS OF STATE 77

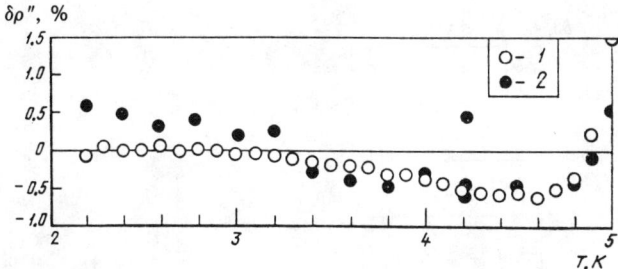

Figure 15 Deviations of the calculated [32] and experimental [74] densities of the saturated vapor from the average LT equation. *(Data from [32] and [74].)*

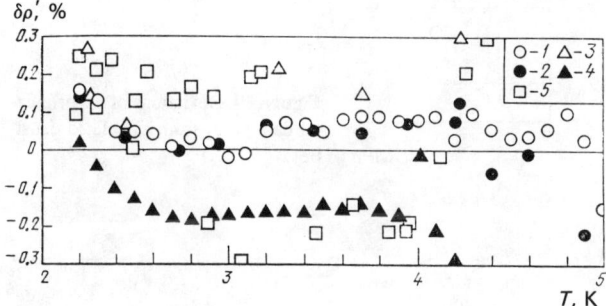

Figure 16 Deviations of the density of boiling liquid from the calculated values. *(Data from 1-[32]; 2-[75]; 3-[101]; 4-[68]; 5-[115].)*

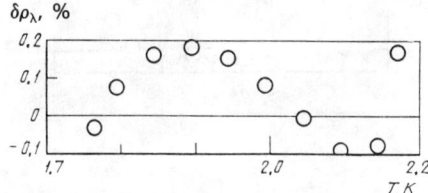

Figure 17 Deviations of the calculated [32] density of liquid He I on the λ-transition line from the average LT equation.

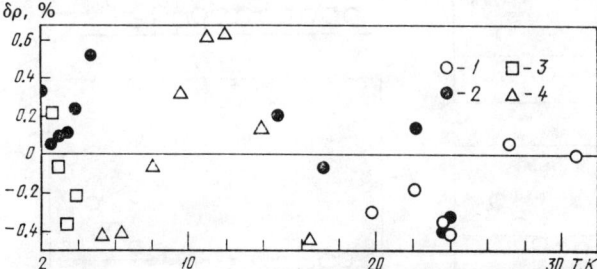

Figure 18 Deviations of the density of liquid helium on the melting line from the calculated values. *(Data from 1-[134] (calculated from the LT equation), 2-[134] (calculated from the HT equation), 3-[67], 4-[81].)*

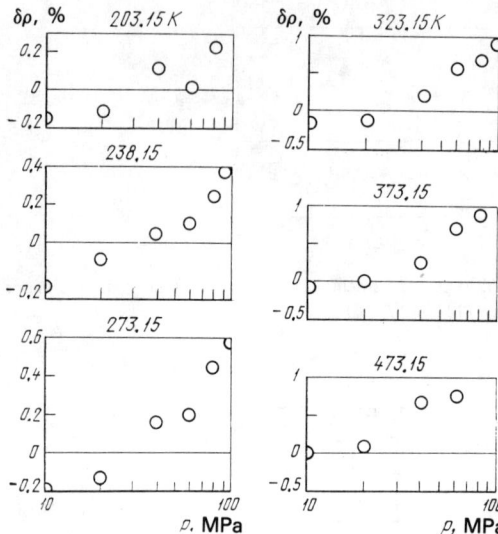

Figure 19 Deviations of experimental data [168] from calculated densities.

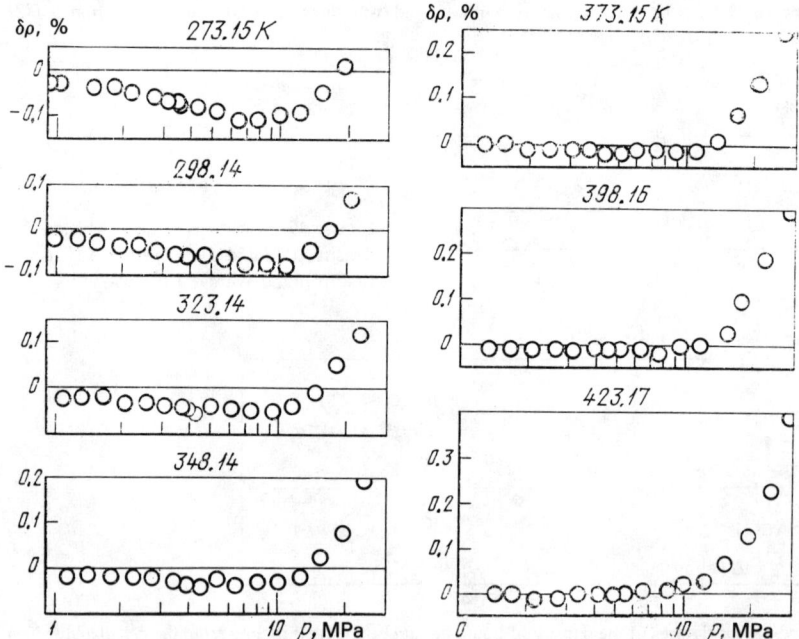

Figure 20 Deviations of experimental data [133] from calculated densities.

EQUATIONS OF STATE 79

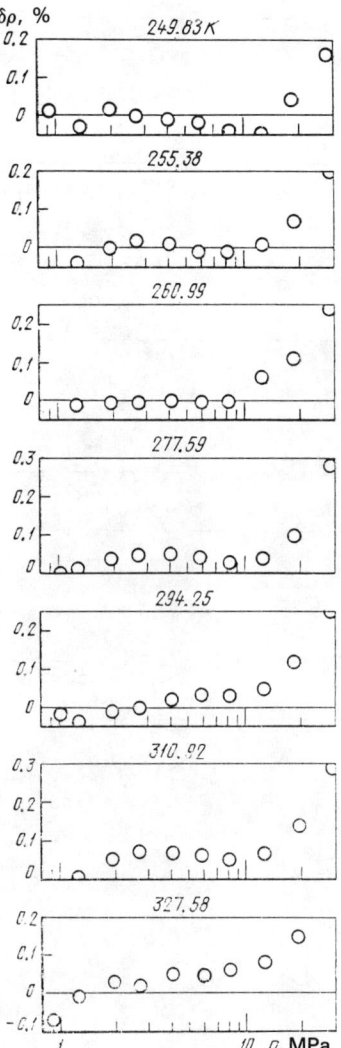

Figure 21 Deviations of experimental data [157] from calculated densities.

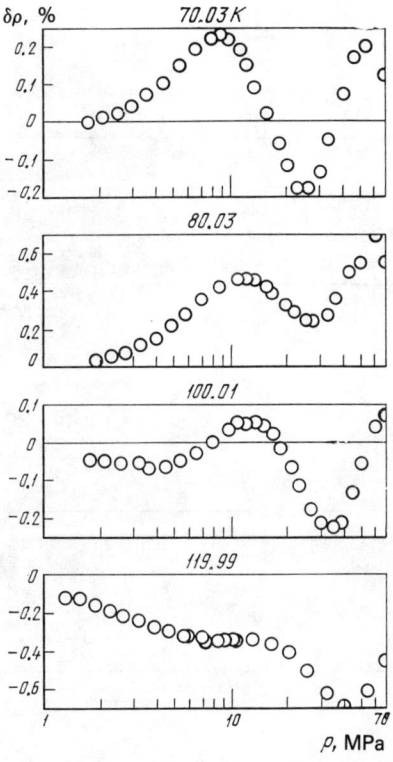

Figure 22 Deviations of the experimental data [158] from calculated densities.

80 THERMODYNAMIC PROPERTIES OF HELIUM

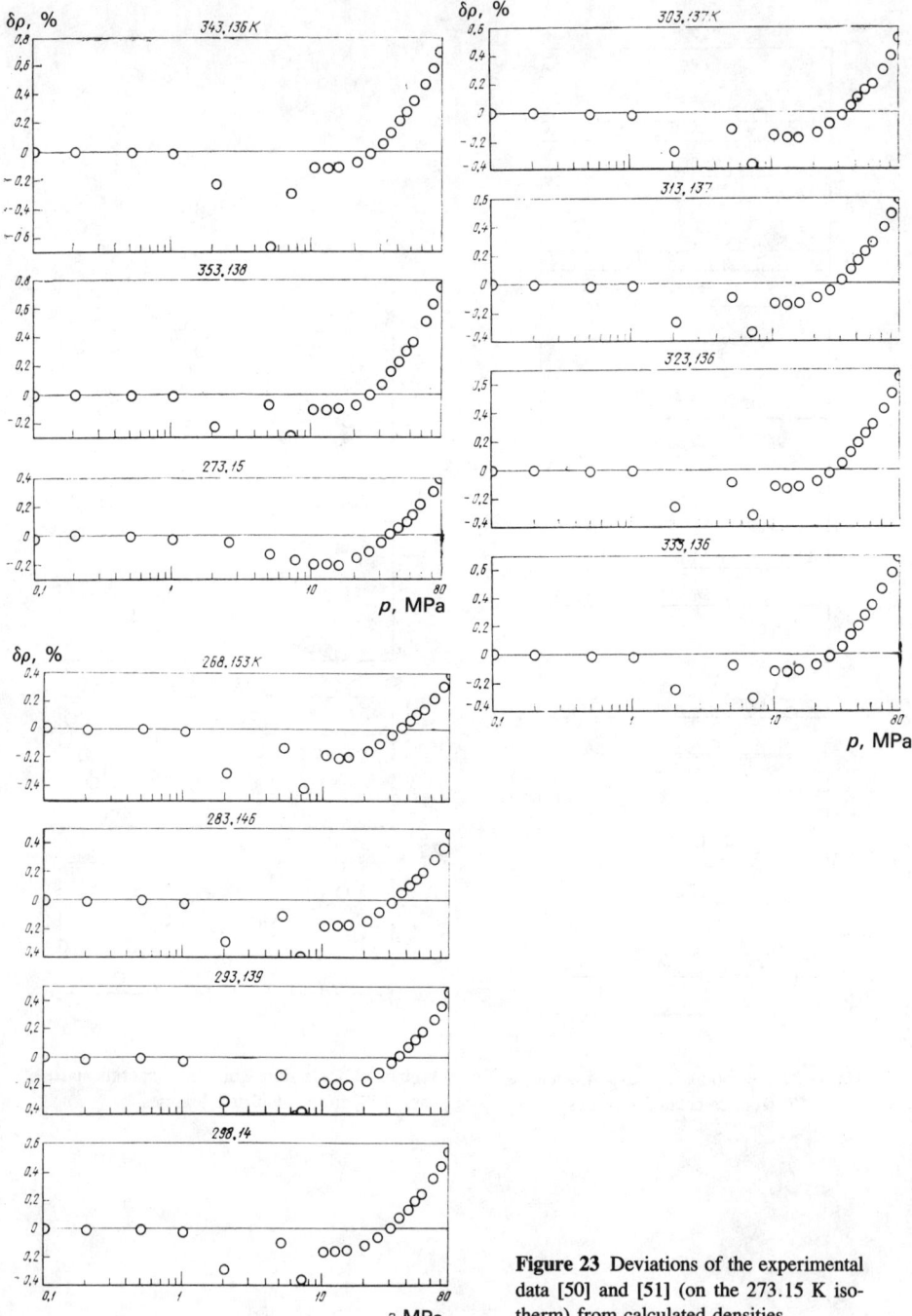

Figure 23 Deviations of the experimental data [50] and [51] (on the 273.15 K isotherm) from calculated densities.

EQUATIONS OF STATE **81**

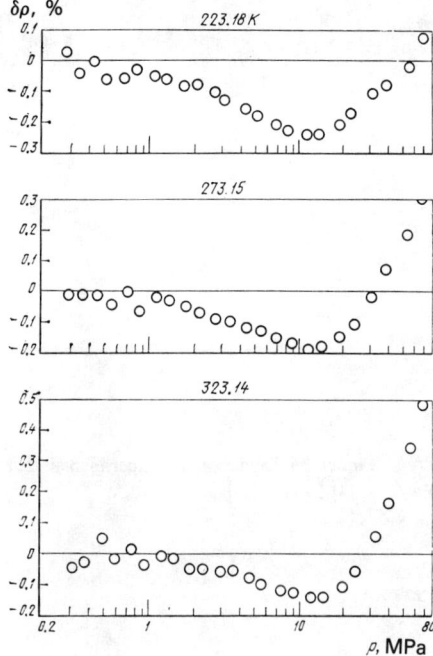

Figure 24 Deviations of experimental data [45] from calculated densities.

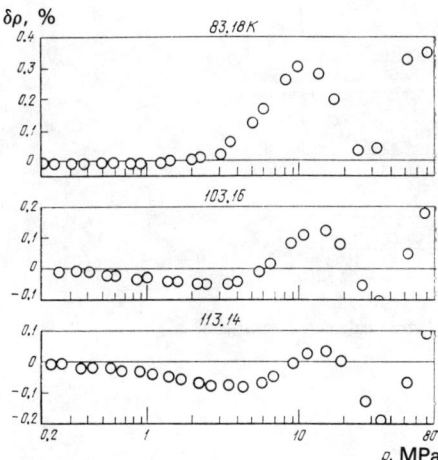

Figure 25 Deviations of the experimental data [89] from calculated densities.

82 THERMODYNAMIC PROPERTIES OF HELIUM

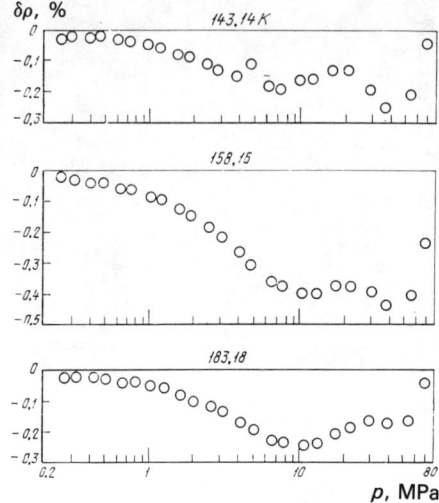

Figure 26 Deviation of experimental data [144] from calculated densities.

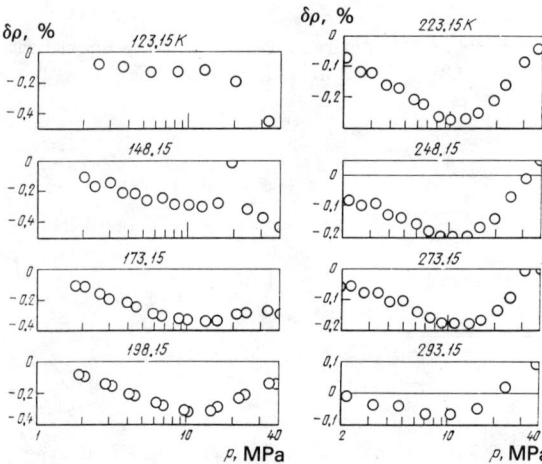

Figure 27 Deviations of experimental data [15] from calculated densities.

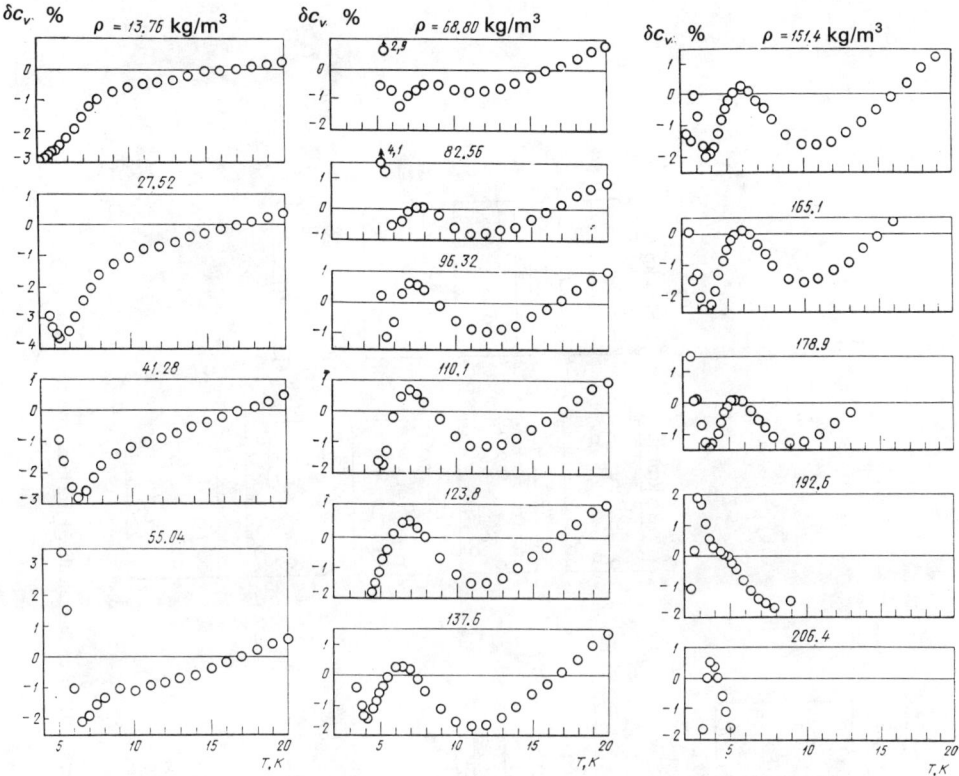

Figure 28 Deviations of smoothed data from [91] from the calculated isochoric specific heats.

In the low-temperature region, we also have several measurements of speed of sound. They are compared with calculations in Fig. 34 and in Tables 3.5 and 3.6.

In the gas phase, the deviations of the data given in [99] from the values calculated by the present authors do not, at low pressures, exceed 0.3% for 60% of the points, and reach 1.2–1.6% only at the lowest temperatures (2.218 and 2.259 K). The specific heats c_p and c_v were calculated in [99] from experimental data. Most of the values obtained agree with those calculated by the present authors to within ±3%, but on the 2.218 and 2.259 K isotherms the data in [99] are systematically lower and the deviations reach 4–6% for many of the points. For a few points on the 3.760 and 4.228 K isotherms near the saturation curve, the deviation δc_v reaches 8–18% and δc_p reaches 6–17%. The wide spread of the deviations may be regarded as an indication that the data in [99] are unreliable.

Tables 3.5 and 3.6 list the results of a comparison of calculated and measured [36, 165] values of the speed of sound in liquid helium. Agreement to

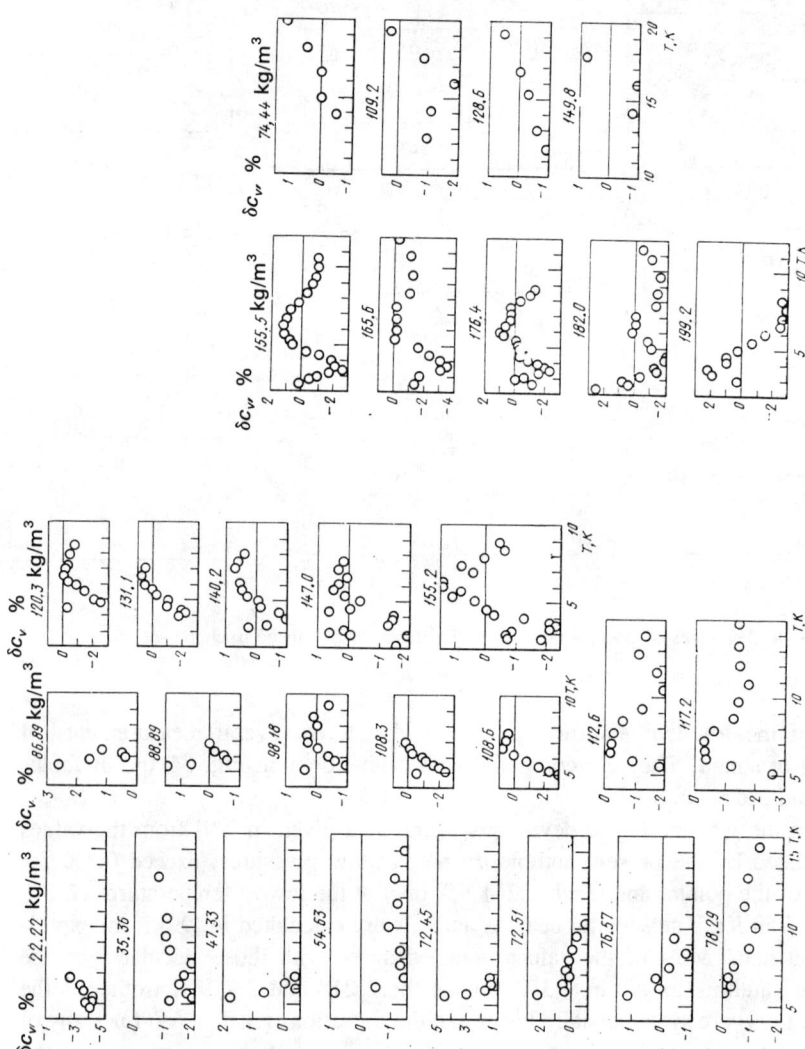

Figure 29 Deviations of experimental data reported in [91] from the calculated isochoric specific heats.

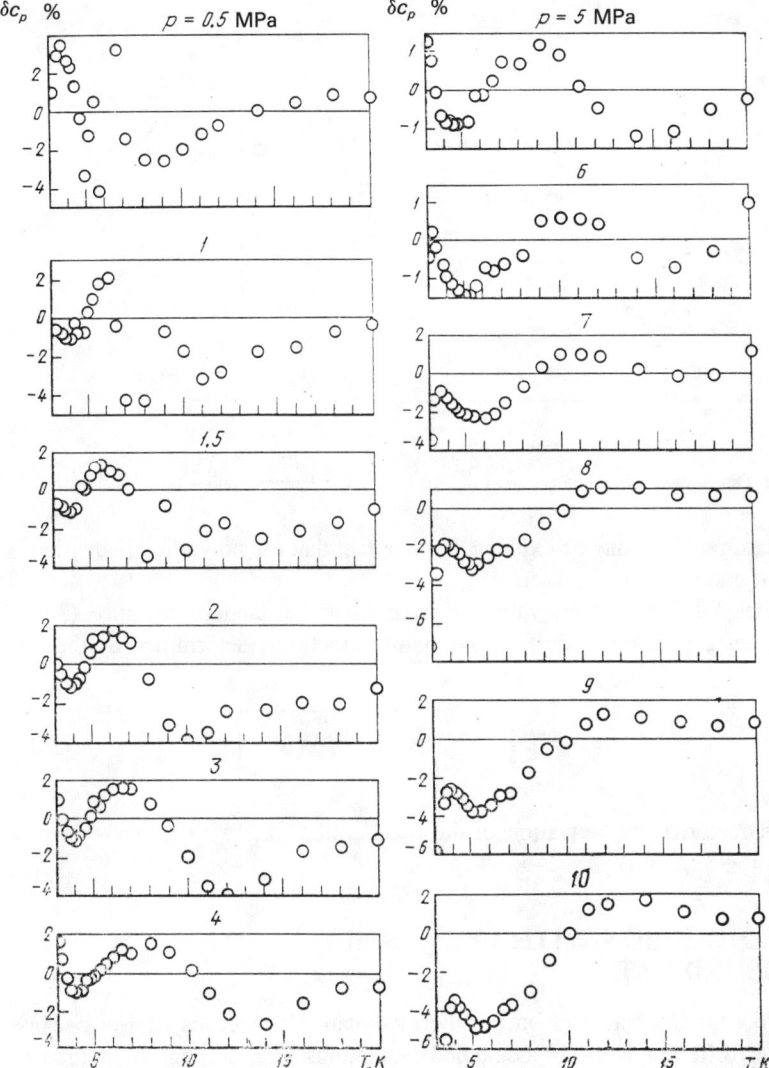

Figure 30 Deviations of the calculated data given in [91] from the specific heats c_p calculated from the average equation of state.

within 2% is apparent. Values of the velocity of sound on the saturation line, reported in [36], are higher than the calculated values by 4–5%. This may be explained by inaccuracies in the graphical extrapolation used to obtain these values. At the lowest temperature (2.19 K), the deviation between the data from [36] and the values calculated by the present authors increases substantially

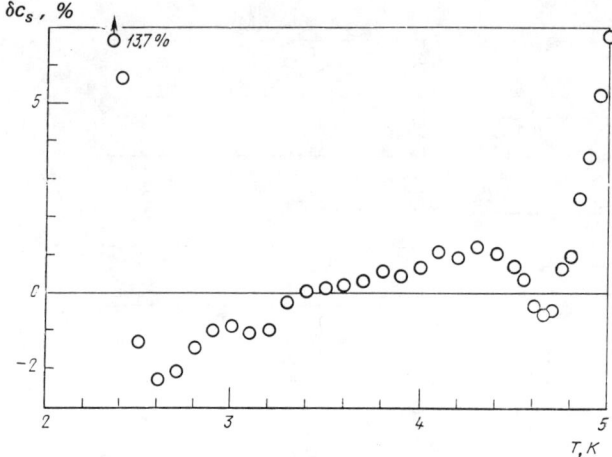

Figure 31 Deviations of the data reported in [90] from the calculated values of c_s.

with pressure. This may be explained by the fact that the polynomial equation is not good enough near the λ-curve.

Tables 3.7–3.15 present values of three times the standard deviation (3σ) of the calculated data for basic thermodynamic functions, determined by the formula:

$$3\sigma_x = 3\sqrt{\sum_{k=1}^{N}(\bar{x}-x_k)^2/N(N-1)} \qquad (3.3a)$$

They characterize the deviation in the mean value of x with a probability of 0.997 for $\pm 3\sigma_x$.[†]

3.4 COMPARISON WITH PREVIOUSLY PUBLISHED DATA

As more experimental data on the thermodynamic properties of He^4 became available, different workers assembled tabulations with a range of degree of rigor, completeness, and parameter values. Our brief review of the best known published calculations has led us to the conclusion that it is useful to compare our calculations with the most recent and rigorous publications published outside the USSR, namely, the tables in [32], based on the results in [132]. In addition, it was decided to compare our values with detailed Soviet calculations [20], bearing in mind that the data in [20, 32] cover a broader range of parameters than the tables in other publications.

[†]*Editor's Note.* Units for $3\sigma_x$ as indicated in the headings for Tables 3.7 through 3.15 are as given. In particular, this applies to Table 3.8 units are not percent as they are in the other tables in the series but in kJ/kg. This procedure appears in the other volumes in the series and is backed up with numerical data to verify the practice and insure correct translation.

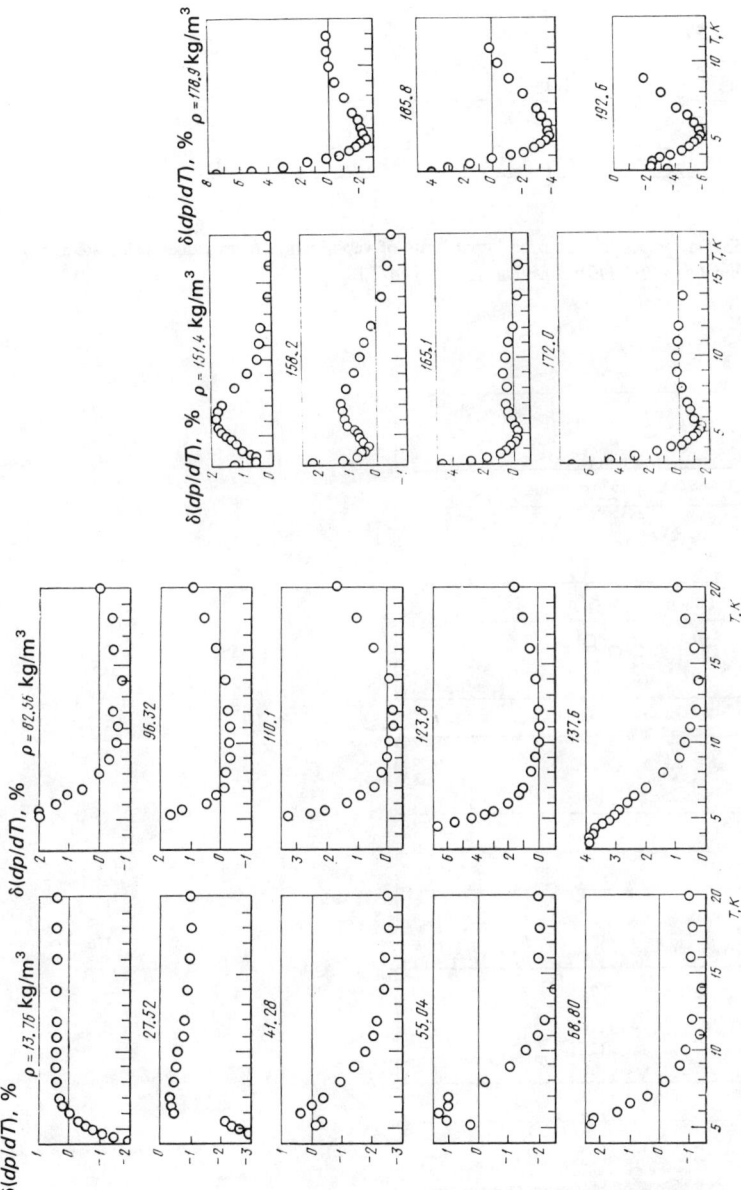

Figure 32 Deviations of the smoothed values of $\partial p/\partial T$ [91] from calculated values.

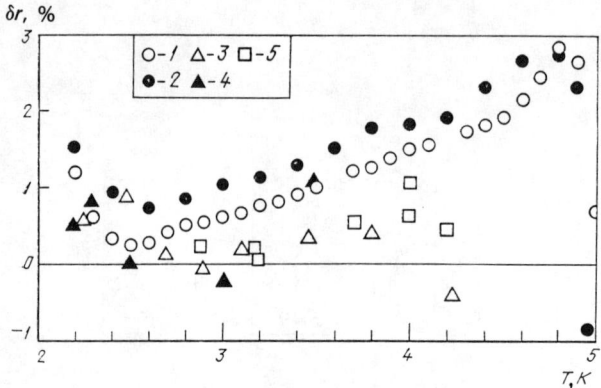

Figure 33 Deviations of measured latent heats of vaporization from values calculated from data. *(Data from 1-[163]; 2-[40]; 3-[59]; 4-[124]; 5-[41].)*

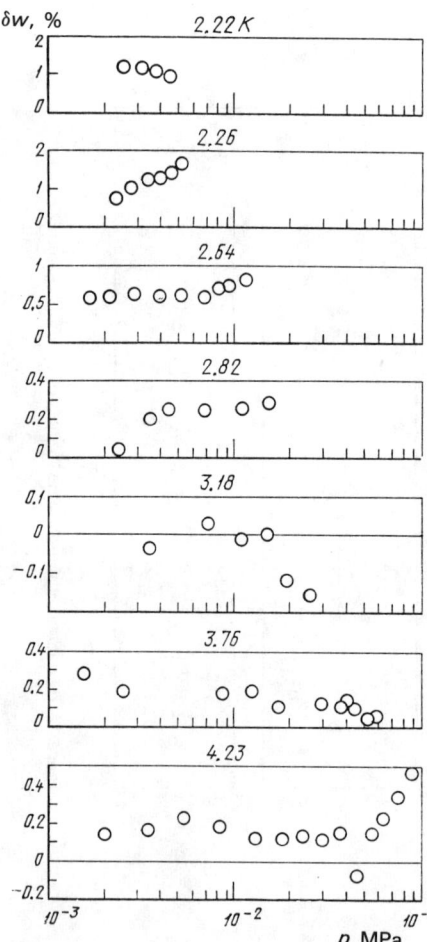

Figure 34 Deviations of data reported in [99] from the calculated speed of sound.

Table 3.5 Deviations of the measured speed of sound in helium [36] from values calculated from the LT equation of state

p, MPa	Deviation δw, % at temperature T, K						
	2.19	2.24	2.50	2.99	3.49	4.00	4.20
p_s	4.68	0.05	4.77	4.25	4.52	5.32	4.41
0.253	0.04	0.83	1.24	1.60	1.10	2.18	1.18
0.507	−2.15	−1.10	0.23	0.53	0.59	1.23	1.35
1.013	−3.75	−1.86	−0.70	−0.43	0.61	1.04	0.53
1.520	−5.29	−1.79	−0.43	−0.15	0.25	0.50	0.22
2.026	−10.0	−2.15	−0.51	−0.51	−0.03	0.35	0.17
2.533	−18.7	−1.71	−0.03	−0.27	−0.08	0.14	0.14
3.040	−30.4	−1.65	0.44	0.03	−0.05	−0.13	−0.23
4.053	−25.2	−1.82	1.27	0.57	0.14	−0.05	−0.31
5.066	—	—	1.44	0.71	0.11	−0.31	−0.42
6.080	—	—	—	1.22	0.59	0.08	−0.06
7.093	—	—	—	1.81	1.31	0.83	0.69

A comparison of specific volumes (Table 3.16) shows that for $T > 100$ K and $p < 20$ MPa the data in [20, 32] agree well with each other and also with our calculations. The deviation for most of the points is 0.1–0.2% and it is only for temperatures of 800–1400 K and pressures of 15–20 MPa that it reaches 0.3–0.4%. At lower temperatures and higher pressures the deviation rises to 0.5–0.7% and, at some points, to 1%.

We note, however, that whenever there are significant deviations between all the values that are being compared, the calculated specific volumes are either intermediate between the values obtained in [20] and [32] or in full agreement with the data in one of these papers.

When the values of enthalpy were compared with the results in [32], the

Table 3.6 Deviations of the measured speed of sound in helium [165] from values calculated from the LT equation of state

p, MPa	Deviation δw, % at temperature T, K			
	2.5	3.0	3.5	4.0
0.253	0.83	0.80	2.11	2.65
0.507	−0.91	−0.53	−0.90	0.82
1.013	−1.70	−1.40	−0.68	−0.35
1.520	−1.65	−1.37	−0.96	−0.75
2.026	−1.09	−1.37	−0.83	−0.81
2.533	−0.84	−0.81	−0.85	−0.95
3.040	−0.34	−0.77	−0.82	−1.15
4.053	0.55	−0.14	−0.57	−0.90
5.066	1.21	0.27	−0.35	−0.75

Table 3.7 Standard deviation (3σ) for the differences in the calculated density of He⁴

T, K	3σ, % at pressure p, MPa									
	0.1	1.0	2.0	3.0	4.0	5.0	6.0	7.0	8.0	9.0
2.5	0.09	0.12	0.06	0.06	0.12	0.14	0.12	—	—	—
3	0.07	0.10	0.05	0.04	0.08	0.10	0.10	0.06	0.06	0.16
4	0.17	0.08	0.07	0.05	0.04	0.05	0.07	0.08	0.09	0.08
5	0.20	0.10	0.08	0.07	0.05	0.04	0.05	0.07	0.09	0.11
6	0.12	0.16	0.10	0.08	0.06	0.05	0.04	0.05	0.07	0.09
8	0.05	0.27	0.17	0.12	0.08	0.06	0.06	0.06	0.06	0.07
10	0.03	0.21	0.19	0.15	0.12	0.09	0.08	0.07	0.07	0.08
12	0.02	0.20	0.17	0.16	0.13	0.11	0.09	0.09	0.08	0.08
14	0.01	0.15	0.11	0.17	0.13	0.11	0.10	0.09	0.09	0.09
16	0.01	0.12	0.13	0.13	0.15	0.11	0.09	0.09	0.09	0.09
18	0.01	0.10	0.16	0.10	0.14	0.13	0.09	0.08	0.08	0.09
20	0.01	0.08	0.17	0.12	0.13	0.15	0.12	0.08	0.08	0.09
30	0.01	0.02	0.05	0.09	0.11	0.12	0.12	0.11	0.09	0.07
40	0.01	0.01	0.02	0.04	0.06	0.08	0.10	0.11	0.11	0.11
50	0.01	0.01	0.01	0.02	0.03	0.04	0.05	0.06	0.07	0.08
100	0.01	0.01	0.02	0.02	0.03	0.04	0.04	0.05	0.06	0.06
200	0.01	0.01	0.01	0.01	0.01	0.01	0.02	0.02	0.02	0.02
300	0.01	0.01	0.01	0.01	0.01	0.01	0.01	0.02	0.02	0.02
400	0.01	0.01	0.01	0.01	0.01	0.01	0.01	0.02	0.02	0.02
500	0.01	0.01	0.01	0.01	0.01	0.01	0.01	0.01	0.02	0.02
600	0.01	0.01	0.01	0.01	0.01	0.01	0.01	0.01	0.02	0.02
700	0.01	0.01	0.01	0.01	0.01	0.01	0.01	0.01	0.01	0.02
800	0.01	0.01	0.01	0.01	0.01	0.01	0.01	0.01	0.01	0.02
900	0.01	0.01	0.01	0.01	0.01	0.01	0.01	0.01	0.01	0.01
1000	0.01	0.01	0.01	0.01	0.01	0.01	0.01	0.01	0.01	0.01
1100	0.01	0.01	0.01	0.01	0.01	0.01	0.01	0.01	0.01	0.01
1200	0.01	0.01	0.01	0.01	0.01	0.01	0.01	0.01	0.01	0.01
1300	0.01	0.01	0.01	0.01	0.01	0.01	0.01	0.01	0.01	0.01
1400	0.01	0.01	0.01	0.01	0.01	0.01	0.01	0.01	0.01	0.01
1500	0.01	0.01	0.01	0.01	0.01	0.01	0.01	0.01	0.01	0.01

Table 3.7 (*Continued*)

T, K	3σ, % at pressure p, MPa									
	10	20	30	40	50	60	70	80	90	100
3	0.32	—	—	—	—	—	—	—	—	—
4	0.06	—	—	—	—	—	—	—	—	—
5	0.12	—	—	—	—	—	—	—	—	—
6	0.11	0.13	—	—	—	—	—	—	—	—
8	0.07	0.16	0.16	0.20	—	—	—	—	—	—
10	0.08	0.10	0.14	0.15	0.16	—	—	—	—	—
12	0.09	0.10	0.12	0.14	0.13	0.13	0.13	—	—	—
14	0.09	0.11	0.11	0.12	0.14	0.14	0.13	0.10	—	—
16	0.09	0.12	0.10	0.10	0.11	0.11	0.11	0.12	0.14	—
18	0.10	0.13	0.12	0.14	0.17	0.16	0.11	0.06	0.10	—
20	0.10	0.14	0.17	0.14	0.32	0.34	0.29	0.21	0.17	0.23
30	0.06	0.15	0.15	0.14	0.13	0.12	0.15	0.23	0.34	0.45
40	0.10	0.15	0.20	0.20	0.19	0.17	0.18	0.30	0.50	0.75
50	0.09	0.10	0.19	0.19	0.17	0.18	0.18	0.26	0.47	0.79
100	0.06	0.06	0.07	0.08	0.09	0.08	0.19	0.37	0.60	0.44
200	0.03	0.06	0.05	0.03	0.04	0.04	0.05	0.09	0.16	0.25
300	0.02	0.04	0.06	0.06	0.05	0.04	0.05	0.06	0.07	0.09
400	0.02	0.03	0.06	0.07	0.07	0.07	0.06	0.06	0.06	0.07
500	0.02	0.03	0.05	0.07	0.08	0.08	0.08	0.07	0.07	0.07
600	0.02	0.03	0.04	0.06	0.07	0.08	0.09	0.09	0.08	0.08
700	0.02	0.03	0.04	0.05	0.07	0.08	0.09	0.09	0.09	0.09
800	0.02	0.03	0.04	0.05	0.06	0.08	0.09	0.09	0.10	0.10
900	0.02	0.03	0.04	0.05	0.06	0.07	0.08	0.09	0.10	0.10
1000	0.01	0.03	0.04	0.05	0.06	0.07	0.08	0.08	0.09	0.10
1100	0.01	0.03	0.04	0.04	0.05	0.06	0.07	0.08	0.09	0.09
1200	0.01	0.02	0.03	0.04	0.05	0.06	0.07	0.08	0.08	0.09
1300	0.01	0.02	0.03	0.04	0.05	0.06	0.06	0.07	0.08	0.09
1400	0.01	0.02	0.03	0.04	0.05	0.05	0.06	0.07	0.08	0.08
1500	0.01	0.02	0.03	0.04	0.04	0.05	0.06	0.07	0.07	0.08

Table 3.8 Standard deviation (3σ) for the differences in calculated enthalpy of He^4

T, K	3σ, kJ/kg at pressure p, MPa									
	0.1	1	2	3	4	5	6	7	8	9
2.5	0.1	0.1	0.1	0.1	0.1	0.1	0.1	—	—	—
3	0.1	0.1	0.1	0.1	0.1	0.1	0.1	0.1	0.1	0.1
4	0.1	0.1	0.1	0.1	0.1	0.1	0.1	0.1	0.1	0.1
5	0.1	0.1	0.1	0.1	0.1	0.1	0.1	0.1	0.1	0.1
6	0.1	0.1	0.1	0.1	0.1	0.1	0.1	0.1	0.1	0.1
8	0.1	0.1	0.1	0.1	0.1	0.1	0.1	0.1	0.1	0.1
10	0.1	0.1	0.1	0.1	0.1	0.1	0.1	0.1	0.1	0.1
12	0.1	0.1	0.2	0.2	0.1	0.1	0.1	0.1	0.1	0.1
14	0.1	0.1	0.1	0.2	0.2	0.1	0.1	0.1	0.1	0.1
16	0.1	0.1	0.1	0.2	0.2	0.2	0.2	0.2	0.2	0.2
18	0.1	0.1	0.1	0.1	0.2	0.2	0.2	0.2	0.2	0.2
20	0.1	0.1	0.1	0.1	0.1	0.2	0.2	0.2	0.2	0.2
30	0.1	0.1	0.1	0.1	0.1	0.1	0.1	0.1	0.1	0.1
40	0.1	0.1	0.1	0.1	0.2	0.2	0.3	0.3	0.3	0.4
50	0.1	0.1	0.1	0.1	0.2	0.2	0.2	0.3	0.3	0.4
100	0.1	0.1	0.1	0.1	0.1	0.1	0.1	0.1	0.2	0.2
200	0.1	0.1	0.1	0.1	0.1	0.1	0.2	0.2	0.2	0.3
300	0.1	0.1	0.1	0.1	0.1	0.1	0.1	0.1	0.2	0.2
400	0.1	0.1	0.1	0.1	0.1	0.1	0.1	0.1	0.1	0.1
500	0.1	0.1	0.1	0.1	0.1	0.1	0.1	0.1	0.1	0.1
600	0.1	0.1	0.1	0.1	0.1	0.1	0.1	0.1	0.1	0.1
700	0.1	0.1	0.1	0.1	0.1	0.1	0.1	0.1	0.1	0.2
800	0.1	0.1	0.1	0.1	0.1	0.1	0.1	0.1	0.2	0.2
900	0.1	0.1	0.1	0.1	0.1	0.1	0.1	0.2	0.2	0.2
1000	0.1	0.1	0.1	0.1	0.1	0.1	0.1	0.2	0.2	0.2
1100	0.1	0.1	0.1	0.1	0.1	0.1	0.1	0.2	0.2	0.2
1200	0.1	0.1	0.1	0.1	0.1	0.1	0.2	0.2	0.2	0.2
1300	0.1	0.1	0.1	0.1	0.1	0.1	0.2	0.2	0.2	0.2
1400	0.1	0.1	0.1	0.1	0.1	0.1	0.2	0.2	0.2	0.2
1500	0.1	0.1	0.1	0.1	0.1	0.1	0.2	0.2	0.2	0.2

Table 3.8 (Continued)

T, K	3σ, kJ/kg at pressure p, MPa									
	10	20	30	40	50	60	70	80	90	100
3	0.1	—	—	—	—	—	—	—	—	—
4	0.1	—	—	—	—	—	—	—	—	—
5	0.1	—	—	—	—	—	—	—	—	—
6	0.1	0.1	—	—	—	—	—	—	—	—
8	0.1	0.2	0.4	0.5	—	—	—	—	—	—
10	0.1	0.1	0.3	0.5	0.6	—	—	—	—	—
12	0.1	0.1	0.2	0.4	0.6	0.7	0.8	—	—	—
14	0.1	0.2	0.3	0.5	0.7	0.8	0.9	0.9	—	—
16	0.2	0.2	0.3	0.6	0.9	1.2	1.5	1.7	1.8	—
18	0.2	0.3	0.3	0.6	1.0	1.6	2.1	2.5	2.9	—
20	0.2	0.3	0.3	0.5	1.0	1.7	2.4	3.1	3.7	4.2
30	0.1	0.2	0.1	0.2	0.3	0.4	0.5	0.7	0.9	1.2
40	0.4	0.3	0.3	0.3	0.4	0.4	0.6	0.8	1.1	1.4
50	0.4	0.4	0.4	0.4	0.4	0.6	0.7	1.1	1.6	2.2
100	0.2	0.3	0.3	0.3	0.4	0.4	0.4	0.6	1.0	1.5
200	0.3	0.5	0.6	0.7	0.8	0.9	0.8	0.8	0.8	0.9
300	0.2	0.4	0.7	0.7	0.8	0.8	0.9	1.0	1.0	0.9
400	0.1	0.3	0.6	0.7	0.8	0.8	0.8	0.8	0.9	1.0
500	0.1	0.2	0.4	0.7	0.8	0.9	0.9	0.8	0.8	0.8
600	0.1	0.2	0.4	0.6	0.8	1.0	1.0	1.0	1.0	0.9
700	0.2	0.3	0.4	0.6	0.8	1.0	1.1	1.2	1.2	1.2
800	0.2	0.2	0.4	0.6	0.8	1.0	1.2	1.3	1.4	1.4
900	0.2	0.4	0.5	0.6	0.8	1.0	1.2	1.4	1.6	1.7
1000	0.2	0.4	0.5	0.7	0.8	1.0	1.3	1.5	1.7	1.8
1100	0.2	0.4	0.6	0.7	0.9	1.1	1.3	1.5	1.7	1.9
1200	0.2	0.5	0.6	0.8	0.9	1.1	1.3	1.6	1.8	2.0
1300	0.3	0.5	0.7	0.8	1.0	1.2	1.4	1.6	1.8	2.1
1400	0.3	0.5	0.7	0.9	1.1	1.2	1.4	1.6	1.9	2.1
1500	0.3	0.5	0.8	0.9	1.1	1.3	1.5	1.7	1.9	2.2

Table 3.9 Standard deviation (3σ) for the differences in calculated entropy of He^4

T, K	3σ, % at pressure p, MPa									
	0.1	1	2	3	4	5	6	7	8	9
2.5	0.60	0.71	0.67	0.65	0.81	1.21	1.69	—	—	—
3	0.66	0.61	0.64	0.73	0.88	1.08	1.23	1.21	0.93	0.80
4	0.39	0.30	0.36	0.45	0.56	0.67	0.78	0.85	0.84	0.73
5	0.07	0.27	0.25	0.28	0.33	0.39	0.46	0.53	0.61	0.67
6	0.03	0.26	0.21	0.19	0.20	0.22	0.25	0.29	0.35	0.41
8	0.02	0.18	0.14	0.13	0.11	0.11	0.12	0.13	0.14	0.15
10	0.01	0.11	0.17	0.11	0.10	0.10	0.10	0.12	0.13	0.14
12	0.01	0.07	0.17	0.17	0.13	0.11	0.11	0.12	0.14	0.15
14	0.01	0.04	0.12	0.16	0.16	0.14	0.14	0.14	0.15	0.16
16	0.01	0.02	0.07	0.14	0.17	0.16	0.15	0.15	0.16	0.17
18	0.01	0.01	0.04	0.10	0.14	0.15	0.15	0.15	0.16	0.17
20	0.01	0.01	0.03	0.07	0.11	0.14	0.14	0.14	0.15	0.17
30	0.01	0.01	0.01	0.01	0.01	0.02	0.02	0.03	0.04	0.05
40	0.01	0.01	0.01	0.02	0.03	0.04	0.04	0.05	0.06	0.06
50	0.01	0.01	0.01	0.02	0.02	0.03	0.04	0.04	0.05	0.06
100	0.01	0.01	0.01	0.01	0.01	0.01	0.01	0.01	0.01	0.01
200	0.01	0.01	0.01	0.01	0.01	0.01	0.01	0.01	0.01	0.01
300	0.01	0.01	0.01	0.01	0.01	0.01	0.01	0.01	0.01	0.01
400	0.01	0.01	0.01	0.01	0.01	0.01	0.01	0.01	0.01	0.01
500	0.01	0.01	0.01	0.01	0.01	0.01	0.01	0.01	0.01	0.01
600	0.01	0.01	0.01	0.01	0.01	0.01	0.01	0.01	0.01	0.01
700	0.01	0.01	0.01	0.01	0.01	0.01	0.01	0.01	0.01	0.01
800	0.01	0.01	0.01	0.01	0.01	0.01	0.01	0.01	0.01	0.01
900	0.01	0.01	0.01	0.01	0.01	0.01	0.01	0.01	0.01	0.01
1000	0.01	0.01	0.01	0.01	0.01	0.01	0.01	0.01	0.01	0.01
1100	0.01	0.01	0.01	0.01	0.01	0.01	0.01	0.01	0.01	0.01
1200	0.01	0.01	0.01	0.01	0.01	0.01	0.01	0.01	0.01	0.01
1300	0.01	0.01	0.01	0.01	0.01	0.01	0.01	0.01	0.01	0.01
1400	0.01	0.01	0.01	0.01	0.01	0.01	0.01	0.01	0.01	0.01
1500	0.01	0.01	0.01	0.01	0.01	0.01	0.01	0.01	0.01	0.01

Table 3.9 (*Continued*)

T, K	3σ, % at pressure p, MPa									
	10	20	30	40	50	60	70	80	90	100
3	2.25	—	—	—	—	—	—	—	—	—
4	0.55	—	—	—	—	—	—	—	—	—
5	0.70	—	—	—	—	—	—	—	—	—
6	0.49	0.92	—	—	—	—	—	—	—	—
8	0.16	0.92	1.70	3.00	—	—	—	—	—	—
10	0.14	0.36	1.16	1.93	2.48	—	—	—	—	—
12	0.16	0.18	0.51	1.06	1.55	1.83	2.37	—	—	—
14	0.17	0.23	0.38	0.75	1.18	1.53	1.70	1.72	—	—
16	0.18	0.26	0.38	0.78	1.36	1.99	2.57	3.01	3.23	—
18	0.19	0.28	0.35	0.72	1.39	2.23	3.13	3.99	4.77	—
20	0.18	0.29	0.34	0.60	1.25	2.17	3.23	4.32	5.41	6.48
30	0.06	0.08	0.07	0.13	0.22	0.32	0.42	0.55	0.72	1.00
40	0.06	0.07	0.08	0.09	0.11	0.13	0.17	0.24	0.33	0.45
50	0.06	0.07	0.07	0.08	0.09	0.10	0.13	0.22	0.34	0.50
100	0.01	0.02	0.02	0.02	0.02	0.02	0.02	0.05	0.07	0.09
200	0.01	0.01	0.02	0.02	0.02	0.03	0.03	0.03	0.03	0.03
300	0.01	0.01	0.01	0.01	0.02	0.02	0.02	0.03	0.03	0.03
400	0.01	0.01	0.01	0.01	0.01	0.01	0.02	0.02	0.02	0.02
500	0.01	0.01	0.01	0.01	0.01	0.01	0.01	0.01	0.02	0.02
600	0.01	0.01	0.01	0.01	0.01	0.01	0.01	0.01	0.01	0.01
700	0.01	0.01	0.01	0.01	0.01	0.01	0.01	0.01	0.01	0.01
800	0.01	0.01	0.01	0.01	0.01	0.01	0.01	0.01	0.01	0.01
900	0.01	0.01	0.01	0.01	0.01	0.01	0.01	0.01	0.01	0.01
1000	0.01	0.01	0.01	0.01	0.01	0.01	0.01	0.01	0.01	0.01
1100	0.01	0.01	0.01	0.01	0.01	0.01	0.01	0.01	0.01	0.01
1200	0.01	0.01	0.01	0.01	0.01	0.01	0.01	0.01	0.01	0.01
1300	0.01	0.01	0.01	0.01	0.01	0.01	0.01	0.01	0.01	0.01
1400	0.01	0.01	0.01	0.01	0.01	0.01	0.01	0.01	0.01	0.01
1500	0.01	0.01	0.01	0.01	0.01	0.01	0.01	0.01	0.01	0.01

96 THERMODYNAMIC PROPERTIES OF HELIUM

Table 3.10 Standard deviation (3σ) for the differences in calculated isochoric specific heat of He^4

T, K	3σ, % at pressure p, MPa									
	0.1	1	2	3	4	5	6	7	8	9
2.5	3.6	1.2	1.4	1.2	1.6	0.9	4.1	—	—	—
3	1.8	1.5	0.9	0.8	0.9	0.7	1.1	2.4	4.2	6.4
4	0.9	0.7	0.7	0.6	0.7	0.9	1.1	1.2	1.1	1.2
5	0.3	0.4	0.5	0.6	0.8	0.9	1.1	1.1	1.1	0.9
6	0.3	0.5	0.6	0.8	0.9	1.0	1.1	1.2	1.2	1.1
8	0.1	0.9	0.8	0.8	0.8	0.8	0.7	0.7	0.7	0.7
10	0.1	0.4	0.6	0.7	0.7	0.7	0.7	0.6	0.6	0.6
12	0.1	0.2	0.3	0.5	0.6	0.6	0.7	0.7	0.7	0.7
14	0.1	0.2	0.2	0.3	0.4	0.5	0.6	0.6	0.7	0.7
16	0.1	0.3	0.3	0.2	0.3	0.4	0.5	0.6	0.6	0.6
18	0.1	0.3	0.3	0.3	0.3	0.3	0.4	0.5	0.5	0.6
20	0.1	0.3	0.3	0.3	0.3	0.3	0.3	0.3	0.4	0.5
30	0.1	0.2	0.5	0.6	0.8	0.9	1.1	1.2	1.3	1.4
40	0.1	0.1	0.2	0.3	0.3	0.4	0.4	0.5	0.5	0.6
50	0.1	0.1	0.1	0.1	0.1	0.1	0.1	0.1	0.1	0.2
100	0.1	0.1	0.1	0.1	0.1	0.1	0.1	0.1	0.1	0.1
200	0.1	0.1	0.1	0.1	0.1	0.1	0.1	0.1	0.1	0.1
300	0.1	0.1	0.1	0.1	0.1	0.1	0.1	0.1	0.1	0.1
400	0.1	0.1	0.1	0.1	0.1	0.1	0.1	0.1	0.1	0.1
500	0.1	0.1	0.1	0.1	0.1	0.1	0.1	0.1	0.1	0.1
600	0.1	0.1	0.1	0.1	0.1	0.1	0.1	0.1	0.1	0.1
700	0.1	0.1	0.1	0.1	0.1	0.1	0.1	0.1	0.1	0.1
800	0.1	0.1	0.1	0.1	0.1	0.1	0.1	0.1	0.1	0.1
900	0.1	0.1	0.1	0.1	0.1	0.1	0.1	0.1	0.1	0.1
1000	0.1	0.1	0.1	0.1	0.1	0.1	0.1	0.1	0.1	0.1
1100	0.1	0.1	0.1	0.1	0.1	0.1	0.1	0.1	0.1	0.1
1200	0.1	0.1	0.1	0.1	0.1	0.1	0.1	0.1	0.1	0.1
1300	0.1	0.1	0.1	0.1	0.1	0.1	0.1	0.1	0.1	0.1
1400	0.1	0.1	0.1	0.1	0.1	0.1	0.1	0.1	0.1	0.1
1500	0.1	0.1	0.1	0.1	0.1	0.1	0.1	0.1	0.1	0.1

Table 3.10 (*Continued*)

T, K	3σ, % at pressure p, MPa									
	10	20	30	40	50	60	70	80	90	100
3	9.4	—	—	—	—	—	—	—	—	—
4	2.3	—	—	—	—	—	—	—	—	—
5	0.8	—	—	—	—	—	—	—	—	—
6	1.0	10.4	—	—	—	—	—	—	—	—
8	0.8	1.1	6.9	24.4	—	—	—	—	—	—
10	0.6	1.3	3.0	5.3	12.5	—	—	—	—	—
12	0.7	1.2	3.2	6.1	9.0	13.3	23.2	—	—	—
14	0.7	1.0	2.6	5.6	9.2	13.1	18.1	26.6	—	—
16	0.7	0.9	2.0	4.6	8.1	12.1	16.5	22.2	30.8	—
18	0.6	0.9	1.5	3.5	6.6	10.2	14.3	18.9	24.7	—
20	0.5	0.8	1.2	0.6	5.2	8.4	11.9	15.8	20.3	25.8
30	1.4	1.6	1.6	1.5	1.6	1.9	2.4	3.2	4.3	5.7
40	0.6	0.8	0.6	0.6	1.0	1.6	2.3	3.1	4.0	4.9
50	0.2	0.2	0.2	0.4	0.9	1.4	2.0	2.6	3.4	4.2
100	0.2	0.3	0.3	0.4	0.5	0.6	0.7	0.9	1.1	1.3
200	0.1	0.1	0.1	0.1	0.1	0.2	0.2	0.2	0.2	0.3
300	0.1	0.1	0.1	0.1	0.1	0.1	0.1	0.1	0.1	0.1
400	0.1	0.1	0.1	0.1	0.1	0.1	0.1	0.1	0.1	0.1
500	0.1	0.1	0.1	0.1	0.1	0.1	0.1	0.1	0.1	0.1
600	0.1	0.1	0.1	0.1	0.1	0.1	0.1	0.1	0.1	0.1
700	0.1	0.1	0.1	0.1	0.1	0.1	0.1	0.1	0.1	0.1
800	0.1	0.1	0.1	0.1	0.1	0.1	0.1	0.1	0.1	0.1
900	0.1	0.1	0.1	0.1	0.1	0.1	0.1	0.1	0.1	0.1
1000	0.1	0.1	0.1	0.1	0.1	0.1	0.1	0.1	0.1	0.1
1100	0.1	0.1	0.1	0.1	0.1	0.1	0.1	0.1	0.1	0.1
1200	0.1	0.1	0.1	0.1	0.1	0.1	0.1	0.1	0.1	0.1
1300	0.1	0.1	0.1	0.1	0.1	0.1	0.1	0.1	0.1	0.1
1400	0.1	0.1	0.1	0.1	0.1	0.1	0.1	0.1	0.1	0.1
1500	0.1	0.1	0.1	0.1	0.1	0.1	0.1	0.1	0.1	0.1

Table 3.11 Standard deviation (3σ) for the differences in calculated isobaric specific heat of He^4

T, K	3σ, % at pressure p, MPa									
	0.1	1	2	3	4	5	6	7	8	9
2.5	3.0	1.0	1.2	1.3	1.4	1.5	5.2	—	—	—
3	1.1	1.1	0.8	0.6	0.7	0.8	1.1	0.7	3.5	7.3
4	0.8	0.7	0.7	0.8	1.1	1.4	1.5	1.4	1.4	2.5
5	0.3	0.8	0.7	0.7	0.8	1.1	1.3	1.3	1.2	1.0
6	0.2	0.7	0.7	0.7	0.7	0.9	1.1	1.2	1.4	1.4
8	0.1	0.9	0.8	0.5	0.5	0.5	0.5	0.6	0.7	0.9
10	0.1	0.4	1.0	0.8	0.6	0.5	0.4	0.5	0.5	0.6
12	0.1	0.3	0.6	0.8	0.7	0.6	0.5	0.5	0.5	0.5
14	0.1	0.3	0.5	0.5	0.6	0.6	0.5	0.5	0.5	0.5
16	0.1	0.2	0.5	0.5	0.5	0.5	0.5	0.4	0.4	0.5
18	0.1	0.2	0.4	0.6	0.5	0.5	0.5	0.4	0.4	0.4
20	0.1	0.2	0.4	0.6	0.6	0.5	0.5	0.5	0.5	0.4
30	0.1	0.1	0.2	0.3	0.4	0.6	0.7	0.8	0.9	0.9
40	0.1	0.1	0.1	0.1	0.1	0.1	0.1	0.2	0.2	0.3
50	0.1	0.1	0.1	0.1	0.1	0.1	0.1	0.1	0.1	0.1
100	0.1	0.1	0.1	0.1	0.1	0.1	0.1	0.1	0.1	0.1
200	0.1	0.1	0.1	0.1	0.1	0.1	0.1	0.1	0.1	0.1
300	0.1	0.1	0.1	0.1	0.1	0.1	0.1	0.1	0.1	0.1
400	0.1	0.1	0.1	0.1	0.1	0.1	0.1	0.1	0.1	0.1
500	0.1	0.1	0.1	0.1	0.1	0.1	0.1	0.1	0.1	0.1
600	0.1	0.1	0.1	0.1	0.1	0.1	0.1	0.1	0.1	0.1
700	0.1	0.1	0.1	0.1	0.1	0.1	0.1	0.1	0.1	0.1
800	0.1	0.1	0.1	0.1	0.1	0.1	0.1	0.1	0.1	0.1
900	0.1	0.1	0.1	0.1	0.1	0.1	0.1	0.1	0.1	0.1
1000	0.1	0.1	0.1	0.1	0.1	0.1	0.1	0.1	0.1	0.1
1100	0.1	0.1	0.1	0.1	0.1	0.1	0.1	0.1	0.1	0.1
1200	0.1	0.1	0.1	0.1	0.1	0.1	0.1	0.1	0.1	0.1
1300	0.1	0.1	0.1	0.1	0.1	0.1	0.1	0.1	0.1	0.1
1400	0.1	0.1	0.1	0.1	0.1	0.1	0.1	0.1	0.1	0.1
1500	0.1	0.1	0.1	0.1	0.1	0.1	0.1	0.1	0.1	0.1

Table 3.11 (*Continued*)

T, K	3σ, % at pressure p, MPa									
	10	20	30	40	50	60	70	80	90	100
3	14.2	—	—	—	—	—	—	—	—	—
4	4.9	—	—	—	—	—	—	—	—	—
5	1.1	—	—	—	—	—	—	—	—	—
6	1.4	12.6	—	—	—	—	—	—	—	—
8	1.1	2.0	7.9	29.4	—	—	—	—	—	—
10	0.7	2.3	4.0	6.3	15.7	—	—	—	—	—
12	0.6	1.5	3.6	6.2	9.1	14.5	27.1	—	—	—
14	0.5	1.0	2.2	4.6	7.8	11.7	17.1	27.3	—	—
16	0.5	0.9	1.4	2.6	5.0	8.5	12.7	18.3	26.9	—
18	0.4	0.8	1.4	1.8	2.7	5.0	8.3	12.4	17.5	—
20	0.4	0.7	1.5	2.1	2.5	3.0	4.9	7.7	11.3	15.7
30	1.0	0.8	0.8	1.0	1.3	1.7	2.2	2.9	3.7	4.7
40	0.3	0.5	0.3	0.3	0.5	0.8	1.2	1.7	2.4	3.3
50	0.1	0.3	0.2	0.2	0.2	0.4	0.6	0.8	1.1	1.5
100	0.1	0.2	0.2	0.2	0.2	0.2	0.3	0.4	0.6	1.0
200	0.1	0.1	0.1	0.1	0.1	0.1	0.2	0.2	0.2	0.3
300	0.1	0.1	0.1	0.1	0.1	0.1	0.1	0.1	0.1	0.1
400	0.1	0.1	0.1	0.1	0.1	0.1	0.1	0.1	0.1	0.1
500	0.1	0.1	0.1	0.1	0.1	0.1	0.1	0.1	0.1	0.1
600	0.1	0.1	0.1	0.1	0.1	0.1	0.1	0.1	0.1	0.1
700	0.1	0.1	0.1	0.1	0.1	0.1	0.1	0.1	0.1	0.1
800	0.1	0.1	0.1	0.1	0.1	0.1	0.1	0.1	0.1	0.1
900	0.1	0.1	0.1	0.1	0.1	0.1	0.1	0.1	0.1	0.1
1000	0.1	0.1	0.1	0.1	0.1	0.1	0.1	0.1	0.1	0.1
1100	0.1	0.1	0.1	0.1	0.1	0.1	0.1	0.1	0.1	0.1
1200	0.1	0.1	0.1	0.1	0.1	0.1	0.1	0.1	0.1	0.1
1300	0.1	0.1	0.1	0.1	0.1	0.1	0.1	0.1	0.1	0.1
1400	0.1	0.1	0.1	0.1	0.1	0.1	0.1	0.1	0.1	0.1
1500	0.1	0.1	0.1	0.1	0.1	0.1	0.1	0.1	0.1	0.1

Table 3.12 Standard deviation (3σ) for the differences in calculated speed of sound in He^4

T, K	3σ, % at pressure p, MPa									
	0.1	1	2	3	4	5	6	7	8	9
2.5	1.4	0.5	1.0	0.9	0.5	0.6	1.5	—	—	—
3	0.8	0.4	0.6	0.5	0.3	0.2	0.6	0.9	1.2	1.3
4	0.9	0.3	0.3	0.3	0.3	0.4	0.4	0.5	0.5	0.5
5	0.1	0.3	0.2	0.2	0.4	0.5	0.5	0.6	0.6	0.6
6	0.1	0.3	0.2	0.3	0.4	0.5	0.5	0.5	0.5	0.5
8	0.1	0.3	0.3	0.3	0.4	0.4	0.4	0.4	0.4	0.4
10	0.1	0.2	0.3	0.3	0.3	0.3	0.4	0.4	0.4	0.3
12	0.1	0.2	0.2	0.3	0.3	0.3	0.3	0.3	0.3	0.3
14	0.1	0.1	0.1	0.2	0.3	0.3	0.3	0.3	0.3	0.3
16	0.1	0.1	0.1	0.2	0.3	0.3	0.3	0.3	0.3	0.3
18	0.1	0.1	0.1	0.2	0.3	0.3	0.3	0.4	0.4	0.4
20	0.1	0.1	0.1	0.2	0.3	0.3	0.3	0.4	0.4	0.4
30	0.1	0.1	0.1	0.1	0.1	0.2	0.2	0.3	0.3	0.4
40	0.1	0.1	0.1	0.1	0.1	0.1	0.1	0.1	0.2	0.2
50	0.1	0.1	0.1	0.1	0.1	0.1	0.1	0.1	0.1	0.1
100	0.1	0.1	0.1	0.1	0.1	0.1	0.1	0.1	0.1	0.1
200	0.1	0.1	0.1	0.1	0.1	0.1	0.1	0.1	0.1	0.1
300	0.1	0.1	0.1	0.1	0.1	0.1	0.1	0.1	0.1	0.1
400	0.1	0.1	0.1	0.1	0.1	0.1	0.1	0.1	0.1	0.1
500	0.1	0.1	0.1	0.1	0.1	0.1	0.1	0.1	0.1	0.1
600	0.1	0.1	0.1	0.1	0.1	0.1	0.1	0.1	0.1	0.1
700	0.1	0.1	0.1	0.1	0.1	0.1	0.1	0.1	0.1	0.1
800	0.1	0.1	0.1	0.1	0.1	0.1	0.1	0.1	0.1	0.1
900	0.1	0.1	0.1	0.1	0.1	0.1	0.1	0.1	0.1	0.1
1000	0.1	0.1	0.1	0.1	0.1	0.1	0.1	0.1	0.1	0.1
1100	0.1	0.1	0.1	0.1	0.1	0.1	0.1	0.1	0.1	0.1
1200	0.1	0.1	0.1	0.1	0.1	0.1	0.1	0.1	0.1	0.1
1300	0.1	0.1	0.1	0.1	0.1	0.1	0.1	0.1	0.1	0.1
1400	0.1	0.1	0.1	0.1	0.1	0.1	0.1	0.1	0.1	0.1
1500	0.1	0.1	0.1	0.1	0.1	0.1	0.1	0.1	0.1	0.1

Table 3.12 (Continued)

T, K	3σ, % at pressure p, MPa									
	10	20	30	40	50	60	70	80	90	100
3	1.3	—	—	—	—	—	—	—	—	—
4	0.5	—	—	—	—	—	—	—	—	—
5	0.6	—	—	—	—	—	—	—	—	—
6	0.5	0.5	—	—	—	—	—	—	—	—
8	0.3	0.5	0.8	1.4	—	—	—	—	—	—
10	0.3	0.5	0.7	0.6	0.8	—	—	—	—	—
12	0.3	0.5	0.6	0.6	0.8	0.6	0.8	—	—	—
14	0.4	0.5	0.8	1.0	1.8	1.2	1.2	1.3	—	—
16	0.4	0.6	1.2	1.7	2.0	2.1	2.1	2.4	3.3	—
18	0.4	0.6	1.5	2.4	3.0	3.1	3.2	3.6	4.7	—
20	0.4	0.7	1.7	3.1	3.9	4.1	4.2	4.6	5.5	7.3
30	0.4	0.7	0.7	0.6	0.5	0.9	1.7	2.7	3.8	5.0
40	0.2	0.5	0.4	0.5	0.6	0.8	1.7	3.0	4.5	6.2
50	0.1	0.3	0.3	0.5	0.9	1.2	1.8	2.8	4.4	6.4
100	0.1	0.1	0.1	0.2	0.4	0.9	1.5	2.3	3.1	4.0
200	0.1	0.1	0.1	0.1	0.1	0.2	0.3	0.4	0.7	1.0
300	0.1	0.1	0.1	0.1	0.1	0.1	0.1	0.2	0.2	0.3
400	0.1	0.1	0.1	0.1	0.1	0.1	0.1	0.1	0.1	0.2
500	0.1	0.1	0.1	0.1	0.1	0.1	0.1	0.1	0.1	0.1
600	0.1	0.1	0.1	0.1	0.1	0.1	0.1	0.1	0.1	0.1
700	0.1	0.1	0.1	0.1	0.1	0.1	0.1	0.1	0.1	0.1
800	0.1	0.1	0.1	0.1	0.1	0.1	0.1	0.1	0.1	0.1
900	0.1	0.1	0.1	0.1	0.1	0.1	0.1	0.1	0.1	0.1
1000	0.1	0.1	0.1	0.1	0.1	0.1	0.1	0.1	0.1	0.1
1100	0.1	0.1	0.1	0.1	0.1	0.1	0.1	0.1	0.1	0.1
1200	0.1	0.1	0.1	0.1	0.1	0.1	0.1	0.1	0.1	0.1
1300	0.1	0.1	0.1	0.1	0.1	0.1	0.1	0.1	0.1	0.1
1400	0.1	0.1	0.1	0.1	0.1	0.1	0.1	0.1	0.1	0.1
1500	0.1	0.1	0.1	0.1	0.1	0.1	0.1	0.1	0.1	0.1

Table 3.13 Standard deviation (3σ) for the differences in calculated adiabatic Joule-Thomson effect in He^4

T, K	3σ, % at pressure p, MPa									
	0.1	1.0	2.0	3.0	4.0	5.0	6.0	7.0	8.0	9.0
2.5	2.9	1.0	1.2	1.4	1.5	1.9	5.4	—	—	—
3	1.1	1.1	0.8	0.6	0.8	1.0	1.2	1.7	3.8	8.2
4	3.0	0.9	0.7	0.9	1.3	1.6	1.7	1.5	1.5	2.8
5	0.7	1.4	1.0	0.8	0.9	1.2	1.5	1.6	1.5	1.2
6	0.8	2.1	1.1	0.9	0.8	1.0	1.2	1.4	1.6	1.7
8	1.0	1.2	2.3	0.8	0.7	0.6	0.6	0.7	0.9	1.1
10	1.1	1.2	11.7	2.4	1.0	0.7	0.6	0.6	0.6	0.7
12	1.1	0.9	3.3	9.2	1.1	0.8	0.7	0.6	0.6	0.6
14	0.1	1.2	3.3	19.4	4.3	1.9	1.1	0.9	0.8	0.7
16	0.9	1.3	2.9	9.9	7.4	3.1	1.8	1.1	0.9	0.8
18	1.0	1.3	2.5	9.8	16.9	4.3	2.8	1.8	1.2	1.0
20	1.3	1.5	2.4	9.9	46.7	5.5	3.6	2.7	1.9	1.4
30	1.0	1.1	4.0	14.8	8.2	2.5	2.4	2.6	2.5	2.4
40	4.9	18.1	16.6	7.6	5.5	4.2	3.1	2.1	1.4	1.0
50	11.4	5.2	3.7	3.2	3.1	3.0	2.9	2.7	2.4	1.9
100	0.9	0.8	0.8	0.9	0.9	0.8	0.8	0.7	0.6	0.6
200	1.0	1.0	1.0	1.0	1.0	1.0	1.0	1.0	1.1	1.1
300	0.8	0.8	0.7	0.7	0.7	0.7	0.7	0.7	0.8	0.8
400	0.6	0.6	0.6	0.6	0.5	0.5	0.5	0.5	0.5	0.5
500	0.6	0.6	0.6	0.6	0.5	0.5	0.5	0.4	0.4	0.4
600	0.6	0.6	0.6	0.6	0.6	0.5	0.5	0.5	0.4	0.4
700	0.7	0.7	0.7	0.7	0	0.6	0.6	0.6	0.5	0.5
800	0.8	0.8	0.8	0.7	0.7	0.7	0.7	0.7	0.6	0.6
900	0.8	0.8	0.8	0.8	0.8	0.8	0.8	0.7	0.7	0.7
1000	0.9	0.9	0.9	0.9	0.9	0.8	0.8	0.8	0.8	0.8
1100	0.9	0.9	0.9	0.9	0.9	0.9	0.9	0.9	0.8	0.8
1200	1.0	1.0	1.0	1.0	1.0	0.9	0.9	0.9	0.9	0.9
1300	1.0	1.0	1.0	1.0	1.0	1.0	1.0	1.0	1.0	0.9
1400	1.1	1.1	1.0	1.0	1.0	1.0	1.0	1.0	1.0	1.0
1500	1.1	1.1	1.1	1.1	1.1	1.1	1.1	1.0	1.0	1.0

Table 3.13 (*Continued*)

T, K	3σ, % at pressure p, MPa									
	10	20	30	40	50	60	70	80	90	100
3	14.7	—	—	—	—	—	—	—	—	—
4	5.5	—	—	—	—	—	—	—	—	—
5	1.1	—	—	—	—	—	—	—	—	—
6	1.7	13.3	—	—	—	—	—	—	—	—
8	1.3	2.5	7.4	29.2	—	—	—	—	—	—
10	0.8	2.8	4.7	6.3	12.8	—	—	—	—	—
12	0.6	1.8	4.0	6.8	10.0	14.5	22.1	—	—	—
14	0.7	1.2	2.4	4.7	8.0	12.7	19.6	31.2	—	—
16	0.8	1.1	1.9	2.7	4.7	8.2	13.2	21.0	34.7	—
18	0.9	1.0	2.1	2.7	3.1	4.5	7.5	12.0	18.7	—
20	1.1	1.0	2.1	3.4	4.0	4.0	4.8	7.0	10.5	15.6
30	2.2	0.9	1.0	1.3	1.6	1.9	2.5	3.3	4.4	5.7
40	1.1	1.1	0.7	0.5	0.5	0.9	1.2	1.5	2.1	3.2
50	1.5	1.4	0.9	0.8	0.6	0.8	1.4	2.0	2.3	2.5
100	0.7	1.0	0.8	1.0	0.6	0.8	1.5	1.8	2.0	2.7
200	1.1	0.6	0.9	0.9	0.6	0.7	0.9	1.1	1.5	2.3
300	0.9	1.0	0.6	0.7	1.0	1.0	0.8	0.7	0.8	1.1
400	0.6	1.0	0.9	0.6	0.6	0.9	1.0	1.0	0.9	0.8
500	0.4	0.8	1.0	0.8	0.6	0.6	0.8	1.0	1.1	1.1
600	0.4	0.7	1.0	1.0	0.8	0.6	0.5	0.7	0.9	1.0
700	0.5	0.6	0.9	1.0	1.0	0.8	0.6	0.5	0.6	0.8
800	0.6	0.5	0.8	1.0	1.0	0.9	0.8	0.6	0.5	0.6
900	0.7	0.6	0.7	0.9	1.1	1.0	0.9	0.8	0.6	0.5
1000	0.7	0.6	0.7	0.9	1.0	1.1	1.0	0.9	0.8	0.6
1100	0.8	0.7	0.7	0.9	1.0	1.1	1.1	1.0	0.9	0
1200	0.9	0.7	0.7	0.8	1.0	1.1	1.1	1.1	1.0	0.9
1300	0.9	0.8	0.7	0.8	0.9	1.1	1.1	1.1	1.1	1.0
1400	1.0	0.8	0.8	0.8	0.9	1.0	1.1	1.1	1.1	1.1
1500	1.0	0.9	0.8	0.8	0.9	1.0	1.1	1.1	1.2	1.1

Table 3.14 Standard deviation (3σ) for the differences in calculated adiabatic index of He[4]

T, K	3σ, % at pressure p, MPa									
	0.1	1.0	2.0	3.0	4.0	5.0	6.0	7.0	8.0	9.0
2.5	2.6	1.2	2.0	1.8	0.8	1.2	3.2	—	—	—
3	1.7	0.9	1.2	1.0	0.6	0.5	1.2	1.9	2.3	2.4
4	1.9	0.6	0.5	0.6	0.7	0.8	0.8	0.9	0.9	0.9
5	0.1	0.6	0.3	0.5	0.8	0.9	1.0	1.1	1.1	1.1
6	0.1	0.5	0.4	0.6	0.8	0.9	1.0	1.0	1.0	1.0
8	0.1	0.7	0.5	0.7	0.8	0.8	0.8	0.8	0.7	0.7
10	0.1	0.5	0.7	0.6	0.6	0.7	0.7	0.7	0.7	0.6
12	0.1	0.3	0.3	0.6	0.6	0.6	0.7	0.7	0.7	0.7
14	0.1	0.2	0.3	0.4	0.6	0.6	0.6	0.6	0.7	0.7
16	0.1	0.2	0.2	0.4	0.5	0.6	0.6	0.7	0.7	0.7
18	0.1	0.1	0.2	0.4	0.4	0.5	0.6	0.7	0.7	0.7
20	0.1	0.1		0.3	0.5	0.5	0.6	0.7	0.7	0.8
30	0.1	0.1	0.2	0.3	0.3	0.4	0.5	0.6	0.7	0.8
40	0.1	0.1	0.1	0.2	0.2	0.2	0.3	0.3	0.4	0.4
50	0.1	0.1	0.1	0.1	0.1	0.1	0.1	0.2	0.2	0.2
100	0.1	0.1	0.1	0.1	0.1	0.1	0.1	0.1	0.1	0.1
200	0.1	0.1	0.1	0.1	0.1	0.1	0.1	0.1	0.1	0.1
300	0.1	0.1	0.1	0.1	0.1	0.1	0.1	0.1	0.1	0.1
400	0.1	0.1	0.1	0.1	0.1	0.1	0.1	0.1	0.1	0.1
500	0.1	0.1	0.1	0.1	0.1	0.1	0.1	0.1	0.1	0.1
600	0.1	0.1	0.1	0.1	0.1	0.1	0.1	0.1	0.1	0.1
700	0.1	0.1	0.1	0.1	0.1	0.1	0.1	0.1	0.1	0.1
800	0.1	0.1	0.1	0.1	0.1	0.1	0.1	0.1	0.1	0.1
900	0.1	0.1	0.1	0.1	0.1	0.1	0.1	0.1	0.1	0.1
1000	0.1	0.1	0.1	0	0.1	0.1	0.1	0.1	0.1	0.1
1100	0.1	0.1	0.1	0.1	0.1	0.1	0.1	0.1	0.1	0.1
1200	0.1	0.1	0.1	0.1	0.1	0.1	0.1	0.1	0.1	0.1
1300	0.1	0.1	0.1	0.1	0.1	0.1	0.1	0.1	0.1	0.1
1400	0.1	0.1	0.1	0.1	0.	0.1	0.1	0.1	0.1	0.1
1500	0.1	0.1	0.1	0.1	0.1	0.1	0.1	0.1	0.1	0.1

Table 3.14 (Continued)

T, K	3σ, % at pressure p, MPa									
	10	20	30	40	50	60	70	80	90	100
3	2.4	—	—	—	—	—	—	—	—	—
4	0.9	—	—	—	—	—	—	—	—	—
5	1.0	—	—	—	—	—	—	—	—	—
6	0.9	1.0	—	—	—	—	—	—	—	—
8	0.6	1.1	1.7	2.7	—	—	—	—	—	—
10	0.6	1.0	1.3	1.3	1.6	—	—	—	—	—
12	0.7	1.0	1.2	1.2	1.2	1.3	1.6	—	—	—
14	0.7	1.0	1.6	0.1	2.3	2.4	2.5	2.6	—	—
16	0.7	1.2	2.3	3.5	4.0	4.2	4.3	4.9	6.7	—
18	0.7	1.3	2.9	4.8	5.8	6.2	8.4	7.3	9.5	—
20	0.8	1.3	3.4	6.0	7.5	8.0	8.2	9.1	11.0	14.7
30	0.9	1.3	1.3	1.2	1.0	1.7	3.2	5.2	7.3	9.6
40	0.5	0.9	0.8	1.0	1.2	1.7	3.3	5.7	8.5	11.6
50	0.2	0.6	0.5	1.1	1.8	2.4	3.5	5.4	8.3	11.9
100	0.1	0.1	0.2	0.4	0.8	1.7	2.9	4.2	5.6	7.2
200	0.1	0.1	0.1	0.2	0.2	0.3	0.5	0.8	1.2	1.8
300	0.1	0.1	0.1	0.1	0.1	0.2	0.2	0.3	0.4	0.5
400	0.1	0.1	0.1	0.1	0.1	0.1	0.2	0.2	0.2	0.2
500	0.1	0.1	0.1	0.1	0.1	0.1	0.1	0.1	0.2	0.2
600	0.1	0.1	0.1	0.1	0.1	0.1	0.1	0.1	0.1	0.1
700	0.1	0.1	0.1	0.1	0.1	0.1	0.1	0.1	0.1	0.1
800	0.1	0.1	0.1	0.1	0.1	0.1	0.1	0.1	0.1	0.1
900	0.1	0.1	0.1	0.1	0.1	0.1	0.1	0.1	0.1	0.1
1000	0.1	0.1	0.1	0.1	0.1	0.1	0.1	0.1	0.1	0.1
1100	0.1	0.1	0.1	0.1	0.1	0.1	0.1	0.1	0.1	0.1
1200	0.1	0.1	0.1	0.1	0.1	0.1	0.1	0.1	0.1	0.1
1300	0.1	0.1	0.1	0.1	0.1	0.1	0.1	0.1	0.1	0.1
1400	0.1	0.1	0.1	1.1	0.1	0.1	0.1	0.1	0.1	0.1
1500	0.1	0.1	0.1	0.1	0.1	0.1	0.1	0.1	0.1	0.1

Table 3.15 Standard deviation (3σ) for the difference in calculated fugacity of He^4

T, K	3σ, % at pressure p, MPa									
	0.1	1.0	2.0	3.0	4.0	5.0	6.0	7.0	8.0	9.0
2.5	0.2	0.2	0.3	0.3	0.3	0.4	0.4	—	—	—
3	0.2	0.2	0.3	0.3	0.3	0.3	0.3	0.4	0.4	0.4
4	0.2	0.2	0.3	0.3	0.4	0.4	0.3	0.3	0.3	0.3
5	0.2	0.3	0.3	0.3	0.4	0.4	0.4	0.4	0.4	0.3
6	0.1	0.3	0.3	0.3	0.4	0.4	0.4	0.4	0.4	0.4
8	0.1	0.3	0.3	0.3	0.3	0.3	0.3	0.3	0.3	0.3
10	0.1	0.2	0.2	0.2	0.2	0.3	0.3	0.3	0.3	0.3
12	0.1	0.1	0.2	0.2	0.2	0.2	0.2	0.2	0.2	0.2
14	0.1	0.1	0.2	0.1	0.1	0.1	0.2	0.2	0.2	0.2
16	0.1	0.1	0.2	0.2	0.1	0.1	0.1	0.2	0.2	0.2
18	0.1	0.1	0.1	0.2	0.2	0.2	0.2	0.2	0.2	0.2
20	0.1	0.1	0.1	0.2	0.2	0.2	0.2	0.2	0.2	0.2
30	0.1	0.1	0.1	0.1	0.1	0.1	0.2	0.2	0.2	0.2
40	0.1	0.1	0.1	0.1	0.1	0.1	0.1	0.1	0.1	0.1
50	0.1	0.1	0.1	0.1	0.1	0.1	0.1	0.1	0.1	0.1
100	0.1	0.1	0.1	0.1	0.1	0.1	0.1	0.1	0.1	0.1
200	0.1	0.1	0.1	0.1	0.1	0.1	0.1	0.1	0.1	0.1
300	0.1	0.1	0.1	0.1	0.1	0.1	0.1	0.1	0.1	0.1
400	0.1	0.1	0.1	0.1	0.1	0.1	0.1	0.1	0.1	0.1
500	0.1	0.1	0.1	0.1	0.1	0.1	0.1	0.1	0.1	0.1
600	0.1	0.1	0.1	0.1	0.1	0.1	0.1	0.1	0.1	0.1
700	0.1	0.1	0.1	0.1	0.1	0.1	0.1	0.1	0.1	0.1
800	0.1	0.1	0.1	0.1	0.1	0.1	0.1	0.1	0.1	0.1
900	0.1	0.1	0.1	0.1	0.1	0.1	0.1	0.1	0.1	0.1
1000	0.1	0.1	0.1	0.1	0.1	0.1	0.1	0.1	0.1	0.1
1100	0.1	0.1	0.1	0.1	0.1	0.1	0.1	0.1	0.1	0.1
1200	0.1	0.1	0.1	0.1	0.1	0.1	0.1	0.1	0.1	0.1
1300	0.1	0.1	0.1	0.1	0.1	0.1	0.1	0.1	0.1	0.1
1400	0.1	0.1	0.1	0.1	0.1	0.1	0.1	0.1	0.1	0.1
1500	0.1	0.1	0.1	0.1	0.1	0.1	0.1	0.1	0.1	0.1

Table 3.15 (*Continued*)

T, K	3σ, % at pressure p, MPa									
	10	20	30	40	50	60	70	80	90	100
3	0.4	—	—	—	—	—	—	—	—	—
4	0.3	—	—	—	—	—	—	—	—	—
5	0.3	—	—	—	—	—	—	—	—	—
6	0.3	0.5	—	—	—	—	—	—	—	—
8	0.3	0.4	0.6	0.8	—	—	—	—	—	—
10	0.3	0.3	0.5	0.7	0.9	—	—	—	—	—
12	0.2	0.3	0.4	0.6	0.7	0.9	1.0	—	—	—
14	0.2	0.3	0.4	0.4	0.5	0.7	0.8	0.9	—	—
16	0.2	0.2	0.3	0.3	0.4	0.4	0.4	0.5	0.5	—
18	0.2	0.2	0.2	0.3	0.3	0.4	0.5	0.6	0.5	—
20	0.2	0.2	0.2	0.3	0.5	0.8	1.0	1.2	1.3	1.3
30	0.2	0.2	0.2	0.2	0.3	0.4	0.4	0.5	0.6	0.8
40	0.1	0.1	0.1	0.2	0.3	0.4	0.4	0.5	0.6	0.8
50	0.1	0.1	0.1	0.2	0.3	0.3	0.3	0.4	0.4	0.6
100	0.1	0.1	0.1	0.2	0.2	0.2	0.2	0.1	0.2	0.3
200	0.1	0.1	0.1	0.1	0.1	0.1	0.1	0.1	0.1	0.1
300	0.1	0.1	0.1	0.1	0.1	0.1	0.1	0.1	0.1	0.1
400	0.1	0.1	0.1	0.1	0.1	0.1	0.1	0.1	0.1	0.1
500	0.1	0.1	0.1	0.1	0.1	0.1	0.1	0.1	0.1	0.1
600	0.1	0.1	0.1	0.1	0.1	0.1	0.1	0.1	0.1	0.1
700	0.1	0.1	0.1	0.1	0.1	0.1	0.1	0.1	0.1	0.1
800	0.1	0.1	0.1	0.1	0.1	0.1	0.1	0.1	0.1	0.1
900	0.1	0.1	0.1	0.1	0.1	0.1	0.1	0.1	0.1	0.1
1000	0.1	0.1	0.1	0.1	0.1	0.1	0.1	0.1	0.1	0.1
1100	0.1	0.1	0.1	0.1	0.1	0.1	0.1	0.1	0.1	0.1
1200	0.1	0.1	0.1	0.1	0.1	0.1	0.1	0.1	0.1	0.1
1300	0.1	0.1	0.1	0.1	0.1	0.1	0.1	0.1	0.1	0.1
1400	0.1	0.1	0.1	0.1	0.1	0.1	0.1	0.1	0.1	0.1
1500	0.1	0.1	0.1	0.1	0.1	0.1	0.1	0.1	0.1	0.1

Table 3.16 Deviations δv of tabulated data [20, 32] from the calculated specific volume

T, K	Reference	δv, % at p, MPa								
		0.1	0.2	0.4	0.6	0.8	1	2	4	6	8
2.4	[20]	0.09	0.11	0.13	0.09	0.07	0.04	−0.12	−0.25	—	—
2.49	[32]	−0.05	−0.06	−0.06	−0.05	−0.04	−0.03	0.03	0.03	—	—
3.0	[20]	0.18	0.18	0.16	0.11	0.10	0.06	−0.08	−0.13	−0.13	—
2.99	[32]	−0.08	−0.12	−0.18	−0.20	−0.21	−0.21	−0.16	−0.06	−0.08	—
3.4	[20]	−0.25	0.24	0.22	0.16	0.11	0.11	−0.05	−0.11	−0.05	−0.17
3.49	[32]	−0.12	−0.18	−0.24	−0.27	−0.28	−0.28	−0.23	−0.09	−0.01	−0.04
4.0	[20]	0.52	0.45	0.36	0.32	0.25	0.20	0.02	−0.10	−0.10	−0.14
3.99	[32]	−0.09	−0.12	−0.17	−0.20	−0.21	−0.21	−0.18	−0.04	0.06	−0.12
4.4	[20]	0.56	0.69	0.53	0.44	0.38	0.31	0.07	−0.12	−0.14	−0.18
4.49	[32]	0.29	0.06	0.08	0.05	0.02	0.01	−0.01	0.07	0.17	0.23
5.0	[20]	0.08	1.66	0.83	0.70	0.61	0.52	0.20	−0.08	−0.19	−0.28
4.99	[32]	0.13	0.40	0.72	0.57	0.48	0.41	0.27	0.23	0.28	0.34
6.0	[20]	−0.04	−0.02	0.02	0.70	0.83	0.80	−0.29	−0.06	−0.27	−0.43
	[32]	−0.23	−0.24	−1.27	0.43	0.68	0.71	0.60	0.45	0.41	0.41
8.0	[20]	−0.02	0.01	0.05	−0.11	−0.60	−0.45	0.80	0.29	−0.20	−0.47
	[32]	−0.14	−0.22	−0.54	−0.73	−0.53	−0.77	0.62	0.70	0.54	0.43
10	[20]	0.01	0.03	0.10	0.06	−0.07	−0.41	0.33	0.63	0.12	−0.31
	[32]	−0.06	−0.13	−0.40	−0.76	−0.84	−0.50	−0.17	0.71	0.60	−0.44

108

Table 3.16 (*Continued*)

| T, K | Reference | δv, % at p, MPa |||||||||
		0.1	0.2	0.4	0.6	0.8	1	2	4	6	8
20	[20]	0.01	0.02	0.01	0.03	−0.01	−0.03	−0.33	−0.49	0.02	0.20
	[32]	0.05	0.08	0.05	−0.03	−0.16	−0.31	−1.01	−1.02	−0.48	−0.34
50	[20]	−0.01	−0.01	−0.03	−0.06	−0.07	−0.09	−0.22	−0.46	−0.56	−0.53
	[32]	0.02	0.03	0.06	0.07	0.07	0.06	−0.07	−0.52	−0.86	−1.01
100	[20]	0	0		−0.01		0	0.01	−0.03	−0.02	−0.09
	[32]	0.01	−0.01	0.03	0.05	0.07	0.07	0.12	0.13	0.07	−0.01
200	[20]	0.01	0.01	0.02	0.03	0.03	0.04	0.07	0.11	0.18	0.21
	[32]	0.01	0.01	0.02	0.03	0.04	0.05	0.09	0.17	0.23	0.26
300	[20]	0	0.01	0.01	0.02	0.02	0.02	0.05	0.09	0.12	0.16
	[32]	0	0	0.01	0.01	0.02	0.02	0.06	0.09	0.13	0.15
400	[32]	0	0	0	0.01	0.01	0.01	0.02	0.04	0.05	0.06
600	[32]	0	0	0	0	0	0	−0.01	−0.02	−0.03	−0.05
800	[32]	0	0	0	−0.01	−0.01	−0.01	−0.02	−0.05	−0.07	−0.09
1000	[32]	0	0	−0.01	−0.01	−0.02	−0.02	−0.03	−0.07	−0.10	−0.13
1200	[32]	0	0	−0.01	−0.01	−0.02	−0.02	−0.04	−0.08	−0.11	−0.15
1400	[32]	0	0	−0.01	−0.02	−0.02	−0.03	−0.04	−0.08	−0.12	−0.16

Table 3.16 (Continued)

T, K	Reference	δv, % at p, MPa									
		10	15	20	30	40	50	60	70	80	100
3.4	[20]	—	—	—	—	—	—	—	—	—	—
3.49	[32]	−0.20	—	—	—	—	—	—	—	—	—
4.0	[20]	—	—	—	—	—	—	—	—	—	—
3.99	[32]	0.08	—	—	—	—	—	—	—	—	—
4.4	[20]	—	—	—	—	—	—	—	—	—	—
4.49	[32]	0.25	0.01	—	—	—	—	—	—	—	—
5.0	[20]	—	—	—	—	—	—	—	—	—	—
4.99	[32]	0.36	0.26	—	—	—	—	—	—	—	—
6.0	[20]	—	—	0.26	—	—	—	—	—	—	—
	[32]	0.42	0.41	—	—	—	—	—	—	—	—
8.0	[20]	−0.71	—	0.25	0.05	−0.44	—	—	—	—	—
	[32]	0.37	0.29	—	—	—	—	—	—	—	—
10	[20]	−0.60	—	0.07	0.03	−0.11	−0.43	—	—	—	—
	[32]	0.31	0.13	−0.76	−0.36	0.02	0.31	0.46	0.46	—	—
20	[20]	0.10	−0.43	−0.66	−0.47	−0.42	−0.17	−0.84	−0.46	—	—
	[32]	−0.42	−0.64	−0.32	−1.29	−1.36	−1.18	0.16	−0.19	—	—
50	[20]	−0.45	−0.29	−0.99	0.23	0.23	0.17	−0.16	0.11	—	—
	[32]	−1.03	−0.94	0.07	0.13	0.11	−0.03	0.17	—	—	—
100	[20]	−0.09	−0.05	−0.04	0.19	0.21	0.21	0.21	—	0.52	1.24
	[32]	−0.07	−0.11	0.20	0.21	0.23	0.25	0.24	−0.55	0.04	−0.08
200	[20]	0.21	0.19	0.24	0.07	−0.04	−0.14	−0.42	−1.08	—	—
	[32]	0.27	0.27	0.19	−0.02	−0.17	−0.30	−0.87	—	−0.44	−0.68
300	[20]	0.18	0.19	0.12	−0.20	−0.42	−0.64	—	—	—	—
	[32]	0.17	0.16	−0.02	—	—	—	—	—	—	—
400	[20]	0.06	0.04	−0.21	—	—	—	—	—	—	—
600	[32]	−0.07	−0.13	−0.31	—	—	—	—	—	—	—
800	[32]	−0.13	−0.21	−0.36	—	—	—	—	—	—	—
1000	[32]	−0.17	−0.26	−0.40	—	—	—	—	—	—	—
1200	[32]	−0.21	−0.29	−0.42	—	—	—	—	—	—	—
1400	[32]	−0.20	−0.30	—	—	—	—	—	—	—	—

EQUATIONS OF STATE

Table 3.17 Deviations Δh of data tabulated in [20, 32] from calculated enthalpy

T, K	Ref.	Δh, kJ/kg, at pressure p, MPa									
		0.1	0.2	0.4	0.6	0.8	1	2	4	6	8
2.4	[20]	0.2	0.1	0.1	0.1	0.1	0.2	0.1	0	—	—
2.49	[32]	0.4	0.3	0.3	0.2	0.2	0.2	0.1	0	—	—
3.0	[20]	0.1	0.2	0.1	0.1	0	0.1	0.2	0.2	0.1	—
2.99	[32]	0.3	0.2	0.2	0.2	0.1	0.1	0	−0.1	−0.1	—
3.4	[20]	0.2	0.2	0.1	0.2	0.1	0.3	0.2	0.1	0	0.1
3.49	[32]	0.2	0.2	0.2	0.1	0.1	0.1	0	−0.1	−0.2	−0.3
4.0	[20]	0.3	0.2	0.1	0.1	0.1	0.3	0.2	0.2	0.2	0.1
3.99	[32]	0.2	0.2	0.1	0.1	0.1	0.1	0	−0.1	−0.2	−0.3
4.4	[20]	0.3	0.3	0.2	0.3	0.1	0.1	0.2	0.1	0.2	0.2
4.49	[32]	0.1	0.1	0.1	0.1	0.1	0.1	0	−0.1	−0.2	−0.3
5.0	[20]	0.1	0.3	0.4	0.2	0.2	0.3	0.2	0.2	0.2	—
4.99	[32]	0.1	0.1	0.2	0.2	0.2	0.2	0.1	−0.1	−0.1	−0.2
6.0	[20]	0	0.2	0.2	0.2	0.2	0.4	0.2	0.2	0.2	0.1
	[32]	0	0	−0.2	0.1	0.2	0.2	0.1	0	−0.1	0
8.0	[20]	0.1	0.1	0.1	0.1	−0.1	−0.2	0.4	0.3	0.2	0
	[32]	−0.1	−0.1	−0.1	−0.1	−0.1	−0.2	0	0	0	0
10	[20]	0.1	0.2	0.1	0.1	0.2	−0.1	0.2	0.6	0.5	0.3
	[32]	−0.1	−0.1	−0.1	−0.2	−0.1	0	−0.1	−0.1	−0.1	−0.1
20	[20]	0.1	0.2	0.3	0.3	0.3	0.2	0.2	0.2	0.3	0.6
	[32]	0.1	0	−0.1	−0.1	−0.3	−0.4	−0.8	−0.8	−0.3	−0.1
50	[20]	0.1	0.1	0	0.1	0	0.1	0	−0.4	−0.7	−0.8
	[32]	0.1	0.1	0.1	0.2	0.2	0.2	0.1	−0.6	−1.3	−1.8
100	[20]	0.1	0.1	0.1	0.1	−0.1	−0.1	−0.2	−0.8	−1.2	−1.7
	[32]	0.1	0.1	0.1	0.1	0.1	0.1	0.1	−0.2	−0.8	−1.3
200	[20]	0.1	0.1	0.3	0.2	0.1	0.2	0.3	0.3	0.2	0.1
	[32]	0.1	0.1	0.1	0.2	0.2	0.3	0.5	0.9	1.2	1.3
300	[20]	0.2	0.2	0.1	0.3	0.2	0.3	0.4	0.7	1.0	1.5
	[32]	0.1	0.1	0.2	0.2	0.3	0.3	0.4	1.3	1.9	2.5
400	[32]	0.1	0.1	0.2	0.2	0.3	0.4	0.8	1.5	2.2	2.9
600	[32]	0	0	0.2	0.2	0.2	0.5	0.7	1.4	2.2	2.9
800	[32]	0	0	0.2	0	0.2	0.2	0.7	1.4	2.0	2.6
1000	[32]	0	0	0.2	0	0.2	0.2	0.5	1.2	1.7	2.4
1200	[32]	0	0	0	0	0.2	0.2	0.5	1.0	1.4	2.2
1400	[32]	−0.2	0	0	0	0.2	0.2	0.5	0.7	1.2	1.7

Table 3.17 (*Continued*)

T, K	Ref.	Δh, kJ/kg, at pressure p, MPa									
		10	15	20	30	40	50	60	70	80	100
3.4 3.49	[20] [32]	— −0.5	— —	— —	— —	— —	— —	— —	— —	— —	— —
4.0 3.99	[20] [32]	— −0.4	— —	— —	— —	— —	— —	— —	— —	— —	— —
4.4 4.49	[20] [32]	— −0.3	— −0.7	— —	— —	— —	— —	— —	— —	— —	— —
5.0 4.99	[20[[32]	— −0.2	— −0.4	— —	— —	— —	— —	— —	— —	— —	— —
6.0	[20] [32]	— 0	— 0.1	— 0	— —	— —	— —	— —	— —	— —	— —
8.0	[20] [32]	0.1 0	— 0.2	— 0.4	— 0.6	— −0.1	— —	— —	— —	— —	— —
10	[20] [32]	0 −0.1	— 0	— 0.2	— 0.4	— −0.3	— —	— —	— —	— —	— —
20	[20] [32]	0.7 0.2	0.7 0.6	0.4 1.0	— 1.7	— 2.1	— 2.2	— 2.2	— 2.2	— —	— —
50	[20] [32]	−0.6 −2.1	−0.8 −2.6	−1.0 −3.2	−1.7 −4.8	−2.3 −6.0	−2.4 −6.5	— −6.7	— −6.5	— —	— —
100	[20] [32]	−2.1 −1.9	−2.7 −3.1	−3.1 −3.8	−3.4 −4.8	−3.7 −6.1	−4.3 −7.5	−4.8 −8.8	— −9.9	−4.2 —	−2.1 —
200	[20] [32]	0.1 1.5	−0.2 1.6	−0.5 1.7	−0.4 2.3	−0.2 3.4	0.4 4.6	0.8 5.8	— 6.8	1.8 —	2.8 —
300	[20] [32]	1.5 2.9	1.8 3.9	2.0 4.7	2.1 6.1	2.4 7.6	2.8 9.4	3.5 11.3	— 13.4	4.9 —	6.0 —
400	[32]	3.5	4.8	6.0	7.8	9.4	11.1	12.8	14.8	—	—
600	[32]	3.6	5.0	6.5	—	—	—	—	—	—	—
800	[32]	2.4	4.6	6.0	—	—	—	—	—	—	—
1000	[32]	2.9	4.1	5.3	—	—	—	—	—	—	—
1200	[32]	2.4	2.4	4.3	—	—	—	—	—	—	—
1400	[32]	1.9	2.6	3.6	—	—	—	—	—	—	—

Table 3.18 Deviations Δs of values tabulated in [20, 32] from calculated entropy

T, K	Ref.	$\Delta s \cdot 10^3$, kJ/(kg·K), at pressure p, MPa								
		0.1	0.2	0.4	0.6	0.8	1	2	4	6	8
2.4	[20]	−64	−67	−70	−72	−74	−73	−77	−94	—	—
2.49	[32]	114	102	84	71	61	53	33	10	—	—
3.0	[20]	−61	−60	−63	−63	−64	−64	−67	−74	−89	—
2.99	[32]	83	74	59	49	41	84	16	0	−27	
3.4	[20]	−31	−48	−50	−52	−53	−55	−62	−68	−73	−80
3.49	[32]	69	62	50	41	33	28	8	15	−37	−71
4.0	[20]	−24	−31	−36	−41	−45	−48	−60	−65	−61	−52
3.99	[32]	53	48	38	33	26	19	2	−30	−51	−75
4.4	[20]	−5	−19	−26	−33	−41	−43	−57	−68	−58	−46
4.49	[32]	24	38	36	31	26	19	−5	−36	−57	−72
5.0	[20]	−31	7	−14	−22	−29	−36	−55	−67	−62	−46
4.99	[32]	17	29	50	43	38	33	7	−26	−45	−58
6.0	[20]	−43	−41	−48	−17	−12	−17	−36	−62	−65	−60
	[32]	−5	0	−29	26	36	36	19	−12	−29	−36
8.0	[20]	−48	−50	−50	−57	−65	−53	−10	31	−53	−53
	[32]	−10	−10	−7	−2	−5	−26	0	−10	−24	−29
10	[20]	−45	−48	−50	−50	−53	−60	−31	−10	−31	−45
	[32]	−7	−12	−14	−12	−5	7	−33	−24	−34	−36
20	[20]	−45	−45	−45	−50	−50	−53	−60	−53	−43	−41
	[32]	−5	−5	−10	−19	−26	−33	−55	−33	−10	0
50	[20]	−43	−43	−43	−45	−45	−43	−43	−45	−45	−45
	[32]	0	0	2	0	0	0	−2	−12	−19	−24
100	[20]	−43	−45	−45	−48	−45	−48	−48	−55	−60	−72
	[32]	−10	0	0	−2	0	−2	−2	−10	−14	−22
200	[20]	−45	−45	−45	−45	−45	−45	−45	−48	−48	−50
	[32]	−12	2	−7	10	10	0	0	0	0	0
300	[20]	−45	−45	−43	−43	−43	−45	−45	−45	−45	−43
	[32]	5	−5	12	5	2	−10	5	0	2	5
400	[32]	10	0	−7	10	7	−5	12	2	2	5
600	[32]	2	−5	10	2	0	−10	5	−5	12	10
800	[32]	7	0	−10	7	5	−5	10	0	−7	14
1000	[32]	−2	−10	5	−2	−5	10	0	−10	7	7
1200	[32]	0	−7	7	0	−2	12	2	−7	12	10
1400	[32]	0	−10	7	0	−2	10	2	−7	10	7

Table 3.18 (*Continued*)

T, K	Ref.	\multicolumn{9}{c}{$\Delta s \cdot 10^3$, kJ/(kg·K), at pressure p, MPa}
		10	15	20	30	40	50	60	70	80	100
4.0	[20]	—	—	—	—	—	—	—	—	—	—
3.99	[32]	−108	—	—	—	—	—	—	—	—	—
4.4	[20]	—	—	—	—	—	—	—	—	—	—
4.49	[32]	−92	−184	—	—	—	—	—	—	—	—
5.0	[20]	—	—	—	—	—	—	—	—	—	—
4.99	[32]	−69	−116	—	—	—	—	—	—	—	—
6.0	[20]	—	—	—	—	—	—	—	—	—	—
	[32]	−38	−43	−63	—	—	—	—	—	—	—
8.0	[20]	−60	—	—	—	—	—	—	—	—	—
	[32]	−31	−19	0	8	−61	—	—	—	—	—
10	[20]	−57	—	—	—	—	—	—	—	—	—
	[32]	−41	−36	−26	−2	−10	−69	—	—	—	—
20	[20]	−43	−48	−50	—	—	—	—	—	—	—
	[32]	10	36	65	120	153	170	172	−163	—	—
50	[20]	−43	−41	−41	−50	−55	−55	—	—	—	—
	[32]	−24	−24	−29	−43	−53	−50	−33	−33	—	—
100	[20]	−67	−74	−74	−81	−86	−93	−98	—	−98	−84
	[32]	−26	−36	−43	−55	−67	−81	−93	−103	—	—
200	[20]	−50	−55	−57	−60	−60	−57	−57	—	−53	−48
	[32]	0	−2	−5	−2	0	7	10	14	—	—
300	[20]	−45	−43	−45	−45	−45	−43	−41	—	−33	−26
	[32]	5	7	7	12	17	24	33	43	—	—
400	[32]	5	10	12	17	24	31	36	48	—	—
600	[32]	−2	10	14	—	—	—	—	—	—	—
800	[32]	5	22	19	—	—	—	—	—	—	—
1000	[32]	−5	12	10	—	—	—	—	—	—	—
1200	[32]	−2	14	12	—	—	—	—	—	—	—
1400	[32]	−2	14	12	—	—	—	—	—	—	—

Table 3.19 Deviations δc_p of data tabulated in [20, 32] from calculated isobaric specific heat

| T, K | Ref. | δc_p, % at pressure p, MPa |||||||||
		0.1	0.2	0.4	0.6	0.8	1	2	4	6	8
2.6	[20]	−1.7	−1.2	−0.5	0	0.3	0.4	1.0	5.4	—	—
2.49	[32]	−7.9	−7.3	−6.6	−5.9	−6.5	−4.9	−6.0	2.5	—	—
3.0	[20]	4.7	4.6	4.5	4.2	4.0	3.7	2.8	4.7	12.1	—
2.99	[32]	−4.7	−3.7	−1.6	0.1	0.9	0.9	2.7	1.4	6.1	—
3.4	[20]	4.4	4.2	3.8	3.4	3.1	2.7	1.3	2.4	8.1	—
3.49	[32]	−3.8	−2.8	−1.5	−0.3	−0.5	0.2	0.2	−2.3	−2.4	5.5
4.2	[20]	3.1	2.8	2.5	2.4	2.0	1.7	0.4	−0.6	0.5	3.8
3.99	[32]	−3.8	−2.4	−1.3	−0.2	−0.3	−0.2	0	−1.1	−1.9	2.9
4.6	[20]	−3.2	2.5	2.4	2.5	2.5	2.3	1.3	−0.3	−0.8	0.1
4.49	[32]	−0.8	−0.7	0.8	1.0	1.9	1.7	2.5	2.0	2.4	5.2
5.0	[20]	−1.9	7.2	1.6	2.3	2.6	2.8	2.2	0.3	−1.1	−1.5
4.99	[32]	−0.7	5.0	1.0	1.4	1.7	2.2	2.9	2.9	4.1	5.9
6.0	[20]	−0.6	−0.9	−2.8	−2.2	0.2	1.5	3.4	2.3	0.1	−2.0
	[32]	−0.3	−0.6	−3.7	−0.1	−3.8	−2.2	0.6	1.3	2.8	2.9
8	[20]	0	0	0.2	0.7	−0.1	−2.6	0.3	3.2	2.4	0.8
	[32]	−0.1	−0.3	−0.5	−0.7	−0.1	−3.1	−3.8	−1.4	−0.7	−1.1
10	[20]	0.1	0.2	0.2	0.3	0.6	0.9	−2.7	1.4	2.6	2.5
	[32]	0.3	0	0	−0.4	−1.1	−1.6	−3.6	−3.0	−1.6	−1.7
20	[20]	0.1	0.1	0.3	0.5	0.6	0.8	1.3	1.1	1.4	2.1
	[32]	0.1	0.2	0.8	1.1	1.3	1.6	1.9	−0.2	−0.3	0
50	[20]	0	0	−0.1	−0.1	−0.2	−0.2	−0.5	−0.8	−1.2	−1.6
	[32]	−0.1	0.2	−0.1	0	−0.3	−0.1	−0.1	−0.4	−0.6	−0.9
100	[20]	0	0	0	0	0	0	0.1	0.2	0.3	0.3
	[32]	0	0	0	−0.1	−0.1	0.2	0	0.1	0.3	0.5
200	[20]	0	0	0	0	0	0	0.1	0.2	0.3	0.4
	[32]	0	0	0	0	0	0	0	0	0	0.5
300	[20]	0	0	0	0	0	0	0.1	0.1	0.2	0.3
	[32]	0	0	0	0	0	0	0.1	0.1	0.1	0.1
400	[32]	0	0	0	0	0	0.1	0.1	0.1	0.1	0.1
600	[32]	0	0	0	0	0	0	0.1	0.1	0.1	0.1
800	[32]	0	0	0	0	0	0	0	0.1	0.1	0.1
1000	[32]	0	0	0	0	0	0	0	0	0.1	0.1
1200	[32]	0	0	0	0	0	0	0	0	0	0.1
1400	[32]	0	0	0	0	0	0	0	0	0	0

116 THERMODYNAMIC PROPERTIES OF HELIUM

Table 3.19 (*Continued*)

T, K	Ref.	δc_p, % at pressure p, MPa									
		10	15	20	30	40	50	60	70	80	100
4.2	[20]	—	—	—	—	—	—	—	—	—	—
3.99	[32]	11.8	—	—	—	—	—	—	—	—	—
4.6	[20]	—	—	—	—	—	—	—	—	—	—
4.99	[32]	11.4	45.8	—	—	—	—	—	—	—	—
5.0	[20]	—	—	—	—	—	—	—	—	—	—
4.99	[32]	9.4	29.6	—	—	—	—	—	—	—	—
6.0	[20]	—	—	—	—	—	—	—	—	—	—
	[32]	4.0	11.2	—	—	—	—	—	—	—	—
8	[20]	−1.0	—	—	—	—	—	—	—	—	—
	[32]	−1.0	−1.3	—	—	—	—	—	—	—	—
10	[20]	1.6	—	—	—	—	—	—	—	—	—
	[32]	−1.6	−2.9	−4.9	−6.7	—	—	—	—	—	—
20	[20]	2.6	1.6	3.4	—	—	—	—	—	—	—
	[32]	−0.1	−1.0	−2.1	−1.5	0.4	3.2	6.7	11.4	—	—
50	[20]	−1.9	−2.1	−2.2	−2.8	−3.3	−3.5	−2.7	—	—	—
	[32]	−1.3	−2.2	−2.5	−3.4	−4.7	−6.4	−7.3	−8.3	—	—
100	[20]	0.3	0.2	0.2	0.6	1.0	1.3	1.3	—	1.0	1.3
	[32]	0.7	1.0	1.0	1.7	2.0	2.9	3.1	2.8	—	—
200	[20]	0.5	0.7	0.9	1.0	1.1	1.3	1.6	—	2.1	2.3
	[32]	0.6	0.6	0.6	1.0	1.4	1.4	2.0	2.2	—	—
300	[20]	0.4	0.6	0.7	1.0	1.2	1.4	1.5	—	2.0	2.4
	[32]	0.2	0.3	0.3	0.4	0.4	0.4	0.4	0.4	—	—
400	[32]	0.1	0.2	0.3	0.4	0	0	0	0	—	—
600	[32]	0.1	0.1	0.2	—	—	—	—	—	—	—
800	[32]	0.1	0.1	0.1	—	—	—	—	—	—	—
1000	[32]	0.1	0.1	0.1	—	—	—	—	—	—	—
1200	[32]	0.1	0.1	0.1	—	—	—	—	—	—	—
1400	[32]	0.1	0.1	0.1	—	—	—	—	—	—	—

heat of phase transition at 0 K of helium (59 kJ/kmol) was added to the former, and a misprint at $T = 3.5$ K and $p = 10$ MPa was corrected. Our values of enthalpy agree well with the values given in [20, 32] (see Table 3.17) in the very important temperature range $T \leq 200$ K at pressures up to 8 MPa. The deviations for the majority of points in this range are a few tenths of kJ/kg. The deviations rise with increasing pressure and temperature. However, where the data deviate significantly from [32], they are in satisfactory agreement with [20].

The tables in [20, 32] show the presence of noticeable discrepancies between values of entropy which, at some points, reach 0.15 kJ/kg K. It is clear from Table 3.18 that our values of entropy are often intermediate between those in [20] and [32] for a number of isotherms. At supercritical temperatures, our values agree better with those in [32] for the majority of isobars. For $T \geq 400$ K, only the data given in [32] are available. The deviations of the calculated values from these data are mostly less than 0.01 kJ/(kg·K), and only at a few points reach 0.014–0.022 kJ/(kg·K). On subcritical isotherms the deviations from the data in [20] all have the same sign and change relatively little with pressure. This is explained by the fact that the deviations are largely determined by the differences between the data in [20] and our data. For example, on the 2.4 and 3 K isotherms the differences are 0.046 and 0.053 kJ/(kg·K), respectively.

Comparison of the calculated isobaric heat capacity with the tabulated data [20, 32] (Table 3.19) shows, as in the case of specific volumes, that there is good agreement for $T \geq 100$ K and $p \leq 20$ MPa: the deviations are in the range 0.1–0.3%, and only for a few cases reach 0.7–1%. More substantial deviations are seen outside this range, but, in many cases, our calculated values of c_p are intermediate (as were the values of s) between those published previously. The largest deviations occur at the highest pressures on the 4.5 and 5 K isotherms where c_p rises after passing through a minimum [32]. This is not observed for other well-investigated media. Further studies of the thermal properties of helium at high pressures and low temperatures would seem to be desirable.

Our tables contain far more data than those in [20], where only the values of v, h, s and c_p are presented. The values of μ, k, a/a_0, and γ/γ_0 presented here are not given in [32]. Figures 35–45 give the temperature and pressure dependence of the thermodynamic functions of He^4.

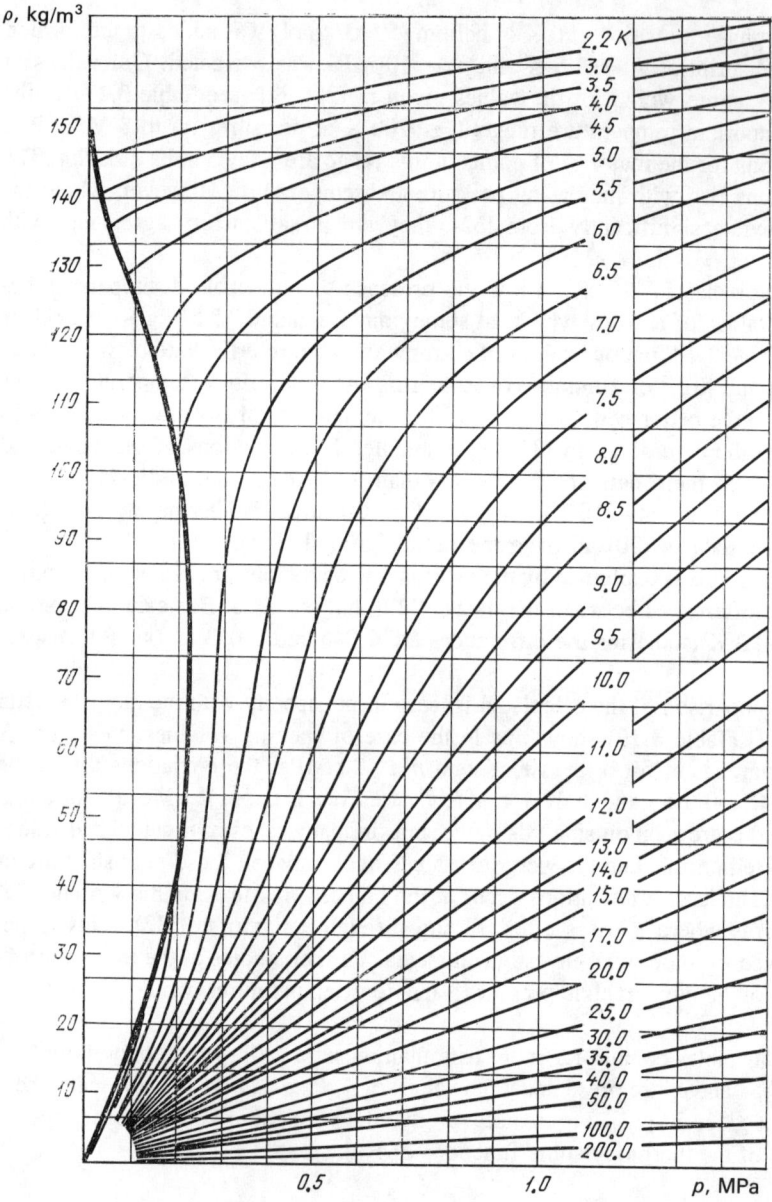

Figure 35 The pressure and temperature dependence of the density of He^4.

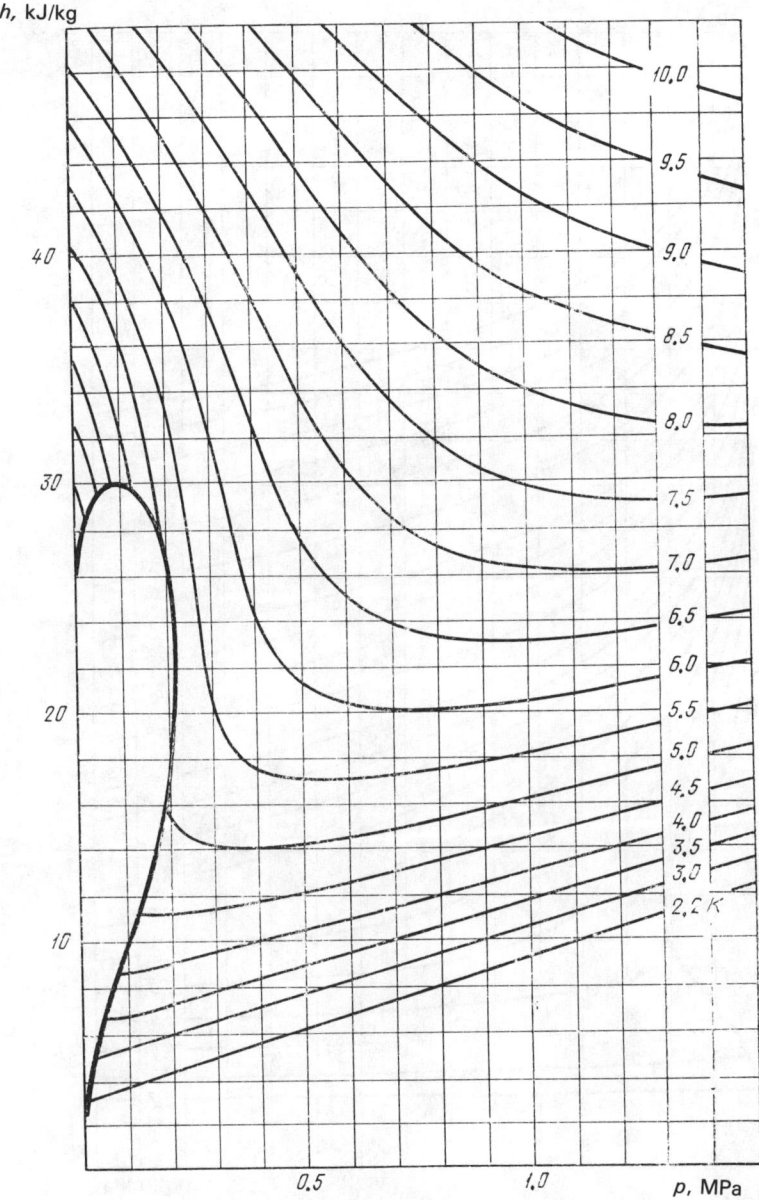

Figure 36 The pressure and temperature dependence of the enthalpy of He^4.

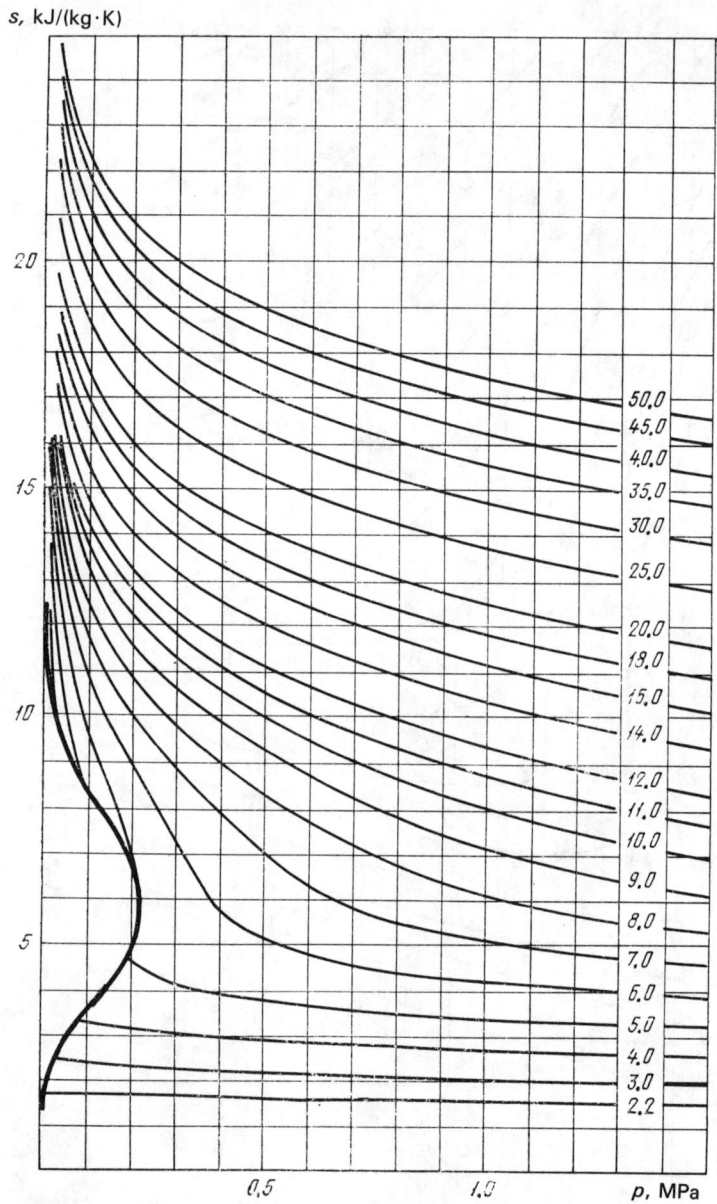

Figure 37 The pressure and temperature dependence of the entropy of He[4].

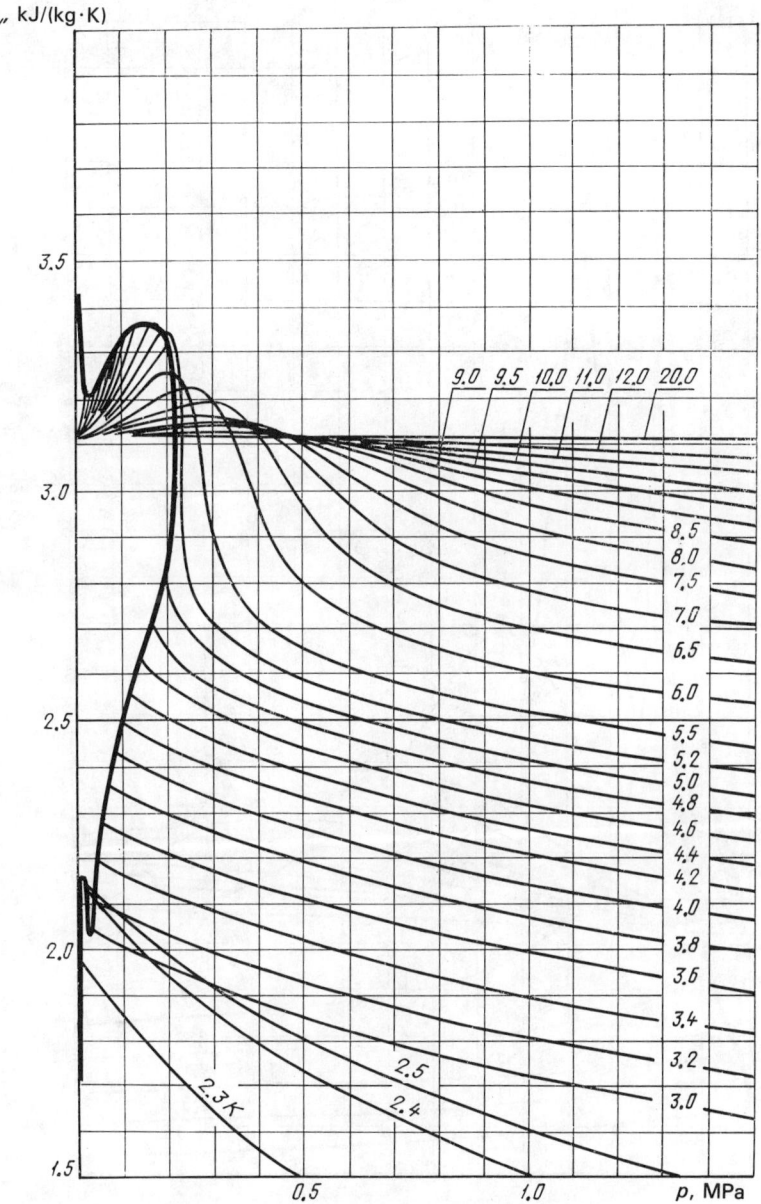

Figure 38 The pressure and temperature dependence of the isochoric specific heat of He^4.

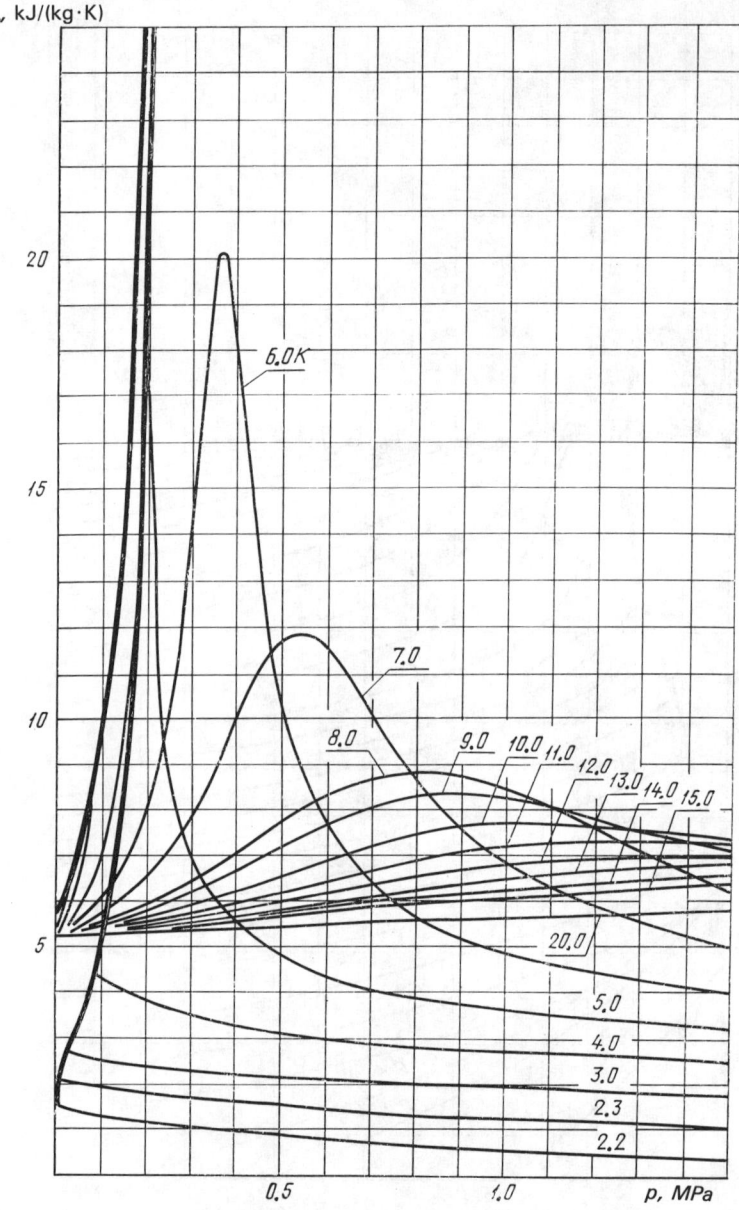

Figure 39 The pressure and temperature dependence of the isobaric specific heat of He[4].

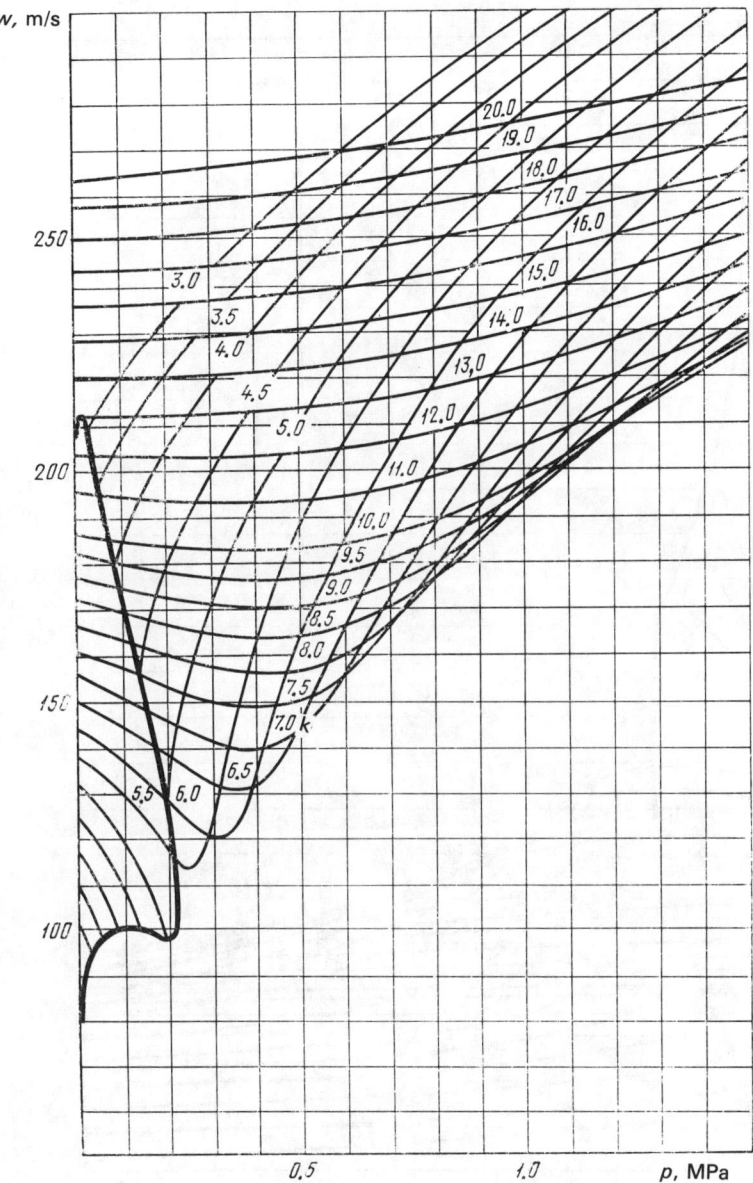

Figure 40 The pressure and temperature dependence of the speed of sound in He^4.

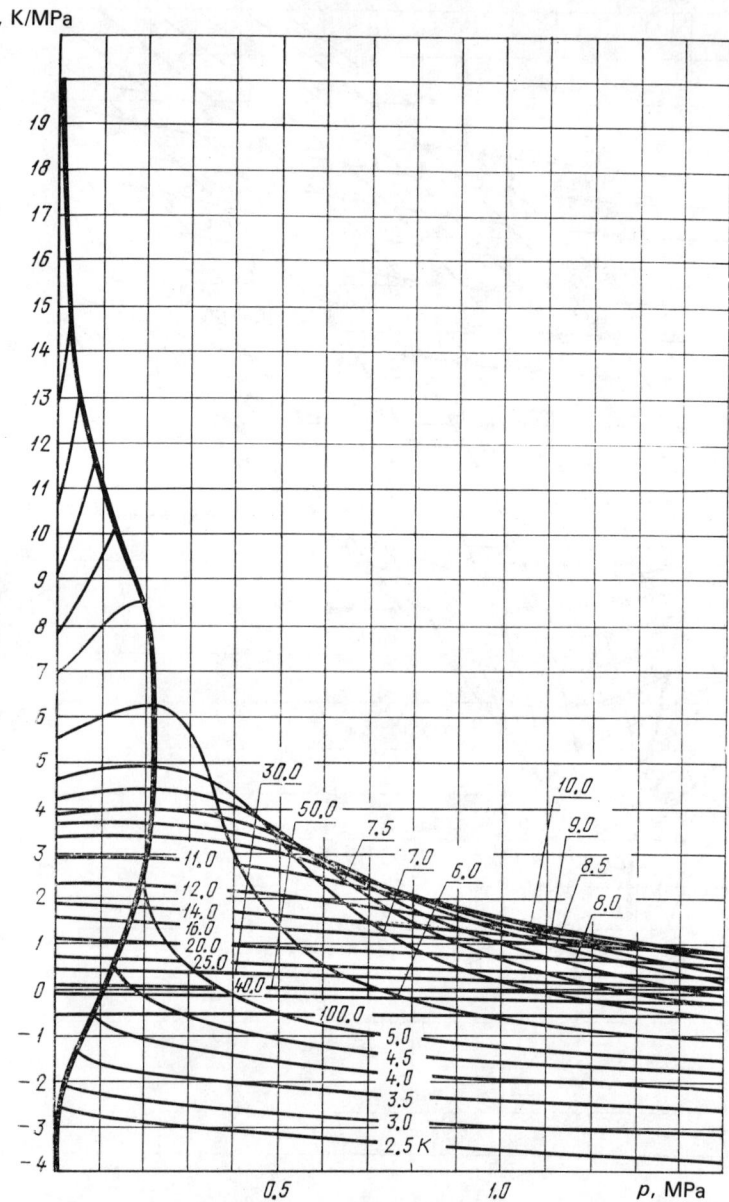

Figure 41 The pressure and temperature dependence of the adiabatic Joule-Thomson coefficient in He^4.

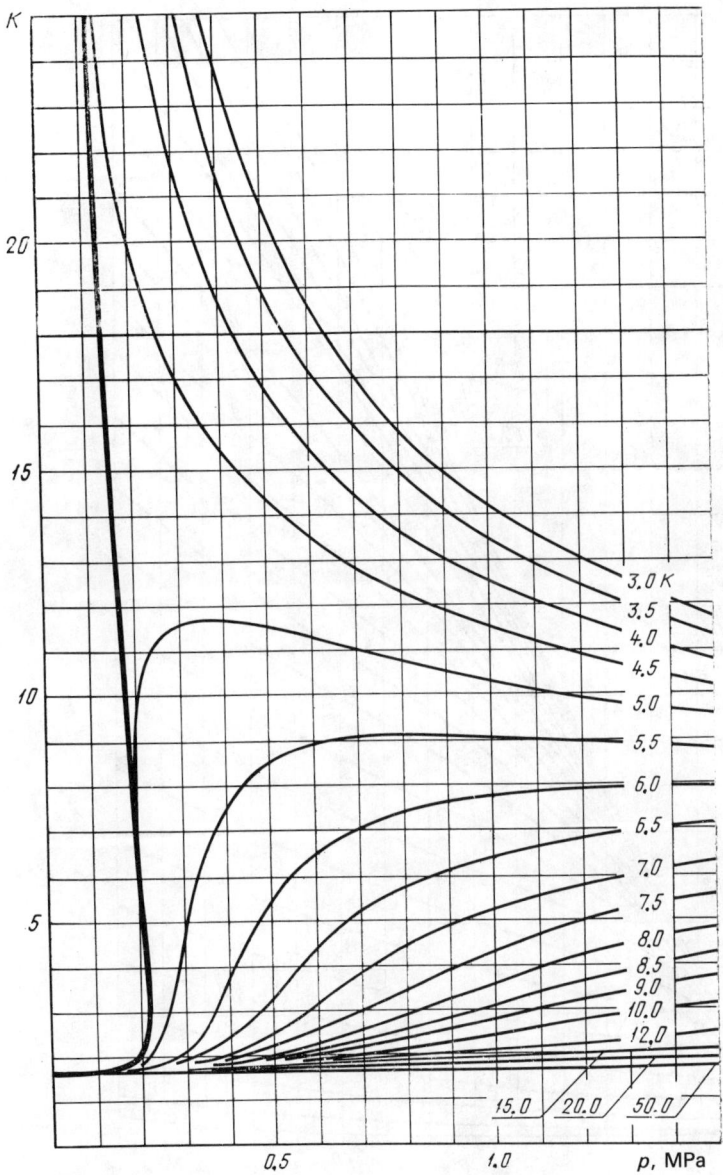

Figure 42 The pressure and temperature dependence of the adiabatic index of He^4.

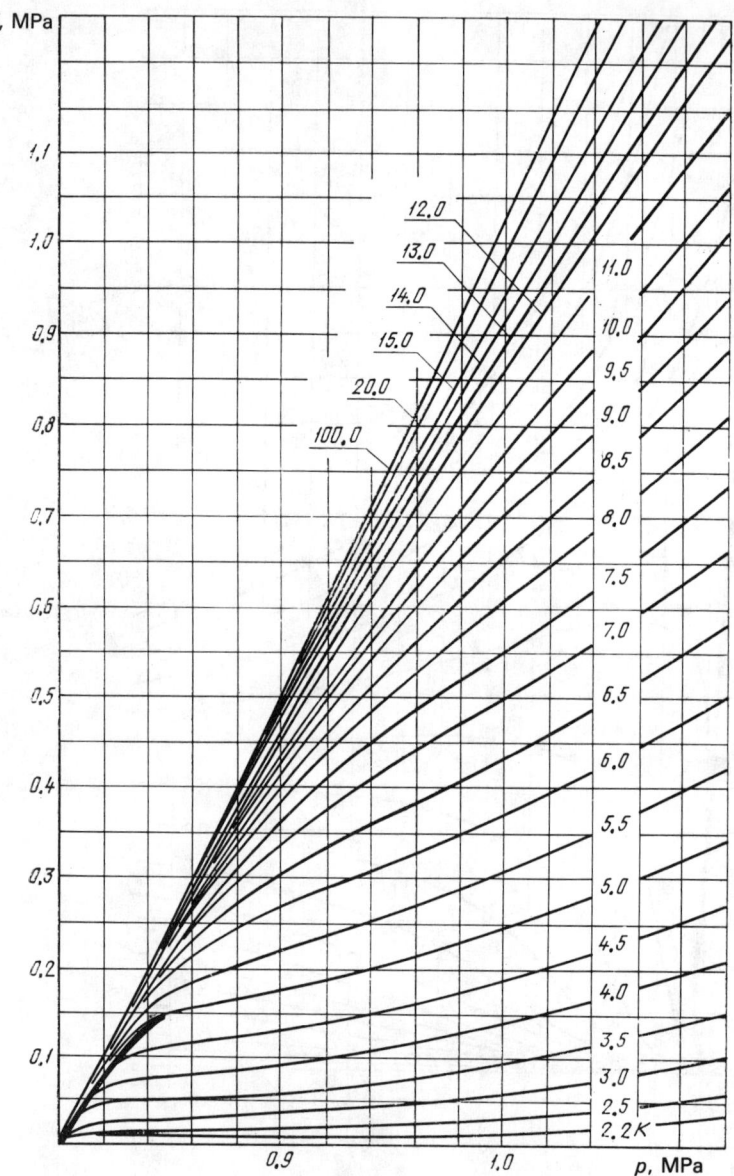

Figure 43 The pressure and temperature dependence of the fugacity of He^4.

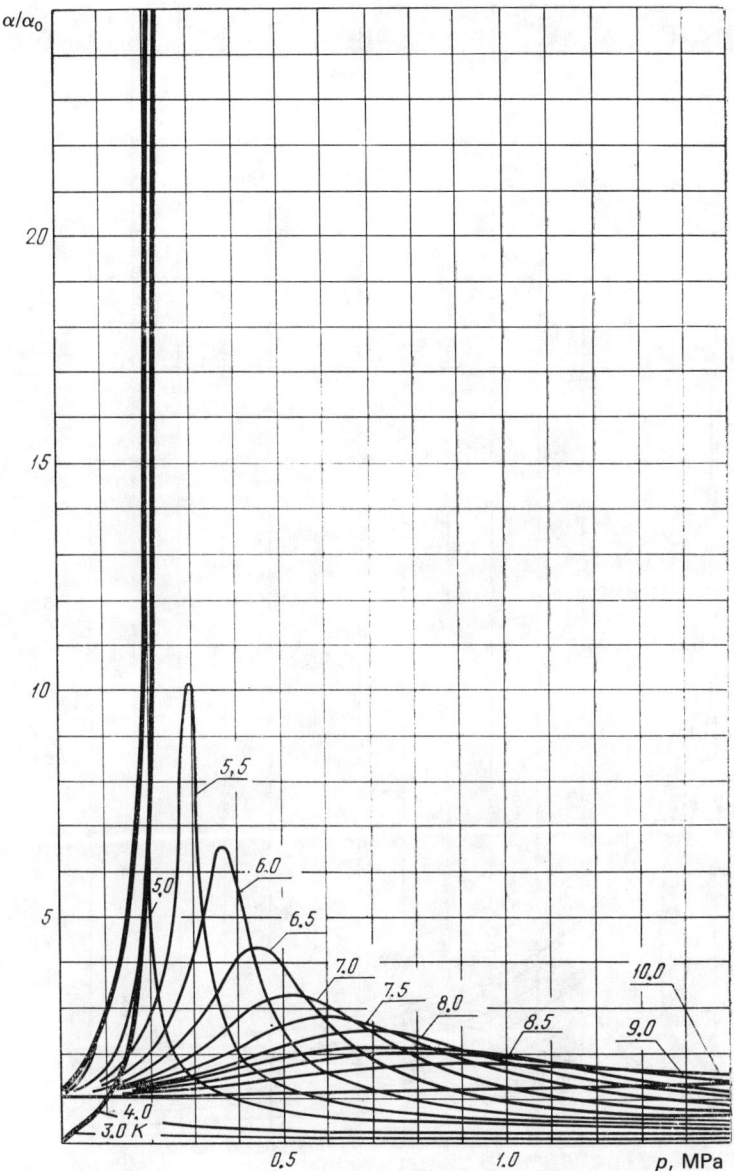

Figure 44 The pressure and temperature dependence of the volume expansion coefficient of He^4.

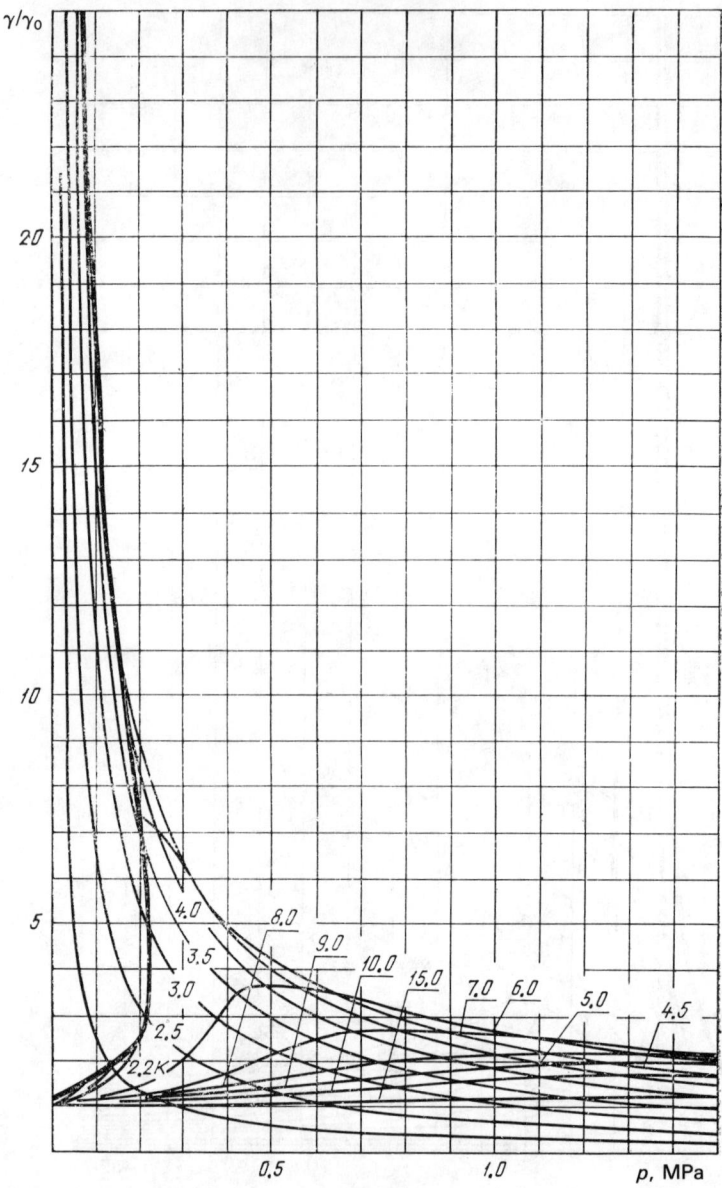

Figure 45 The pressure and temperature dependence of the thermal coefficient of pressure.

PART TWO

TABLES OF THERMODYNAMIC PROPERTIES OF HELIUM-4

A. BASIC NUMERICAL VALUES

Relative atomic mass	μ = 4.0026 kg/kmol
Gas constant	R = 2077.252 J/(kg·K)
Temperature at upper λ point	$T_\lambda^{(U)}$ = 1.7633 K
Pressure at upper λ point	$P_\lambda^{(U)}$ = 3.013 MPa
Temperature at lower λ point	$T_\lambda^{(L)}$ = 2.1720 K
Pressure at lower λ point	$p_\lambda^{(L)}$ = 5040 Pa
Normal boiling point	$T_{n,b}$ = 4.224 K
Critical point temperature	T_{crit} = 5.190 K
Critical point pressure	p_{crit} = 0.22746 MPa
Critical point density	ρ_{crit} = 69.64 kg/m^3
Heat of phase transition at 0 K	h_0^0 = 14740.4 J/kg

B. DEFINITION OF SYMBOLS AND UNITS

T	temperature, K	τ	reduced temperature = T/T_{crit}†
p	pressure, MPa	π	reduced pressure = P/P_{crit}†
ρ	density, kg/m^3	ω	reduced density = ρ/ρ_{crit}†

†Changed in proof.—T.B.S.

132 THERMODYNAMIC PROPERTIES OF HELIUM

Z	coefficient of compressibility
h	enthalpy, kJ/kg
s	entropy, kJ/(kg·K)
c_v	isochoric specific heat, kJ/(kg·K)
c_p	isobaric specific heat, kJ/(kg·K)
c_s	specific heat on the saturation curve, kJ/(kg·K)
c_m	specific heat on the melting curve, kJ/(kg·K)
c_λ	specific heat on the λ line, kJ/(kg·K)
w	speed of sound, m/s
μ	adiabatic Joule-Thomson coefficient K/MPa
k	adiabatic index
f	fugacity, MPa
α/α_0	volume expansion coefficient $= [(1/v)(\partial v/\partial T)_p]T$ [†]
γ/γ_0	thermal coefficient of pressure $= [(1/p)(\partial P/\partial T)_v]T$ [†]
$d\pi/d\tau$	first derivative of reduced pressure with respect to reduced temperature
$d\pi^2/d\tau^2$	second derivative of reduced pressure with respect to reduced temperature
ϕ	Gibbs potential, kJ/kg
r	latent heat vaporization, kJ/kg
$(')$	property at saturation in the liquid phase
$('')$	property at saturation in the gas phase
$\lg_{10}$	$\log_{10}$
$\ln$	$\ln_e$

[†]Changed in proof.—T.B.S.

Thermodynamic properties of liquid He⁴ on the freezing curve (at the indicated temperature)

Table II.1

T	p	Φ	$d\pi/d\tau$	$d^2\pi/d\tau^2$
2.0	3.761	22.0	78.8	295.4
2.1	4.118	23.8	83.8	232.4
2.2	4.494	25.6	87.9	192.7
2.3	4.887	27.6	91.4	170.1
2.4	5.295	29.5	94.5	159.2
2.5	5.715	31.6	97.5	155.3
2.6	6.149	33.7	100.5	154.4
2.7	6.596	35.8	103.5	153.3
2.8	7.056	37.9	106.4	149.5
2.9	7.529	40.1	109.2	141.2
3.0	8.013	42.4	111.8	127.5
3.1	8.509	44.7	114.1	108.1
3.2	9.013	47.0	115.9	83.4
3.3	9.524	49.3	117.3	54.7
3.4	10.040	51.6	118.0	24.0
3.5	10.558	53.9	118.2	−6.0
3.6	11.075	56.2	117.8	−31.9
3.7	11.590	58.4	117.0	−49.5
3.8	12.101	60.6	116.0	−53.9
3.9	12.607	62.8	115.1	−39.4
4.0	13.110	65.0	114.7	0.5
4.1	13.614	67.1	115.3	72.8
4.2	14.124	69.3	117.7	185.5
4.3	14.650	71.5	122.8	347.3
4.4	15.205	73.8	131.5	567.4
4.5	15.809	76.4	145.1	856.0
4.6	16.486	79.2	165.0	1224.0
4.7	17.267	82.5	192.8	1683.0

Table II.2

T	ρ	h	s	c_v	c_p	c_m
2.0	184.57	25.15	1.585	−4.247	−4.160	−4.788
2.1	186.55	26.77	1.435	−1.176	−1.103	−1.709
2.2	188.56	28.68	1.391	0.193	0.267	−0.366
2.3	190.56	30.75	1.389	0.782	0.863	0.183
2.4	192.52	32.91	1.402	1.024	1.114	0.383
2.5	194.44	35.12	1.419	1.120	1.220	0.437
2.6	196.33	37.39	1.436	1.159	1.268	0.437
2.7	198.17	39.70	1.452	1.179	1.296	0.419
2.8	199.98	42.05	1.467	1.194	1.319	0.400
2.9	201.75	44.44	1.481	1.210	1.342	0.385

Table II.2 (*Continued*)

T	ρ	h	s	c_v	c_p	c_m
3.0	203.48	46.87	1.494	1.226	1.365	0.374
3.1	205.17	49.33	1.506	1.243	1.388	0.368
3.2	206.81	51.81	1.518	1.259	1.410	0.367
3.3	208.41	54.31	1.529	1.275	1.431	0.371
3.4	209.95	56.82	1.540	1.290	1.451	0.382
3.5	211.43	59.31	1.551	1.304	1.470	0.397
3.6	212.85	61.79	1.563	1.318	1.487	0.418
3.7	214.21	64.25	1.575	1.332	1.505	0.442
3.8	215.51	66.67	1.587	1.346	1.522	0.469
3.9	216.74	69.06	1.599	1.361	1.540	0.495
4.0	217.93	71.43	1.612	1.377	1.558	0.517
4.1	219.08	73.78	1.625	1.393	1.577	0.531
4.2	220.21	76.16	1.638	1.410	1.596	0.529
4.3	221.35	78.59	1.650	1.428	1.616	0.506
4.4	222.53	81.14	1.661	1.446	1.635	0.452
4.5	223.79	83.89	1.670	1.464	1.653	0.360
4.6	225.19	86.94	1.677	1.481	1.669	0.221
4.7	226.77	90.40	1.680	1.497	1.681	0.030

Table II.3

T	w	μ	k	f	α/α_0	γ/γ_0
2.0	387.1	1.26	7.35	0.67	0.034	0.25
2.1	388.9	4.71	6.85	1.05	0.031	0.22
2.2	484.4	−19.25	9.84	1.61	0.031	0.22
2.3	443.4	−5.89	7.67	2.44	0.032	0.22
2.4	451.0	−4.50	7.40	3.61	0.034	0.23
2.5	462.0	−4.07	7.26	5.25	0.036	0.24
2.6	474.1	−3.87	7.17	7.51	0.037	0.24
2.7	486.4	−3.74	7.11	10.59	0.038	0.25
2.8	498.8	−3.64	7.05	14.72	0.039	0.25
2.9	511.2	−3.54	7.00	20.19	0.040	0.25
3.0	523.6	−3.45	6.96	27.36	0.041	0.26
3.1	535.8	−3.36	6.92	36.64	0.042	0.26
3.2	548.0	−3.28	6.89	48.46	0.042	0.26
3.3	560.0	−3.21	6.86	63.29	0.043	0.26
3.4	571.7	−3.14	6.83	81.59	0.043	0.26
3.5	583.2	−3.08	6.81	103.75	0.044	0.26
3.6	594.3	−3.02	6.79	130.09	0.044	0.27
3.7	605.1	−2.96	6.77	160.82	0.044	0.27
3.8	615.4	−2.91	6.74	196.07	0.045	0.27
3.9	625.3	−2.86	6.72	236.00	0.045	0.27
4.0	634.9	−2.81	6.70	280.91	0.045	0.27
4.1	644.1	−2.76	6.68	331.56	0.045	0.27

Table II.3 (*Continued*)

T	w	μ	k	f	α/α_0	γ/γ_0
4.2	653.2	−2.72	6.65	389.55	0.046	0.27
4.3	662.3	−2.67	6.63	457.92	0.046	0.27
4.4	671.6	−2.62	6.60	542.15	0.046	0.27
4.5	681.4	−2.58	6.57	651.82	0.046	0.26
4.6	692.1	−2.54	6.54	803.84	0.045	0.26
4.7	704.1	−2.51	6.51	1028.98	0.044	0.26

Thermodynamic properties of liquid He^4 on the freezing curve (at the indicated pressure)

Table II.4

p	T	Φ	$d\pi/d\tau$	$d^2\pi/d\tau^2$
4.0	2.068	23.2	82.3	249.9
4.5	2.202	25.6	87.9	192.2
5.0	2.328	28.1	92.3	166.1
5.5	2.449	30.5	96.0	156.7
6.0	2.566	32.9	99.5	154.6
6.5	2.679	35.3	102.9	153.7
7.0	2.788	37.7	106.1	150.1
7.5	2.894	40.0	109.1	141.9
8.0	2.997	42.3	111.7	127.9
8.5	3.098	44.6	114.1	108.4
9.0	3.197	46.9	115.9	84.1
9.5	3.295	49.2	117.2	56.2
10.0	3.392	51.4	118.0	26.4
10.5	3.489	53.6	118.2	−2.8
11.0	3.585	55.8	117.9	−28.5
11.5	3.682	58.0	117.2	−47.2
12.0	3.780	60.2	116.2	−54.3
12.5	3.879	62.4	115.3	−44.3
13.0	3.978	64.5	114.7	−10.8
13.5	4.077	66.6	115.0	53.3
14.0	4.176	68.8	116.9	154.3
14.5	4.272	70.9	121.0	296.5
15.0	4.364	73.0	127.8	480.5
15.5	4.450	75.1	137.6	702.6
16.0	4.529	77.2	150.2	955.8
16.5	4.602	79.3	165.4	1232.0
17.0	4.668	81.4	182.8	1523.4
17.5	4.727	83.5	201.9	1823.9

Table II.5

p	ρ	h	s	c_v	c_p	c_m
4.0	185.91	26.20	1.467	−1.916	−1.841	−2.447
4.5	188.59	28.71	1.391	0.206	0.280	−0.354
5.0	191.11	31.35	1.392	0.873	0.956	0.263
5.5	193.47	33.99	1.410	1.082	1.177	0.420
6.0	195.69	36.61	1.430	1.149	1.255	0.440
6.5	197.78	39.20	1.449	1.176	1.291	0.423
7.0	199.76	41.76	1.465	1.193	1.317	0.402
7.5	201.64	44.29	1.480	1.209	1.341	0.385
8.0	203.43	46.80	1.493	1.226	1.365	0.374
8.5	205.14	49.29	1.506	1.243	1.388	0.358
9.0	206.77	51.75	1.517	1.259	1.410	0.367
9.5	208.34	54.20	1.528	1.274	1.430	0.371
10.0	209.83	56.62	1.539	1.289	1.450	0.381
10.5	211.27	59.04	1.550	1.303	1.468	0.395
11.0	212.65	61.43	1.561	1.316	1.485	0.414
11.5	213.98	63.82	1.573	1.330	1.502	0.438
12.0	215.25	66.19	1.584	1.344	1.519	0.464
12.5	216.48	68.56	1.597	1.358	1.536	0.490
13.0	217.67	70.91	1.609	1.373	1.554	0.513
13.5	218.82	73.25	1.622	1.389	1.572	0.529
14.0	219.94	75.58	1.635	1.406	1.591	0.531
14.5	221.03	77.90	1.647	1.423	1.610	0.515
15.0	222.09	80.20	1.657	1.439	1.628	0.475
15.5	223.15	82.49	1.666	1.455	1.644	0.411
16.0	224.19	84.75	1.673	1.469	1.658	0.324
16.5	225.21	87.00	1.677	1.481	1.669	0.218
17.0	226.23	89.22	1.679	1.492	1.678	0.098
17.5	227.24	91.43	1.680	1.500	1.683	−0.031

Table II.6

p	w	μ	k	f	α/α_0	γ/γ_0
4.0	390.3	2.83	7.08	0.091	0.031	0.23
4.5	480.3	−18.34	9.67	1.62	0.031	0.22
5.0	444.8	−5.29	7.56	2.73	0.033	0.23
5.5	456.2	−4.24	7.32	4.35	0.035	0.23
6.0	469.9	−3.92	7.20	6.66	0.037	0.24
6.5	483.8	−3.77	7.12	9.85	0.038	0.25
7.0	497.3	−3.65	7.06	14.15	0.039	0.25
7.5	510.5	−3.55	7.01	19.81	0.040	0.25
8.0	523.2	−3.45	6.96	27.14	0.041	0.26
8.5	535.6	−3.37	6.92	36.45	0.042	0.26
9.0	547.7	−3.28	6.89	48.12	0.042	0.26
9.5	559.4	−3.21	6.85	62.52	0.043	0.26
10.0	570.8	−3.14	6.84	80.04	0.043	0.26

Table II.6 (*Continued*)

p	w	μ	k	f	α/α_0	γ/γ_0
10.5	581.9	−3.08	6.81	101.07	0.044	0.26
11.0	592.7	−3.03	6.79	125.97	0.044	0.27
11.5	603.2	−2.97	6.77	155.09	0.044	0.27
12.0	613.4	−2.92	6.75	188.71	0.045	0.27
12.5	623.2	−2.87	6.73	227.11	0.045	0.27
13.0	632.8	−2.82	6.71	270.58	0.045	0.27
13.5	642.1	−2.77	6.68	319.57	0.045	0.27
14.0	651.0	−2.73	6.66	374.79	0.046	0.27
14.5	659.7	−2.68	6.63	437.46	0.046	0.27
15.0	668.2	−2.64	6.61	509.40	0.046	0.27
15.5	676.4	−2.60	6.59	593.01	0.046	0.27
16.0	684.4	−2.57	6.56	691.21	0.045	0.26
16.5	692.3	−2.54	6.54	807.34	0.045	0.26
17.0	700.0	−2.52	6.52	945.12	0.045	0.26
17.5	707.6	−2.50	6.50	1108.75	0.044	0.26

Thermodynamic properties of He^4 on the saturation curve (at the indicated temperature)

Table II.7

T	p	Φ	r	$d\pi/d\tau$	$d^2\pi/d\tau^2$
2.000	0.0032	−0.343	21.5	0.201	2.274
2.100	0.0042	−0.507	22.1	0.246	2.419
2.200	0.0054	−0.667	22.5	0.294	2.597
2.300	0.0068	−0.832	22.7	0.346	2.797
2.400	0.0084	−1.003	22.9	0.402	3.011
2.500	0.0103	−1.183	23.0	0.462	3.234
2.600	0.0125	−1.369	23.2	0.527	3.462
2.700	0.0150	−1.563	23.3	0.596	3.691
2.800	0.0177	−1.762	23.3	0.669	3.920
2.900	0.0208	−1.968	23.4	0.747	4.147
3.000	0.0243	−2.180	23.4	0.829	4.373
3.100	0.0281	−2.397	23.4	0.915	4.597
3.200	0.0323	−2.620	23.4	1.006	4.818
3.300	0.0369	−2.848	23.3	1.101	5.039
3.400	0.0420	−3.081	23.2	1.200	5.260
3.500	0.0474	−3.320	23.0	1.303	5.481
3.600	0.0534	−3.563	22.8	1.411	5.703
3.700	0.0598	−3.812	22.6	1.523	5.929
3.800	0.0667	−4.065	22.3	1.640	6.158
3.900	0.0742	−4.324	21.9	1.761	6.394
4.000	0.0822	−4.587	21.5	1.886	6.638
4.100	0.0907	−4.855	21.1	2.016	6.893

Table II.7 (Continued)

T	p	Φ	r	$d\pi/d\tau$	$d^2\pi/d\tau^2$
4.200	0.0999	−5.127	20.5	2.152	7.162
4.400	0.1200	−5.686	19.2	2.439	7.759
4.500	0.1310	−5.972	18.4	2.592	8.101
4.600	0.1427	−6.263	17.5	2.752	8.486
4.700	0.1551	−6.557	16.5	2.919	8.929
4.800	0.1683	−6.856	15.2	3.096	9.455
4.900	0.1823	−7.159	13.7	3.284	10.108
5.000	0.1971	−7.466	11.8	3.487	10.964
5.100	0.2129	−7.776	9.3	3.709	12.185

Table II.8

T	ρ'	ρ''	h'	h''	s'	s''
2.000	145.87	0.81	3.2	24.7	1.764	12.523
2.100	145.97	1.02	3.0	25.1	1.674	12.215
2.200	145.86	1.25	3.1	25.6	1.706	11.928
2.300	145.59	1.52	3.3	26.0	1.784	11.662
2.400	145.20	1.82	3.5	26.4	1.876	11.415
2.500	144.70	2.17	3.7	26.8	1.970	11.184
2.600	144.11	2.55	4.0	27.1	2.061	10.968
2.700	143.45	2.97	4.2	27.5	2.149	10.763
2.800	142.71	3.43	4.5	27.8	2.234	10.569
2.900	141.92	3.94	4.8	28.1	2.319	10.383
3.000	141.06	4.51	5.0	28.4	2.402	10.206
3.100	140.14	5.12	5.3	28.7	2.486	10.035
3.200	139.16	5.79	5.6	29.0	2.571	9.870
3.300	138.12	6.52	5.9	29.2	2.656	9.710
3.400	137.01	7.31	6.2	29.4	2.743	9.554
3.500	135.84	8.17	6.6	29.6	2.831	9.403
3.600	134.59	9.10	6.9	29.8	2.920	9.254
3.700	133.27	10.12	7.3	29.9	3.012	9.108
3.800	131.85	11.22	7.7	30.0	3.105	8.964
3.900	130.35	12.42	8.2	30.1	3.200	8.821
4.000	128.75	13.73	8.6	30.1	3.298	8.678
4.100	127.03	15.15	9.1	30.1	3.398	8.535
4.200	125.17	16.72	9.6	30.1	3.502	8.391
4.400	120.99	20.34	10.7	29.9	3.722	8.095
4.500	118.61	22.46	11.3	29.8	3.840	7.940
4.600	115.96	24.85	12.0	29.5	3.965	7.778
4.700	112.99	27.59	12.7	29.2	4.100	7.605
4.800	109.58	30.77	13.5	28.8	4.247	7.419
4.900	105.58	34.58	14.5	28.2	4.412	7.211
5.000	100.66	39.35	15.6	27.4	4.605	6.970
5.100	94.10	45.86	17.0	26.2	4.850	6.668

Table II.9

T	c_v'	c_v''	c_p'	c_p''	c_s'	c_s''
2.000	−3.977	3.331	−3.962	5.696	−3.960	−6.355
2.100	−0.218	3.345	−0.217	5.746	−0.218	−6.257
2.200	1.372	3.339	1.394	5.781	1.391	−6.076
2.300	1.972	3.322	2.036	5.811	2.030	−5.888
2.400	2.143	3.300	2.260	5.844	2.251	−5.720
2.500	2.149	3.276	2.324	5.884	2.310	−5.580
2.600	2.106	3.255	2.343	5.936	2.323	−5.467
2.700	2.064	3.237	2.366	6.000	2.339	−5.378
2.800	2.040	3.222	2.409	6.080	2.375	−5.310
2.900	2.036	3.213	2.476	6.177	2.433	−5.259
3.000	2.049	3.207	2.566	6.290	2.512	−5.223
3.100	2.074	3.206	2.675	6.421	2.608	−5.202
3.200	2.109	3.208	2.801	6.571	2.717	−5.194
3.300	2.149	3.214	2.941	6.741	2.838	−5.199
3.400	2.191	3.222	3.094	6.933	2.967	−5.218
3.500	2.234	3.233	3.260	7.150	3.106	−5.252
3.600	2.276	3.245	3.442	7.393	3.254	−5.301
3.700	2.318	3.258	3.641	7.668	3.411	−5.368
3.800	2.357	3.273	3.862	7.979	3.580	−5.456
3.900	2.396	3.288	4.108	8.333	3.764	−5.567
4.000	2.432	3.302	4.387	8.741	3.964	−5.707
4.100	2.467	3.317	4.708	9.215	4.186	−5.879
4.200	2.501	3.331	5.086	9.774	4.437	−6.094
4.400	2.567	3.356	6.093	11.267	5.057	−6.693
4.500	2.600	3.365	6.796	12.303	5.458	−7.116
4.600	2.634	3.373	7.719	13.651	5.951	−7.661
4.700	2.669	3.377	8.991	15.488	6.582	−8.386
4.800	2.707	3.376	10.864	18.150	7.431	−9.387
4.900	2.749	3.370	13.896	22.370	8.661	−10.856
5.000	2.800	3.353	19.630	30.115	10.663	−13.237
5.100	2.866	3.319	34.288	49.003	14.723	−17.937

Table II.10

T	w'	w''	μ'	μ''	k'	k''
2.000	203.6	80.7	1.78	24.60	1835.66	1.64
2.100	206.4	82.3	31.36	22.43	1474.86	1.63
2.200	209.4	83.9	−4.74	20.77	1183.90	1.63
2.300	210.6	85.4	−3.17	19.49	949.27	1.63
2.400	211.4	86.9	−2.80	18.50	768.73	1.63
2.500	211.9	88.3	−2.67	17.70	628.73	1.63
2.600	212.1	89.6	−2.60	17.05	518.72	1.64
2.700	212.0	90.9	−2.52	16.49	430.96	1.64
2.800	211.4	92.1	−2.43	15.59	359.96	1.64
2.900	210.5	93.3	−2.31	15.54	301.89	1.65

Table II.10 (*Continued*)

T	w'	w''	μ'	μ''	k'	k''
3.000	209.1	94.3	−2.18	15.11	254.02	1.65
3.100	207.3	95.2	−2.04	14.71	214.36	1.65
3.200	205.2	96.1	−1.89	14.32	181.37	1.65
3.300	202.8	96.9	−1.74	13.95	153.84	1.66
3.400	200.1	97.6	−1.58	13.59	130.79	1.66
3.500	197.3	98.2	−1.43	13.24	111.42	1.66
3.600	194.2	98.7	−1.27	12.91	95.09	1.66
3.700	191.0	99.2	−1.11	12.58	81.28	1.66
3.800	187.6	99.5	−0.95	12.26	69.54	1.67
3.900	184.1	99.8	−0.78	11.96	59.53	1.67
4.000	180.4	100.0	−0.60	11.66	50.96	1.67
4.100	176.5	100.1	−0.42	11.37	43.60	1.67
4.200	172.4	100.2	−0.22	11.09	37.26	1.68
4.400	163.7	100.0	0.23	10.55	27.02	1.70
4.500	159.0	99.9	0.49	10.28	22.88	1.71
4.600	154.0	99.6	0.78	10.00	19.26	1.73
4.700	148.7	99.3	1.11	9.71	16.10	1.75
4.800	143.0	98.9	1.49	9.39	13.31	1.79
4.900	136.8	98.5	1.94	9.03	10.84	1.84
5.000	130.0	98.2	2.48	8.58	8.64	1.93
5.100	122.3	98.2	3.20	7.93	6.61	2.08

Table II.11

T	f'	f''	α'/σ_0	α''/σ_0	γ'/γ_0	γ''/γ_0
2.000	0.003	0.003	−0.027	1.114	−50.40	1.066
2.100	0.004	0.004	0.006	1.131	8.55	1.073
2.200	0.005	0.005	0.034	1.150	39.50	1.082
2.300	0.006	0.006	0.059	1.172	54.02	1.091
2.400	0.008	0.008	0.081	1.197	59.33	1.102
2.500	0.010	0.010	0.103	1.226	59.70	1.115
2.600	0.011	0.011	0.123	1.258	57.52	1.129
2.700	0.014	0.014	0.144	1.294	54.16	1.145
2.800	0.016	0.016	0.165	1.334	50.36	1.162
2.900	0.019	0.019	0.187	1.379	46.50	1.181
3.000	0.021	0.021	0.211	1.428	42.78	1.202
3.100	0.025	0.025	0.236	1.484	39.30	1.224
3.200	0.028	0.028	0.264	1.545	36.09	1.248
3.300	0.032	0.032	0.295	1.613	33.15	1.274
3.400	0.036	0.036	0.329	1.689	30.47	1.302
3.500	0.040	0.040	0.367	1.774	28.03	1.331
3.600	0.044	0.044	0.410	1.869	25.80	1.363
3.700	0.049	0.049	0.459	1.976	23.76	1.397
3.800	0.054	0.054	0.516	2.098	21.89	1.433
3.900	0.059	0.059	0.581	2.238	20.18	1.472

Table II.11 (*Continued*)

T	f'	f''	α'/α_0	$\alpha'''\alpha_0$	γ'/γ_0	γ''/γ_0
4.000	0.065	0.065	0.658	2.399	18.60	1.515
4.100	0.071	0.071	0.750	2.588	17.14	1.560
4.200	0.077	0.077	0.862	2.812	15.79	1.611
4.400	0.090	0.090	1.172	3.417	13.35	1.727
4.500	0.097	0.097	1.397	3.840	12.23	1.796
4.600	0.105	0.105	1.700	4.392	11.18	1.875
4.700	0.112	0.112	2.128	5.148	10.17	1.966
4.800	0.120	0.120	2.772	6.244	9.20	2.076
4.900	0.128	0.128	3.840	7.981	8.24	2.213
5.000	0.137	0.137	5.907	11.163	7.28	2.393
5.100	0.146	0.146	11.326	18.883	6.26	2.658

Thermodynamic properties of He^4 on the saturation curve (at the indicated pressure)

Table II.12

p	T	Φ	r	$d\pi/d\tau$	$d^2\pi/d\tau^2$
0.0060	2.24	−0.740	22.6	0.317	2.684
0.0070	2.31	−0.854	22.7	0.353	2.824
0.0080	2.37	−0.959	22.9	0.387	2.956
0.0090	2.43	−1.058	22.9	0.420	3.080
0.0100	2.48	−1.153	23.0	0.452	3.197
0.0200	2.87	−1.915	23.4	0.726	4.090
0.0300	3.15	−2.500	23.4	0.957	4.700
0.0400	3.36	−2.992	23.2	1.162	5.176
0.0500	3.54	−3.426	22.9	1.350	5.578
0.0600	3.70	−3.819	22.5	1.526	5.935
0.0700	3.84	−4.180	22.1	1.693	6.262
0.0800	3.97	−4.516	21.6	1.852	6.572
0.0900	4.09	−4.832	21.1	2.005	6.871
0.1000	4.20	−5.131	20.5	2.154	7.165
0.1100	4.30	−5.416	19.9	2.298	7.460
0.1200	4.40	−5.687	19.2	2.439	7.760
0.1300	4.49	−5.947	18.5	2.578	8.070
0.1400	4.58	−6.197	17.8	2.715	8.395
0.1500	4.66	−6.438	16.9	2.850	8.741
0.1600	4.74	−6.670	16.0	2.985	9.116
0.1700	4.81	−6.894	15.1	3.119	9.529
0.1800	4.88	−7.111	14.0	3.254	9.993
0.1900	4.95	−7.321	12.8	3.389	10.528
0.2000	5.02	−7.524	11.4	3.527	11.159
0.2100	5.08	−7.721	9.8	3.668	11.929

Table II.13

p	ρ'	ρ''	h'	h''	s'	s''
0.0060	145.76	1.37	3.2	25.8	1.738	11.806
0.0070	145.55	1.56	3.3	26.0	1.796	11.629
0.0080	145.31	1.74	3.4	26.3	1.853	11.476
0.0090	145.05	1.93	3.6	26.5	1.906	11.341
0.0100	144.78	2.11	3.7	26.7	1.954	11.221
0.0200	142.13	3.81	4.7	28.1	2.297	10.430
0.0300	139.69	5.42	5.4	28.8	2.525	9.957
0.0400	137.44	7.00	6.1	29.3	2.710	9.613
0.0500	135.30	8.57	6.7	29.7	2.870	9.337
0.0600	133.23	10.15	7.3	29.9	3.014	9.104
0.0700	131.20	11.74	7.9	30.0	3.147	8.900
0.0800	129.19	13.37	8.5	30.1	3.272	8.716
0.0900	127.17	15.03	9.0	30.1	3.390	8.547
0.1000	125.15	16.74	9.6	30.1	3.504	8.389
0.1100	123.09	18.51	10.1	30.0	3.614	8.239
0.1200	120.99	20.34	10.7	29.9	3.723	8.094
0.1300	118.82	22.27	11.3	29.8	3.830	7.953
0.1400	116.58	24.29	11.8	29.6	3.936	7.815
0.1500	114.24	26.43	12.4	29.3	4.044	7.677
0.1600	111.76	28.72	13.0	29.0	4.153	7.537
0.1700	109.12	31.20	13.6	28.7	4.266	7.394
0.1800	106.27	33.92	14.3	28.3	4.384	7.246
0.1900	103.13	36.95	15.0	27.8	4.509	8.089
0.2000	99.59	40.41	15.8	27.2	4.646	6.919
0.2100	95.45	44.51	16.7	26.5	4.801	6.729

Table II.14

p	c_v'	c_v''	c_p'	c_p''	c_s'	c_s''
0.0060	1.720	3.332	1.760	5.795	1.755	−5.990
0.0070	2.011	3.319	2.082	5.815	2.075	−5.865
0.0080	2.123	3.306	2.226	5.835	2.217	−5.760
0.0090	2.155	3.292	2.290	5.855	2.279	−5.674
0.0100	2.153	3.280	2.319	4.877	2.305	−5.601
0.0200	2.035	3.215	2.457	6.150	2.416	−5.270
0.0300	2.090	3.207	2.732	6.488	2.657	−5.197
0.0400	2.175	3.219	3.034	6.858	2.917	−5.209
0.0500	2.253	3.238	3.339	7.253	3.170	−5.271
0.0600	2.319	3.259	3.647	7.676	3.416	−5.371
0.0700	2.375	3.279	3.968	8.131	3.660	−5.503
0.0800	2.422	3.298	4.309	8.626	3.909	−5.666
0.0900	2.464	3.316	4.680	9.173	4.167	−5.863
0.1000	2.502	3.331	5.091	9.783	4.440	−6.097
0.1100	2.536	3.345	5.557	10.474	4.735	−6.371
0.1200	2.567	3.356	6.095	11.270	5.058	−6.694
0.1300	2.597	3.365	6.727	12.201	5.419	−7.074

Table II. 14 (*Continued*)

p	c_v'	c_v''	c_p'	c_p''	c_s'	c_s''
0.1400	2.626	3.371	7.486	13.312	5.829	−7.525
0.1500	2.655	3.375	8.422	14.668	6.305	−8.066
0.1600	2.683	3.377	9.607	16.368	6.870	−8.723
0.1700	2.712	3.376	11.162	18.569	7.559	−9.539
0.1800	2.742	3.371	11.295	21.541	8.429	−10.578
0.1900	2.775	3.363	16.401	25.786	9.578	−11.950
0.2000	2.811	3.348	21.325	32.356	11.196	−13.866
0.2100	2.852	3.327	30.242	43.888	13.708	−16.779

Table II. 15

p	w'	w''	μ'	μ''	k'	k''
0.0060	210.0	84.5	−3.72	20.15	1071.00	1.63
0.0070	210.7	85.6	−3.10	19.35	923.11	1.63
0.0080	211.2	86.5	−2.86	18.73	810.23	1.63
0.0090	211.6	87.3	−2.75	18.23	721.34	1.63
0.0100	211.8	88.1	−2.68	17.82	649.55	1.63
0.0200	210.8	93.0	−2.34	15.65	315.64	1.65
0.0300	206.4	95.6	−1.97	14.53	198.28	1.65
0.0400	201.2	97.3	−1.64	13.73	139.03	1.66
0.0500	195.9	98.4	−1.36	13.09	103.90	1.66
0.0600	190.9	99.2	−1.11	12.57	80.93	1.66
0.0700	186.1	99.7	−0.88	12.12	64.88	1.67
0.0800	181.4	100.0	−0.65	11.74	53.12	1.67
0.0900	176.8	100.1	−0.43	11.40	44.18	1.67
0.1000	172.4	100.2	−0.21	11.09	37.18	1.68
0.1100	168.0	100.2	0.01	10.81	31.58	1.69
0.1200	163.7	100.0	0.23	10.55	27.01	1.70
0.1300	159.4	99.9	0.47	10.30	23.22	1.71
0.1400	155.1	99.7	0.71	10.06	20.04	1.72
0.1500	150.9	99.4	0.97	9.83	17.33	1.74
0.1600	146.6	99.1	1.25	9.59	15.01	1.76
0.1700	142.2	98.8	1.54	9.35	12.99	1.79
0.1800	137.8	98.6	1.86	9.09	11.22	1.83
0.1900	133.3	98.3	2.21	8.80	9.65	1.88
0.2000	128.7	98.2	2.60	8.48	8.24	1.95
0.2100	123.7	98.2	3.05	8.09	6.96	2.04

Table II.16

p	f'	f''	a'/a_0	a''/a_0	γ'/γ_0	γ''/γ_0
0.0060	0.006	0.006	0.045	1.160	47.53	1.086
0.0070	0.007	0.007	0.062	1.175	55.11	1.093
0.0080	0.007	0.007	0.076	1.190	58.58	1.099
0.0090	0.008	0.008	0.088	1.206	59.84	1.106
0.0100	0.009	0.009	0.099	1.221	59.86	1.113

Table II.16 (*Continued*)

p	f'	f''	a'/a_0	a''/a_0	γ'/γ_0	γ''/γ_0
0.0200	0.018	0.018	0.182	1.367	47.48	1.176
0.0300	0.026	0.026	0.249	1.511	37.78	1.235
0.0400	0.034	0.034	0.316	1.659	31.46	1.291
0.0500	0.042	0.042	0.385	1.814	27.02	1.345
0.0600	0.049	0.049	0.461	1.979	23.71	1.398
0.0700	0.056	0.056	0.544	2.158	21.11	1.450
0.0800	0.063	0.063	0.637	2.354	19.01	1.503
0.0900	0.070	0.070	0.742	2.571	17.26	1.556
0.1000	0.077	0.077	0.863	2.816	15.77	1.611
0.1100	0.084	0.084	1.005	3.095	14.48	1.668
0.1200	0.090	0.090	1.173	3.418	13.34	1.727
0.1300	0.097	0.097	1.375	3.798	12.33	1.790
0.1400	0.103	0.103	1.623	4.253	11.41	1.856
0.1500	0.109	0.109	1.935	4.810	10.57	1.927
0.1600	0.115	0.115	2.338	5.510	9.80	2.005
0.1700	0.121	0.121	2.876	6.416	9.08	2.092
0.1800	0.127	0.127	3.626	7.640	8.39	2.189
0.1900	0.133	0.133	4.737	9.386	7.73	2.301
0.2000	0.139	0.139	6.526	12.082	7.09	2.434
0.2100	0.144	0.144	9.818	16.797	6.44	2.601

Thermodynamic properties of He^4 in the single phase region

Table II. 17

p	ρ	z	h	s	c_v	c_p
\multicolumn{7}{c}{$T = 2.2$ K}
0.01	145.97	0.0150	3.1	1.706	1.364	1.387
0.02	146.19	0.0299	3.2	1.705	1.347	1.370
0.03	146.42	0.0448	3.2	1.704	1.331	1.354
0.04	146.64	0.0597	3.3	1.703	1.314	1.338
0.05	146.86	0.0745	3.4	1.702	1.299	1.322
0.06	147.08	0.0893	3.4	1.700	1.283	1.307
0.07	147.30	0.1040	3.5	1.699	1.268	1.292
0.08	147.51	0.1187	3.6	1.698	1.253	1.277
0.09	147.72	0.1333	3.6	1.697	1.238	1.262
0.1	147.93	0.1479	3.7	1.696	1.224	1.248
0.2	149.89	0.2920	4.4	1.686	1.093	1.119
0.3	151.68	0.4328	5.0	1.677	0.983	1.009
0.4	153.33	0.5709	5.6	1.668	0.888	0.914
0.5	154.86	0.7065	6.3	1.659	0.805	0.832
0.6	156.30	0.8400	6.9	1.650	0.732	0.759
0.7	157.66	0.9715	7.5	1.642	0.668	0.694
0.8	158.95	1.1013	8.1	1.634	0.610	0.637
0.9	160.18	1.2295	8.7	1.627	0.558	0.585
1.0	161.36	1.3561	9.3	1.619	0.512	0.538
1.5	166.64	1.9697	12.3	1.585	0.337	0.364
2.0	171.18	2.5566	15.2	1.553	0.229	0.257
2.5	175.24	3.1218	18.0	1.523	0.165	0.198
3.0	178.94	3.6687	20.8	1.492	0.135	0.173
3.5	182.36	4.1998	23.5	1.460	0.132	0.179
4.0	185.57	4.7167	26.1	1.427	0.152	0.211
\multicolumn{7}{c}{$T = 2.3$ K}
0.01	145.67	0.0144	3.3	1.784	1.967	2.032
0.02	145.90	0.0287	3.4	1.782	1.953	2.017
0.03	146.12	0.0430	3.4	1.780	1.939	2.003
0.04	146.35	0.0572	3.5	1.779	1.926	1.989
0.05	146.57	0.0714	3.6	1.777	1.913	1.976
0.06	146.79	0.0856	3.6	1.775	1.900	1.963
0.07	147.01	0.0997	3.7	1.774	1.887	1.950
0.08	147.22	0.1137	3.7	1.772	1.874	1.937
0.09	147.43	0.1278	3.8	1.770	1.862	1.924
0.1	147.64	0.1418	3.9	1.769	1.850	1.912
0.2	149.62	0.2798	4.5	1.753	1.739	1.799
0.3	151.41	0.4147	5.1	1.739	1.645	1.702
0.4	153.07	0.5469	5.8	1.726	1.562	1.617
0.5	154.62	0.6769	6.4	1.714	1.490	1.542
0.6	156.06	0.8047	7.0	1.702	1.425	1.476
0.7	157.43	0.9307	7.6	1.691	1.367	1.416
0.8	158.73	1.0549	8.2	1.681	1.315	1.362
0.9	159.96	1.1776	8.8	1.671	1.267	1.313
1.0	161.15	1.2989	9.4	1.662	1.223	1.268
1.5	166.44	1.8863	12.4	1.619	1.052	1.092

Table II.17 (*Continued*)

p	ρ	z	h	s	c_v	c_p
2.0	170.99	2.4482	15.3	1.583	0.935	0.973
2.5	175.04	2.9894	18.1	1.549	0.854	0.893
3.0	178.73	3.5132	20.8	1.517	0.800	0.843
3.5	182.15	4.0219	23.5	1.484	0.769	0.818
4.0	185.34	4.5174	26.2	1.451	0.758	0.816
4.5	188.34	5.0021	28.8	1.417	0.765	0.834

$T = 2.4$ K

p	ρ	z	h	s	c_v	c_p
0.01	145.23	0.0138	3.5	1.876	2.142	2.258
0.02	145.47	0.0276	3.6	1.874	2.130	2.246
0.03	145.70	0.0413	3.6	1.871	2.118	2.233
0.04	145.92	0.0550	3.7	1.869	2.107	2.221
0.05	146.15	0.0686	3.8	1.867	2.096	2.209
0.06	146.37	0.0822	3.8	1.865	2.085	2.197
0.07	146.59	0.0958	3.9	1.863	2.074	2.185
0.08	146.81	0.1093	4.0	1.860	2.063	2.174
0.09	147.02	0.1228	4.0	1.858	2.053	2.162
0.10	147.24	0.1362	4.1	1.856	2.043	2.151
0.2	149.24	0.2688	4.7	1.836	1.949	2.050
0.3	151.06	0.3984	5.3	1.818	1.859	1.963
0.4	152.74	0.5253	5.9	1.802	1.799	1.887
0.5	154.30	0.6500	6.6	1.786	1.736	1.819
0.6	155.76	0.7727	7.2	1.772	1.680	1.759
0.7	157.14	0.8936	7.8	1.759	1.630	1.704
0.8	158.44	1.0128	8.4	1.746	1.583	1.654
0.9	159.69	1.1305	9.0	1.734	1.541	1.609
1.0	160.88	1.2468	9.6	1.723	1.502	1.567
1.5	166.21	1.8103	12.5	1.673	1.345	1.399
2.0	170.77	2.3491	15.4	1.632	1.231	1.281
2.5	174.83	2.8682	18.2	1.594	1.148	1.196
3.0	178.52	3.3707	20.9	1.560	1.087	1.137
3.5	181.93	3.8589	23.6	1.526	1.045	1.099
4.0	185.10	4.3345	26.3	1.493	1.020	1.081
4.5	188.09	4.7989	28.9	1.458	1.010	1.080
5.0	190.92	5.2531	31.4	1.423	1.015	1.097

$T = 2.5$ K

p	ρ	z	h	s	c_v	c_p
0.01	2.09	0.9217	26.8	11.260	3.270	5.854
0.02	144.93	0.0266	3.8	1.967	2.139	2.312
0.03	145.16	0.0398	3.9	1.964	2.129	2.301
0.04	145.40	0.0530	3.9	1.962	2.120	2.289
0.05	145.63	0.0661	4.0	1.959	2.110	2.278
0.06	145.85	0.0792	4.0	1.956	2.101	2.267
0.07	146.08	0.0923	4.1	1.953	2.092	2.256
0.08	146.30	0.1053	4.2	1.951	2.083	2.246
0.09	146.52	0.1183	4.2	1.948	2.074	2.235
0.10	146.74	0.1312	4.3	1.946	2.065	2.225
0.2	148.78	0.2589	4.9	1.922	1.986	2.132
0.3	150.63	0.3835	5.5	1.900	1.917	2.051
0.4	152.33	0.5056	6.1	1.881	1.857	1.981
0.5	153.91	0.6255	6.7	1.863	1.803	1.919
0.6	155.40	0.7435	7.4	1.846	1.755	1.863
0.7	156.79	0.8597	8.0	1.831	1.711	1.813
0.8	158.11	0.9743	8.6	1.816	1.670	1.767

Table II.17 (*Continued*)

p	ρ	z	h	s	c_v	c_p
0.9	159.37	1.0874	9.1	1.802	1.633	1.724
1.0	160.58	1.1992	9.7	1.790	1.598	1.685
1.5	165.94	1.7406	12.7	1.733	1.457	1.528
2.0	170.54	2.2583	15.5	1.687	1.352	1.414
2.5	174.61	2.7571	18.3	1.646	1.271	1.330
3.0	178.30	3.2400	21.1	1.609	1.209	1.268
3.5	181.70	3.7092	23.7	1.574	1.164	1.224
4.0	184.87	4.1664	26.4	1.539	1.132	1.198
4.5	187.85	4.6129	29.0	1.505	1.114	1.187
5.0	190.66	5.0498	31.5	1.470	1.108	1.190
5.5	193.33	5.4781	34.1	1.435	1.114	1.208

$T = 2.6$ K

p	ρ	z	h	s	c_v	c_p
0.01	1.99	0.9290	27.4	11.487	3.219	5.736
0.02	144.29	0.0257	4.0	2.058	2.100	2.334
0.03	144.54	0.0384	4.1	2.055	2.091	2.323
0.04	144.78	0.0512	4.2	2.052	2.083	2.312
0.05	145.01	0.0638	4.2	2.049	2.075	2.301
0.06	145.25	0.0765	4.3	2.046	2.067	2.290
0.07	145.48	0.0891	4.3	2.042	2.059	2.280
0.08	145.70	0.1017	4.4	2.039	2.051	2.270
0.09	145.93	0.1142	4.5	2.037	2.044	2.260
0.1	146.15	0.1267	4.5	2.034	2.036	2.250
0.2	148.24	0.2498	5.1	2.006	1.967	2.160
0.3	150.13	0.3700	5.7	1.982	1.908	2.084
0.4	151.87	0.4877	6.3	1.959	1.856	2.018
0.5	153.48	0.6032	6.9	1.939	1.809	1.959
0.6	154.98	0.7168	7.5	1.920	1.767	1.906
0.7	156.40	0.8287	8.1	1.903	1.728	1.859
0.8	157.74	0.9390	8.7	1.886	1.692	1.815
0.9	159.02	1.0479	9.3	1.871	1.659	1.776
1.0	160.23	1.1555	9.9	1.857	1.629	1.739
1.5	165.65	1.6766	12.8	1.795	1.501	1.590
2.0	170.27	2.1748	15.7	1.744	1.404	1.480
2.5	174.36	2.6547	18.4	1.700	1.327	1.397
3.0	178.07	3.1195	21.2	1.660	1.267	1.334
3.5	181.47	3.5711	23.9	1.623	1.220	1.288
4.0	184.64	4.0113	26.5	1.588	1.185	1.257
4.5	187.61	4.4412	29.1	1.553	1.162	1.239
5.0	190.41	4.8621	31.7	1.518	1.150	1.234
5.5	193.07	5.2747	34.2	1.483	1.147	1.241
6.0	195.59	5.6798	36.7	1.447	1.155	1.260

$T = 2.8$ K

p	ρ	z	h	s	c_v	c_p
0.01	1.83	0.9407	28.5	11.906	3.130	5.585
0.02	142.77	0.0241	4.5	2.233	2.038	2.405
0.03	143.03	0.0361	4.6	2.229	2.032	2.394
0.04	143.29	0.0480	4.6	2.225	2.025	2.383
0.05	143.55	0.0599	4.7	2.221	2.019	2.371
0.06	143.80	0.0717	4.7	2.217	2.012	2.360
0.07	144.04	0.0836	4.8	2.214	2.006	2.349
0.08	144.29	0.0953	4.9	2.210	2.000	2.339
0.09	144.53	0.1071	4.9	2.206	1.994	2.328
0.1	144.76	0.1188	5.0	2.202	1.988	2.318

Table II.17 (*Continued*)

p	ρ	z	h	s	c_v	c_p
0.2	146.98	0.2339	5.6	2.168	1.933	2.227
0.3	148.97	0.3462	6.2	2.138	1.886	2.152
0.4	150.79	0.4561	6.7	2.111	1.845	2.087
0.5	152.47	0.5638	7.3	2.087	1.807	2.030
0.6	154.03	0.6697	7.9	2.064	1.773	1.979
0.7	155.50	0.7740	8.5	2.043	1.742	1.934
0.8	156.89	0.8767	9.1	2.024	1.713	1.893
0.9	158.20	0.9781	9.7	2.006	1.686	1.855
1.0	159.45	1.0783	10.3	1.988	1.660	1.820
1.5	164.99	1.5631	13.1	1.916	1.552	1.678
2.0	169.69	2.0264	16.0	1.857	1.467	1.573
2.5	173.83	2.4727	18.7	1.807	1.396	1.491
3.0	177.56	2.9049	21.5	1.762	1.338	1.427
3.5	180.97	3.3251	24.1	1.722	1.290	1.376
4.0	184.14	3.7347	26.8	1.684	1.252	1.338
4.5	187.11	4.1350	29.4	1.647	1.223	1.311
5.0	189.90	4.5268	31.9	1.611	1.201	1.294
5.5	192.54	4.9112	34.4	1.576	1.188	1.287
6.0	195.05	5.2887	36.9	1.541	1.183	1.289
6.5	197.44	5.6601	39.4	1.506	1.184	1.299

$T = 3.0$ K

p	ρ	z	h	s	c_v	c_p
0.01	1.69	0.9497	29.6	12.289	3.135	5.499
0.02	3.60	0.8921	28.8	10.689	3.179	5.990
0.03	141.22	0.0341	5.1	2.399	2.045	2.558
0.04	141.50	0.0454	5.1	2.395	2.039	2.544
0.05	141.78	0.0566	5.2	2.390	2.034	2.531
0.06	142.05	0.0678	5.2	2.385	2.028	2.518
0.07	142.32	0.0789	5.3	2.380	2.023	2.505
0.08	142.59	0.0900	5.3	2.376	2.017	2.493
0.09	142.85	0.1011	5.4	2.371	2.012	2.481
0.1	143.10	0.1121	5.5	2.367	2.007	2.469
0.2	145.49	0.2206	6.0	2.326	1.960	2.366
0.3	147.61	0.3261	6.6	2.291	1.920	2.283
0.4	149.53	0.4292	7.2	2.259	1.884	2.213
0.5	151.30	0.5303	7.8	2.230	1.852	2.153
0.6	152.93	0.6296	8.3	2.204	1.822	2.100
0.7	154.46	0.7272	8.9	2.180	1.795	2.053
0.8	155.90	0.8234	9.5	2.158	1.769	2.010
0.9	157.26	0.9184	10.1	2.137	1.745	1.971
1.0	158.55	1.0121	10.6	2.118	1.723	1.936
1.5	164.25	1.4655	13.5	2.035	1.625	1.792
2.0	169.04	1.8986	16.3	1.969	1.545	1.684
2.5	173.23	2.3158	19.0	1.913	1.477	1.600
3.0	177.00	2.7198	21.8	1.864	1.420	1.532
3.5	180.44	3.1125	24.4	1.820	1.370	1.477
4.0	183.62	3.4956	27.0	1.779	1.329	1.433
4.5	186.59	3.8699	29.6	1.740	1.294	1.398
5.0	189.38	4.2366	32.2	1.703	1.267	1.372
5.5	192.02	4.5962	34.7	1.667	1.245	1.355
6.0	194.52	4.9496	37.2	1.632	1.230	1.344
6.5	196.90	5.2972	39.6	1.597	1.221	1.341
7.0	199.18	5.6396	42.0	1.563	1.218	1.343
7.5	201.35	5.9773	44.4	1.529	1.219	1.352
8.0	203.43	6.3107	46.8	1.495	1.226	1.365

Table II.17 *(Continued)*

p	ρ	z	h	s	c_v	c_p
			$T = 3.2$ K			
0.01	1.57	0.9568	30.7	12.642	3.125	5.445
0.02	3.31	0.9085	30.0	11.070	3.150	5.819
0.03	5.29	0.8535	29.2	10.066	3.195	6.349
0.04	139.40	0.0432	5.6	2.566	2.105	2.787
0.05	139.71	0.0538	5.7	2.560	2.099	2.770
0.06	140.01	0.0645	5.8	2.555	2.094	2.753
0.07	140.31	0.0751	5.8	2.549	2.089	2.737
0.08	140.60	0.0856	5.9	2.544	2.084	2.721
0.09	140.89	0.0961	5.9	2.538	2.079	2.706
0.1	141.17	0.1066	6.0	2.533	2.074	2.692
0.2	143.77	0.2093	6.5	2.485	2.031	2.566
0.3	146.06	0.3090	7.1	2.444	1.993	2.468
0.4	148.11	0.4063	7.6	2.407	1.960	2.388
0.5	149.97	0.5015	8.2	2.375	1.930	2.320
0.6	151.70	0.5950	8.8	2.345	1.903	2.261
0.7	153.30	0.6869	9.3	2.318	1.878	2.209
0.8	154.80	0.7775	9.9	2.293	1.854	2.163
0.9	156.21	0.8667	10.5	2.269	1.832	2.121
1.0	157.55	0.9548	11.0	2.247	1.810	2.082
1.5	163.42	1.3809	13.9	2.155	1.717	1.929
2.0	168.32	1.7876	16.6	2.082	1.640	1.815
2.5	172.58	2.1792	19.4	2.020	1.573	1.726
3.0	176.40	2.5585	22.1	1.967	1.514	1.654
3.5	179.87	2.9273	24.7	1.919	1.463	1.593
4.0	183.07	3.2870	27.3	1.875	1.418	1.544
4.5	186.05	3.6386	29.9	1.834	1.380	1.503
5.0	188.85	3.9830	32.5	1.795	1.347	1.470
5.5	191.49	4.3209	35.0	1.757	1.320	1.444
6.0	193.99	4.6530	37.4	1.721	1.298	1.424
6.5	196.37	4.9797	39.9	1.686	1.281	1.411
7.0	198.63	5.3016	42.3	1.651	1.268	1.402
7.5	200.80	5.6190	44.7	1.617	1.260	1.399
8.0	202.87	5.9323	47.1	1.584	1.256	1.399
8.5	204.86	6.2420	49.4	1.551	1.256	1.403
9.0	206.77	6.5482	51.8	1.518	1.259	1.410
			$T = 3.4$ K			
0.01	1.47	0.9625	31.8	12.971	3.122	5.408
0.02	3.07	0.9214	31.2	11.419	3.139	5.707
0.03	4.85	0.8757	30.4	10.445	3.168	6.132
0.04	6.87	0.8239	29.6	9.689	3.211	6.770
0.05	137.29	0.0516	6.3	2.737	2.187	3.075
0.06	137.64	0.0617	6.3	2.730	2.181	3.052
0.07	137.97	0.0718	6.4	2.724	2.176	3.030
0.08	138.30	0.0819	6.4	2.717	2.171	3.009
0.09	138.62	0.0919	6.5	2.711	2.166	2.989
0.10	138.94	0.1019	6.5	2.704	2.162	2.969
0.2	141.82	0.1997	7.1	2.648	2.119	2.807
0.3	144.31	0.2944	7.6	2.600	2.082	2.686
0.4	146.51	0.3866	8.1	2.558	2.050	2.590
0.5	148.50	0.4767	8.7	2.521	2.021	2.510
0.6	150.32	0.5651	9.2	2.487	1.995	2.443
0.7	152.01	0.6520	9.8	2.457	1.970	2.384

Table II.17 (*Continued*)

p	ρ	z	h	s	c_v	c_p
0.8	153.58	0.7375	10.4	2.429	1.947	2.332
0.9	155.06	0.8218	10.9	2.403	1.926	2.285
1.0	156.45	0.9050	11.5	2.378	1.905	2.243
1.5	162.51	1.3069	14.3	2.276	1.814	2.076
2.0	167.53	1.6903	17.0	2.196	1.737	1.955
2.5	171.88	2.0595	19.7	2.129	1.670	1.859
3.0	175.75	2.4169	22.4	2.071	1.611	1.781
3.5	179.26	2.7645	25.1	2.019	1.558	1.716
4.0	182.49	3.1036	27.7	1.976	1.511	1.661
4.5	185.49	3.4351	30.2	1.928	1.470	1.615
5.0	188.29	3.7598	32.8	1.887	1.433	1.576
5.5	190.94	4.0785	35.3	1.848	1.402	1.544
6.0	193.44	4.3916	37.7	1.810	1.375	1.517
6.5	195.82	4.6998	40.2	1.774	1.352	1.496
7.0	198.09	5.0034	42.6	1.738	1.333	1.480
7.5	200.26	5.3028	45.0	1.704	1.318	1.468
8.0	202.33	5.5984	47.4	1.670	1.307	1.459
8.5	204.32	5.8905	49.7	1.637	1.298	1.454
9.0	206.22	6.1793	52.0	1.605	1.293	1.451
9.5	208.05	6.4652	54.3	1.573	1.290	1.450
10.0	209.81	6.7484	56.6	1.543	1.290	1.451

$T = 3.5$ K

p	ρ	z	h	s	c_v	c_p
0.01	1.43	0.9650	32.3	13.127	3.122	5.394
0.02	2.97	0.9268	31.7	11.584	3.136	5.665
0.03	4.66	0.8849	31.0	10.621	3.160	6.038
0.04	6.57	0.8380	30.3	9.882	3.196	6.574
0.05	135.93	0.0506	6.6	2.829	2.232	3.253
0.06	136.30	0.0605	6.6	2.821	2.227	3.226
0.07	136.66	0.0705	6.7	2.814	2.222	3.200
0.08	137.01	0.0803	6.7	2.807	2.217	3.175
0.09	137.36	0.0901	6.8	2.800	2.212	3.151
0.1	137.70	0.0999	6.8	2.793	2.207	3.128
0.2	140.75	0.1955	7.3	2.731	2.164	2.941
0.3	143.35	0.2879	7.9	2.679	2.127	2.805
0.4	145.64	0.3778	8.4	2.635	2.095	2.698
0.5	147.70	0.4656	8.9	2.595	2.067	2.611
0.6	149.58	0.5517	9.5	2.559	2.040	2.538
0.7	151.32	0.6363	10.0	2.527	2.016	2.475
0.8	152.93	0.7195	10.6	2.497	1.993	2.419
0.9	154.44	0.8015	11.2	2.470	1.972	2.370
1.0	155.87	0.8824	11.7	2.445	1.951	2.325
1.5	162.03	1.2733	14.5	2.338	1.861	2.150
2.0	167.12	1.6461	17.2	2.254	1.784	2.024
2.5	171.51	2.0050	19.9	2.184	1.717	1.925
3.0	175.40	2.3525	22.6	2.124	1.657	1.844
3.5	178.94	2.6904	25.2	2.070	1.604	1.777
4.0	182.18	3.0200	27.8	2.021	1.556	1.719
4.5	185.19	3.3422	30.4	1.976	1.514	1.671
5.0	188.01	3.6580	32.9	1.934	1.476	1.630
5.5	190.66	3.9678	35.4	1.893	1.443	1.595
6.0	193.16	4.2723	37.9	1.855	1.414	1.566
6.5	195.55	4.5720	40.3	1.818	1.389	1.542
7.0	197.82	4.8672	42.8	1.782	1.368	1.523
7.5	199.98	5.1584	45.1	1.747	1.350	1.507

Table II.17 (*Continued*)

p	ρ	z	h	s	c_v	c_p
8.0	202.06	5.4458	47.5	1.713	1.336	1.495
8.5	204.04	5.7298	49.9	1.680	1.325	1.486
9.0	205.95	6.0107	52.2	1.647	1.316	1.479
9.5	207.78	6.2886	54.5	1.616	1.310	1.475
10.0	209.54	6.5640	56.8	1.585	1.306	1.472

$T = 3.6$ K

p	ρ	z	h	s	c_v	c_p
0.01	1.38	0.9672	32.9	13.279	3.121	5.381
0.02	2.87	0.9317	32.3	11.743	3.134	5.628
0.03	4.49	0.8931	31.6	10.790	3.155	5.960
0.04	6.29	0.8503	30.9	10.065	3.185	6.419
0.05	8.34	0.8021	30.1	9.449	3.227	7.089
0.06	134.85	0.0595	7.0	2.915	2.273	3.420
0.07	135.24	0.0692	7.0	2.907	2.267	3.388
0.08	135.62	0.0789	7.1	2.899	2.262	3.358
0.09	135.99	0.0885	7.1	2.891	2.257	3.330
0.1	136.36	0.0981	7.2	2.883	2.252	3.302
0.2	139.60	0.1916	7.6	2.816	2.208	3.083
0.3	142.34	0.2818	8.2	2.760	2.171	2.928
0.4	144.73	0.3696	8.7	2.712	2.139	2.810
0.5	146.87	0.4553	9.2	2.670	2.110	2.714
0.6	148.81	0.5392	9.7	2.632	2.084	2.635
0.7	150.59	0.6216	10.3	2.598	2.060	2.567
0.8	152.25	0.7027	10.8	2.567	2.037	2.507
0.9	153.80	0.7825	11.4	2.538	2.016	2.454
1.0	155.25	0.8613	11.9	2.511	1.995	2.407
1.5	161.53	1.2418	14.7	2.399	1.906	2.223
2.0	166.69	1.6045	17.4	2.312	1.829	2.092
2.5	171.12	1.9537	20.1	2.239	1.762	1.990
3.0	175.05	2.2918	22.8	2.176	1.702	1.906
3.5	178.60	2.6205	25.4	2.121	1.648	1.837
4.0	181.86	2.9412	28.0	2.070	1.600	1.777
4.5	184.89	3.2547	30.6	2.024	1.557	1.727
5.0	187.71	3.5619	33.1	1.980	1.518	1.684
5.5	190.37	3.8635	35.6	1.939	1.483	1.647
6.0	192.88	4.1598	38.1	1.900	1.453	1.616
6.5	195.27	4.4514	40.5	1.862	1.427	1.589
7.0	197.54	4.7387	42.9	1.825	1.404	1.567
7.5	199.70	5.0220	45.3	1.790	1.384	1.549
8.0	201.78	5.3018	47.7	1.756	1.367	1.534
8.5	203.77	5.5781	50.0	1.722	1.354	1.521
9.0	205.68	5.8514	52.3	1.690	1.343	1.512
9.5	207.51	6.1219	54.6	1.658	1.334	1.504
10.0	209.28	6.3898	56.9	1.627	1.327	1.498

$T = 3.8$ K

p	ρ	z	h	s	c_v	c_p
0.01	1.30	0.9711	34.0	13.569	3.122	5.359
0.02	2.70	0.9401	33.4	12.045	3.132	5.569
0.03	4.19	0.9069	32.8	11.109	3.149	5.838
0.04	5.82	0.8709	32.2	10.406	3.172	6.189
0.05	7.62	0.8313	31.4	9.820	3.202	6.662
0.06	9.66	0.7870	30.6	9.298	3.240	7.334
0.07	132.01	0.0672	7.7	3.101	2.355	3.846
0.08	132.46	0.0765	7.8	3.091	2.350	3.799

Table II.17 (*Continued*)

p	ρ	z	h	s	c_v	c_p
0.09	132.90	0.0858	7.8	3.082	2.344	3.755
0.1	133.33	0.0950	7.9	3.072	2.339	3.713
0.2	137.07	0.1849	8.3	2.991	2.291	3.401
0.3	140.13	0.2712	8.8	2.925	2.253	3.196
0.4	142.75	0.3550	9.3	2.870	2.221	3.046
0.5	145.07	0.4366	9.8	2.822	2.192	2.929
0.6	147.15	0.5166	10.3	2.780	2.165	2.833
0.7	149.05	0.5950	10.8	2.742	2.141	2.753
0.8	150.80	0.6721	11.4	2.707	2.118	2.684
0.9	152.43	0.7480	11.9	2.675	2.097	2.624
1.0	153.95	0.8229	12.4	2.646	2.077	2.571
1.5	160.48	1.1841	15.2	2.523	1.988	2.367
2.0	165.78	1.5284	17.9	2.428	1.912	2.225
2.5	170.31	1.8596	20.5	2.350	1.845	2.116
3.0	174.31	2.1804	23.2	2.283	1.785	2.027
3.5	177.91	2.4923	25.8	2.223	1.731	1.953
4.0	181.21	2.7965	28.4	2.170	1.682	1.891
4.5	184.26	3.0940	30.9	2.120	1.638	1.837
5.0	187.10	3.3855	33.4	2.074	1.598	1.790
5.5	189.77	3.6716	35.9	2.031	1.562	1.750
6.0	192.30	3.9528	38.4	1.990	1.530	1.715
6.5	194.69	4.2296	40.8	1.950	1.501	1.684
7.0	196.97	4.5022	43.2	1.913	1.476	1.658
7.5	199.14	4.7712	45.6	1.876	1.453	1.636
8.0	201.22	5.0366	48.0	1.841	1.433	1.616
8.5	203.22	5.2989	50.3	1.807	1.416	1.600
9.0	205.13	5.5583	52.6	1.773	1.401	1.585
9.5	206.97	5.8149	54.9	1.741	1.388	1.572
10.0	208.74	6.0691	57.2	1.709	1.377	1.561

$T = 4.0$ K

p	ρ	z	h	s	c_v	c_p
0.01	1.24	0.9743	35.0	13.844	3.122	5.341
0.02	2.54	0.9471	34.5	12.330	3.132	5.523
0.03	3.93	0.9182	34.0	11.406	3.145	5.746
0.04	5.43	0.8873	33.4	10.719	3.163	6.026
0.05	7.05	0.8540	32.7	10.154	3.186	6.382
0.06	8.83	0.8176	32.0	9.661	3.214	6.852
0.07	10.84	0.7773	31.2	9.209	3.249	7.499
0.08	13.16	0.7316	30.3	8.773	3.292	8.462
0.09	129.17	0.0839	8.6	3.288	2.427	4.329
0.1	129.70	0.0928	8.7	3.276	2.421	4.260
0.2	134.15	0.1794	9.0	3.175	2.368	3.781
0.3	137.64	0.2623	9.4	3.097	2.327	3.497
0.4	140.55	0.3425	9.9	3.033	2.293	3.302
0.5	143.08	0.4206	10.4	2.978	2.264	3.156
0.6	145.33	0.4969	10.9	2.931	2.237	3.040
0.7	147.36	0.5717	11.4	2.888	2.213	2.945
0.8	149.22	0.6452	11.9	2.849	2.190	2.864
0.9	150.94	0.7176	12.4	2.814	2.169	2.795
1.0	152.55	0.7889	13.0	2.782	2.149	2.734
1.5	159.35	1.1329	15.6	2.648	2.060	2.507
2.0	164.81	1.4605	18.3	2.546	1.985	2.353
2.5	169.45	1.7757	21.0	2.462	1.920	2.237
3.0	173.52	2.0808	23.6	2.390	1.860	2.143
3.5	177.18	2.3774	26.2	2.326	1.807	2.065
4.0	180.51	2.6668	28.8	2.269	1.758	2.000

Table II.17 (*Continued*)

p	ρ	z	h	s	c_v	c_p
4.5	183.59	2.9499	31.3	2.217	1.714	1.943
5.0	185.46	3.2272	33.8	2.169	1.674	1.894
5.5	189.15	3.4994	36.3	2.123	1.637	1.851
6.0	191.69	3.7670	38.7	2.080	1.604	1.813
6.5	194.10	4.0304	41.2	2.039	1.574	1.780
7.0	196.38	4.2899	43.6	2.000	1.547	1.751
7.5	198.57	4.5458	46.0	1.962	1.523	1.726
8.0	200.65	4.7984	48.3	1.926	1.501	1.703
8.5	202.65	5.0480	50.7	1.891	1.482	1.683
9.0	204.57	5.2948	53.0	1.857	1.464	1.666
9.5	206.42	5.5390	55.3	1.823	1.449	1.650
10.0	208.19	5.7807	57.6	1.791	1.435	1.635
			$T = 4.2$ K			
0.01	1.17	0.9771	36.1	14.104	3.122	5.326
0.02	2.41	0.9530	35.6	12.598	3.131	5.485
0.03	3.71	0.9276	35.1	11.685	3.143	5.675
0.04	5.09	0.9008	34.6	11.010	3.157	5.904
0.05	6.57	0.8722	34.0	10.461	3.175	6.185
0.06	8.17	0.8415	33.4	9.988	3.197	6.537
0.07	9.93	0.8083	32.7	9.561	3.223	6.988
0.08	11.88	0.7719	31.9	9.163	3.253	7.592
0.09	14.11	0.7312	31.1	8.777	3.289	8.450
0.1	125.18	0.0916	9.6	3.502	2.501	5.084
0.2	130.74	0.1753	9.8	3.370	2.439	4.263
0.3	134.81	0.2551	10.2	3.276	2.395	3.851
0.4	138.10	0.3320	10.6	3.201	2.359	3.589
0.5	140.89	0.4068	11.0	3.138	2.328	3.402
0.6	143.34	0.4798	11.5	3.084	2.301	3.259
0.7	145.53	0.5513	12.0	3.036	2.276	3.145
0.8	147.52	0.6216	12.5	2.994	2.254	3.050
0.9	149.35	0.6907	13.0	2.955	2.232	2.969
1.0	151.04	0.7589	13.5	2.919	2.212	2.899
1.5	158.14	1.0872	16.2	2.774	2.125	2.645
2.0	163.78	1.3997	18.8	2.664	2.051	2.478
2.5	168.53	1.7003	21.4	2.574	1.987	2.354
3.0	172.69	1.9912	24.0	2.497	1.929	2.255
3.5	176.41	2.2741	26.6	2.430	1.876	2.174
4.0	179.79	2.5501	29.2	2.369	1.828	2.105
4.5	182.90	2.8200	31.7	2.314	1.784	2.046
5.0	185.80	3.0845	34.2	2.263	1.744	1.995
5.5	188.51	3.3442	36.7	2.216	1.708	1.950
6.0	191.07	3.5994	39.1	2.171	1.675	1.910
6.5	193.48	3.8506	41.5	2.128	1.644	1.875
7.0	195.78	4.0981	43.9	2.088	1.617	1.844
7.5	197.97	4.3423	46.3	2.049	1.592	1.817
8.0	200.07	4.5832	48.7	2.011	1.569	1.792
8.5	202.08	4.8213	51.0	1.975	1.549	1.770
9.0	204.00	5.0567	53.3	1.940	1.530	1.750
9.5	205.86	5.2896	55.6	1.906	1.513	1.732
10.0	207.64	5.5201	57.9	1.873	1.498	1.715
			$T = 4.4$ K			
0.01	1.12	0.9794	37.2	14.351	3.122	5.313

Table II.17 (Continued)

p	ρ	z	h	s	c_v	c_p
0.02	2.28	0.9579	36.7	12.853	3.130	5.453
0.03	3.51	0.9355	36.2	11.947	3.140	5.617
0.04	4.80	0.9119	35.7	11.282	3.153	5.810
0.05	6.17	0.8870	35.2	10.745	3.167	6.038
0.06	7.63	0.8607	34.7	10.283	3.185	6.314
0.07	9.20	0.8327	34.1	9.878	3.204	6.652
0.08	10.91	0.8025	33.4	9.503	3.227	7.076
0.09	12.79	0.7698	32.7	9.149	3.253	7.625
0.10	14.91	0.7337	31.9	8.804	3.282	8.373
0.2	126.65	0.1728	10.7	3.583	2.507	4.931
0.3	131.57	0.2495	11.0	3.465	2.457	4.286
0.4	135.35	0.3233	11.3	3.375	2.418	3.921
0.5	138.47	0.3951	11.7	3.303	2.383	3.677
0.6	141.16	0.4650	12.2	3.241	2.358	3.497
0.7	143.54	0.5336	12.7	3.188	2.333	3.357
0.8	145.67	0.6008	13.1	3.140	2.310	3.244
0.9	147.62	0.6670	13.6	3.097	2.289	3.149
1.0	149.42	0.7322	14.1	3.058	2.269	3.068
1.5	156.86	1.0462	16.7	2.900	2.183	2.783
2.0	162.69	1.3450	19.3	2.782	2.111	2.601
2.5	167.57	1.6323	21.9	2.683	2.048	2.468
3.0	171.82	1.9104	24.5	2.604	1.991	2.364
3.5	175.60	2.1808	27.1	2.533	1.940	2.279
4.0	179.03	2.4445	29.6	2.470	1.893	2.208
4.5	182.18	2.7025	32.1	2.412	1.851	2.146
5.0	185.11	2.9553	34.6	2.359	1.811	2.093
5.5	187.84	3.2035	37.1	2.309	1.775	2.047
6.0	190.42	3.4475	39.5	2.262	1.743	2.006
6.5	192.85	3.6876	41.9	2.218	1.713	1.970
7.0	195.16	3.9243	44.3	2.176	1.685	1.937
7.5	197.37	4.1576	46.7	2.135	1.660	1.909
8.0	199.47	4.3880	49.0	2.097	1.637	1.883
8.5	201.49	4.6156	51.4	2.059	1.616	1.859
9.0	203.42	4.8406	53.7	2.023	1.597	1.838
9.5	205.28	5.0532	56.0	1.988	1.579	1.818
10.0	207.08	5.2836	58.2	1.955	3.563	1.800
15.0	222.01	7.3923	80.3	1.671	1.450	1.642

$T = 4.5$ K

p	ρ	z	h	s	c_v	c_p
0.01	1.09	0.9805	37.7	14.471	3.122	5.308
0.02	2.23	0.9601	37.3	12.975	3.130	5.439
0.03	3.42	0.9389	36.8	12.073	3.139	5.592
0.04	4.67	0.9168	36.3	11.412	3.150	5.770
0.05	5.99	0.8935	35.8	10.880	3.164	5.978
0.06	7.39	0.8690	35.3	10.427	3.179	6.226
0.07	8.88	0.8430	34.7	10.026	3.197	6.523
0.08	10.50	0.8153	34.1	9.660	3.217	6.888
0.09	12.26	0.7855	33.4	9.317	3.239	7.347
0.1	14.20	0.7532	32.7	8.987	3.264	7.944
0.2	124.25	0.1722	11.2	3.699	2.542	5.388
0.3	129.75	0.2473	11.4	3.564	2.487	4.550
0.4	133.85	0.3197	11.7	3.466	2.446	4.111
0.5	137.17	0.3900	12.1	3.387	2.413	3.828
0.6	140.00	0.4585	12.5	3.321	2.385	3.625

Table II.17 (Continued)

p	ρ	z	h	s	c_v	c_p
0.7	142.48	0.5256	13.0	3.264	2.360	3.470
0.8	144.70	0.5915	13.5	3.214	2.337	3.346
0.9	146.72	0.6562	13.9	3.169	2.315	3.243
1.0	148.57	0.7201	14.4	3.128	2.296	3.155
1.5	156.19	1.0274	17.0	2.964	2.210	2.852
2.0	162.13	1.3197	19.6	2.841	2.139	2.662
2.5	167.07	1.6008	22.2	2.742	2.077	2.525
3.0	171.36	1.8728	24.7	2.658	2.021	2.418
3.5	175.18	2.1374	27.3	2.585	1.970	2.331
4.0	178.64	2.3954	29.8	2.520	1.924	2.258
4.5	181.81	2.6478	32.3	2.461	1.882	2.196
5.0	184.75	2.8952	34.8	2.406	1.844	2.142
5.5	187.50	3.1380	37.3	2.355	1.808	2.095
6.0	190.09	3.3768	39.7	2.308	1.776	2.053
6.5	192.53	3.6117	42.1	2.263	1.746	2.016
7.0	194.85	3.8433	44.5	2.220	1.718	1.984
7.5	197.06	4.0716	46.9	2.179	1.693	1.954
8.0	199.17	4.2971	49.2	2.139	1.671	1.928
8.5	201.19	4.5198	51.5	2.102	1.650	1.904
9.0	203.13	4.7399	53.8	2.065	1.630	1.882
9.5	204.99	4.9577	56.1	2.030	1.613	1.862
10.0	206.79	5.1733	58.4	1.996	1.597	1.843
15.0	221.77	7.2357	80.4	1.708	1.480	1.682

$T = 4.6$ K

p	ρ	z	h	s	c_v	c_p
0.01	1.07	0.9814	38.2	14.587	3.122	5.302
0.02	2.18	0.9622	37.8	13.095	3.129	5.427
0.03	3.33	0.9421	37.4	12.196	3.138	5.569
0.04	4.54	0.9213	36.9	11.538	3.148	5.734
0.05	5.82	0.8994	36.4	11.011	3.161	5.925
0.06	7.16	0.8765	35.9	10.563	3.175	6.148
0.07	8.59	0.8523	35.4	10.168	3.190	6.413
0.08	10.13	0.8268	34.8	9.810	3.208	6.732
0.09	11.78	0.7995	34.2	9.476	3.228	7.122
0.1	13.59	0.7701	33.5	9.158	3.249	7.614
0.2	121.52	0.1722	11.8	3.823	2.577	5.986
0.3	127.78	0.2457	11.9	3.667	2.516	4.856
0.4	132.24	0.3166	12.2	3.558	2.473	4.321
0.5	135.79	0.3854	12.5	3.473	2.439	3.991
0.6	138.77	0.4525	12.9	3.402	2.410	3.761
0.7	141.37	0.5182	13.3	3.342	2.385	3.588
0.8	143.68	0.5827	13.8	3.289	2.362	3.451
0.9	145.77	0.6461	14.3	3.241	2.341	3.339
1.0	147.69	0.7086	14.8	3.198	2.321	3.244
1.5	155.51	1.0095	17.3	3.027	2.236	2.922
2.0	161.55	1.2956	19.8	2.900	2.166	2.723
2.5	166.56	1.5708	22.4	2.798	2.104	2.581
3.0	170.90	1.8371	25.0	2.712	2.050	2.471
3.5	174.75	2.0960	27.5	2.637	2.000	2.382
4.0	178.24	2.3486	30.1	2.570	1.955	2.308
4.5	181.43	2.5957	32.6	2.509	1.913	2.245
5.0	184.39	2.8378	35.0	2.454	1.875	2.190
5.5	187.15	3.0755	37.5	2.402	1.840	2.142
6.0	189.75	3.3092	39.9	2.353	1.808	2.100
6.5	192.20	3.5393	42.3	2.307	1.778	2.063

Table II.17 (Continued)

p	ρ	z	h	s	c_v	c_p
7.00	194.53	3.7659	44.7	2.264	1.751	2.030
7.5	196.74	3.9895	47.1	2.222	1.727	2.000
8.0	198.86	4.2101	49.4	2.182	1.704	1.973
8.5	200.89	4.4282	51.7	2.144	1.683	1.948
9.0	202.83	4.6437	54.0	2.107	1.664	1.926
9.5	204.70	4.8569	56.3	2.071	1.646	1.905
10.0	206.50	5.0679	58.6	2.037	1.630	1.886
15.5	221.54	7.0860	80.6	1.746	1.512	1.723

$T = 4.8$ K

p	ρ	z	h	s	c_v	c_p
0.01	1.02	0.9832	39.3	14.813	3.121	5.292
0.02	2.08	0.9658	38.9	13.325	3.128	5.404
0.03	3.17	0.9479	38.5	12.432	3.136	5.530
0.04	4.32	0.9292	38.0	11.781	3.145	5.672
0.05	5.51	0.9099	37.6	11.261	3.155	5.834
0.06	6.76	0.8897	37.1	10.822	3.166	6.020
0.07	8.08	0.8686	36.6	10.437	3.179	6.235
0.08	9.48	0.8465	36.1	10.091	3.193	6.485
0.09	10.96	0.8232	35.6	9.771	3.209	6.781
0.1	12.56	0.7986	35.0	9.472	3.225	7.136
0.2	114.47	0.1752	13.2	4.117	2.658	8.122
0.3	123.23	0.2442	12.9	3.890	2.576	5.665
0.4	128.68	0.3118	13.1	3.752	2.526	4.822
0.5	132.79	0.3776	13.3	3.650	2.490	4.361
0.6	136.15	0.4420	13.7	3.569	2.459	4.060
0.7	139.01	0.5050	14.1	3.500	2.433	3.842
0.8	141.53	0.5669	14.5	3.440	2.410	3.675
0.9	143.78	0.6278	15.0	3.387	2.388	3.541
1.0	145.83	0.6877	15.4	3.340	2.369	3.430
1.5	154.07	0.9764	17.9	3.154	2.285	3.064
2.0	160.34	1.2510	20.4	3.019	2.217	2.846
2.5	165.51	1.5149	22.9	2.910	2.158	2.693
3.0	169.95	1.7704	25.5	2.819	2.104	2.577
3.5	173.88	2.0188	28.0	2.741	2.056	2.484
4.0	177.42	2.2612	30.5	2.670	2.013	2.407
4.5	180.65	2.4982	33.0	2.607	1.973	2.342
5.0	183.65	2.7306	35.5	2.549	1.936	2.286
5.5	186.44	2.9587	37.9	2.495	1.902	2.236
6.0	189.06	3.1829	40.3	2.445	1.871	2.193
6.5	191.53	3.4036	42.7	2.397	1.842	2.155
7.0	193.87	3.6212	45.1	2.352	1.816	2.121
7.5	196.10	3.8357	47.5	2.309	1.792	2.090
8.0	198.23	4.0475	49.8	2.268	1.769	2.062
8.5	200.27	4.2567	52.1	2.229	1.749	2.037
9.0	202.23	4.4635	54.4	2.191	1.730	2.014
9.5	204.11	4.6681	56.7	2.154	1.712	1.993
10.0	205.92	4.8706	59.0	2.119	1.696	1.973
15.0	221.05	6.8057	81.0	1.821	1.578	1.808

$T = 5.0$ K

p	ρ	z	h	s	c_v	c_p
0.01	0.98	0.9847	40.3	15.029	3.121	5.284
0.02	1.99	0.9690	40.0	13.545	3.127	5.384
0.03	3.03	0.9528	39.6	12.657	3.134	5.496
0.04	4.11	0.9361	39.2	12.012	3.141	5.621

Table II.17 (*Continued*)

p	ρ	z	h	s	c_v	c_p
0.05	5.24	0.9188	38.7	11.498	3.150	5.761
0.06	6.41	0.9008	38.3	11.066	3.159	5.918
0.07	7.64	0.8822	37.9	10.689	3.170	6.097
0.08	8.93	0.8629	37.4	10.352	3.181	6.300
0.09	10.28	0.8427	36.9	10.043	3.193	6.534
0.10	11.72	0.8215	36.4	9.756	3.207	6.805
0.2	101.78	0.1892	15.4	4.573	2.788	17.664
0.3	117.50	0.2458	14.2	4.145	2.639	6.959
0.4	124.52	0.3093	14.1	3.962	2.579	5.486
0.5	129.40	0.3720	14.3	3.837	2.538	4.812
0.6	133.24	0.4336	14.5	3.741	2.506	4.406
0.7	136.44	0.4940	14.9	3.662	2.478	4.128
0.8	139.20	0.5533	15.3	3.595	2.455	3.921
0.9	141.64	0.6118	15.7	3.536	2.433	3.759
1.0	143.85	0.6693	16.1	3.484	2.414	3.628
1.5	152.56	0.9467	18.5	3.282	2.332	3.209
2.0	159.08	1.2104	21.0	3.137	2.265	2.970
2.5	164.41	1.4641	23.5	3.023	2.208	2.806
3.0	168.96	1.7096	26.0	2.927	2.157	2.683
3.5	172.96	1.9483	28.5	2.844	2.110	2.585
4.0	176.57	2.1812	31.0	2.771	2.068	2.505
4.5	179.85	2.4090	33.5	2.705	2.030	2.438
5.0	182.88	2.6323	35.9	2.644	1.994	2.380
5.5	185.71	2.8515	38.4	2.588	1.962	2.329
6.0	188.35	3.0671	40.8	2.536	1.932	2.285
6.5	190.84	3.2793	43.2	2.487	1.904	2.246
7.0	193.21	3.4883	45.6	2.441	1.878	2.211
7.5	195.45	3.6946	47.9	2.396	1.855	2.179
8.0	197.59	3.8981	50.2	2.354	1.833	2.151
8.5	199.64	4.0992	52.6	2.314	1.813	2.125
9.0	201.61	4.2980	54.9	2.275	1.794	2.102
9.5	203.50	4.4947	57.1	2.237	1.777	2.080
10.0	205.32	4.6893	59.4	2.201	1.761	2.060
15.0	220.55	6.5482	81.3	1.896	1.645	1.895

$T = 5.2$ K

p	ρ	z	h	s	c_v	c_p
0.01	0.94	0.9861	41.4	15.236	3.120	5.276
0.02	1.91	0.9718	41.0	13.756	3.126	5.367
0.03	2.90	0.9571	40.7	12.872	3.132	5.467
0.04	3.93	0.9420	40.3	12.231	3.138	5.578
0.05	5.00	0.9264	39.9	11.722	3.146	5.700
0.06	6.10	0.9104	39.5	11.296	3.153	5.835
0.07	7.25	0.8938	39.1	10.926	3.162	5.987
0.08	8.45	0.8767	38.6	10.596	3.171	6.156
0.09	9.70	0.8590	38.2	10.295	3.181	6.346
0.1	11.01	0.8405	37.7	10.018	3.192	6.563
0.2	31.73	0.5836	30.7	7.614	3.313	14.200
0.3	109.70	0.2532	15.8	4.450	2.715	9.493
0.4	119.55	0.3098	15.3	4.194	2.633	6.423
0.5	125.54	0.3687	15.3	4.037	2.585	5.379
0.6	130.00	0.4273	15.5	3.922	2.550	4.817
0.7	133.61	0.4850	15.7	3.830	2.522	4.454
0.8	136.67	0.5419	16.1	3.753	2.497	4.194
0.9	139.34	0.5980	16.5	3.688	2.476	3.998
1.0	141.72	0.6532	16.9	3.630	2.456	3.841

Table II.17 (*Continued*)

p	ρ	z	h	s	c_v	c_p
1.5	150.97	0.9198	19.2	3.411	2.376	3.360
2.0	157.77	1.1736	21.6	3.256	2.311	3.095
2.5	163.26	1.4177	24.1	3.135	2.256	2.919
3.0	167.93	1.6539	26.6	3.034	2.206	2.788
3.5	172.02	1.8836	29.1	2.947	2.162	2.686
4.0	175.69	2.1078	31.5	2.871	2.122	2.602
4.5	179.02	2.3271	34.0	2.802	2.085	2.532
5.0	182.10	2.5420	36.4	2.739	2.051	2.473
5.5	184.95	2.7530	38.9	2.681	2.019	2.421
6.0	187.62	2.9605	41.3	2.627	1.990	2.375
6.5	190.14	3.1648	43.6	2.577	1.964	2.335
7.0	192.52	3.3661	46.0	2.529	1.939	2.300
7.5	194.79	3.5646	48.4	2.484	1.916	2.268
8.0	196.94	3.7606	50.7	2.440	1.895	2.239
8.5	199.01	3.9542	53.0	2.399	1.876	2.212
9.0	200.99	4.1456	55.3	2.359	1.858	2.189
9.5	202.89	4.3349	57.6	2.321	1.841	2.167
10.0	204.72	4.5223	59.8	2.284	1.826	2.146
15.0	220.04	6.3109	81.7	1.972	1.712	1.982

$T = 5.5$ K

p	ρ	z	h	s	c_v	c_p
0.01	0.89	0.9878	43.0	15.531	3.120	5.266
0.02	1.79	0.9753	42.6	14.056	3.124	5.345
0.03	2.73	0.9625	42.3	13.178	3.129	5.431
0.04	3.69	0.9495	42.0	12.543	3.134	5.524
0.05	4.68	0.9361	41.6	12.040	3.140	5.626
0.06	5.69	0.9224	41.2	11.620	3.146	5.737
0.07	6.75	0.9083	40.8	11.258	3.152	5.858
0.08	7.83	0.8938	40.4	10.936	3.159	5.991
0.09	8.96	0.8789	40.0	10.645	3.167	6.138
0.1	10.14	0.8636	39.6	10.378	3.174	6.300
0.2	26.01	0.6730	34.2	8.254	3.260	9.895
0.3	85.52	0.3071	20.5	5.333	2.935	28.711
0.4	119.65	0.3193	17.5	4.615	2.725	8.907
0.5	118.55	0.3692	17.1	4.369	2.658	6.581
0.6	124.39	0.4222	17.0	4.213	2.616	5.603
0.7	128.83	0.4756	17.2	4.096	2.585	5.043
0.8	132.46	0.5286	17.4	4.002	2.559	4.671
0.9	135.55	0.5812	17.7	3.924	2.537	4.402
1.0	138.25	0.6331	18.1	3.855	2.517	4.195
1.5	148.42	0.8846	20.2	3.606	2.438	3.597
2.0	155.68	1.1245	22.5	3.435	2.376	3.289
2.5	161.46	1.3553	25.0	3.303	2.324	3.090
3.0	166.31	1.5788	27.4	3.195	2.278	2.947
3.5	170.55	1.7963	29.9	3.102	2.236	2.836
4.0	174.32	2.0085	32.3	3.021	2.198	2.747
4.5	177.74	2.2161	34.8	2.948	2.163	2.673
5.0	180.88	2.4196	37.2	2.882	2.131	2.610
5.5	183.79	2.6194	39.6	2.821	2.102	2.556
6.0	186.50	2.8159	42.0	2.764	2.074	2.508
6.5	189.06	3.0093	44.4	2.711	2.049	2.467
7.0	191.47	3.2000	46.7	2.662	2.026	2.430
7.5	193.76	3.3880	49.1	2.614	2.004	2.397
8.0	195.94	3.5736	51.4	2.569	1.985	2.367
8.5	198.03	3.7570	53.7	2.526	1.966	2.340

Table II.17 (*Continued*)

p	ρ	z	h	s	c_v	c_p
9.0	200.03	3.9382	56.0	2.485	1.949	2.316
9.5	201.95	4.1175	58.2	2.446	1.933	2.293
10.0	203.79	4.2950	60.5	2.407	1.919	2.272
15.0	219.26	5.9879	82.3	2.087	1.812	2.110
20.0	230.99	7.5784	103.2	1.853	1.719	1.935
			$T=6.0$ K			
0.01	0.81	0.9901	45.6	15.989	3.119	5.253
0.02	1.64	0.9799	45.3	14.520	3.122	5.317
0.03	2.48	0.9697	45.0	13.648	3.125	5.385
0.04	3.35	0.9592	44.7	13.020	3.129	5.458
0.05	4.23	0.9483	44.4	12.525	3.132	5.535
0.06	5.13	0.9377	44.1	12.114	3.136	5.618
0.07	6.06	0.9267	43.7	11.761	3.141	5.707
0.08	7.01	0.9154	43.4	11.449	3.145	5.802
0.09	7.99	0.9039	43.0	11.169	3.150	5.909
0.1	8.99	0.8922	42.7	10.913	3.154	6.015
0.2	21.16	0.7582	38.5	9.019	3.205	7.821
0.3	41.71	0.5771	32.4	7.427	3.208	13.526
0.4	79.81	0.4021	24.3	5.778	2.964	18.242
0.5	101.71	0.3944	21.2	5.084	2.798	10.382
0.6	112.19	0.4291	20.3	4.780	2.729	7.664
0.7	118.99	0.4720	20.0	4.590	2.687	6.426
0.8	124.07	0.5174	20.0	4.451	2.656	5.712
0.9	128.15	0.5635	20.1	4.341	2.631	5.242
1.0	131.59	0.6097	20.4	4.250	2.611	4.905
1.5	143.74	0.8373	22.1	3.937	2.534	4.028
2.0	151.92	1.0563	24.3	3.736	2.477	3.627
2.5	158.23	1.2677	26.6	3.585	2.429	3.382
3.0	163.45	1.4726	29.0	3.463	2.387	3.212
3.5	167.94	1.6721	31.4	3.360	2.350	3.085
4.0	171.91	1.8669	33.8	3.270	2.316	2.984
4.5	175.48	2.0575	36.2	3.190	2.284	2.902
5.0	178.74	2.2444	38.6	3.119	2.256	2.833
5.5	181.75	2.4280	40.9	3.053	2.229	2.774
6.0	184.55	2.6085	43.3	2.992	2.205	2.723
6.5	187.18	2.7862	45.6	2.935	2.182	2.679
7.0	189.65	2.9614	48.0	2.882	2.161	2.639
7.5	192.00	3.1342	50.3	2.832	2.141	2.604
8.0	194.22	3.3048	52.6	2.784	2.123	2.573
8.5	196.35	3.4734	54.9	2.739	2.107	2.544
9.0	198.38	3.6400	57.1	2.695	2.091	2.519
9.5	200.33	3.8048	59.4	2.654	2.077	2.495
10.0	202.21	3.9678	61.7	2.614	2.063	2.473
15.0	217.92	5.5227	83.4	2 279	1.968	2.313
20.0	229.91	6.9796	104.3	2.031	1.892	2.160
25.0	239.47	8.3761	124.5	1.853	1.783	1.950
			$T=6.5$ K			
0.01	0.75	0.9918	48.2	16.409	3.118	5.243
0.02	1.51	0.9835	48.0	14.945	3.120	5.296
0.03	2.28	0.9750	47.7	14.078	3.122	5.351
0.04	3.07	0.9665	47.4	13.455	3.125	5.409
0.05	3.87	0.9579	47.1	12.966	3.127	5.471
0.06	4.68	0.9491	46.8	12.560	3.130	5.536
0.07	5.51	0.9402	46.6	12.213	3.132	5.604

Table II.17 (*Continued*)

p	ρ	z	h	s	c_v	c_p
0.08	6.36	0.9312	46.3	11.908	3.135	5.677
0.09	7.23	0.9221	46.0	11.635	3.138	5.753
0.10	8.11	0.9128	45.6	11.387	3.141	5.834
0.20	18.26	0.8113	42.2	9.608	3.172	6.982
0.3	32.18	0.6905	37.9	8.304	3.184	9.389
0.4	53.07	0.5582	32.4	7.087	3.120	13.327
0.5	77.72	0.4765	27.5	6.088	2.967	13.722
0.6	95.16	0.4670	24.9	5.510	2.853	10.618
0.7	106.02	0.4830	23.7	5.180	2.789	8.428
0.8	113.49	0.5221	23.2	4.932	2.749	7.150
0.9	119.12	0.5596	23.0	4.802	2.720	6.349
1.0	123.65	0.5990	23.0	4.676	2.697	5.803
1.5	138.48	0.8023	24.2	4.278	2.621	4.514
2.0	147.78	1.0024	26.2	4.040	2.568	3.986
2.5	154.73	1.1966	28.4	3.857	2.525	3.684
3.0	160.37	1.3855	30.6	3.730	2.487	3.481
3.5	165.16	1.5695	33.0	3.616	2.453	3.333
4.0	169.35	1.7493	35.3	3.518	2.422	3.218
4.5	173.09	1.9254	37.7	3.432	2.394	3.126
5.0	176.49	2.0982	40.0	3.354	2.368	3.049
5.5	179.62	2.2678	42.4	3.283	2.344	2.985
6.0	182.51	2.4348	44.7	3.218	2.322	2.929
6.5	185.22	2.5991	47.0	3.158	2.301	2.881
7.0	187.76	2.7612	49.4	3.101	2.282	2.838
7.5	190.16	2.9210	51.7	3.048	2.265	2.800
8.0	192.44	3.0789	53.9	2.998	2.248	2.767
8.5	194.61	3.2348	56.2	2.950	2.233	2.736
9.0	196.69	3.3889	58.5	2.905	2.219	2.709
9.5	198.68	3.5414	60.7	2.861	2.206	2.684
10.0	200.59	3.6923	63.0	2.819	2.194	2.661
15.0	216.54	5.1303	84.6	2.472	2.108	2.500
20.0	228.73	6.4751	105.4	2.212	2.046	2.363
25.0	238.59	7.7603	125.5	2.019	1.962	2.183
			$T=7.0$ K			
0.01	0.69	0.9931	50.9	16.797	3.117	5.235
0.02	1.39	0.9862	50.6	15.337	3.119	5.280
0.03	2.11	0.9792	50.4	14.473	3.120	5.326
0.04	2.83	0.9721	50.1	13.855	3.122	5.374
0.05	3.56	0.9650	49.9	13.369	3.123	5.424
0.06	4.31	0.9578	49.6	12.968	3.125	5.476
0.07	5.06	0.9505	49.3	12.626	3.127	5.531
0.08	5.83	0.9431	49.1	12.325	3.129	5.588
0.09	6.61	0.9357	48.8	12.057	3.130	5.648
0.1	7.41	0.9282	48.5	11.814	3.132	5.710
0.2	16.22	0.8481	45.6	10.107	3.151	6.524
0.3	27.20	0.7586	42.1	8.933	3.160	7.856
0.4	41.47	0.6633	38.1	7.930	3.139	9.893
0.5	59.10	0.5819	33.9	7.041	3.066	11.677
0.6	76.53	0.5392	30.5	6.351	2.969	11.515
0.7	90.41	0.5325	28.4	5.875	2.892	10.132
0.8	100.63	0.5467	27.2	5.551	2.839	8.720
0.9	108.28	0.5716	26.5	5.320	2.804	7.645
1.0	114.25	0.6019	26.2	5.145	2.777	6.870
1.5	132.57	0.7781	26.6	4.633	2.699	5.058

Table II.17 (Continued)

p	ρ	z	h	s	c_v	c_p
2.0	143.25	0.9602	28.3	4.349	2.650	4.370
2.5	150.96	1.1389	30.3	4.151	2.610	3.996
3.0	157.08	1.3135	32.4	3.998	2.576	3.753
3.5	162.20	1.4839	34.7	3.872	2.546	3.581
4.0	166.65	1.6507	37.0	3.765	2.518	3.449
4.5	170.58	1.8142	39.3	3.671	2.492	3.345
5.0	174.14	1.9746	41.6	3.588	2.469	3.259
5.5	177.39	2.1323	43.9	3.512	2.447	3.187
6.0	180.39	2.2875	46.2	3.442	2.427	3.126
6.5	183.18	2.4403	48.5	3.378	2.408	3.073
7.0	185.80	2.5910	50.8	3.319	2.391	3.027
7.5	188.26	2.7397	53.1	3.263	2.375	2.986
8.0	190.60	2.8865	55.4	3.210	2.360	2.950
8.5	192.82	3.0316	57.6	3.160	2.346	2.917
9.0	194.94	3.1750	59.9	3.112	2.333	2.887
9.5	196.97	3.3169	62.1	3.067	2.321	2.861
10.0	198.92	3.4573	64.3	3.023	2.310	2.836
15.0	215.13	4.7951	85.9	2.664	2.232	2.669
20.0	227.56	6.0443	106.6	2.394	2.181	2.543
25.0	237.63	7.2353	126.6	2.188	2.116	2.388
30.0	245.99	8.3871	146.3	2.038	2.009	2.179

$T = 7.5$ K

p	ρ	z	h	s	c_v	c_p
0.01	0.65	0.9942	53.5	17.158	3.117	5.229
0.02	1.30	0.9884	53.2	15.701	3.118	5.267
0.03	1.96	0.9825	53.0	14.840	3.119	5.306
0.04	2.63	0.9766	52.8	14.224	3.120	5.346
0.05	3.31	0.9706	52.6	13.742	3.121	5.388
0.06	3.99	0.9646	52.3	13.345	3.122	5.432
0.07	4.69	0.9585	52.1	13.005	3.123	5.476
0.08	5.39	0.9524	51.9	12.708	3.124	5.523
0.09	6.10	0.9463	51.6	12.444	3.125	5.571
0.1	6.83	0.9401	51.4	12.205	3.126	5.621
0.2	14.67	0.8752	48.8	10.547	3.138	6.232
0.3	23.91	0.8054	45.9	9.448	3.144	7.132
0.4	35.03	0.7330	42.6	8.554	3.134	8.362
0.5	48.19	0.6660	39.2	7.774	3.098	9.684
0.6	62.38	0.6174	36.0	7.111	3.039	10.397
0.7	75.72	0.5934	33.5	6.585	2.975	10.225
0.8	87.06	0.5899	31.8	6.186	2.920	9.522
0.9	96.25	0.6002	30.6	5.886	2.880	8.668
1.0	103.65	0.6193	29.9	5.655	2.849	7.878
1.5	126.01	0.7641	29.3	5.002	2.768	5.649
2.0	138.33	0.9280	30.5	4.665	2.722	4.776
2.5	146.91	1.0923	32.3	4.438	2.686	4.317
3.0	153.58	1.2538	34.4	4.267	2.655	4.029
3.5	159.08	1.4122	36.5	4.128	2.628	3.828
4.0	163.80	1.5674	38.8	4.011	2.603	3.677
4.5	167.96	1.7198	41.0	3.909	2.580	3.559
5.0	171.68	1.8694	43.3	3.819	2.559	3.463
5.5	175.07	2.0166	45.6	3.738	2.539	3.383
6.0	178.18	2.1614	47.8	3.665	2.520	3.315
6.5	181.07	2.3042	50.1	3.597	2.503	3.257
7.0	183.77	2.4450	52.4	3.534	2.487	3.206
7.5	186.31	2.5839	54.6	3.475	2.473	3.162

Table II.17 (*Continued*)

p	ρ	z	h	s	c_v	c_p
8.0	188.71	2.7211	56.9	3.419	2.459	3.122
8.5	190.99	2.8567	59.1	3.367	2.446	3.087
9.0	193.16	2.9908	61.4	3.317	2.434	3.055
9.5	195.23	3.1234	63.6	3.270	2.423	3.026
10.0	197.22	3.2547	65.8	3.224	2.412	2.999
15.0	213.70	4.5054	87.3	2.853	2.340	2.823
20.0	226.32	5.6722	107.9	2.575	2.297	2.703
25.0	236.60	6.7823	127.9	2.359	2.247	2.567
30.0	245.20	7.8534	147.4	2.195	2.164	2.386
35.0	252.50	8.8973	166.6	2.079	2.031	2.155

$T = 8.0$ K

p	ρ	z	h	s	c_v	c_p
0.01	0.60	0.9951	56.1	17.495	3.116	5.225
0.02	1.22	0.9901	55.9	16.040	3.117	5.257
0.03	1.83	0.9851	55.7	15.182	3.118	5.291
0.04	2.46	0.9801	55.5	14.569	3.118	5.325
0.05	3.09	0.9750	55.2	14.089	3.119	5.361
0.06	3.72	0.9700	55.0	13.694	3.120	5.397
0.07	4.37	0.9649	54.8	13.357	3.120	5.435
0.08	5.02	0.9597	54.6	13.063	3.121	5.474
0.09	5.67	0.9546	54.4	12.802	3.122	5.513
0.10	6.34	0.9494	54.2	12.566	3.122	5.555
0.2	13.43	0.8959	51.8	10.943	3.129	6.042
0.3	21.50	0.8398	49.3	9.893	3.132	6.696
0.4	30.76	0.7826	46.6	9.065	3.127	7.533
0.5	41.30	0.7285	43.7	8.357	3.107	8.464
0.6	52.78	0.6840	40.9	7.742	3.072	9.208
0.7	64.34	0.6546	38.5	7.224	3.027	9.508
0.8	75.11	0.6409	36.5	6.800	2.982	9.380
0.9	84.62	0.6400	35.1	6.459	2.944	8.982
1.0	92.78	0.6486	34.0	6.185	2.912	8.458
1.5	118.86	0.7594	32.3	5.385	2.828	6.246
2.0	133.03	0.9047	33.0	4.986	2.785	5.197
2.5	142.60	1.0550	34.6	4.727	2.752	4.647
3.0	149.88	1.2045	36.5	4.535	2.725	4.307
3.5	155.80	1.3518	38.5	4.383	2.700	4.074
4.0	160.83	1.4966	40.7	4.255	2.678	3.902
4.5	165.21	1.6390	42.8	4.146	2.657	3.768
5.0	169.12	1.7791	45.1	4.049	2.638	3.661
5.5	172.66	1.9169	47.3	3.963	2.620	3.572
6.0	175.90	2.0526	49.5	3.884	2.603	3.497
6.5	178.89	2.1865	51.8	3.813	2.587	3.433
7.0	181.68	2.3185	54.0	3.746	2.572	3.377
7.5	184.30	2.4488	56.3	3.684	2.559	3.328
8.0	186.77	2.5776	58.5	3.626	2.546	3.285
8.5	189.10	2.7048	60.7	3.571	2.534	3.246
9.0	191.33	2.8307	62.9	3.519	2.523	3.212
9.5	193.45	2.9552	65.1	3.470	2.512	3.180
10.0	195.48	3.0784	67.3	3.423	2.502	3.152
15.0	212.25	4.2527	88.7	3.040	2.435	2.963
20.0	225.06	5.3475	109.3	2.754	2.396	2.845
25.0	235.52	6.3876	129.2	2.530	2.358	2.723
30.0	244.32	7.3889	148.7	2.355	2.294	2.565
35.0	251.85	8.3626	167.8	2.225	2.187	2.360
40.0	258.36	9.3165	186.6	2.136	2.028	2.114

Table II.17 (Continued)

p	ρ	z	h	s	c_v	c_p
			$T=8.5$ K			
0.01	0.57	0.9958	58.7	17 812	3.116	5.221
0.02	1.14	0.9915	58.5	16.359	3.117	5.249
0.03	1.72	0.9872	58.3	15 503	3.117	5.278
0.04	2.30	0.9830	58.1	14.891	3.117	5.308
0.05	2.89	0.9787	57.9	14.413	3.118	5.339
0.06	3.49	0.9743	57.7	14.020	3.118	5.370
0.07	4.09	0.9700	57.5	13.686	3.118	5.402
0.08	4.69	0.9657	57.3	13.394	3.119	5.435
0.09	5.30	0.9613	57.1	13.135	3.119	5.469
0.1	5.92	0.9569	56.9	12.901	3.120	5.503
0.2	12.42	0.9122	54.8	11.305	3.124	5.901
0.3	19.62	0.8661	52.6	10.290	3.125	6.406
0.4	27.63	0.8198	50.2	9.505	3.121	7.021
0.5	36.51	0.7756	47.7	8.845	3.109	7.707
0.6	46.09	0.7373	45.3	8.272	3.087	8.326
0.7	55.97	0.7083	43.0	7.777	3.057	8.739
0.8	65.62	0.6905	41.1	7.354	3.024	8.877
0.9	74.63	0.6830	39.5	6.999	2.992	8.786
1.0	82.79	0.6841	38.3	6.702	2.963	8.541
1.5	111.32	0.7631	35.5	5.780	2.879	6.772
2.0	127.38	0.8892	35.7	5.314	2.839	5.618
2.5	138.03	1.0258	37.0	5.019	2.810	4.980
3.0	145.98	1.1639	38.7	4.805	2 786	4.587
3.5	152.37	1.3010	40.6	4.637	2.764	4.319
4.0	157.72	1.4363	42.7	4.499	2.743	4.124
4.5	162.36	1.5697	44.8	4.380	2.725	3.973
5.0	166.47	1.7011	46.9	4.277	2.707	3.853
5.5	170.17	1.8305	49.1	4.185	2.691	3.755
6.0	173.54	1.9581	51.3	4.102	2.675	3.672
6.5	176.65	2.0840	53.5	4.026	2.661	3.602
7.0	179.54	2.2082	55.8	3.956	2.647	3.541
7.5	182.24	2.3309	58.0	3.891	2.634	3.487
8.0	184.78	2.4521	60.2	3.830	2.622	3.440
8.5	187.18	2.5719	62.4	3.772	2.611	3.398
9.0	189.46	2.6904	64.6	3.718	2.601	3.360
9.5	191.63	2.8077	66.8	3.667	2.591	3.326
10.0	193.70	2.9238	68.9	3.618	2.581	3.295
15.0	210.78	4.0305	90.3	3.223	2.517	3.091
20.0	223.77	5.0619	110.8	2.930	2.482	2.971
25.0	234.40	6.0405	130.6	2.699	2.451	2.859
30.0	243.39	6.9809	150.0	2 516	2.402	2.720
35.0	251.12	7.8937	169.0	2.374	2.317	2.540
40.0	257.84	8.7861	187.7	2.270	2.186	2.316
			$T=9.0$ K			
0.01	0.54	0.9963	61.3	18.110	3.116	5.218
0.02	1.08	0.9927	61.1	16.658	3.116	5.243
0.03	1.62	0.9890	60.9	15.804	3.116	5.268
0.04	2.17	0.9853	60.8	15.194	3.117	5.294
0.05	2.72	0.9816	60.6	14.718	3.117	5.321
0.06	3.28	0 9779	60.4	14.327	3.117	5.348
0.07	3.84	0.9742	60.2	13.994	3.117	5.376
0.08	4.41	0.9705	60.0	13.704	3.117	5.404

Table II.17 (Continued)

p	ρ	z	h	s	c_v	c_p
0.09	4.98	0.9668	59.9	13.446	3.118	5.433
0.1	5.55	0.9630	59.7	13.214	3.118	5.463
0.2	11.56	0.9252	57.7	11.639	3.120	5.796
0.3	18.10	0.8868	55.7	10.650	3.120	6.200
0.4	25.21	0.8486	53.6	9.896	3.117	6.676
0.5	32.93	0.8121	51.4	9.270	3.109	7.198
0.6	41.17	0.7796	49.3	8.730	3.094	7.702
0.7	49.71	0.7532	47.3	8.257	3.074	8.103
0.8	58.25	0.7346	45.4	7.846	3.050	8.340
0.9	66.50	0.7240	43.8	7.491	3.025	8.409
1.0	74.23	0.7106	42.5	7.185	3.002	8.344
1.5	103.71	0.7736	39.0	6.179	2.923	7.151
2.0	121.47	0.8807	38.7	5.646	2.885	6.012
2.5	133.24	1.0037	39.6	5.313	2.859	5.308
3.0	141.91	1.1308	41.1	5.075	2.838	4.864
3.5	148.78	1.2583	42.8	4.891	2.819	4.562
4.0	154.50	1.3848	44.8	4.740	2.801	4.342
4.5	159.41	1.5099	46.8	4.613	2.784	4.174
5.0	163.73	1.6334	48.9	4.503	2.768	4.041
5.5	167.60	1.7553	51.1	4.405	2.753	3.933
6.0	171.12	1.8755	53.2	4.316	2.738	3.842
6.5	174.35	1.9942	55.4	4.236	2.725	3.765
7.0	177.34	2.1114	57.6	4.163	2.712	3.698
7.5	180.12	2.2272	59.7	4.094	2.701	3.640
8.0	182.74	2.3417	61.9	4.031	2.689	3.588
8.5	185.21	2.4548	64.1	3.971	2.679	3.542
9.0	187.55	2.5668	66.3	3.915	2.669	3.501
9.5	189.78	2.6776	68.5	3.861	2.660	3.464
10.0	191.90	2.7874	70.6	3.811	2.651	3.430
15.0	209.29	3.8336	91.8	3.403	2.589	3.210
20.0	222.47	4.8087	112.3	3.104	2.556	3.085
25.0	233.26	5.7329	132.1	2.866	2.531	2.979
30.0	242.41	6.6198	151.4	2.675	2.493	2.856
35.0	250.32	7.4791	170.3	2.523	2.425	2.696
40.0	257.23	8.3177	189.0	2.408	2.318	2.496
45.0	263.32	9.1409	207.4	2.326	2.162	2.258
50.0	268.73	9.9524	225.7	2.274	1.956	1.991
			$T = 9.5$ K			
0.01	0.51	0.9968	63.9	18.392	3.116	5.215
0.02	1.02	0.9936	63.7	16.942	3.116	5.237
0.03	1.53	0.9905	63.6	16.089	3.116	5.260
0.04	2.05	0.9873	63.4	15.480	3.116	5.283
0.05	2.57	0.9841	63.2	15.005	3.116	5.306
0.06	3.10	0.9809	63.1	14.615	3.116	5.330
0.07	3.63	0.9777	62.9	14.284	3.116	5.355
0.08	4.16	0.9745	62.7	13.995	3.117	5.380
0.09	4.70	0.9713	62.6	13.739	3.117	5.405
0.1	5.23	0.9681	62.4	13.509	3.117	5.430
0.2	10.83	0.9358	60.6	11.950	3.118	5.714
0.3	16.83	0.9034	58.8	10.981	3.117	6.048
0.4	23.26	0.8715	56.9	10.250	3.115	6.430
0.5	30.13	0.8411	55.0	9.649	3.109	6.843
0.6	37.38	0.8135	53.0	9.133	3.098	7.253
0.7	44.89	0.7902	51.2	8.682	3.084	7.610

Table II.17 (*Continued*)

p	ρ	z	h	s	c_v	c_p
0.8	52.48	0.7724	49.5	8.284	3.067	7.869
0.9	69.95	0.7608	47.9	7.935	3.049	8.013
1.0	67.13	0.7549	46.6	7.629	3.030	8.050
1.5	96.37	0.7888	42.7	6.572	2.959	7.355
2.0	115.42	0.8781	41.7	5.981	2.924	6.349
2.5	128.26	0.9877	42.3	5.608	2.902	5.618
3.0	137.68	1.1042	43.6	5.345	2.883	5.134
3.5	145.07	1.2226	45.2	5.144	2.866	4.799
4.0	151.16	1.3409	47.0	4.981	2.850	4.556
4.5	156.36	1.4584	49.0	4.844	2.835	4.371
5.0	160.91	1.5746	51.0	4.726	2.820	4.225
5.5	164.97	1.6895	53.1	4.622	2.807	4.105
6.0	168.63	1.8030	55.2	4.529	2.794	4.006
6.5	171.98	1.9152	57.3	4.444	2.781	3.922
7.0	175.08	2.0260	59.5	4.367	2.770	3.849
7.5	177.96	2.1356	61.6	4.295	2.759	3.786
8.0	180.66	2.2439	63.8	4.228	2.748	3.730
8.5	183.20	2.3511	65.9	4.166	2.738	3.680
9.0	185.61	2.4572	68.1	4.107	2.729	3.636
9.5	187.89	2.5622	70.2	4.052	2.720	3.595
10.0	190.06	2.6662	72.4	3.999	2.712	3.559
15.0	207.79	3.6582	93.5	3.580	2.652	3.320
20.0	221.15	4.5827	113.8	3.273	2.621	3.189
25.0	232.09	5.4585	133.6	3.030	2.599	3.086
30.0	241.39	6.2979	152.8	2.833	2.569	2.974
35.0	249.46	7.1099	171.7	2.673	2.516	2.833
40.0	256.54	7.9011	190.2	2.547	2.428	2.654
45.0	262.82	8.6765	208.6	2.453	2.297	2.437
50.0	268.41	9.4398	226.8	2.387	2.119	2.187
			$T=10$ K			
0.01	0.48	0.9972	66.5	18.660	3.116	5.213
0.02	0.97	0.9945	66.4	17.210	3.116	5.233
0.03	1.46	0.9917	66.2	16.358	3.116	5.253
0.04	1.95	0.9889	66.0	15.751	3.116	5.274
0.05	2.44	0.9862	65.9	15.277	3.116	5.294
0.06	2.94	0.9834	65.7	14.888	3.116	5.316
0.07	3.44	0.9806	65.6	14.558	3.116	5.337
0.08	3.94	0.9778	65.4	14.271	3.116	5.359
0.09	4.44	0.9751	65.3	14.016	3.116	5.381
0.10	4.95	0.9723	65.1	13.787	3.116	5.404
0.2	10.19	0.9446	63.5	12.241	3.116	5.649
0.3	15.75	0.9170	61.8	11.288	3.115	5.931
0.4	21.63	0.8901	60.1	10.575	3.113	6.246
0.5	27.85	0.8644	58.3	9.994	3.108	6.583
0.6	34.35	0.8410	56.6	9.497	3.101	6.922
0.7	41.06	0.8207	54.9	9.062	3.091	7.231
0.8	47.87	0.8045	53.3	8.678	3.079	7.481
0.9	54.64	0.7929	51.8	8.337	3.064	7.652
1.0	61.25	0.7860	50.5	8.034	3.050	7.744
1.5	89.55	0.8064	46.4	6.951	2.989	7.413
2.0	109.39	0.8802	45.0	6.313	2.957	6.606
2.5	123.18	0.9771	45.2	5.904	2.937	5.898
3.0	133.32	1.0833	46.2	5.615	2.921	5.390
3.5	141.24	1.1930	47.6	5.396	2.906	5.029

Table II.17 (*Continued*)

p	ρ	z	h	s	c_v	c_p
4.0	147.73	1.3035	49.3	5.220	2.892	4.764
4.5	153.23	1.4138	51.2	5.073	2.879	4.562
5.0	158.01	1.5233	53.1	4.947	2.866	4.403
5.5	162.26	1.6318	55.2	4.837	2.854	4.273
6.0	166.08	1.7392	57.2	4.738	2.842	4.165
6.5	169.57	1.8454	59.3	4.649	2.830	4.074
7.0	172.78	1.9504	61.4	4.568	2.820	3.995
7.5	175.76	2.0543	63.5	4.493	2.809	3.922
8.0	178.54	2.1571	65.7	4.423	2.799	3.866
8.5	181.16	2.2588	67.8	4.358	2.790	3.813
9.0	183.63	2.3595	69.9	4.297	2.781	3.765
9.5	185.97	2.4592	72.1	4.240	2.773	3.721
10.0	188.20	2.5580	74.2	4.185	2.765	3.682
15.0	206.27	3.5009	95.2	3.753	2.707	3.424
20.0	219.83	4.3799	115.5	3.439	2.676	3.284
25.0	230.90	5.2122	135.2	3.191	2.657	3.182
30	240.34	6.0091	154.3	2.988	2.633	3.078
35	248.55	6.7790	173.1	2.821	2.591	2.952
40	255.79	7.5281	191.6	2.687	2.519	2.793
45	262.23	8.2612	209.9	2.582	2.410	2.597
50	267.99	8.9818	227.9	2.504	2.256	2.366
55	273.17	9.6927	245.9	2.452	2.057	2.106

$T = 11$ K

p	ρ	z	h	s	c_v	c_p
0.01	0.44	0.9979	71.7	19.157	3.116	5.209
0.02	0.88	0.9957	71.6	17.709	3.116	5.226
0.03	1.32	0.9936	71.5	16.858	3.116	5.242
0.04	1.77	0.9915	71.3	16.253	3.116	5.259
0.05	2.21	0.9894	71.2	15.781	3.116	5.276
0.06	2.66	0.9873	71.0	15.394	3.115	5.293
0.07	3.11	0.9852	70.9	15.065	3.115	5.310
0.08	3.56	0.9831	70.8	14.780	3.115	5.328
0.09	4.02	0.9810	70.6	14.527	3.115	5.345
0.1	4.47	0.9789	70.5	14.300	3.115	5.363
0.2	9.14	0.9581	69.1	12.775	3.115	5.553
0.3	14.00	0.9377	67.6	11.845	3.113	5.764
0.4	19.07	0.9181	66.2	11.158	3.112	5.991
0.5	24.33	0.8995	64.7	10.603	3.109	6.231
0.6	29.76	0.8824	63.2	10.134	3.104	6.473
0.7	35.33	0.8672	61.8	9.724	3.099	6.714
0.8	40.97	0.8545	60.5	9.362	3.092	6.911
0.9	46.64	0.8446	59.2	9.038	3.084	7.082
1.0	52.25	0.8376	58.0	8.746	3.075	7.209
1.5	77.88	0.8429	53.7	7.653	3.033	7.268
2.0	98.95	0.8936	51.7	6.958	3.007	6.867
2.5	112.99	0.9683	51.3	6.487	2.992	6.320
3.0	124.39	1.0555	51.8	6.151	2.980	5.833
3.5	133.33	1.1488	52.9	5.895	2.970	5.449
4.0	140.61	1.2449	54.3	5.693	2.960	5.154
4.5	146.73	1.3421	55.9	5.525	2.950	4.924
5.0	152.01	1.4395	57.7	5.383	2.940	4.741
5.5	156.66	1.5364	59.6	5.259	2.930	4.591
6.0	160.82	1.6328	61.5	5.149	2.920	4.467
6.5	164.59	1.7284	63.5	5.051	2.911	4.362

Table II.17 (*Continued*)

p	ρ	z	h	s	c_v	c_p
7.0	168.03	1.8231	65.5	4.962	2.901	4.272
7.5	171.22	1.9170	67.6	4.880	2.892	4.193
8.0	174.18	2.0100	69.7	4.804	2.884	4.124
8.5	176.96	2.1022	71.7	4.733	2.876	4.063
9.0	179.57	2.1934	73.8	4.668	2.868	4.008
9.5	182.04	2.2839	75.9	4.606	2.860	3.958
10.0	184.38	2.3736	78.0	4.547	2.853	3.913
15.0	203.18	3.2309	98.7	4.088	2.798	3.617
20.0	217.14	4.0310	118.8	3.761	2.768	3.457
25.0	228.50	4.7883	138.4	3.502	2.751	3.348
30	238.17	5.5124	157.5	3.290	2.735	3.254
35	246.63	6.2106	176.2	3.112	2.708	3.150
40	254.13	6.8883	194.5	2.964	2.661	3.024
45	260.85	7.5498	212.6	2.842	2.584	2.866
50	266.90	8.1985	230.5	2.745	2.472	2.675
55	272.39	8.8369	248.2	2.669	2.320	2.450
60	277.37	9.4670	265.8	2.615	2.126	2.194

$$T = 12 \text{ K}$$

p	ρ	z	h	s	c_v	c_p
0.01	0.40	0.9983	76.9	19.610	3.116	5.207
0.02	0.80	0.9967	76.8	18.163	3.116	5.220
0.03	1.21	0.9951	76.7	17.314	3.116	5.234
0.04	1.62	0.9934	76.6	16.710	3.115	5.248
0.05	2.02	0.9918	76.4	16.239	3.115	5.262
0.06	2.43	0.9902	76.3	15.854	3.115	5.276
0.07	2.84	0.9885	76.2	15.527	3.115	5.290
0.08	3.25	0.9869	76.1	15.242	3.115	5.305
0.09	3.66	0.9853	75.9	14.991	3.115	5.319
0.10	4.08	0.9837	75.8	14.765	3.115	5.334
0.2	8.29	0.9678	74.6	13.255	3.114	5.486
0.3	12.63	0.9525	73.3	12.341	3.113	5.651
0.4	17.11	0.9379	72.1	11.671	3.112	5.825
0.5	21.70	0.9242	70.8	11.135	3.110	6.006
0.6	26.41	0.9116	69.6	10.684	3.107	6.189
0.7	31.19	0.9003	68.4	10.292	3.103	6.367
0.8	36.03	0.8906	67.2	9.946	3.099	6.532
0.9	40.90	0.8828	66.1	9.636	3.095	6.680
1.0	45.75	0.8769	65.0	9.355	3.089	6.804
1.5	68.66	0.8765	60.9	8.278	3.062	7.052
2.0	87.91	0.9127	58.6	7.557	3.042	6.886
2.5	103.33	0.9706	57.8	7.048	3.030	6.538
3.0	115.54	1.0417	57.8	6.672	3.022	6.146
3.5	125.31	1.1205	58.5	6.385	3.015	5.789
4.0	133.32	1.2037	59.6	6.156	3.009	5.491
4.5	140.04	1.2891	61.0	5.968	3.002	5.247
5.0	145.82	1.3756	62.6	5.809	2.994	5.049
5.5	150.88	1.4624	64.3	5.671	2.986	4.884
6.0	155.38	1.5491	66.1	5.550	2.979	4.747
6.5	159.44	1.6355	68.0	5.442	2.971	4.630
7.0	163.14	1.7214	70.0	5.345	2.963	4.529
7.5	166.54	1.8066	71.9	5.255	2.956	4.442
8.0	169.69	1.8913	73.9	5.173	2.948	4.364
8.5	172.64	1.9752	75.9	5.097	2.941	4.296
9.0	175.40	2.0585	77.9	5.026	2.934	2.234

Table II.17 (*Continued*)

p	ρ	z	h	s	c_v	c_p
9.5	178.00	2.1411	80.0	4.960	2.927	4.179
10.0	180.46	2.2230	82.0	4.897	2.921	4.128
15.0	200.04	3.0082	102.3	4.411	2.869	3.794
20.0	214.40	3.7422	122.4	4.068	2.839	3.612
25.0	226.05	4.4368	141.8	3.800	2.822	3.493
30	235.95	5.1007	160.8	3.579	2.810	3.398
35	244.62	5.7399	179.4	3.393	2.792	3.307
40	252.33	6.3594	197.6	3.235	2.761	3.204
45	259.27	6.9628	215.6	3.101	2.708	3.079
50	265.56	7.5534	233.3	2.988	2.627	2.925
55	271.28	8.1333	250.8	2.895	2.518	2.739
60	276.52	8.7047	268.1	2.821	2.362	2.521
65	281.33	9.2688	285.4	2.763	2.172	2.270
70	285.77	9.8269	302.5	2.722	1.942	1.992
75	289.87	10.3798	319.6	2.697	1.671	1.687

$$T = 13 \text{ K}$$

p	ρ	z	h	s	c_v	c_p
0.01	0.37	0.9987	82.1	20.026	3.116	5.205
0.02	0.74	0.9974	82.0	18.581	3.116	5.216
0.03	1.12	0.9961	81.9	17.733	3.116	5.228
0.04	1.49	0.9949	81.8	17.129	3.115	5.240
0.05	1.86	0.9936	81.7	16.660	3.115	5.251
0.06	2.24	0.9923	81.6	16.275	3.115	5.263
0.07	2.62	0.9911	81.5	15.949	3.115	5.275
0.08	2.99	0.9898	81.4	15.666	3.115	5.287
0.09	3.37	0.9885	81.3	15.416	3.115	5.299
0.1	3.75	0.9873	81.1	15.191	3.115	5.312
0.2	7.60	0.9751	80.0	13.692	3.114	5.438
0.3	11.53	0.9635	78.9	12.790	3.113	5.571
0.4	15.55	0.9525	77.8	12.133	3.112	5.710
0.5	19.65	0.9422	76.7	11.609	3.111	5.852
0.6	23.82	0.9328	75.7	11.171	3.109	5.996
0.7	28.04	0.9245	74.6	10.792	3.107	6.137
0.8	32.30	0.9173	73.6	10.458	3.104	6.271
0.9	36.57	0.9114	72.6	10.158	3.101	6.394
1.0	40.84	0.9068	71.6	9.887	3.098	6.505
1.5	61.37	0.9051	67.8	8.832	3.080	6.811
2.0	79.43	0.9324	65.5	8.105	3.065	6.792
2.5	94.60	0.9786	64.3	7.574	3.056	6.599
3.0	107.12	1.0371	64.1	7.172	3.051	6.323
3.5	117.44	1.1036	64.4	6.858	3.047	6.030
4.0	126.03	1.1753	65.3	6.607	3.043	5.757
4.5	133.28	1.2503	66.4	6.399	3.039	5.519
5.0	139.53	1.3270	67.8	6.224	3.034	5.317
5.5	144.98	1.4048	69.4	6.073	3.028	5.145
6.0	149.82	1.4830	71.0	5.940	3.022	5.000
6.5	154.18	1.5612	72.8	5.823	3.016	4.874
7.0	158.13	1.6393	74.6	5.717	3.010	4.766
7.5	161.75	1.7170	76.5	5.620	3.004	4.671
8.0	165.10	1.7944	78.4	5.531	2.998	4.587
8.5	168.21	1.8712	80.3	5.450	2.991	4.512
9.0	171.13	1.9476	82.3	5.373	2.985	4.445
9.5	173.87	2.0234	84.2	5.302	2.979	4.384
10.0	176.46	2.0986	86.2	5.235	2.973	4.329
15.0	196.84	2.8220	106.3	4.721	2.925	3.959

Table II.17 (*Continued*)

p	ρ	z	h	s	c_v	c_p
20.0	211.64	3.4995	126.1	4.363	2.895	3.755
25.0	223.56	4.1410	145.4	4.085	2.878	3.622
30	233.69	4.7539	164.3	3.856	2.867	3.523
35	242.55	5.3437	182.8	3.663	2.854	3.437
40	250.44	5.9146	200.9	3.498	1.833	3.349
45	257.56	6.4700	218.7	3.355	2.797	3.248
50	264.03	7.0127	236.3	3.231	2.739	3.126
55	269.94	7.5449	253.6	3.124	2.654	2.978
60	275.38	8.0683	270.8	3.034	2.538	2.799
65	280.40	8.5844	287.8	2.958	2.388	2.589
70	285.04	9.0940	304.7	2.896	2.202	2.346
75	289.36	9.5982	321.5	2.847	1.979	2.072
80	293.39	10.0976	338.1	2.811	1.717	1.768
85	297.16	10.5926	354.8	2.788	1.416	1.436

$T = 14$ K

p	ρ	z	h	s	c_v	c_p
0.01	0.34	0.9990	87.3	20.412	3.116	5.203
0.02	0.69	0.9980	87.2	18.967	3.116	5.213
0.03	1.03	0.9970	87.1	18.120	3.116	5.223
0.04	1.38	0.9960	87.0	17.517	3.116	5.233
0.05	1.73	0.9950	86.9	17.049	3.115	5.243
0.06	2.08	0.9940	86.8	16.665	3.115	5.253
0.07	2.42	0.9930	86.7	16.340	3.115	5.264
0.08	2.77	0.9920	86.6	16.057	3.115	5.274
0.09	3.12	0.9910	86.5	15.808	3.115	5.284
0.10	3.47	0.9901	86.4	15.584	3.115	5.294
0.2	7.01	0.9806	85.5	14.094	3.114	5.401
0.3	10.62	0.9718	84.5	13.201	3.113	5.511
0.4	14.28	0.9634	83.5	12.553	3.113	5.625
0.5	17.99	0.9558	82.5	12.039	3.112	5.741
0.6	21.74	0.9488	81.6	11.610	3.110	5.858
0.7	25.53	0.9426	80.6	11.241	3.109	5.972
0.8	29.35	0.9374	79.7	10.915	3.107	6.082
0.9	33.17	0.9330	78.9	10.624	3.105	6.186
1.0	36.99	0.9297	78.0	10.361	3.103	6.282
1.5	55.54	0.9287	74.5	9.329	3.092	6.596
2.0	72.34	0.9507	72.2	8.604	3.082	6.660
2.5	86.94	0.9888	70.9	8.063	3.075	6.570
3.0	99.40	1.0378	70.4	7.644	3.072	6.394
3.5	119.97	1.0944	70.5	7.311	3.069	6.174
4.0	118.94	1.1565	71.1	7.041	3.068	5.946
4.5	126.61	1.2222	72.0	6.816	3.065	5.731
5.0	133.24	1.2903	73.2	6.626	3.062	5.538
5.5	139.06	1.3600	74.6	6.463	3.059	5.368
6.0	144.22	1.4306	76.1	6.319	3.054	5.220
6.5	148.85	1.5015	77.8	6.192	3.050	5.091
7.0	153.05	1.5727	79.5	6.078	3.045	4.978
7.5	156.89	1.6438	81.2	5.974	3.040	4.879
8.0	160.43	1.7147	83.1	5.879	3.035	4.790
8.5	163.72	1.7852	84.9	5.791	3.030	4.711
9.0	166.79	1.8555	86.8	5.710	3.025	4.639
9.5	169.67	1.9253	88.7	5.634	3.019	4.574
10.0	172.38	1.9948	90.7	5.563	3.014	4.515
15.0	193.59	2.6644	110.3	5.020	2.970	4.115
20.0	208.83	3.2932	129.9	4.646	2.939	3.890

Table II.17 (*Continued*)

p	ρ	z	h	s	c_v	c_p
25.0	221.05	3.8890	149.1	4.357	2.922	3.742
30	231.39	4.4582	167.9	4.122	2.911	3.635
35	240.43	5.0056	186.3	3.922	2.901	3.548
40	248.48	5.5353	204.3	3.751	2.887	3.468
45	255.75	6.0503	222.1	3.601	2.862	3.384
50	262.37	6.5530	239.5	3.469	2.821	3.287
55	268.44	7.0454	256.7	3.352	2.758	3.171
60	274.02	7.5291	273.7	3.250	2.670	3.031
65	279.20	8.0055	290.5	3.160	2.553	2.863
70	284.00	8.4754	307.2	3.082	2.406	2.663
75	288.48	8.9397	323.7	3.014	2.224	2.432
80	292.68	9.3990	340.1	2.957	2.008	2.168
85	296.62	9.8537	356.4	2.910	1.755	1.870

$$T = 15 \text{ K}$$

p	ρ	z	h	s	c_v	c_p
0.01	0.32	0.9992	92.5	20.771	3.116	5.202
0.02	0.64	0.9984	92.5	19.327	3.116	5.210
0.03	0.97	0.9976	92.4	18.480	3.116	5.219
0.04	1.29	0.9968	92.3	17.878	3.116	5.228
0.05	1.61	0.9961	92.2	17.410	3.115	5.237
0.06	1.93	0.9953	92.1	17.027	3.115	5.245
0.07	2.26	0.9945	92.0	16.703	3.115	5.254
0.08	2.58	0.9937	91.9	16.421	3.115	5.263
0.09	2.91	0.9930	91.8	16.172	3.115	5.272
0.1	3.23	0.9922	91.7	15.949	3.115	5.281
0.2	6.52	0.9849	90.8	14.465	3.114	5.372
0.3	9.84	0.9782	90.0	13.579	3.114	5.465
0.4	13.21	0.9719	89.1	12.938	3.113	5.561
0.5	16.61	0.9662	88.2	12.432	3.112	5.658
0.6	20.04	0.9610	87.4	12.011	3.112	5.754
0.7	23.49	0.9566	86.6	11.648	3.111	5.849
0.8	26.95	0.9528	85.7	11.330	3.110	5.942
0.9	30.41	0.9497	85.0	11.045	3.108	6.030
1.0	33.87	0.9475	84.2	10.788	3.107	6.112
1.5	50.72	0.9481	81.0	9.778	3.100	6.412
2.0	66.38	0.9669	78.8	9.058	3.093	6.522
2.5	80.29	0.9993	77.5	8.514	3.088	6.499
3.0	92.46	1.0413	76.8	8.085	3.086	6.396
3.5	103.04	1.0901	76.7	7.740	3.085	6.242
4.0	112.20	1.1442	77.1	7.455	3.085	6.065
4.5	120.14	1.2021	77.9	7.217	3.084	5.883
5.0	127.09	1.2627	78.9	7.014	3.082	5.709
5.5	133.20	1.3252	80.1	6.839	3.080	5.549
6.0	138.64	1.3889	81.4	6.686	3.078	5.406
6.5	143.53	1.4534	83.0	6.550	3.075	5.278
7.0	147.96	1.5184	84.5	6.428	3.071	5.165
7.5	152.01	1.5835	86.2	6.317	3.067	5.063
8.0	155.74	1.6486	88.0	6.216	3.063	4.972
8.5	159.19	1.7137	89.7	6.123	3.059	4.890
9.0	162.41	1.7785	91.5	6.036	3.055	4.816
9.5	165.42	1.8431	93.4	5.956	3.050	4.749
10.0	168.26	1.9073	95.3	5.880	3.046	4.687
15.0	190.29	2.5298	114.5	5.309	3.005	4.264
20.0	205.99	3.1160	133.8	4.919	2.975	4.019
25.0	218.51	3.6719	152.9	4.620	2.957	3.856

Table II.17 (*Continued*)

p	ρ	z	h	s	c_v	c_p
30	229.07	4.2031	171.6	4.376	2.946	3.739
35	238.29	4.7140	189.9	4.171	2.938	3.647
40	246.49	5.2081	207.8	3.993	2.928	3.570
45	253.89	5.6983	225.5	3.838	2.910	3.496
50	260.63	6.1569	242.9	3.700	2.881	3.417
55	266.81	6.6157	260.0	3.577	2.835	3.327
60	272.52	7.0661	276.8	3.466	2.769	3.220
65	277.80	7.5093	293.5	3.365	2.679	3.090
70	282.72	7.9462	310.0	3.275	2.563	2.936
75	287.32	8.3776	326.3	3.193	2.418	2.753
80	291.63	8.8039	342.5	3.120	2.242	2.539
85	295.70	9.2255	358.5	3.054	2.035	2.292
90	299.54	9.6428	374.4	2.995	1.794	2.012
$T = 16$ K						
0.01	0.30	0.9994	97.7	21.107	3.116	5.201
0.02	0.60	0.9988	97.7	19.663	3.116	5.208
0.03	0.90	0.9981	97.6	18.817	3.116	5.216
0.04	1.21	0.9975	97.5	18.215	3.116	5.224
0.05	1.51	0.9969	97.4	17.748	3.116	5.231
0.06	1.81	0.9963	97.3	17.366	3.115	5.239
0.07	2.12	0.9957	97.3	17.042	3.115	5.247
0.08	2.42	0.9951	97.2	16.760	3.115	5.255
0.09	2.72	0.9945	97.1	16.512	3.115	5.262
0.10	3.03	0.9939	97.0	16.289	3.115	5.270
0.2	6.09	0.9883	96.2	14.811	3.115	5.349
0.3	9.18	0.9832	95.4	13.931	3.114	5.429
0.4	12.30	0.9785	94.6	13.296	3.114	5.511
0.5	15.44	0.9743	93.9	12.795	3.113	5.593
0.6	18.60	0.9706	93.1	12.379	3.113	5.675
0.7	21.77	0.9674	92.4	12.023	3.112	5.755
0.8	24.95	0.9648	91.6	11.710	3.111	5.834
0.9	28.13	0.9628	90.9	11.431	3.111	5.909

Table II.17 (*Continued*)

p	ρ	z	h	s	c_v	c_p
1.0	31.30	0.9614	90.3	11.178	3.110	5.981
1.5	46.83	0.9638	87.4	10.183	3.105	6.258
2.0	61.35	0.9808	85.3	9.475	3.100	6.391
2.5	74.53	1.0092	83.9	8.931	3.098	6.412
3.0	86.30	1.0460	83.2	8.497	3.096	6.360
3.5	96.72	1.0888	83.0	8.143	3.096	6.259
4.0	105.90	1.1364	83.2	7.849	3.097	6.127
4.5	114.00	1.1877	83.7	7.600	3.097	5.981
5.0	121.14	1.2418	84.6	7.387	3.097	5.832
5.5	127.49	1.2980	85.6	7.202	3.096	5.689
6.0	133.16	1.3557	86.9	7.040	3.095	5.556
6.5	138.27	1.4144	88.2	6.896	3.093	5.434
7.0	142.90	1.4738	89.7	6.766	3.090	5.323
7.5	147.14	1.5336	91.3	6.649	3.088	5.223
8.0	151.04	1.5936	92.9	6.542	3.085	5.132
8.5	154.65	1.6537	94.6	6.443	3.081	5.050
9.0	158.02	1.7137	96.4	6.352	3.078	4.975
9.5	161.16	1.7736	98.2	6.267	3.074	4.906
10.0	164.12	1.8332	100.0	6.188	3.070	4.843
15.0	186.97	2.4139	118.8	5.589	3.034	4.404
20.0	203.12	2.9626	137.9	5.182	3.005	4.144
25.0	215.94	3.4834	156.8	4.872	2.986	3.967
30	226.73	3.9812	175.3	4.620	2.974	3.838
35	236.12	4.4600	193.6	4.409	2.967	3.739
40	244.47	4.9233	211.5	4.227	2.959	3.660
45	251.99	5.3730	229.0	4.067	2.947	3.590
50	258.84	5.8120	246.3	3.924	2.925	3.523
55	265.12	6.2418	263.4	3.795	2.892	3.452
60	270.91	6.6637	280.2	3.678	2.842	3.371
65	276.28	7.0788	296.7	3.571	2.773	3.275
70	281.27	7.4879	313.1	3.472	2.683	3.161
75	285.95	7.8917	329.3	3.380	2.569	3.024

Table II.17 (*Continued*)

p	ρ	z	h	s	c_v	c_p
80	290.33	8.2905	345.2	3.294	2.428	2.863
85	294.47	8.6848	361.0	3.214	2.261	2.673
90	298.40	9.0749	376.7	3.139	2.064	2.454

$T = 17$ K

p	ρ	z	h	s	c_v	c_p
0.01	0.28	0.9995	102.9	21.422	3.116	5.200
0.02	0.57	0.9990	102.9	19.979	3.116	5.207
0.03	0.85	0.9986	102.8	19.133	3.116	5.213
0.04	1.13	0.9981	102.7	18.532	3.116	5.220
0.05	1.42	0.9976	102.6	18.065	3.116	5.227
0.06	1.70	0.9971	102.6	17.683	3.116	5.234
0.07	1.99	0.9967	102.5	17.359	3.116	5.241
0.08	2.27	0.9962	102.4	17.079	3.115	5.247
0.09	2.56	0.9958	102.4	16.831	3.115	5.254
0.10	2.85	0.9953	102.3	16.608	3.115	5.261
0.2	5.71	0.9910	101.5	15.135	3.115	5.330
0.3	8.61	0.9872	100.8	14.259	3.115	5.400
0.4	11.51	0.9837	100.1	13.629	3.114	5.471
0.5	14.44	0.9807	99.4	13.132	3.114	5.542
0.6	17.37	0.9781	98.7	12.721	3.113	5.612
0.7	20.31	0.9760	98.1	12.369	3.113	5.681
0.8	23.25	0.9743	97.4	12.061	3.113	5.749
0.9	26.19	0.9731	96.8	11.786	3.112	5.814
1.0	29.12	0.9725	96.2	11.537	3.112	5.877
1.5	47.50	0.9766	93.5	10.562	3.109	6.130
2.0	57.06	0.9926	91.6	9.859	3.106	6.273
2.5	69.53	1.0182	90.3	9.317	3.104	6.321
3.0	80.84	1.0509	89.5	8.881	3.104	6.305
3.5	91.00	1.0891	89.3	8.522	3.104	6.243
4.0	100.10	1.1316	89.4	8.221	3.105	6.149
4.5	108.23	1.1774	89.8	7.965	3.106	6.036
5.0	115.49	1.2260	90.5	7.743	3.107	5.914
5.5	121.99	1.2768	91.4	7.550	3.107	5.791
6.0	127.83	1.3291	92.5	7.380	3.107	5.671
6.5	133.12	1.3827	93.8	7.229	3.106	5.558
7.0	137.93	1.4371	95.2	7.093	3.104	5.454
7.5	142.33	1.4922	96.7	6.970	3.103	5.358
8.0	146.39	1.5475	98.2	6.857	3.100	5.269
8.5	150.15	1.6031	99.8	6.754	3.098	5.188
9.0	153.64	1.6588	101.5	6.658	3.095	5.114
9.5	156.92	1.7144	103.2	6.569	3.092	5.046
10.0	159.99	1.7700	104.9	6.486	3.089	4.982
15.0	183.62	2.3133	123.3	5.860	3.057	4.535
20.0	200.22	2.8287	142.1	5.437	3.029	4.264
25.0	213.34	3.3184	160.8	5.116	3.010	4.074
30	224.36	3.7865	179.2	4.856	2.998	3.933
35	233.93	4.2369	197.4	4.638	2.990	3.826
40	242.42	4.6725	215.2	4.451	2.984	3.742
45	250.07	5.0959	232.7	4.287	2.974	3.673
50	257.02	5.5090	249.9	4.140	2.959	3.612
55	263.38	5.9135	266.9	4.008	2.935	3.553
60	269.24	6.3106	283.6	3.887	2.897	3.492
65	274.67	6.7014	300.0	3.774	2.845	3.422
70	279.72	7.0867	316.3	3.669	2.775	3.341
75	284.43	7.4669	332.3	3.570	2.685	3.245

Table II.17 (*Continued*)

p	ρ	z	h	s	c_v	c_p
80	288.86	7.8426	348.2	3.476	2.574	3.1
85	293.04	8.2139	363.8	3.385	2.440	2.997
90	297.00	8.5813	379.3	3.300	2.282	2.839

$T = 18$ K

p	ρ	z	h	s	c_v	c_p
0.01	0.27	0.9996	108.1	21.719	3.116	5.199
0.02	0.54	0.9993	108.1	20.276	3.116	5.205
0.03	0.80	0.9989	108.0	19.431	3.116	5.211
0.04	1.07	0.9985	107.9	18.830	3.116	5.217
0.05	1.34	0.9982	107.9	18.364	3.116	5.223
0.06	1.61	0.9978	107.8	17.982	3.116	5.229
0.07	1.88	0.9975	107.7	17.659	3.116	5.235
0.08	2.15	0.9971	107.7	17.378	3.116	5.242
0.09	2.41	0.9968	107.6	17.131	3.115	5.248
0.10	2.68	0.9964	107.5	16.909	3.115	5.254
0.2	5.39	0.9932	106.9	15.439	3.115	5.315
0.3	8.10	0.9904	106.2	14.567	3.115	5.376
0.4	10.83	0.9880	105.6	13.940	3.115	5.438
0.5	13.55	0.9859	104.9	13.448	3.114	5.500
0.6	16.30	0.9842	104.3	13.041	3.114	5.561
0.7	19.05	0.9829	103.7	12.692	3.114	5.622
0.8	21.79	0.9819	103.1	12.387	3.114	5.681
0.9	24.53	0.9814	102.6	12.116	3.113	5.738
1.0	27.25	0.9813	102.0	11.871	3.113	5.793
1.5	40.64	0.9870	99.6	10.909	3.111	6.023
2.0	53.36	1.0025	97.8	10.214	3.110	6.167
2.5	65.16	1.0260	96.6	9.675	3.109	6.233
3.0	75.99	1.0558	95.8	9.240	3.109	6.241
3.5	85.85	1.0903	95.5	8.878	3.110	6.208
4.0	94.79	1.1285	95.5	8.573	3.111	6.145
4.5	102.87	1.1700	95.9	8.310	3.112	6.061
5.0	110.16	1.2139	96.4	8.083	3.114	5.963
5.5	116.75	1.2600	97.3	7.883	3.114	5.860
6.0	122.71	1.3077	98.3	7.707	3.115	5.756
6.5	128.13	1.3567	99.4	7.550	3.115	5.655
7.0	133.08	1.4067	100.7	7.408	3.114	5.558
7.5	137.63	1.4575	102.1	7.279	3.113	5.468
8.0	141.82	1.5087	103.5	7.162	3.112	5.384
8.5	145.70	1.5602	105.1	7.054	3.110	5.306
9.0	149.32	1.6120	106.7	6.954	3.108	5.234
9.5	152.70	1.6638	108.3	6.861	3.106	5.167
10.0	155.88	1.7157	110.0	6.774	3.104	5.105
15.0	180.27	2.2254	127.9	6.123	3.075	4.658
20.0	197.30	2.7110	146.4	5.684	3.049	4.379
25.0	210.73	3.1729	164.9	5.352	3.030	4.179
30	221.97	3.6147	183.2	5.083	3.017	4.027
35	231.72	4.0397	201.2	4.859	3.009	3.911
40	240.36	4.4508	218.9	4.667	3.003	3.820
45	248.13	4.8504	235.4	4.499	2.996	3.748
50	255.17	5.2405	253.5	4.349	2.985	3.689
55	261.61	5.6226	270.5	4.213	2.967	3.638
60	267.54	5.9980	287.1	4.089	2.939	3.589
65	273.01	6.3675	303.5	3.973	2.899	3.539
70	278.10	6.7320	319.7	3.864	2.845	3.483
75	282.84	7.0918	335.7	3.761	2.775	3.420

Table II.17 (*Continued*)

p	ρ	z	h	s	c_v	c_p
80	287.29	7.4475	351.4	3.661	2.688	3.345
85	291.48	7.7998	367.0	3.565	2.581	3.257
90	259.44	8.1473	382.3	3.471	2.455	3.154
95	299.20	8.4918	397.5	3.380	2.308	3.034
			$T = 19$ K			
0.01	0.25	0.9997	113.3	22.000	3.116	5.199
0.02	0.51	0.9994	113.3	20.558	3.116	5.204
0.03	0.76	0 9992	113.2	19.713	3.116	5.209
0.04	1.01	0.9989	113.2	19.112	3.116	5.215
0.05	1.27	0.9986	113.1	18.646	3.116	5.220
0.06	1.52	0.9984	113.0	18.265	3.116	5.226
0.07	1.78	0.9981	113.0	17.942	3.116	5.231
0.08	2.03	0.9978	112.9	17.662	3.116	5.237
0.09	2.29	0.9976	112.8	17.414	3.116	5.242
0.1	2.54	0.9973	112.8	17.193	3.116	5.247
0.2	5.09	0 9950	112.2	15.726	3.115	5.302
0.3	7.65	0.9930	111.6	14.857	3.115	5.357
0.4	10.22	0.9914	111.0	14.234	3.115	5.411
0.5	12.80	0.9901	110.4	13.744	3.115	5.466
0.6	15.37	0.9891	109.9	13.340	3.115	5.520
0.7	17.94	0.9884	109.3	12.995	3.114	5.573
0.8	20.51	0.9881	108.8	12.693	3.114	5.625
0.9	23.08	0.9881	108.3	12.424	3.114	5.675
1.0	25.63	0.9885	107.8	12.182	3.114	5.724
1.5	38.17	0.9956	105.6	11.232	3.113	5.933
2.0	50.13	1.0108	103.9	10.545	3.112	6.074
2.5	61.33	1.0328	102.7	10.010	3.112	6.150
3.0	71.69	1.0603	102.0	9.576	3.113	6.176
3.5	81.22	1.0918	101.7	9.213	3.114	6.165
4.0	89.95	1.1268	101.7	8.904	3.115	6.125
4.5	97.91	1.1645	101.9	8.638	3.117	6.064
5.0	105.17	1 2045	102.4	8.406	3.118	5.988
5.5	111.79	1.2466	103.1	8.202	3.120	5.903
6.0	117.82	1.2903	104.1	8.020	3.121	5.815
6.5	123.34	1.3353	105.1	7.857	3.121	5.726
7.0	128.39	1 3814	106.3	7.711	3.121	5.639
7.5	133.05	1.4283	107.6	7.577	3.121	5.556
8.0	137.35	1.4758	109.0	7.456	3.120	5.478
8.5	141 35	1.5237	110.4	7.343	3.119	5.404
9.0	145.07	1.5719	111.9	7.240	3.118	5.335
9.5	148.56	1.6203	113.5	7.143	3.116	5.271
10.0	151.83	1.6688	115.1	7.053	3.114	5.211
15.0	176.91	2.1483	132.6	6.378	3.090	4.771
20.0	194.37	2.6071	150.9	5.924	3.066	4.489
25.0	208.09	3.0440	169.2	5.580	3.047	4.281
30	219.56	3.4620	187.3	5.304	3.034	4.120
35	229.49	3.8642	205.2	5.073	3.025	3.994
40	238.28	4.2533	222.8	4.876	3.020	3.896
45	246 17	4 6316	240.2	4.703	3.014	3.819
50	253.32	5.0011	257.3	4.556	3.006	3.758
55	259.84	5.3632	274.1	4.412	2.992	3.710
60	265.82	5.7191	290.7	4.285	2.971	3.669
65	271.33	6.0697	307.1	4.167	2.941	3.632
70	276 44	6.4158	323.3	4.056	2.899	3.595

Table II.17 (*Continued*)

p	ρ	z	h	s	c_v	c_p
75	281.20	6.7577	339.2	3.949	2.844	3.555
80	285.65	7.0959	354.9	3.847	2.775	3.511
85	289.84	7.4306	370.3	3.747	2.691	3.460
90	293.78	7.7620	385.6	3.649	2.591	3.400
95	297.53	8.0901	400.7	3.552	2.474	3.331
100	301.09	8.4150	415.5	3.456	2.338	3.251

$T = 20$ K

p	ρ	z	h	s	c_v	c_p
0.01	0.24	0.9998	118.5	22.267	3.116	5.198
0.02	0.48	0.9996	118.5	20.825	3.116	5.203
0.03	0.72	0.9994	118.4	19.980	3.116	5.208
0.04	0.96	0.9992	118.4	19.380	3.116	5.213
0.05	1.20	0.9990	118.3	18.914	3.116	5.218
0.06	1.45	0.9988	118.3	18.533	3.116	5.223
0.07	1.69	0.9986	118.2	18.210	3.116	5.227
0.08	1.93	0.9984	118.1	17.930	3.116	5.232
0.09	2.17	0.9982	118.1	17.683	3.116	5.237
0.1	2.41	0.9981	118.0	17.462	3.116	5.242
0.2	4.83	0.9964	117.5	15.998	3.115	5.291
0.3	7.26	0.9951	116.9	15.132	3.115	5.340
0.4	9.69	0.9941	116.4	14.510	3.115	5.389
0.5	12.12	0.9933	115.9	14.024	3.115	5.438
0.6	14.55	0.9929	115.3	13.622	3.115	5.486
0.7	16.97	0.9928	114.9	13.279	3.115	5.533
0.8	19.39	0.9929	114.4	12.980	3.115	5.579
0.9	21.81	0.9934	113.9	12.714	3.115	5.625
1.0	24.21	0.9941	113.5	12.474	3.115	5.668
1.5	36.02	1.0023	111.5	11.534	3.114	5.859
2.0	47.32	1.0173	109.9	10.854	3.114	5.995
2.5	57.96	1.0382	108.8	10.322	3.114	6.077
3.0	67.89	1.0637	108.1	9.889	3.115	6.115
3.5	77.08	1.0929	107.7	9.526	3.116	6.119
4.0	85.57	1.1251	107.7	9.216	3.118	6.096
4.5	93.39	1.1598	107.9	8.948	3.120	6.054
5.0	100.58	1.1966	108.4	8.712	3.121	5.996
5.5	107.18	1.2352	109.0	8.504	3.123	5.928
6.0	113.23	1.2754	109.8	8.318	3.124	5.854
6.5	118.80	1.3170	110.8	8.151	3.125	5.777
7.0	123.93	1.3596	111.9	8.000	3.126	5.700
7.5	128.66	1.4031	113.1	7.863	3.126	5.625
8.0	133.06	1.4472	114.4	7.737	3.126	5.552
8.5	137.14	1.4919	115.8	7.622	3.125	5.484
9.0	140.95	1.5369	117.3	7.515	3.125	5.419
9.5	144.52	1.5822	118.8	7.415	3.123	5.358
10.0	147.88	1.6277	120.3	7.322	3.122	5.300
15.0	173.58	2.0800	137.4	6.625	3.102	4.871
20.0	191.43	2.5148	155.4	6.157	3.079	4.590
25.0	205.41	2.9295	173.5	5.803	3.060	4.379
30	217.09	3.3262	191.4	5.518	3.047	4.210
35	227.20	3.7080	209.2	5.281	3.038	4.075
40	236.13	4.0774	226.7	5.079	3.033	3.968
45	244.14	4.4366	244.0	4.903	3.028	3.885
50	251.38	4.7876	261.1	4.746	3.021	3.821
55	257.98	5.1318	277.9	4.606	3.011	3.772
60	264.01	5.4703	294.5	4.477	2.995	3.734

Table II.17 (*Continued*)

p	ρ	z	h	s	c_v	c_p
65	269.57	5.8040	310.8	4.357	2.972	3.705
70	274.70	6.1336	326.9	4.245	2.940	3.679
75	279.47	6.4596	342.8	4.137	2.897	3.656
80	283.92	6.7823	358.5	4.033	2.843	3.633
85	288.09	7.1019	373.9	3.931	2.776	3.608
90	292.01	7.4186	389.2	3.831	2.696	3.580
95	295.73	7.7324	404.2	3.732	2.603	3.549
100	299.26	8.0434	419.0	3.633	2.495	3.512

$T = 25$ K

p	ρ	z	h	s	c_v	c_p
0.01	0.19	1.0000	144.5	23.427	3.115	5.196
0.02	0.39	1.0001	144.5	21.985	3.115	5.199
0.03	0.58	1.0001	144.5	21.141	3.115	5.201
0.04	0.77	1.0001	144.4	20.542	3.114	5.204
0.05	0.96	1.0001	144.4	20.077	3.114	5.207
0.06	1.16	1.0002	144.3	19.697	3.114	5.210
0.07	1.35	1.0002	144.3	19.375	3.113	5.213
0.08	1.54	1.0002	144.3	19.096	3.113	5.215
0.09	1.73	1.0003	144.2	18.850	3.112	5.218
0.1	1.92	1.0003	144.2	18.629	3.112	5.221
0.2	3.85	1.0008	143.8	17.174	3.108	5.248
0.3	5.77	1.0015	143.5	16.317	3.105	5.275
0.4	7.68	1.0024	143.1	15.704	3.102	5.302
0.5	9.59	1.0035	142.8	15.226	3.098	5.328
0.6	11.50	1.0047	142.5	14.833	3.095	5.354
0.7	13.40	1.0061	142.2	14.498	3.092	5.379
0.8	15.29	1.0077	141.9	14.207	3.089	5.404
0.9	17.17	1.0095	141.6	13.949	3.086	5.428
1.0	19.04	1.0114	141.3	13.717	3.083	5.451
1.5	28.23	1.0233	140.1	12.813	3.071	5.558
2.0	37.09	1.0384	139.2	12.160	3.060	5.648
2.5	45.58	1.0563	138.5	11.647	3.051	5.721
3.0	53.67	1.0763	138.0	11.225	3.043	5.776
3.5	61.37	1.0982	137.8	10.866	3.037	5.815
4.0	68.67	1.1217	137.7	10.556	3.031	5.836
4.5	75.57	1.1467	137.8	10.282	3.027	5.842
5.0	82.07	1.1732	138.1	10.040	3.023	5.833
5.5	88.19	1.2009	138.5	9.822	3.021	5.813
6.0	93.95	1.2298	139.1	9.625	3.018	5.783
6.5	99.36	1.2597	139.8	9.447	3.017	5.745
7.0	104.45	1.2905	140.6	9.284	3.016	5.703
7.5	109.23	1.3221	141.6	9.135	3.015	5.657
8.0	113.74	1.3544	142.6	8.997	3.015	5.610
8.5	117.99	1.3873	143.7	8.869	3.015	5.563
9.0	122.00	1.4206	144.9	8.751	3.016	5.516
9.5	125.79	1.4543	146.2	8.640	3.017	5.470
10.0	129.39	1.4883	147.5	8.537	3.017	5.426
15.0	157.51	1.8339	162.8	7.758	3.031	5.081
20.0	177.14	2.1741	179.6	7.239	3.048	4.865
25.0	192.35	2.5028	196.9	6.846	3.063	4.716
30	204.88	2.8196	214.1	6.528	3.075	4.602
35	215.61	3.1259	231.2	6.261	3.085	4.508
40	225.04	3.4227	248.1	6.029	3.090	4.426
45	233.48	3.7113	264.8	5.825	3.093	4.352
50	241.15	3.9926	281.3	5.643	3.092	4.284

Table II.17 (*Continued*)

p	ρ	z	h	s	c_v	c_p
55	248.18	4.2674	297.6	5.477	3.087	4.220
60	254.69	4.5364	313.7	5.326	3.079	4.159
65	260.75	4.8002	329.6	5.188	3.068	4.099
70	266.43	5.0593	345.4	5.059	3.052	4.040
75	271.77	5.3142	361.0	4.940	3.034	3.981
80	276.81	5.5652	376.4	4.828	3.012	3.922
85	281.58	5.8128	391.7	4.724	2.986	3.863
90	286.12	6.0571	406.8	4.625	2.958	3.802
95	290.44	6.2985	421.9	4.532	2.926	3.741
100	294.56	6.5372	436.8	4.445	2.891	3.678

$T = 30$ K

p	ρ	z	h	s	c_v	c_p
0.01	0.16	1.0001	170.5	24.374	3.116	5.195
0.02	0.32	1.0003	170.5	22.933	3.116	5.197
0.03	0.48	1.0004	170.5	22.089	3.116	5.200
0.04	0.64	1.0005	170.4	21.491	3.115	5.202
0.05	0.80	1.0007	170.4	21.026	3.115	5.204
0.06	0.96	1.0008	170.4	20.646	3.115	5.206
0.07	1.12	1.0009	170.4	20.325	3.115	5.208
0.08	1.28	1.0011	170.3	20.046	3.115	5.210
0.09	1.44	1.0012	170.3	19.801	3.115	5.212
0.1	1.60	1.0013	170.3	19.581	3.115	5.214
0.2	3.20	1.0028	170.0	18.130	3.114	5.236
0.3	4.79	1.0043	169.8	17.276	3.113	5.256
0.4	6.38	1.0060	169.6	16.668	3.112	5.277
0.5	7.96	1.0078	169.3	16.194	3.111	5.296
0.6	9.54	1.0097	169.1	15.805	3.110	5.316
0.7	11.10	1.0117	168.9	15.475	3.109	5.335
0.8	12.66	1.0139	168.7	15.188	3.109	5.354
0.9	14.21	1.0161	168.6	14.933	3.108	5.372
1.0	15.76	1.0184	168.4	14.705	3.108	5.390
1.5	23.34	1.0314	167.7	13.818	3.106	5.473
2.0	30.67	1.0465	167.1	13.180	3.106	5.546
2.5	37.74	1.0631	166.8	12.679	3.107	5.609
3.0	44.53	1.0811	166.6	12.267	3.108	5.663
3.5	51.06	1.1000	166.5	11.916	3.110	5.708
4.0	57.31	1.1199	166.6	11.611	3.113	5.745
4.5	63.31	1.1406	166.8	11.341	3.115	5.773
5.0	69.05	1.1620	167.1	11.100	3.119	5.793
5.5	74.54	1.1840	167.6	10.882	3.122	5.805
6.0	79.79	1.2067	168.1	10.683	3.125	5.809
6.5	84.80	1.2300	168.7	10.501	3.129	5.806
7.0	89.58	1.2539	169.4	10.334	3.133	5.797
7.5	94.15	1.2783	170.2	10.179	3.137	5.784
8.0	98.50	1.3033	171.7	10.036	3.141	5.766
8.5	102.66	1.3286	172.1	9.902	3.145	5.745
9.0	106.63	1.3544	173.1	9.777	3.149	5.722
9.5	110.42	1.3806	174.2	9.660	3.153	5.697
10.0	114.04	1.4071	175.3	9.550	3.158	5.671
15.0	143.12	1.6818	189.1	8.716	3.200	5.423
20.0	163.81	1.9593	205.0	8.161	3.240	5.246
25.0	179.86	2.2304	221.5	7.743	3.277	5.125
30	193.08	2.4933	238.2	7.407	3.311	5.036
35	204.37	2.7481	254.9	7.125	3.342	4.966
40	214.28	2.9954	271.4	6.881	3.371	4.908

Table II.17 (*Continued*)

p	ρ	z	h	s	c_v	c_p
45	223.15	3.2360	287.8	6.665	3.396	4.859
50	231.19	3.4705	304.1	6.473	3.418	4.816
55	238.55	3.6997	320.1	6.298	3.438	4.777
60	245.36	3.9240	336.0	6.139	3.455	4.742
65	251.70	4.1440	351.7	5.992	3.469	4.710
70	257.63	4.3601	367.2	5.855	3.480	4.680
75	263.20	4.5726	382.6	5.728	3.488	4.651
80	268.46	4.7819	397.9	5.609	3.494	4.624
85	273.43	4.9884	412.9	5.497	3.497	4.597
90	278.16	5.1921	427.9	5.391	3.497	4.572
95	282.65	5.3934	442.7	5.290	3.494	4.546
100	286.94	5.5925	457.4	5.194	3.489	4.520

$T = 35$ K

p	ρ	z	h	s	c_v	c_p
0.01	0.14	1.0002	196.5	25.175	3.116	5.195
0.02	0.27	1.0004	196.5	23.734	3.116	5.197
0.03	0.41	1.0005	196.5	22.891	3.116	5.198
0.04	0.55	1.0007	196.4	22.292	3.116	5.200
0.05	0.69	1.0009	196.4	21.828	3.116	5.202
0.06	0.82	1.0011	196.4	21.449	3.116	5.204
0.07	0.96	1.0013	196.4	21.127	3.116	5.205
0.08	1.10	1.0014	196.4	20.849	3.116	5.207
0.09	1.24	1.0016	196.4	20.604	3.117	5.209
0.1	1.37	1.0018	196.3	20.384	3.117	5.210
0.2	2.74	1.0037	196.2	18.936	3.117	5.228
0.3	4.10	1.0056	196.0	18.086	3.118	5.244
0.4	5.46	1.0077	195.9	17.480	3.119	5.261
0.5	6.81	1.0098	195.8	17.009	3.120	5.277
0.6	8.16	1.0120	195.7	16.623	3.121	5.293
0.7	9.49	1.0142	195.5	16.295	3.122	5.308
0.8	10.82	1.0165	195.4	16.010	3.124	5.323
0.9	12.15	1.0189	195.3	15.759	3.125	5.338
1.0	13.47	1.0214	195.2	15.533	3.126	5.353
1.5	19.94	1.0345	194.9	14.657	3.132	5.420
2.0	26.22	1.0491	194.7	14.029	3.139	5.481
2.5	32.30	1.0646	194.6	13.538	3.147	5.536
3.0	38.17	1.0811	194.7	13.134	3.155	5.584
3.5	43.84	1.0982	194.9	12.790	3.163	5.628
4.0	49.31	1.1158	195.1	12.491	3.170	5.666
4.5	54.58	1.1339	195.5	12.226	3.178	5.699
5.0	59.68	1.1524	195.9	11.988	3.186	5.728
5.5	64.59	1.1713	196.5	11.773	3.194	5.751
6.0	69.32	1.1905	197.0	11.576	3.202	5.769
6.5	73.88	1.2101	197.7	11.395	3.210	5.783
7.0	78.28	1.2299	198.4	11.228	3.218	5.792
7.5	82.52	1.2501	199.2	11.073	3.225	5.797
8.0	86.60	1.2706	200.1	10.928	3.233	5.799
8.5	90.54	1.2913	201.0	10.793	3.240	5.797
9.0	94.32	1.3124	201.9	10.666	3.247	5.792
9.5	97.97	1.3337	202.9	10.546	3.254	5.784
10.0	101.49	1.3553	204.0	10.433	3.261	5.774
15.0	130.53	1.5806	216.8	9.570	3.327	5.629
20.0	151.78	1.8124	231.9	8.990	3.384	5.496
25.0	168.41	2.0418	247.9	8.556	3.436	5.400

Table II.17 (*Continued*)

p	ρ	z	h	s	c_v	c_p
30	182.12	2.2657	264.2	8.207	3.483	5.330
35	193.84	2.4835	280.6	7.915	3.527	5.276
40	204.13	2.6952	296.9	7.664	3.568	5.234
45	213.33	2.9014	313.1	7.442	3.606	5.199
50	221.67	3.1025	329.1	7.244	3.642	5.169
55	229.31	3.2990	345.0	7.065	3.675	5.144
60	236.37	3.4915	360.7	6.901	3.705	5.122
65	242.93	3.6802	376.3	6.750	3.734	5.103
70	249.07	3.8656	391.8	6.610	3.759	5.087
75	254.83	4.0480	407.0	6.480	3.783	5.073
80	260.27	4.2278	422.2	6.358	3.804	5.061
85	265.41	4.4050	437.2	6.243	3.823	5.051
90	270.28	4.5800	452.0	6.134	3.839	5.043
95	274.91	4.7530	466.8	6.031	3.854	5.035
100	279.33	4.9241	481.4	5.933	3.866	5.029

$T = 40$ K

p	ρ	z	h	s	c_v	c_p
0.01	0.12	1.0002	222.5	25.868	3.116	5.195
0.02	0.24	1.0004	222.4	24.428	3.116	5.196
0.03	0.36	1.0006	222.4	23.585	3.116	5.197
0.04	0.48	1.0008	222.4	22.987	3.117	5.199
0.05	0.60	1.0010	222.4	22.523	3.117	5.200
0.06	0.72	1.0012	222.4	22.143	3.117	5.202
0.07	0.84	1.0014	222.4	21.822	3.117	5.203
0.08	0.96	1.0016	222.4	21.544	3.117	5.204
0.09	1.08	1.0018	222.4	21.299	3.117	5.206
0.1	1.20	1.0020	222.4	21.080	3.118	5.207
0.2	2.40	1.0041	222.3	19.634	3.120	5.221
0.3	3.59	1.0062	222.2	18.785	3.121	5.235
0.4	4.77	1.0083	222.2	18.182	3.123	5.248
0.5	5.95	1.0106	222.1	17.713	3.125	5.261
0.6	7.13	1.0128	222.1	17.328	3.127	5.274
0.7	8.30	1.0151	222.0	17.002	3.129	5.287
0.8	9.46	1.0175	222.0	16.720	3.131	5.299
0.9	10.62	1.0199	222.0	16.470	3.133	5.311
1.0	11.77	1.0223	221.9	16.245	3.135	5.323
1.5	17.44	1.0352	221.9	15.378	3.145	5.379
2.0	22.95	1.0490	222.0	14.758	3.156	5.430
2.5	28.29	1.0636	222.2	14.273	3.167	5.476
3.0	33.47	1.0787	222.5	13.875	3.177	5.518
3.5	38.49	1.0943	222.8	13.537	3.188	5.555
4.0	43.36	1.1103	223.3	13.242	3.198	5.591
4.5	48.07	1.1266	223.8	12.982	3.209	5.623
5.0	52.64	1.1431	224.4	12.748	3.219	5.652
5.5	57.07	1.1599	225.0	12.536	3.229	5.677
6.0	61.36	1.1768	225.7	12.342	3.239	5.700
6.5	65.52	1.1939	226.5	12.164	3.249	5.720
7.0	69.56	1.2112	227.3	11.998	3.259	5.736
7.5	73.47	1.2287	228.1	11.844	3.268	5.749
8.0	77.26	1.2463	229.0	11.700	3.277	5.760
8.5	80.93	1.2641	229.9	11.565	3.286	5.768
9.0	84.49	1.2820	230.9	11.438	3.295	5.773
9.5	87.94	1.3001	231.9	11.319	3.304	5.776
10.0	91.28	1.3184	232.9	11.205	3.312	5.777
15.0	119.66	1.5087	245.2	10.328	3.390	5.715

Table II.17 (*Continued*)

p	ρ	z	h	s	c_v	c_p
20.0	141.07	1.7063	259.7	9.733	3.456	5.621
25.0	158.03	1.9039	275.3	9.287	3.515	5.546
30	172.08	2.0982	291.3	8.931	3.569	5.490
35	184.11	2.2879	307.4	8.632	3.618	5.448
40	194.68	2.4728	323.5	8.376	3.665	5.415
45	204.14	2.6530	339.6	8.150	3.708	5.388
50	212.72	2.8289	355.5	7.948	3.750	5.365
55	220.58	3.0009	371.3	7.766	3.789	5.346
60	227.85	3.1692	386.9	7.600	3.825	5.330
65	234.61	3.3344	402.4	7.447	3.860	5.316
70	240.93	3.4967	417.8	7.305	3.893	5.305
75	246.86	3.6564	433.0	7.174	3.924	5.296
80	252.46	3.8138	448.1	7.050	3.953	5.290
85	257.74	3.9690	463.1	6.935	3.980	5.285
90	262.75	4.1224	477.9	6.825	4.004	5.283
95	267.51	4.2741	492.7	6.722	4.028	5.282
100	272.03	4.4241	507.3	6.624	4.049	5.283

$T = 45$ K

p	ρ	z	h	s	c_v	c_p
0.01	0.11	1.0002	248.4	26.480	3.116	5.194
0.02	0.21	1.0004	248.4	25.040	3.116	5.195
0.03	0.32	1.0006	248.4	24.197	3.117	5.197
0.04	0.43	1.0008	248.4	23.599	3.117	5.198
0.05	0.53	1.0010	248.4	23.135	3.117	5.199
0.06	0.64	1.0012	248.4	22.756	3.117	5.200
0.07	0.75	1.0014	248.4	22.435	3.117	5.201
0.08	0.85	1.0017	248.4	22.157	3.118	5.202
0.09	0.96	1.0019	248.4	21.912	3.118	5.204
0.1	1.07	1.0021	248.4	21.693	3.118	5.205
0.2	2.13	1.0042	248.4	20.248	3.120	5.216
0.3	3.19	1.0063	248.4	19.402	3.123	5.227
0.4	4.24	1.0085	248.4	18.800	3.125	5.238
0.5	5.29	1.0107	248.4	18.332	3.127	5.249
0.6	6.34	1.0130	248.4	17.949	3.129	5.259
0.7	7.38	1.0152	248.4	17.624	3.132	5.270
0.8	8.41	1.0176	248.4	17.343	3.134	5.280
0.9	9.44	1.0199	248.5	17.094	3.136	5.290
1.0	10.46	1.0223	248.5	16.871	3.139	5.299
1.5	15.51	1.0346	248.7	16.010	3.150	5.345
2.0	20.42	1.0477	249.0	15.395	3.162	5.387
2.5	25.20	1.0612	249.4	14.915	3.174	5.426
3.0	29.85	1.0752	249.9	14.522	3.185	5.461
3.5	34.36	1.0896	250.5	14.187	3.197	5.494
4.0	38.75	1.1042	251.1	13.897	3.208	5.525
4.5	43.02	1.1191	251.7	13.640	3.219	5.553
5.0	47.16	1.1341	252.5	13.410	3.230	5.579
5.5	51.20	1.1492	253.2	13.201	3.241	5.603
6.0	55.12	1.1645	254.0	13.009	3.251	5.626
6.5	58.94	1.1799	254.9	12.833	3.262	5.646
7.0	62.65	1.1953	255.8	12.670	3.272	5.664
7.5	66.26	1.2108	256.7	12.518	3.282	5.680
8.0	69.78	1.2264	257.6	12.375	3.291	5.694
8.5	73.20	1.2422	258.6	12.241	3.301	5.706
9.0	76.54	1.2580	259.6	12.115	3.310	5.717
9.5	79.78	1.2738	260.6	11.996	3.319	5.725

Table II.17 (*Continued*)

p	ρ	z	h	s	c_v	c_p
10.0	82.94	1.2898	261.7	11.883	3.328	5.731
15.0	110.31	1.4547	273.8	11.002	3.410	5.726
20.0	131.58	1.6261	287.9	10.398	3.479	5.666
25.0	148.68	1.7989	303.2	9.945	3.540	5.608
30	162.92	1.9698	318.9	9.582	3.595	5.563
35	175.17	2.1375	334.9	9.279	3.646	5.528
40	185.94	2.3014	350.9	9.019	3.693	5.501
45	195.59	2.4613	366.8	8.791	3.738	5.479
50	204.35	2.6175	382.6	8.587	3.781	5.460
55	212.39	2.7703	398.3	8.402	3.822	5.444
60	219.83	2.9199	413.9	8.234	3.860	5.430
65	226.75	3.0666	429.3	8.080	3.897	5.418
70	233.23	3.2108	444.6	7.937	3.932	5.408
75	239.32	3.3526	459.8	7.804	3.966	5.400
80	245.05	3.4924	474.9	7.680	3.998	5.393
85	250.47	3.6304	489.8	7.564	4.028	5.389
90	255.61	3.7667	504.7	7.455	4.057	5.387
95	260.48	3.9016	519.4	7.351	4.084	5.387
100	265.12	4.0351	534.0	7.253	4.109	5.389

$T = 50$ K

p	ρ	z	h	s	c_v	c_p
0.01	0.10	1.0002	274.4	27.027	3.116	5.194
0.02	0.19	1.0004	274.4	25.587	3.116	5.195
0.03	0.29	1.0006	274.4	24.745	3.117	5.196
0.04	0.38	1.0008	274.4	24.147	3.117	5.197
0.05	0.48	1.0010	274.4	23.683	3.117	5.198
0.06	0.58	1.0012	274.4	23.304	3.117	5.199
0.07	0.67	1.0014	274.4	22.983	3.117	5.200
0.08	0.77	1.0016	274.4	22.705	3.118	5.201
0.09	0.86	1.0019	274.4	22.460	3.118	5.202
0.1	0.96	1.0021	274.4	22.241	3.118	5.203
0.2	1.92	1.0041	274.5	20.798	3.120	5.212
0.3	2.87	1.0063	274.5	19.952	3.123	5.221
0.4	3.82	1.0084	274.6	19.351	3.125	5.230
0.5	4.76	1.0106	274.6	18.884	3.127	5.239
0.6	5.70	1.0127	274.7	18.502	3.130	5.247
0.7	6.64	1.0150	274.7	18.179	3.132	5.256
0.8	7.57	1.0172	274.8	17.898	3.134	5.264
0.9	8.50	1.0195	274.9	17.650	3.137	5.272
1.0	9.42	1.0217	274.9	17.428	3.139	5.280
1.5	13.97	1.0335	275.3	16.572	3.151	5.318
2.0	18.41	1.0457	275.8	15.961	3.162	5.353
2.5	22.74	1.0584	276.4	15.485	3.174	5.384
3.0	26.96	1.0714	277.1	15.094	3.185	5.414
3.5	31.07	1.0846	277.8	14.764	3.196	5.442
4.0	35.07	1.0981	278.6	14.476	3.207	5.468
4.5	38.97	1.1117	279.4	14.222	3.218	5.492
5.0	42.77	1.1255	280.2	13.994	3.229	5.515
5.5	46.48	1.1394	281.1	13.787	3.240	5.537
6.0	50.09	1.1533	282.0	13.598	3.250	5.557
6.5	53.61	1.1673	282.9	13.424	3.260	5.576
7.0	57.05	1.1813	283.9	13.263	3.270	5.593
7.5	60.41	1.1954	284.9	13.112	3.290	5.609
8.0	63.68	1.2095	285.9	12.972	3.289	5.624

Table II.17 (*Continued*)

p	ρ	z	h	s	c_v	c_p
8.5	66.88	1.2236	286.9	12.839	3.298	5.637
9.0	70.00	1.2378	288.0	12.714	3.308	5.649
9.5	73.05	1.2521	289.1	12.596	3.317	5.660
10.0	76.03	1.2664	290.2	12.484	3.325	5.669
15.0	102.26	1.4123	302.4	11.604	3.406	5.698
20.0	123.18	1.5632	316.3	10.996	3.474	5.665
25.0	140.25	1.7162	331.3	10.537	3.534	5.622
30	154.59	1.8684	346.8	10.170	3.588	5.584
35	166.96	2.0184	362.6	9.864	3.637	5.555
40	177.86	2.1653	378.5	9.601	3.683	5.532
45	187.65	2.3089	394.3	9.370	3.727	5.514
50	196.55	2.4493	410.0	9.164	3.769	5.498
55	204.72	2.5867	425.6	8.978	3.809	5.483
60	212.29	2.7212	441.1	8.809	3.847	5.471
65	219.35	2.8531	456.5	8.653	3.884	5.459
70	225.96	2.9827	471.8	8.509	3.919	5.448
75	232.17	3.1102	486.9	8.376	3.952	5.439
80	238.04	3.2358	502.0	8.251	3.985	5.431
85	243.58	3.3598	516.9	8.134	4.016	5.425
90	248.84	3.4823	531.7	8.024	4.045	5.420
95	253.83	3.6035	546.4	7.921	4.073	5.418
100	258.57	3.7236	561.0	7.823	4.100	5.417

$$T = 55 \text{ K}$$

p	ρ	z	h	s	c_v	c_p
0.01	0.09	1.0002	300.4	27.522	3.116	5.194
0.02	0.17	1.0004	300.4	26.082	3.116	5.195
0.03	0.26	1.0006	300.4	25.240	3.117	5.195
0.04	0.35	1.0008	300.4	24.642	3.117	5.196
0.05	0.44	1.0010	300.4	24.178	3.117	5.197
0.06	0.52	1.0012	300.4	23.799	3.117	5.198
0.07	0.61	1.0014	300.4	23.479	3.117	5.199
0.08	0.70	1.0016	300.4	23.201	3.118	5.199
0.09	0.79	1.0018	300.4	22.956	3.118	5.200
0.1	0.87	1.0020	300.4	22.737	3.118	5.201
0.2	1.74	1.0040	300.5	21.294	3.120	5.208
0.3	2.61	1.0061	300.6	20.449	3.122	5.216
0.4	3.47	1.0082	300.7	19.849	3.125	5.223
0.5	4.33	1.0102	300.8	19.383	3.127	5.230
0.6	5.19	1.0123	300.9	19.002	3.129	5.237
0.7	6.04	1.0145	301.0	18.679	3.131	5.244
0.8	6.89	1.0166	301.1	18.399	3.134	5.251
0.9	7.73	1.0188	301.2	18.152	3.136	5.258
1.0	8.57	1.0209	301.3	17.931	3.138	5.264
1.5	12.72	1.0321	301.9	17.077	3.149	5.296
2.0	16.78	1.0435	302.5	16.469	3.160	5.324
2.5	20.73	1.0554	303.3	15.996	3.171	5.351
3.0	24.60	1.0674	304.1	15.609	3.181	5.375
3.5	28.37	1.0797	304.9	15.280	3.192	5.399
4.0	32.06	1.0922	305.8	14.995	3.202	5.421
4.5	35.65	1.1048	306.7	14.743	3.213	5.441
5.0	39.16	1.1175	307.6	14.517	3.223	5.461
5.5	42.59	1.1303	308.6	14.312	3.232	5.480
6.0	45.94	1.1431	309.6	14.125	3.242	5.497
6.5	49.22	1.1559	310.7	13.953	3.252	5.514

Table II.17 (*Continued*)

p	ρ	z	h	s	c_v	c_p
7.0	52.42	1.1688	311.7	13.793	3.261	5.530
7.5	55.55	1.1817	312.8	13.644	3.270	5.545
8.0	58.61	1.1946	313.9	13.504	3.279	5.559
8.5	61.61	1.2076	315.0	13.373	3.288	5.572
9.0	64.54	1.2205	316.1	13.250	3.296	5.584
9.5	67.41	1.2335	317.2	13.132	3.305	5.595
10.0	70.22	1.2465	318.4	13.021	3.313	5.604
15.0	95.29	1.3778	330.8	12.145	3.389	5.653
20.0	115.73	1.5126	344.6	11.535	3.454	5.641
25.0	132.66	1.6495	359.4	11.072	3.511	5.609
30	146.99	1.7864	374.7	10.702	3.563	5.578
35	159.42	1.9217	390.4	10.393	3.610	5.553
40	170.40	2.0547	406.1	10.128	3.654	5.533
45	180.27	2.1849	421.9	9.896	3.695	5.517
50	189.26	2.3124	437.5	9.689	3.735	5.503
55	197.53	2.4371	453.1	9.501	3.773	5.490
60	205.20	2.5593	468.5	9.331	3.809	5.478
65	212.36	2.6791	483.8	9.174	3.844	5.467
70	219.08	2.7967	499.0	9.029	3.878	5.456
75	225.40	2.9124	514.1	8.895	3.910	5.445
80	231.37	3.0264	529.1	8.769	3.942	5.435
85	237.03	3.1388	544.0	8.651	3.972	5.427
90	242.39	3.2499	558.8	8.541	4.001	5.419
95	247.49	3.3598	573.5	8.437	4.029	5.413
100	252.34	3.4687	588.1	8.339	4.056	5.409

$T = 60$ K

p	ρ	z	h	s	c_v	c_p
0.01	0.08	1.0002	326.3	27.974	3.116	5.194
0.02	0.16	1.0004	326.4	26.534	3.116	5.194
0.03	0.24	1.0006	326.4	25.692	3.116	5.195
0.04	0.32	1.0008	326.4	25.094	3.117	5.196
0.05	0.40	1.0010	326.4	24.630	3.117	5.196
0.06	0.48	1.0012	326.4	24.251	3.117	5.197
0.07	0.56	1.0014	326.4	23.931	3.117	5.198
0.08	0.64	1.0016	326.4	23.653	3.118	5.198
0.09	0.72	1.0017	326.4	23.408	3.118	5.199
0.1	0.80	1.0019	326.4	23.189	3.118	5.199
0.2	1.60	1.0039	326.6	21.747	3.120	5.206
0.3	2.39	1.0059	326.7	20.903	3.122	5.212
0.4	3.18	1.0078	326.8	20.303	3.124	5.218
0.5	3.97	1.0098	326.9	19.838	3.126	5.224
0.6	4.76	1.0119	327.0	19.457	3.128	5.230
0.7	5.54	1.0139	327.2	19.135	3.130	5.235
0.8	6.32	1.0159	327.3	18.856	3.132	5.241
0.9	7.09	1.0180	327.4	18.609	3.134	5.246
1.0	7.87	1.0200	327.6	18.388	3.136	5.252
1.5	11.68	1.0305	328.3	17.537	3.146	5.278
2.0	15.41	1.0413	329.1	16.932	3.156	5.301
2.5	19.06	1.0524	329.9	16.461	3.166	5.323
3.0	22.63	1.0637	330.8	16.075	3.176	5.344
3.5	26.12	1.0751	331.8	15.748	3.186	5.363
4.0	29.53	1.0867	332.8	15.465	3.195	5.382
4.5	32.87	1.0984	333.8	15.215	3.205	5.399
5.0	36.14	1.1101	334.8	14.990	3.214	5.416
5.5	39.33	1.1220	335.9	14.787	3.223	5.432

Table II.17 (*Continued*)

p	ρ	z	h	s	c_v	c_p
6.0	42.46	1.1338	337.0	14.601	3.232	5.447
6.5	45.52	1.1457	338.1	14.430	3.240	5.462
7.0	48.52	1.1576	339.2	14.272	3.249	5.475
7.5	51.45	1.1696	340.4	14.124	3.257	5.489
8.0	54.33	1.1815	341.5	13.986	3.266	5.501
8.5	57.15	1.1934	342.7	13.856	3.274	5.513
9.0	59.91	1.2053	343.8	13.733	3.282	5.524
9.5	62.62	1.2173	345.0	13.617	3.290	5.535
10.0	65.27	1.2292	346.2	13.506	3.297	5.545
15.0	89.22	1.3490	358.9	12.635	3.368	5.603
20.0	109.10	1.4708	372.7	12.024	3.429	5.605
25.0	125.80	1.5945	387.3	11.559	3.482	5.584
30	140.06	1.7186	402.6	11.186	3.530	5.558
35	152.48	1.8417	418.1	10.876	3.574	5.536
40	163.49	1.9630	433.8	10.609	3.615	5.518
45	173.41	2.0821	449.4	10.375	3.654	5.503
50	182.45	2.1988	465.0	10.167	3.691	5.491
55	190.79	2.3130	480.5	9.979	3.726	5.480
60	198.53	2.4249	495.9	9.807	3.760	5.469
65	205.76	2.5346	511.2	9.649	3.793	5.458
70	212.56	2.6423	526.3	9.503	3.825	5.447
75	218.97	2.7481	541.3	9.368	3.855	5.435
80	225.03	2.8524	556.3	9.242	3.885	5.424
85	230.78	2.9551	571.1	9.123	3.914	5.414
90	236.24	3.0567	585.9	9.012	3.942	5.404
95	241.43	3.1571	600.5	8.907	3.969	5.395
100	246.38	3.2565	615.1	8.809	3.994	5.387
$T=65$ K						
0.01	0.07	1.0002	352.3	28.390	3.116	5.194
0.02	0.15	1.0004	352.3	26.950	3.116	5.194
0.03	0.22	1.0006	352.3	26.108	3.116	5.195
0.04	0.30	1.0007	352.3	25.510	3.117	5.195
0.05	0.37	1.0009	352.4	25.046	3.117	5.196
0.06	0.44	1.0011	352.4	24.667	3.117	5.196
0.07	0.52	1.0013	352.4	24.347	3.117	5.197
0.08	0.59	1.0015	352.4	24.069	3.117	5.197
0.09	0.67	1.0017	352.4	23.824	3.118	5.198
0.1	0.74	1.0019	352.4	23.605	3.118	5.198
0.2	1.48	1.0037	352.6	22.164	3.120	5.204
0.3	2.21	1.0056	352.7	21.320	3.121	5.209
0.4	2.94	1.0075	352.9	20.721	3.123	5.214
0.5	3.67	1.0094	353.0	20.256	3.125	5.218
0.6	4.39	1.0113	353.2	19.875	3.127	5.223
0.7	5.12	1.0133	353.3	19.554	3.129	5.228
0.8	5.84	1.0152	353.5	19.275	3.131	5.233
0.9	6.55	1.0171	353.6	19.029	3.133	5.237
1.0	7.27	1.0191	353.8	18.808	3.134	5.242
1.5	10.80	1.0290	354.7	17.959	3.144	5.263
2.0	14.25	1.0392	355.6	17.355	3.153	5.283
2.5	17.64	1.0495	356.5	16.886	3.162	5.301
3.0	20.96	1.0601	357.5	16.502	3.170	5.318
3.5	24.21	1.0708	358.5	16.176	3.179	5.335
4.0	27.39	1.0815	359.6	15.894	3.188	5.350
4.5	30.51	1.0924	360.7	15.645	3.196	5.365

Table II.17 (*Continued*)

p	ρ	z	h	s	c_v	c_p
5.0	33.56	1.1034	361.8	15.422	3.205	5.379
5.5	36.55	1.1144	362.9	15.220	3.213	5.392
6.0	39.48	1.1255	364.1	15.036	3.221	5.405
6.5	42.36	1.1365	365.3	14.866	3.229	5.418
7.0	45.18	1.1476	366.5	14.708	3.236	5.430
7.5	47.94	1.1587	367.7	14.561	3.244	5.441
8.0	50.65	1.1698	368.9	14.424	3.252	5.452
8.5	53.31	1.1808	370.1	14.295	3.259	5.463
9.0	55.92	1.1919	371.3	14.173	3.266	5.473
9.5	58.49	1.2030	372.6	14.057	3.273	5.483
10.0	61.01	1.2140	373.8	13.948	3.280	5.492
15.0	83.89	1.3243	386.8	13.081	3.345	5.553
20.0	103.18	1.4357	400.6	12.471	3.401	5.565
25.0	119.58	1.5484	415.2	12.005	3.451	5.552
30	133.71	1.6617	430.3	11.630	3.495	5.531
35	146.09	1.7744	445.7	11.318	3.536	5.512
40	157.09	1.8859	461.3	11.050	3.574	5.495
45	167.02	1.9955	476.9	10.815	3.610	5.482
50	176.09	2.1030	492.4	10.606	3.644	5.471
55	184.46	2.2083	507.9	10.417	3.676	5.460
60	192.24	2.3116	523.2	10.244	3.708	5.451
65	199.52	2.4128	538.4	10.086	3.738	5.440
70	206.38	2.5121	553.5	9.939	3.768	5.430
75	212.85	2.6097	568.5	9.802	3.797	5.419
80	218.98	2.7057	583.4	9.675	3.824	5.407
85	224.81	2.8003	598.1	9.556	3.851	5.395
90	230.35	2.8937	612.8	9.444	3.878	5.384
95	235.63	2.9860	627.4	9.338	3.903	5.372
100	240.66	3.0774	642.0	9.239	3.927	5.361
$T=70$ K						
0.01	0.07	1.0002	378.3	28.775	3.116	5.194
0.02	0.14	1.0004	378.3	27.335	3.116	5.194
0.03	0.21	1.0005	378.3	26.493	3.116	5.194
0.04	0.27	1.0007	378.3	25.895	3.117	5.195
0.05	0.34	1.0009	378.3	25.431	3.117	5.195
0.06	0.41	1.0011	378.4	25.052	3.117	5.196
0.07	0.48	1.0012	378.4	24.732	3.117	5.196
0.08	0.55	1.0014	378.4	24.454	3.117	5.197
0.09	0.62	1.0016	378.4	24.210	3.117	5.197
0.1	0.69	1.0018	378.4	23.991	3.118	5.197
0.2	1.37	1.0036	378.6	22.549	3.119	5.202
0.3	2.05	1.0054	378.8	21.706	3.121	5.206
0.4	2.73	1.0072	378.9	21.107	3.123	5.210
0.5	3.41	1.0090	379.1	20.642	3.124	5.214
0.6	4.08	1.0108	379.3	20.262	3.126	5.218
0.7	4.75	1.0126	379.5	19.941	3.128	5.222
0.8	5.42	1.0145	379.6	19.662	3.129	5.226
0.9	6.09	1.0163	379.8	19.416	3.131	5.230
1.0	6.75	1.0182	380.0	19.196	3.133	5.234
1.5	10.04	1.0276	380.9	18.349	3.141	5.251
2.0	13.26	1.0371	381.9	17.746	3.149	5.268
2.5	16.42	1.0469	383.0	17.278	3.157	5.283
3.0	19.52	1.0567	384.0	16.895	3.165	5.298
3.5	22.56	1.0667	385.1	16.571	3.173	5.311

Table II.17 (*Continued*)

p	ρ	z	h	s	c_v	c_p
4.0	25.55	1.0768	386.3	16.290	3.180	5.324
4.5	28.47	1.0870	387.4	16.042	3.188	5.336
5.0	31.34	1.0973	388.6	15.820	3.196	5.348
5.5	34.15	1.1075	389.8	15.619	3.203	5.360
6.0	36.91	1.1179	391.0	15.435	3.210	5.371
6.5	39.62	1.1282	392.3	15.266	3.217	5.381
7.0	42.28	1.1385	393.5	15.109	3.224	5.392
7.5	44.89	1.1489	394.8	14.963	3.231	5.402
8.0	47.46	1.1592	396.0	14.826	3.238	5.411
8.5	49.98	1.1696	397.3	14.698	3.245	5.420
9.0	52.46	1.1799	398.6	14.577	3.251	5.429
9.5	54.89	1.1902	399.9	14.462	3.258	5.438
10.0	57.29	1.2005	401.2	14.353	3.264	5.446
15.0	79.17	1.3030	414.4	13.491	3.323	5.506
20.0	97.86	1.4056	428.3	12.882	3.375	5.525
25.0	113.93	1.5091	442.9	12.415	3.420	5.519
30	127.89	1.6132	457.9	12.039	3.461	5.502
35	140.18	1.7171	473.2	11.725	3.499	5.485
40	151.14	1.8200	488.7	11.456	3.534	5.469
45	161.06	1.9215	504.2	11.220	3.567	5.457
50	170.13	2.0212	519.7	11.010	3.598	5.447
55	178.51	2.1189	535.1	10.820	3.628	5.437
60	186.31	2.2148	550.4	10.647	3.657	5.429
65	193.62	2.3087	565.6	10.488	3.685	5.420
70	200.51	2.4009	580.6	10.341	3.712	5.410
75	207.03	2.4914	595.5	10.203	3.738	5.399
80	213.21	2.5804	610.3	10.075	3.764	5.388
85	219.10	2.6681	625.1	9.955	3.789	5.376
90	224.70	2.7545	639.7	9.842	3.813	5.363
95	230.05	2.8399	654.2	9.736	3.837	5.350
100	235.16	2.9244	668.7	9.635	3.860	5.337

$T = 75$ K

p	ρ	z	h	s	c_v	c_p
0.01	0.06	1.0002	404.2	29.133	3.116	5.193
0.02	0.13	1.0003	404.3	27.693	3.116	5.194
0.03	0.19	1.0005	404.3	26.851	3.116	5.194
0.04	0.26	1.0007	404.3	26.253	3.116	5.195
0.05	0.32	1.0008	404.3	25.790	3.117	5.195
0.06	0.38	1.0010	404.3	25.411	3.117	5.195
0.07	0.45	1.0012	404.4	25.090	3.117	5.196
0.08	0.51	1.0014	404.4	24.813	3.117	5.196
0.09	0.58	1.0015	404.4	24.568	3.117	5.196
0.1	0.64	1.0017	404.4	24.349	3.117	5.197
0.2	1.28	1.0034	404.6	22.908	3.119	5.200
0.3	1.92	1.0051	404.8	22.065	3.120	5.204
0.4	2.55	1.0068	405.0	21.466	3.122	5.207
0.5	3.18	1.0086	405.2	21.002	3.123	5.211
0.6	3.81	1.0103	405.4	20.622	3.125	5.214
0.7	4.44	1.0120	405.6	20.301	3.126	5.217
0.8	5.07	1.0138	405.8	20.023	3.128	5.221
0.9	5.69	1.0155	405.9	19.777	3.129	5.224
1.0	6.31	1.0173	406.1	19.557	3.131	5.227
1.5	9.38	1.0262	407.2	18.711	3.138	5.242
2.0	12.40	1.0352	408.2	18.109	3.145	5.256
2.5	15.37	1.0444	409.3	17.642	3.153	5.269

Table II.17 (Continued)

p	ρ	z	h	s	c_v	c_p
3.0	18.28	1.0536	410.5	17.260	3.160	5.281
3.5	21.13	1.0630	411.7	16.937	3.167	5.292
4.0	23.94	1.0725	412.8	16.656	3.174	5.303
4.5	26.69	1.0821	414.1	16.409	3.180	5.313
5.0	29.40	1.0917	415.3	16.188	3.187	5.323
5.5	32.06	1.1013	416.6	15.987	3.194	5.333
6.0	34.67	1.1110	417.8	15.804	3.200	5.342
6.5	37.23	1.1207	419.1	15.636	3.207	5.351
7.0	39.75	1.1304	420.4	15.480	3.213	5.360
7.5	42.23	1.1401	421.7	15.335	3.219	5.368
8.0	44.66	1.1498	423.0	15.199	3.225	5.377
8.5	47.06	1.1595	424.3	15.071	3.231	5.385
9.0	49.41	1.1691	425.6	14.950	3.237	5.393
9.5	51.73	1.1788	427.0	14.836	3.243	5.400
10.0	54.01	1.1884	428.3	14.728	3.249	5.408
15.0	74.97	1.2842	441.9	13.870	3.302	5.464
20.0	93.06	1.3794	455.9	13.262	3.349	5.487
25.0	108.78	1.4752	470.4	12.795	3.391	5.486
30	122.54	1.5715	485.3	12.418	3.429	5.473
35	134.71	1.6677	500.6	12.103	3.463	5.458
40	145.61	1.7632	516.0	11.833	3.495	5.444
45	155.49	1.8576	531.5	11.596	3.526	5.432
50	164.54	1.9505	546.9	11.385	3.554	5.422
55	172.91	2.0417	562.2	11.195	3.582	5.413
60	180.72	2.1311	577.5	11.021	3.608	5.406
65	188.04	2.2188	592.6	10.861	3.634	5.398
70	194.94	2.3048	607.6	10.713	3.659	5.389
75	201.48	2.3893	622.5	10.575	3.683	5.380
80	207.70	2.4723	637.2	10.446	3.707	5.369
85	213.63	2.5540	651.9	10.325	3.730	5.357
90	219.28	2.6344	666.5	10.211	3.752	5.345
95	224.69	2.7139	680.9	10.104	3.774	5.331
100	229.86	2.7924	695.3	10.003	3.796	5.317

$T = 80$ K

p	ρ	z	h	s	c_v	c_p
0.01	0.06	1.0002	430.2	29.468	3.116	5.193
0.02	0.12	1.0003	430.2	28.028	3.116	5.194
0.03	0.18	1.0005	430.3	27.186	3.116	5.194
0.04	0.24	1.0006	430.3	26.588	3.116	5.194
0.05	0.30	1.0008	430.3	26.125	3.117	5.195
0.06	0.36	1.0010	430.3	25.746	3.117	5.195
0.07	0.42	1.0011	430.3	25.426	3.117	5.195
0.08	0.48	1.0013	430.4	25.148	3.117	5.196
0.09	0.54	1.0015	430.4	24.904	3.117	5.196
0.1	0.60	1.0016	430.4	24.685	3.117	5.196
0.2	1.20	1.0033	430.6	23.244	3.119	5.199
0.3	1.80	1.0049	430.8	22.401	3.120	5.202
0.4	2.39	1.0065	431.0	21.802	3.121	5.205
0.5	2.98	1.0082	431.2	21.338	3.123	5.208
0.6	3.58	1.0098	431.4	20.959	3.124	5.211
0.7	4.16	1.0115	431.6	20.638	3.125	5.213
0.8	4.75	1.0131	431.8	20.359	3.127	5.216
0.9	5.34	1.0148	432.1	20.114	3.128	5.219
1.0	5.92	1.0165	432.3	19.894	3.129	5.221
1.5	8.81	1.0249	433.4	19.049	3.136	5.234

Table II.17 (*Continued*)

p	ρ	z	h	s	c_v	c_p
2.0	11.65	1.0334	434.5	18.448	3.142	5.246
2.5	14.44	1.0421	435.7	17.982	3.149	5.257
3.0	17.18	1.0508	436.9	17.600	3.155	5.267
3.5	19.88	1.0596	438.1	17.278	3.161	5.276
4.0	22.53	1.0686	439.3	16.998	3.168	5.285
4.5	25.13	1.0775	440.6	16.751	3.174	5.294
5.0	27.69	1.0866	441.9	16.531	3.180	5.302
5.5	30.21	1.0956	443.2	16.331	3.186	5.310
6.0	32.68	1.1047	444.5	16.148	3.191	5.318
6.5	35.12	1.1138	445.8	15.980	3.197	5.326
7.0	37.51	1.1230	447.1	15.825	3.203	5.333
7.5	39.87	1.1321	448.5	15.680	3.209	5.341
8.0	42.18	1.1412	449.8	15.545	3.214	5.348
8.5	44.46	1.1503	451.2	15.417	3.219	5.355
9.0	46.71	1.1594	452.5	15.297	3.225	5.362
9.5	48.92	1.1685	453.9	15.183	3.230	5.368
10.0	51.10	1.1776	455.3	15.076	3.235	5.375
15.0	71.21	1.2675	469.1	14.221	3.284	5.427
20.0	88.72	1.3565	483.2	13.615	3.326	5.452
25.0	104.06	1.4456	497.7	13.148	3.365	5.455
30	117.60	1.5352	512.6	12.770	3.399	5.446
35	129.63	1.6247	527.8	12.454	3.431	5.432
40	140.45	1.7138	543.1	12.183	3.460	5.419
45	150.28	1.8019	558.5	11.946	3.488	5.407
50	159.30	1.8888	573.9	11.734	3.514	5.398
55	167.64	1.9742	589.2	11.543	3.539	5.390
60	175.43	2.0581	604.4	11.369	3.564	5.383
65	182.75	2.1404	619.5	11.209	3.587	5.376
70	189.65	2.2211	634.5	11.060	3.610	5.369
75	196.20	2.3003	649.3	10.922	3.632	5.361
80	202.44	2.3780	664.0	10.792	3.654	5.351
85	208.39	2.4545	678.6	10.670	3.675	5.340
90	214.08	2.5298	693.1	10.556	3.696	5.328
95	219.53	2.6041	707.6	10.448	3.716	5.315
100	224.75	2.6775	721.9	10.345	3.736	5.301
			$T=85$ K			
0.01	0.06	1.0002	456.2	29.783	3.116	5.193
0.02	0.11	1.0003	456.2	28.343	3.116	5.194
0.03	0.17	1.0005	456.2	27.501	3.116	5.194
0.04	0.23	1.0006	456.2	26.903	3.116	5.194
0.05	0.28	1.0008	456.3	26.440	3.116	5.194
0.06	0.34	1.0009	456.3	26.061	3.117	5.195
0.07	0.40	1.0011	456.3	25.741	3.117	5.195
0.08	0.45	1.0012	456.3	25.463	3.117	5.195
0.09	0.51	1.0014	456.3	25.219	3.117	5.195
0.1	0.57	1.0015	456.4	25.000	3.117	5.196
0.2	1.13	1.0031	456.6	23.559	3.118	5.198
0.3	1.69	1.0047	456.8	22.716	3.119	5.201
0.4	2.25	1.0062	457.0	22.118	3.121	5.203
0.5	2.81	1.0078	457.2	21.654	3.122	5.206
0.6	3.37	1.0094	457.5	21.274	3.123	5.208
0.7	3.92	1.0109	457.7	20.954	3.124	5.210
0.8	4.47	1.0125	457.9	20.676	3.125	5.213
0.9	5.03	1.0141	458.1	20.430	3.127	5.215

Table II.17 (Continued)

p	ρ	z	h	s	c_v	c_p
1.0	5.58	1.0157	458.4	20.211	3.128	5.217
1.5	8.30	1.0237	459.5	19.366	3.134	5.228
2.0	10.98	1.0317	460.7	18.766	3.139	5.238
2.5	13.62	1.0399	461.9	18.300	3.145	5.247
3.0	16.21	1.0482	463.2	17.919	3.151	5.255
3.5	18.76	1.0565	464.4	17.597	3.157	5.263
4.0	21.27	1.0650	465.7	17.318	3.162	5.271
4.5	23.74	1.0734	467.0	17.072	3.168	5.278
5.0	26.17	1.0819	468.3	16.852	3.173	5.285
5.5	28.57	1.0905	469.7	16.652	3.178	5.292
6.0	30.92	1.0991	471.0	16.470	3.184	5.299
6.5	33.24	1.1077	472.4	16.303	3.189	5.305
7.0	35.52	1.1163	473.7	16.148	3.194	5.311
7.5	37.76	1.1249	475.1	16.003	3.199	5.318
8.0	39.97	1.1335	476.5	15.868	3.204	5.324
8.5	42.15	1.1421	477.9	15.741	3.209	5.330
9.0	44.30	1.1507	479.3	15.621	3.214	5.336
9.5	46.41	1.1592	480.7	15.508	3.218	5.341
10.0	48.50	1.1678	482.1	15.401	3.223	5.347
15.0	67.82	1.2526	496.1	14.549	3.267	5.394
20.0	84.77	1.3362	510.4	13.944	3.306	5.420
25.0	99.74	1.4196	524.9	13.478	3.341	5.427
30	113.03	1.5032	539.8	13.099	3.372	5.420
35	124.91	1.5869	554.9	12.783	3.401	5.408
40	135.63	1.6703	570.2	12.511	3.428	5.396
45	145.39	1.7529	585.5	12.273	3.454	5.385
50	154.36	1.8345	600.9	12.061	3.478	5.375
55	162.67	1.9149	616.1	11.870	3.501	5.368
60	170.44	1.9938	631.3	11.695	3.523	5.361
65	177.73	2.0713	646.4	11.534	3.544	5.355
70	184.63	2.1473	661.3	11.385	3.565	5.349
75	191.17	2.2219	676.1	11.246	3.586	5.342
80	197.41	2.2952	690.8	11.116	3.605	5.334
85	203.37	2.3671	705.3	10.994	3.625	5.325
90	209.08	2.4380	719.7	10.878	3.644	5.314
95	214.55	2.5077	734.1	10.769	3.663	5.301
100	219.81	2.5766	748.3	10.666	3.681	5.287
			$T=90$ K			
0.01	0.05	1.0001	482.1	30.080	3.116	5.193
0.02	0.11	1.0003	482.2	28.640	3.116	5.194
0.03	0.16	1.0004	482.2	27.798	3.116	5.194
0.04	0.21	1.0006	482.2	27.200	3.116	5.194
0.05	0.27	1.0007	482.2	26.737	3.116	5.194
0.06	0.32	1.0009	482.3	26.358	3.117	5.194
0.07	0.37	1.0010	482.3	26.038	3.117	5.195
0.08	0.43	1.0012	482.3	25.760	3.117	5.195
0.09	0.48	1.0013	482.3	25.515	3.117	5.195
0.1	0.53	1.0015	482.3	25.297	3.117	5.195
0.2	1.07	1.0030	482.6	23.856	3.118	5.197
0.3	1.60	1.0044	482.8	23.013	3.119	5.200
0.4	2.13	1.0059	483.0	22.415	3.120	5.202
0.5	2.65	1.0074	483.3	21.951	3.121	5.204
0.6	3.18	1.0089	483.5	21.572	3.122	5.206
0.7	3.71	1.0104	483.7	21.251	3.123	5.208
0.8	4.23	1.0119	484.0	20.973	3.124	5.210

Table II.17 (*Continued*)

p	ρ	z	h	s	c_v	c_p
0.9	4.75	1.0134	484.3	20.728	3.125	5.211
1.0	5.27	1.0149	484.4	20.509	3.127	5.213
1.5	7.85	1.0225	485.6	19.665	3.132	5.222
2.0	10.38	1.0302	486.9	19.065	3.137	5.231
2.5	12.88	1.0380	488.1	18.600	3.142	5.238
3.0	15.34	1.0458	489.4	18.219	3.147	5.246
3.5	17.77	1.0537	490.7	17.898	3.152	5.253
4.0	20.15	1.0617	492.0	17.619	3.157	5.259
4.5	22.50	1.0697	493.4	17.373	3.162	5.265
5.0	24.82	1.0777	494.7	17.153	3.167	5.271
5.5	27.09	1.0858	496.1	16.954	3.172	5.277
6.0	29.34	1.0939	497.5	16.773	3.177	5.282
6.5	31.55	1.1020	498.9	16.605	3.181	5.288
7.0	33.73	1.1101	500.2	16.451	3.186	5.293
7.5	35.87	1.1183	501.7	16.307	3.190	5.298
8.0	37.99	1.1264	503.1	16.172	3.195	5.304
8.5	40.07	1.1346	504.5	16.045	3.199	5.309
9.0	42.13	1.1427	505.9	15.926	3.204	5.314
9.5	44.16	1.1508	507.3	15.813	3.208	5.319
10.0	46.15	1.1589	508.7	15.705	3.212	5.323
15.0	64.75	1.2392	523.0	14.857	3.252	5.366
20.0	81.16	1.3180	537.4	14.253	3.287	5.392
25.0	95.76	1.3965	552.0	13.787	3.319	5.400
30	108.80	1.4749	566.8	13.409	3.348	5.396
35	120.51	1.5535	581.9	13.091	3.375	5.386
40	131.12	1.6318	597.1	12.819	3.399	5.375
45	140.80	1.7095	612.4	12.580	3.423	5.364
50	149.71	1.7864	627.7	12.368	3.445	5.355
55	157.98	1.8622	642.9	12.176	3.466	5.348
60	165.71	1.9368	658.1	12.001	3.486	5.342
65	172.97	2.0100	673.1	11.840	3.506	5.336
70	179.85	2.0819	688.0	11.690	3.525	5.331
75	186.38	2.1525	702.8	11.551	3.544	5.325
80	192.61	2.2217	717.7	11.420	3.562	5.318
85	198.56	2.2897	431.9	11.298	3.580	5.310
90	204.28	2.3566	746.3	11.182	3.597	5.301
95	209.77	2.4225	760.6	11.072	3.614	5.290
100	215.05	2.4874	774.8	10.968	3.631	5.277

$$T = 95 \text{ K}$$

p	ρ	z	h	s	c_v	c_p
0.01	0.05	1.0001	508.1	30.361	3.116	5.193
0.02	0.10	1.0003	508.1	28.921	3.116	5.193
0.03	0.15	1.0004	508.2	28.079	3.116	5.194
0.04	0.20	1.0006	508.2	27.481	3.116	5.194
0.05	0.25	1.0007	508.2	27.018	3.116	5.194
0.06	0.30	1.0008	508.2	26.639	3.116	5.194
0.07	0.35	1.0010	508.3	26.318	3.117	5.194
0.08	0.40	1.0011	568.3	26.041	3.117	5.195
0.09	0.46	1.0013	508.3	25.796	3.117	5.195
0.1	0.51	1.0014	508.3	25.577	3.117	5.195
0.2	1.01	1.0028	508.6	24.137	3.118	5.197
0.3	1.51	1.0042	508.8	23.294	3.119	5.199
0.4	2.02	1.0057	509.0	22.696	3.120	5.200
0.5	2.52	1.0071	509.3	22.233	3.121	5.202
0.6	3.01	1.0085	509.5	21.853	3.122	5.204

Table II.17 (*Continued*)

p	ρ	z	h	s	c_v	c_p
0.7	3.51	1.0099	509.8	21.533	3.123	5.205
0.8	4.01	1.0114	510.0	21.255	3.124	5.207
0.9	4.50	1.0128	510.3	21.010	3.124	5.209
1.0	5.00	1.0143	510.5	20.791	3.125	5.210
1.5	7.44	1.0215	511.7	19.947	3.130	5.218
2.0	9.85	1.0288	513.0	19.348	3.135	5.225
2.5	12.23	1.0362	514.3	18.883	3.139	5.232
3.0	14.57	1.0436	515.6	18.503	3.144	5.238
3.5	16.87	1.0511	516.9	18.181	3.149	5.244
4.0	19.15	1.0586	518.3	17.903	3.153	5.249
4.5	21.39	1.0662	519.7	17.658	3.157	5.254
5.0	23.59	1.0738	521.0	17.438	3.162	5.259
5.5	25.77	1.0815	522.4	17.239	3.166	5.264
6.0	27.92	1.0892	523.8	17.058	3.170	5.269
6.5	30.03	1.0969	525.3	16.891	3.175	5.273
7.0	32.11	1.1046	526.7	16.736	3.179	5.278
7.5	34.17	1.1123	528.1	16.593	3.183	5.282
8.0	36.20	1.1200	529.5	16.458	3.187	5.287
8.5	38.19	1.1277	531.0	16.332	3.191	5.291
9.0	40.17	1.1355	532.4	16.213	3.195	5.295
9.5	42.11	1.1432	533.9	16.100	3.199	5.299
10.0	44.03	1.1508	535.3	15.993	3.202	5.304
15.0	61.95	1.2270	549.8	15.146	3.238	5.341
20.0	77.86	1.3017	564.3	14.544	3.271	5.366
25.0	92.08	1.3758	578.9	14.078	3.300	5.376
30	104.86	1.4497	593.8	13.700	3.326	5.375
35	116.40	1.5237	608.8	13.382	3.351	5.366
40	126.89	1.5975	623.9	13.109	3.374	5.356
45	136.48	1.6708	639.2	12.870	3.395	5.345
50	145.32	1.7435	654.4	12.657	3.415	5.336
55	153.54	1.8152	669.6	12.464	3.435	5.329
60	161.22	1.8859	684.7	12.289	3.453	5.323
65	168.46	1.9553	699.7	12.128	3.471	5.318
70	175.30	2.0235	714.6	11.978	3.489	5.314
75	181.81	2.0905	729.3	11.838	3.506	5.309
80	188.02	2.1562	743.9	11.708	3.522	5.304
85	193.96	2.2207	758.4	11.584	3.539	5.297
90	199.67	2.2841	772.8	11.468	3.555	5.289
95	205.16	2.3465	787.0	11.358	3.571	5.280
100	210.45	2.4079	801.1	11.253	3.586	5.269

$T = 100$ K

p	ρ	z	h	s	c_v	c_p
0.01	0.05	1.0001	534.1	30.627	3.116	5.193
0.02	0.10	1.0003	534.1	29.187	3.116	5.193
0.03	0.14	1.0004	534.1	28.345	3.116	5.194
0.04	0.19	1.0005	534.2	27.747	3.116	5.194
0.05	0.24	1.0007	534.2	27.284	3.116	5.194
0.06	0.29	1.0008	534.2	26.905	3.116	5.194
0.07	0.34	1.0009	534.2	26.585	3.116	5.194
0.08	0.38	1.0011	534.2	26.307	3.117	5.194
0.09	0.43	1.0012	534.3	26.063	3.117	5.195
0.1	0.48	1.0014	534.3	25.844	3.117	5.195
0.2	0.96	1.0027	534.5	24.404	3.118	5.196
0.3	1.44	1.0041	534.8	23.561	3.118	5.198

Table II.17 (*Continued*)

p	ρ	z	h	s	c_v	c_p
0.4	1.92	1.0054	535.0	22.963	3.119	5.199
0.5	2.39	1.0068	535.3	22.499	3.120	5.201
0.6	2.87	1.0081	535.5	22.120	3.121	5.202
0.7	3.34	1.0095	535.8	21.800	3.122	5.204
0.8	3.81	1.0109	536.0	21.522	3.123	5.205
0.9	4.28	1.0122	536.3	21.277	3.124	5.206
1.0	4.75	1.0136	536.5	21.058	3.124	5.208
1.5	7.08	1.0205	537.8	20.214	3.129	5.214
2.0	9.37	1.0275	539.1	19.615	3.133	5.220
2.5	11.63	1.0345	540.4	19.151	3.137	5.226
3.0	13.87	1.0416	541.8	18.771	3.141	5.231
3.5	16.07	1.0487	543.1	18.450	3.145	5.236
4.0	18.24	1.0558	544.5	18.172	3.149	5.241
4.5	20.38	1.0631	545.9	17.927	3.153	5.245
5.0	22.49	1.0703	547.3	17.707	3.157	5.249
5.5	24.57	1.0776	548.7	17.509	3.161	5.253
6.0	26.62	1.0849	550.2	17.328	3.165	5.257
6.5	28.65	1.0922	551.6	17.161	3.169	5.261
7.0	30.65	1.0995	553.0	17.007	3.172	5.265
7.5	32.62	1.1068	554.5	16.863	3.176	5.269
8.0	34.57	1.1142	555.9	16.729	3.180	5.272
8.5	36.49	1.1215	557.4	16.603	3.183	5.276
9.0	38.38	1.1288	558.8	16.484	3.187	5.280
9.5	40.25	1.1362	560.3	16.371	3.190	5.283
10.0	42.10	1.1435	561.8	16.264	3.194	5.287
15.0	59.39	1.2159	576.4	15.419	3.227	5.320
20.0	74.82	1.2869	591.1	14.819	3.256	5.344
25.0	88.68	1.3571	605.8	14.354	3.283	5.355
30	101.20	1.4270	620.6	13.975	3.307	5.355
35	112.56	1.4969	635.6	13.657	3.330	5.348
40	122.91	1.5667	650.7	13.383	3.350	5.338
45	132.41	1.6361	665.9	13.143	3.370	5.328
50	141.18	1.7050	681.1	12.930	3.389	5.320
55	149.34	1.7730	696.2	12.737	3.407	5.312
60	156.97	1.8401	711.3	12.562	3.424	5.307
65	164.16	1.9061	726.3	12.400	3.440	5.302
70	170.97	1.9710	741.1	12.250	3.456	5.298
75	177.44	2.0347	755.8	12.110	3.472	5.294
80	183.63	2.0973	770.4	11.979	3.487	5.290
85	189.55	2.1587	784.9	11.856	3.502	5.285
90	195.25	2.2190	799.2	11.739	3.517	5.279
95	200.73	2.2783	813.4	11.628	3.532	5.271
100	206.02	2.3367	827.4	11.523	3.546	5.261

$T = 110$ K

p	ρ	z	h	s	c_v	c_p
0.01	0.04	1.0001	586.0	31.122	3.116	5.193
0.02	0.09	1.0002	586.0	29.682	3.116	5.193
0.03	0.13	1.0004	586.1	28.840	3.116	5.193
0.04	0.17	1.0005	586.1	28.242	3.116	5.194
0.05	0.22	1.0006	586.1	27.779	3.116	5.194
0.06	0.26	1.0007	586.1	27.400	3.116	5.194
0.07	0.31	1.0009	586.2	27.080	3.116	5.194
0.08	0.35	1.0010	586.2	26.803	3.116	5.194
0.09	0.39	1.0011	586.2	26.558	3.117	5.194
0.1	0.44	1.0012	586.2	26.339	3.117	5.194

Table II.17 (*Continued*)

p	ρ	z	h	s	c_v	c_p
0.2	0.87	1.0025	586.5	24.899	3.117	5.195
0.3	1.31	1.0037	586.8	24.056	3.118	5.197
0.4	1.74	1.0050	587.0	23.459	3.119	5.198
0.5	2.17	1.0062	587.3	22.995	3.119	5.199
0.6	2.61	1.0075	587.5	22.616	3.120	5.200
0.7	3.04	1.0087	587.8	22.296	5.121	5.201
0.8	3.47	1.0100	588.1	22.018	3.121	5.202
0.9	3.90	1.0112	588.3	21.773	3.122	5.203
1.0	4.32	1.0125	588.6	21.554	3.123	5.204
1.5	6.44	1.0188	589.9	20.711	3.126	5.209
2.0	8.54	1.0251	591.3	20.113	3.130	5.213
2.5	10.61	1.0315	592.7	19.649	3.133	5.217
3.0	12.65	1.0380	594.0	19.269	3.136	5.221
3.5	14.67	1.0444	595.4	18.949	3.140	5.225
4.0	16.66	1.0509	596.9	18.671	3.143	5.228
4.5	18.62	1.0575	598.3	18.426	3.146	5.231
5.0	20.56	1.0640	599.7	18.207	3.150	5.234
5.5	22.48	1.0706	601.2	18.009	3.153	5.237
6.0	24.38	1.0773	602.6	17.828	3.156	5.240
6.5	26.24	1.0839	604.1	17.662	3.159	5.242
7.0	28.09	1.0905	605.6	17.508	3.162	5.245
7.5	29.92	1.0972	607.1	17.364	3.165	5.248
8.0	31.72	1.1039	608.5	17.230	3.168	5.250
8.5	33.50	1.1105	610.0	17.104	3.171	5.253
9.0	35.26	1.1172	611.5	16.986	3.174	5.255
9.5	36.99	1.1239	613.0	16.873	3.177	5.258
10.0	38.71	1.1305	614.5	16.767	3.180	5.261
15.0	54.87	1.1965	629.5	15.925	3.207	5.286
20.0	69.40	1.2611	644.3	15.326	3.231	5.306
25.0	82.59	1.3248	659.1	14.862	3.254	5.318
30	94.59	1.3880	674.0	14.484	3.274	5.321
35	105.57	1.4509	688.9	14.165	3.293	5.317
40	115.64	1.5138	703.9	13.891	3.311	5.309
45	124.93	1.5764	719.0	13.650	3.328	5.300
50	133.54	1.6387	734.1	13.436	3.344	5.292
55	141.56	1.7003	749.2	13.242	3.359	5.285
60	149.09	1.7613	764.2	13.066	3.373	5.279
65	156.18	1.8214	779.2	12.904	3.387	5.274
70	162.91	1.8805	794.0	12.754	3.401	5.271
75	169.31	1.9387	808.7	12.614	3.414	5.269
80	175.42	1.9958	823.2	12.482	3.427	5.266
85	181.29	2.0519	837.6	12.358	3.440	5.263
90	186.93	2.1070	851.9	12.241	3.452	5.260
95	192.38	2.1612	866.0	12.130	3.465	5.255
100	197.64	2.2143	880.0	12.024	3.477	5.249
			$T = 120$ K			
0.01	0.04	1.0001	637.9	31.574	3.116	5.193
0.02	0.08	1.0002	638.0	30.134	3.116	5.193
0.03	0.12	1.0003	638.0	29.292	3.116	5.193
0.04	0.16	1.0005	638.0	28.694	3.116	5.193
0.05	0.20	1.0006	638.0	28.231	3.116	5.194
0.06	0.24	1.0007	638.1	27.852	3.116	5.194
0.07	0.28	1.0008	638.1	27.532	3.116	5.194
0.08	0.32	1.0009	638.1	27.254	3.116	5.194
0.09	0.36	1.0010	638.2	27.010	3.116	5.194

Table II.17 (*Continued*)

p	ρ	z	h	s	c_v	c_p
0.10	0.40	1.0011	638.2	26.791	3.116	5.194
0.2	0.80	1.0023	638.5	25.351	3.117	5.195
0.3	1.20	1.0034	638.7	24.509	3.118	5.196
0.4	1.60	1.0046	639.0	23.911	3.118	5.196
0.5	1.99	1.0057	639.3	23.447	3.119	5.197
0.6	2.39	1.0069	639.5	23.068	3.119	5.198
0.7	2.79	1.0080	639.8	22.748	3.120	5.199
0.8	3.18	1.0092	640.1	22.471	3.120	5.200
0.9	3.57	1.0103	640.4	22.226	3.121	5.200
1.0	3.97	1.0115	640.6	22.007	3.122	5.201
1.5	5.92	1.0173	642.0	21.164	3.124	5.205
2.0	7.84	1.0231	643.4	20.566	3.127	5.208
2.5	9.75	1.0290	644.8	20.102	3.130	5,211
3.0	11.63	1.0349	646.2	19.723	3.133	5.214
3.5	13.49	1.0408	647.6	19.403	3.136	5.216
4.0	15.33	1.0467	649.1	19.125	3.138	5.219
4.5	17.15	1.0527	650.5	18.881	3.141	5.221
5.0	18.95	1.0587	652.0	18.662	3.144	5.223
5.5	20.72	1.0648	653.5	18.464	3.146	5.225
6.0	22.48	1.0708	655.0	18.283	3.149	5.227
6.5	24.21	1.0769	656.5	18.117	3.151	5.229
7.0	25.93	1.0829	658.0	17.963	3.154	5.231
7.5	27.63	1.0890	659.5	17.820	3.156	5.232
8.0	29.31	1.0951	661.0	17.686	3.259	5.234
8.5	30.97	1.1012	662.5	17.561	3.161	5.236
9.0	32.61	1.1073	664.0	17.442	3.164	5.238
9.5	34.23	1.1134	665.5	17.330	3.166	5.240
10.0	35.83	1.1195	667.0	17.224	3.169	5.241
15.0	50.99	1.1801	682.2	16.384	3.191	5.260
20.0	64.74	1.2394	697.2	15.787	3.212	5.277
25.0	77.28	1.2977	712.2	15.324	3.231	5.289
30	88.79	1.3554	727.0	14.945	3.248	5.293
35	99.39	1.4128	741.9	14.627	3.265	5.291
40	109.17	1.4700	756.9	14.352	3.280	5.286
45	118.23	1.5270	771.9	14.110	3.294	5.278
50	126.66	1.5837	786.9	13.895	3.308	5.270
55	134.54	1.6400	801.9	13.701	3.321	5.263
60	141.94	1.6958	816.9	13.524	3.333	5.257
65	148.93	1.7509	831.8	13.362	3.345	5.252
70	155.56	1.8053	846.6	13.212	3.357	5.249
75	161.87	1.8588	861.2	13.071	3.368	5.247
80	167.90	1.9114	875.8	12.940	3.379	5.246
85	173.70	1.9631	890.1	12.816	3.390	5.244
90	179.28	2.0140	904.4	12.698	3.400	5.243
95	184.66	2.0639	918.5	12.587	3.411	5.241
100	189.87	2.1129	932.4	12.480	3.421	5.238
			$T=130$ K			
0.01	0.04	1.0001	689.9	31.990	3.116	5.193
0.02	0.07	1.0002	689.9	30.550	3.116	5.193
0.03	0.11	1.0003	689.9	29.708	3.116	5.193
0.04	0.15	1.0004	690.0	29.110	3.116	5.193
0.05	0.19	1.0005	690.0	28.647	3.116	5.193
0.06	0.22	1.0006	690.0	28.268	3.116	5.194
0.07	0.26	1.0007	690.0	27.948	3.116	5.194

Table II.17 (*Continued*)

p	ρ	z	h	s	c_v	c_p
0.08	0.30	1.0008	690.1	27.670	3.116	5.194
0.09	0.33	1.0010	690.1	27.426	3.116	5.194
0.10	0.37	1.0011	690.1	27.207	3.116	5.194
0.2	0.74	1.0021	690.4	25.767	3.117	5.194
0.3	1.11	1.0032	690.7	24.924	3.117	5.195
0.4	1.47	1.0042	691.0	24.327	3.118	5.196
0.5	1.84	1.0053	691.2	23.863	3.118	5.196
0.6	2.21	1.0064	691.5	23.484	3.119	5.197
0.7	2.57	1.0074	691.8	23.164	3.119	5.197
0.8	2.94	1.0085	692.1	22.887	3.120	5.198
0.9	3.30	1.0096	692.3	22.642	3.120	5.199
1.0	3.66	1.0106	692.6	22.423	3.121	5.199
1.5	5.47	1.0160	694.0	21.581	3.123	5.202
2.0	7.25	1.0214	695.5	20.983	3.125	5.204
2.5	9.02	1.0268	696.9	20.519	3.128	5.206
3.0	10.76	1.0322	698.3	20.140	3.130	5.208
3.5	12.49	1.0377	699.8	19.820	3.132	5.210
4.0	14.20	1.0431	701.2	19.543	3.134	5.212
4.5	15.89	1.0486	702.7	19.298	3.137	5.213
5.0	17.56	1.0542	704.2	19.080	3.139	5.215
5.5	19.22	1.0597	705.7	18.882	3.141	5.216
6.0	20.86	1.0653	707.2	18.701	3.143	5.218
6.5	22.48	1.0709	708.7	18.535	3.145	5.219
7.0	24.08	1.0764	710.2	18.382	3.148	5.220
7.5	25.67	1.0820	711.7	18.239	3.150	5.221
8.0	27.24	1.0876	713.2	18.105	3.152	5.223
8.5	28.79	1.0933	714.8	17.979	3.154	5.224
9.0	30.33	1.0989	716.3	17.861	3.156	5.225
9.5	31.85	1.1045	717.8	17.749	3.158	5.226
10.0	33.36	1.1101	719.4	17.643	3.160	5.227
15.0	47.64	1.1660	734.7	16.804	3.179	5.241
20.0	60.66	1.2209	749.9	16.208	3.197	5.255
25.0	72.63	1.2747	764.9	15.746	3.213	5.265
30	83.66	1.3279	779.8	15.368	3.228	5.270
35	93.88	1.3806	794.7	15.049	3.241	5.271
40	103.36	1.4330	809.6	14.774	3.255	5.267
45	112.19	1.4853	824.6	14.532	3.267	5.260
50	120.43	1.5374	839.5	14.316	3.279	5.253
55	128.16	1.5892	854.5	14.122	3.290	5.246
60	135.43	1.6406	869.4	13.945	3.300	5.240
65	142.31	1.6914	884.2	13.782	3.311	5.235
70	148.83	1.7417	899.0	13.631	3.321	5.232
75	155.05	1.7913	913.6	13.491	3.330	5.230
80	161.00	1.8401	928.1	13.359	3.340	5.229
85	166.71	1.8881	942.5	13.235	3.349	5.228
90	172.21	1.9353	956.7	13.117	3.358	5.228
95	177.52	1.9817	970.8	13.006	3.367	5.227
100	182.66	2.0273	984.7	12.899	3.376	5.227

$$T = 140 \text{ K}$$

p	ρ	z	h	s	c_v	c_p
0.01	0.03	1.0001	741.8	32.375	3.116	5.193
0.02	0.07	1.0002	741.8	30.935	3.116	5.193
0.03	0.10	1.0003	741.9	30.093	3.116	5.193
0.04	0.14	1.0004	741.9	29.495	3.116	5.193
0.05	0.17	1.0005	741.9	29.031	3.116	5.193

Table II.17 (*Continued*)

p	ρ	z	h	s	c_v	c_p
0.06	0.21	1.0006	741.9	28.653	3.116	5.193
0.07	0.24	1.0007	742.0	28.332	3.116	5.193
0.08	0.27	1.0008	742.0	28.055	3.116	5.194
0.09	0.31	1.0009	742.0	27.810	3.116	5.194
0.10	0.34	1.0010	742.1	27.592	3.116	5.194
0.2	0.69	1.0020	743.3	26.152	3.117	5.194
0.3	1.03	1.0030	742.6	25.309	3.117	5.195
0.4	1.37	1.0040	742.9	24.712	3.117	5.195
0.5	1.71	1.0049	743.2	24.248	3.118	5.195
0.6	2.05	1.0059	743.5	23.869	3.118	5.196
0.7	2.39	1.0069	743.8	23.549	3.119	5.196
0.8	2.73	1.0079	744.0	23.272	3.119	5.197
0.9	3.07	1.0089	744.3	23.027	3.119	5.197
1.0	3.40	1.0099	744.6	22.808	3.120	5.198
1.5	5.08	1.0149	746.0	21.966	3.122	5.200
2.0	6.74	1.0199	747.5	21.368	3.124	5.201
2.5	8.39	1.0249	748.9	20.905	3.126	5.203
3.0	10.02	1.0299	750.4	20.526	3.128	5.204
3.5	11.63	1.0350	751.9	20.206	3.130	5.206
4.0	13.23	1.0400	753.3	19.929	3.131	5.207
4.5	14.81	1.0451	754.8	19.684	3.133	5.208
5.0	16.37	1.0502	756.3	19.466	3.135	5.209
5.5	17.92	1.0554	757.8	19.268	3.137	5.210
6.0	19.45	1.0605	759.3	19.088	3.139	5.211
6.5	20.97	1.0656	760.8	18.922	3.141	5.212
7.0	22.48	1.0708	762.4	18.768	3.142	5.212
7.5	23.97	1.0760	763.9	18.625	3.144	5.213
8.0	25.44	1.0812	765.4	18.492	3.146	5.214
8.5	26.90	1.0864	767.0	18.366	3.148	5.215
9.0	28.35	1.0916	768.5	18.248	3.150	5.216
9.5	29.78	1.0968	770.0	18.136	3.151	5.216
10.0	31.20	1.1020	771.6	18.030	3.153	5.217
15.0	44.70	1.1538	787.0	17.192	3.169	5.226
20.0	57.08	1.2048	802.3	16.597	3.184	5.237
25.0	68.50	1.2549	817.5	16.135	3.198	5.246
30	79.10	1.3042	832.4	15.758	3.211	5.252
35	88.95	1.3530	847.3	15.439	3.223	5.253
40	98.14	1.4015	862.2	15.164	3.234	5.251
45	106.73	1.4498	877.1	14.921	3.245	5.246
50	114.78	1.4980	892.0	14.705	3.255	5.239
55	122.34	1.5458	906.9	14.510	3.264	5.233
60	129.48	1.5934	921.7	14.332	3.274	5.227
65	136.24	1.6406	936.5	14.169	3.283	5.222
70	142.65	1.6873	951.2	14.018	3.291	5.218
75	148.78	1.7334	965.8	13.878	3.300	5.216
80	154.64	1.7790	980.3	13.746	3.308	5.215
85	160.26	1.8238	994.7	13.622	3.316	5.214
90	165.68	1.8679	1008.9	13.504	3.324	5.215
95	170.91	1.9113	1023.0	13.392	3.331	5.215
100	175.98	1.9540	1037.0	13.286	3.339	5.216

$$T = 150 \text{ K}$$

p	ρ	z	h	s	c_v	c_p
0.01	0.03	1.0001	793.7	32.733	3.116	5.193
0.02	0.06	1.0002	793.8	31.293	3.116	5.193
0.03	0.10	1.0003	793.8	30.451	3.116	5.193

Table II.17 (*Continued*)

p	ρ	z	h	s	c_v	c_p
0.04	0.13	1.0004	793.8	29.853	3.116	5.193
0.05	0.16	1.0005	793.9	29.390	3.116	5.193
0.06	0.19	1.0006	793.9	29.011	3.116	5.193
0.07	0.22	1.0006	793.9	28.691	3.116	5.193
0.08	0.26	1.0007	793.9	28.413	3.116	5.193
0.09	0.29	1.0008	794.0	28.169	3.116	5.193
0.10	0.32	1.0009	794.0	27.950	3.116	5.193
0.2	0.64	1.0018	794.3	26.510	3.117	5.194
0.3	0.96	1.0028	794.6	25.668	3.117	5.194
0.4	1.28	1.0037	794.9	25.070	3.117	5.195
0.5	1.60	1.0046	795.1	24.607	3.118	5.195
0.6	1.92	1.0055	795.4	24.228	3.118	5.195
0.7	2.23	1.0065	795.7	23.908	3.118	5.196
0.8	2.55	1.0074	796.0	23.630	3.119	5.196
0.9	2.86	1.0083	796.3	23.386	3.119	5.196
1.0	3.18	1.0092	796.6	23.167	3.119	5.196
1.5	4.75	1.0139	798.0	22.325	3.121	5.198
2.0	6.30	1.0185	799.5	21.727	3.123	5.199
2.5	7.84	1.0232	800.9	21.264	3.124	5.200
3.0	9.37	1.0279	802.4	20.885	3.126	5.201
3.5	10.88	1.0326	803.9	20.565	3.127	5.202
4.0	12.38	1.0373	805.4	20.288	3.129	5.203
4.5	13.86	1.0420	806.9	20.044	3.131	5.204
5.0	15.33	1.0468	808.4	19.825	3.132	5.205
5.5	16.79	1.0515	809.9	19.627	3.134	5.205
6.0	18.23	1.0563	811.4	19.447	3.135	5.206
6.5	19.66	1.0611	812.9	19.281	3.137	5.206
7.0	21.08	1.0659	814.5	19.128	3.138	5.207
7.5	22.48	1.0707	816.0	18.985	3.140	5.207
8.0	23.87	1.0755	817.5	18.851	3.141	5.208
8.5	25.25	1.0804	819.1	18.726	3.143	5.208
9.0	26.62	1.0852	820.6	18.607	3.144	5.209
9.5	27.97	1.0900	822.2	18.496	3.146	5.209
10.0	29.31	1.0949	823.7	18.389	3.147	5.209
15.0	42.11	1.1432	839.2	17.552	3.161	5.215
20.0	53.90	1.1908	854.6	16.958	3.174	5.223
25.0	64.83	1.2376	869.9	16.497	3.186	5.231
30	75.01	1.2836	884.9	16.120	3.197	5.237
35	84.51	1.3291	899.8	15.801	3.207	5.239
40	93.41	1.3743	914.7	15.525	3.217	5.238
45	101.76	1.4192	929.5	15.283	3.227	5.234
50	109.62	1.4639	944.3	15.066	3.235	5.229
55	117.02	1.5084	959.1	14.871	3.244	5.223
60	124.02	1.5527	973.9	14.693	3.252	5.217
65	130.65	1.5967	988.7	14.529	3.260	5.212
70	136.96	1.6403	1003.4	14.378	3.267	5.208
75	142.99	1.6834	1017.9	14.237	3.274	5.205
80	148.75	1.7260	1032.4	14.105	3.282	5.204
85	154.29	1.7681	1046.8	13.981	3.289	5.203
90	159.63	1.8095	1061.0	13.863	3.295	5.204
95	164.78	1.8503	1075.1	13.752	3.302	5.205
100	169.77	1.8904	1089.1	13.646	3.309	5.206

$T = 160$ K

0.01	0.03	1.0001	845.7	33.068	3.116	5.193

Table II.17 (*Continued*)

p	ρ	z	h	s	c_v	c_p
0.02	0.06	1.0002	845.7	31.628	3.116	5.193
0.03	0.09	1.0003	845.7	30 786	3.116	5.193
0.04	0.12	1.0003	845.8	30 188	3.116	5.193
0.05	0.15	1.0004	845.8	29 725	3.116	5.193
0.06	0.18	1.0005	845.8	29 346	3.116	5.193
0.07	0.21	1.0006	845.8	29 026	3.116	5.193
0.08	0.24	1.0007	845.9	28 749	3.116	5.193
0.09	0.27	1.0008	845.9	28 504	3.116	5.193
0.10	0.30	1.0009	845.9	28 285	3.116	5.193
0.2	0.60	1.0017	846.2	26 845	3.116	5.194
0.3	0.90	1.0026	846.5	26 003	3.117	5.194
0.4	1.20	1.0035	846.8	25 405	3.117	5.194
0.5	1.50	1.0043	847.1	24 942	3.117	5.194
0.6	1.80	1.0052	847.4	24 563	3.118	5.195
0.7	2.09	1.0061	847.7	24 243	3.118	5.195
0.8	2.39	1.0069	848.0	23 966	3.118	5.195
0.9	2.69	1.0078	848.2	23 721	3.118	5.195
1.0	2.98	1.0087	848.5	23 502	3.119	5.196
1.5	4.46	1.0130	850.0	22 660	3.120	5.197
2.0	5.91	1.0174	851.5	22 063	3.122	5.198
2.5	7.36	1.0217	852.9	21 599	3.123	5.198
3.0	8.80	1.0261	854.4	21 221	3.124	5.199
3.5	10.22	1.0305	855.9	20 901	3.126	5.200
4.0	11.63	1.0349	857.4	20 624	3.127	5.200
4.5	13.03	1.0393	858.9	20 379	3.128	5.201
5.0	14.41	1.0438	860.4	20 161	3.130	5.201
5.5	15.79	1.0482	861.9	19 963	3.131	5.202
6.0	17.15	1.0527	863.4	19 783	3.132	5.202
6.5	18.50	1.0571	865.0	19 617	3.134	5.202
7.0	19.84	1.0616	866.5	19 464	3.135	5.202
7.5	21.17	1.0661	868.0	19 321	3.136	5.203
8.0	22.48	1.0706	869.6	19 187	3.138	5.203
8.5	23.79	1.0751	871.1	19 062	3.139	5.203
9.0	25.08	1.0796	872.7	18 943	3.140	5.203
9.5	26.37	1.0841	874.2	18 832	3.142	5.203
10.0	27.64	1.0886	875.8	18 725	3.143	5.204
15.0	39.81	1.1338	891.3	17 888	3.155	5.207
20.0	51.06	1.1785	906.8	17 295	3.166	5.213
25.0	61.53	1.2224	922.1	16 834	3.176	5.219
30	71.32	1.2656	937.2	16 457	3.186	5.224
35	80.50	1.3082	952.1	16 139	3.195	5.227
40	89.12	1.3505	967.0	15 863	3.203	5.227
45	97.24	1.3925	981.8	15 620	3 212	5.224
50	104.89	1.4342	996.6	15 403	3.219	5.220
55	112.13	1.4758	1011.3	15 207	3 227	5.215
60	118.99	1.5172	1026.1	15 029	3.234	5.209
65	125.50	1.5584	1040.8	14 865	3.241	5.204
70	131.70	1.5992	1055.4	14 714	3.247	5.200
75	137.62	1.6397	1070.0	14 573	3.254	5.197
80	143.30	1.6797	1084.4	14 441	3.260	5.195
85	148.75	1.7193	1098.8	14 316	3.266	5.195
90	154.00	1.7583	1113.0	14 199	3.272	5.195
95	159.08	1.7968	1127.1	14 088	3.278	5.196
100	163.99	1.8347	1141.1	13.981	3.283	5.198

Table II.17 (*Continued*)

p	ρ	z	h	s	c_v	c_p
\multicolumn{7}{c}{$T=170$ K}						
0.01	0.03	1.0001	897.6	33.383	3.116	5.193
0.02	0.06	1.0002	897.6	31.943	3.116	5.193
0.03	0.08	1.0002	897.7	31.101	3.116	5.193
0.04	0.11	1.0003	897.7	30.503	3.116	5.193
0.05	0.14	1.0004	897.7	30.040	3.116	5.193
0.06	0.17	1.0005	897.7	29.661	3.116	5.193
0.07	0.20	1.0006	897.8	29.341	3.116	5.193
0.08	0.23	1.0007	897.8	29.063	3.116	5.193
0.09	0.25	1.0007	897.8	28.819	3.116	5.193
0.10	0.28	1.0008	897.9	28.600	3.116	5.193
0.2	0.57	1.0016	898.2	27.160	3.116	5.194
0.3	0.85	1.0024	898.4	26.318	3.117	5.194
0.4	1.13	1.0033	898.7	25.720	3.117	5.194
0.5	1.41	1.0041	899.0	25.257	3.117	5.194
0.6	1.69	1.0049	899.3	24.878	3.117	5.194
0.7	1.97	1.0057	899.6	24.558	3.118	5.194
0.8	2.25	1.0065	899.9	24.281	3.118	5.195
0.9	2.53	1.0073	900.2	24.036	3.118	5.195
1.0	2.81	1.0081	900.5	23.817	3.118	5.195
1.5	4.20	1.0122	902.0	22.975	3.120	5.196
2.0	5.57	1.0163	903.4	22.378	3.121	5.196
2.5	6.94	1.0204	904.9	21.914	3.122	5.197
3.0	8.29	1.0245	906.4	21.536	3.123	5.197
3.5	9.64	1.0286	907.9	21.216	3.124	5.198
4.0	10.97	1.0328	909.4	20.939	3.126	5.198
4.5	12.29	1.0369	910.9	20.695	3.127	5.199
5.0	13.60	1.0411	912.4	20.476	3.128	5.199
5.5	14.90	1.0452	913.9	20.278	3.129	5.199
6.0	16.19	1.0494	915.5	20.098	3.130	5.199
6.5	17.47	1.0536	917.0	19.932	3.131	5.199
7.0	18.74	1.0578	918.5	19.779	3.132	5.199
7.5	20.00	1.0620	920.0	19.636	3.134	5.199
8.0	21.25	1.0662	921.6	19.502	3.135	5.199
8.5	22.49	1.0705	923.1	19.377	3.136	5.199
9.0	23.72	1.0747	924.7	19.259	3.137	5.199
9.5	24.93	1.0789	926.2	19.147	3.138	5.199
10.0	26.14	1.0831	927.8	19.041	3.139	5.199
15.0	37.74	1.1255	943.4	18.204	3.149	5.201
20.0	48.51	1.1676	958.9	17.610	3.159	5.205
25.0	58.56	1.2090	974.2	17.150	3.168	5.210
30	67.98	1.2497	989.4	16.774	3.177	5.214
35	76.84	1.2898	1004.3	16.455	3.184	5.217
40	85.20	1.3295	1019.2	16.180	3.192	5.218
45	93.09	1.3689	1034.0	15.937	3.199	5.216
50	100.55	1.4081	1048.7	15.719	3.206	5.213
55	107.62	1.4472	1063.4	15.523	3.213	5.208
60	114.34	1.4860	1078.1	15.345	3.219	5.203
65	120.73	1.5247	1092.8	15.181	3.225	5.198
70	126.82	1.5631	1107.4	15.029	3.231	5.194
75	132.64	1.6012	1121.9	14.888	3.236	5.191
80	138.23	1.6389	1136.3	14.755	3.242	5.189
85	143.59	1.6763	1150.7	14.631	3.247	5.188
90	148.77	1.7132	1164.9	14.514	3.252	5.188
95	153.76	1.7496	1179.0	14.402	3.258	5.188

Table II.17 (*Continued*)

p	ρ	z	h	s	c_v	c_p
100	158.60	1.7855	1193.0	14.296	3.263	5.190

$T = 180$ K

p	ρ	z	h	s	c_v	c_p
0.01	0.03	1.0001	949.5	33.680	3.116	5.193
0.02	0.05	1.0002	949.6	32.240	3.116	5.193
0.03	0.08	1.0002	949.6	31.398	3.116	5.193
0.04	0.11	1.0003	949.6	30.800	3.116	5.193
0.05	0.13	1.0004	949.6	30.337	3.116	5.193
0.06	0.16	1.0005	949.7	29.958	3.116	5.193
0.07	0.19	1.0005	949.7	29.638	3.116	5.193
0.08	0.21	1.0006	949.7	29.360	3.116	5.193
0.09	0.24	1.0007	949.8	29.116	3.116	5.193
0.10	0.27	1.0008	949.8	28.897	3.116	5.193
0.2	0.53	1.0015	950.1	27.457	3.116	5.193
0.3	0.80	1.0023	950.4	26.615	3.117	5.194
0.4	1.07	1.0031	950.7	26.017	3.117	5.194
0.5	1.33	1.0038	951.0	25.554	3.117	5.194
0.6	1.60	1.0046	951.3	25.175	3.117	5.194
0.7	1.86	1.0054	951.6	24.855	3.117	5.194
0.8	2.13	1.0061	951.9	24.577	3.118	5.194
0.9	2.39	1.0069	952.1	24.333	3.118	5.194
1.0	2.65	1.0077	952.4	24.114	3.118	5.194
1.5	3.97	1.0115	953.9	23.272	3.119	5.195
2.0	5.27	1.0154	955.4	22.675	3.120	5.195
2.5	6.56	1.0192	956.9	22.211	3.121	5.196
3.0	7.84	1.0231	958.4	21.833	3.122	5.196
3.5	9.11	1.0270	959.9	21.513	3.123	5.196
4.0	10.38	1.0309	961.4	21.236	3.124	5.197
4.5	11.63	1.0348	962.9	20.992	3.125	5.197
5.0	12.87	1.0387	964.4	20.773	3.126	5.197
5.5	14.11	1.0426	965.9	20.576	3.127	5.197
6.0	15.33	1.0465	967.4	20.395	3.128	5.197
6.5	16.55	1.0505	969.0	20.229	3.129	5.197
7.0	17.75	1.0544	970.5	20.076	3.130	5.197
7.5	18.95	1.0584	972.0	19.933	3.131	5.197
8.0	20.14	1.0623	973.6	19.799	3.132	5.197
8.5	21.32	1.0663	975.1	19.674	3.133	5.196
9.0	22.49	1.0703	976.7	19.556	3.134	5.196
9.5	23.65	1.0743	978.2	19.444	3.135	5.196
10.0	24.80	1.0783	979.8	19.338	3.136	5.196
15.0	35.88	1.1182	995.4	18.501	3.145	5.196
20.0	46.20	1.1579	1010.9	17.908	3.153	5.198
25.0	55.86	1.1970	1026.3	17.448	3.161	5.202
30	64.94	1.2354	1041.5	17.072	3.169	5.206
35	73.51	1.2734	1056.5	16.753	3.176	5.209
40	81.61	1.3109	1071.3	16.478	3.182	5.210
45	89.28	1.3480	1086.1	16.234	3.189	5.209
50	96.55	1.3850	1100.8	16.017	3.195	5.207
55	103.46	1.4218	1115.5	15.821	3.201	5.203
60	110.03	1.4584	1130.1	15.642	3.206	5.198
65	116.30	1.4948	1144.7	15.478	3.212	5.194
70	122.28	1.5310	1159.3	15.326	3.217	5.190
75	128.01	1.5670	1173.8	15.184	3.222	5.186
80	133.50	1.6027	1188.2	15.052	3.227	5.184
85	138.78	1.6380	1202.5	14.928	3.231	5.182

Table II.17 (*Continued*)

p	ρ	z	h	s	c_v	c_p
90	143.87	1.6730	1216.8	14.810	3.236	5.182
95	148.79	1.7076	1230.9	14.699	3.241	5.182
100	153.55	1.7417	1244.9	14.593	3.245	5.184

$T = 190$ K

p	ρ	z	h	s	c_v	c_p
0.01	0.03	1.0001	1001.5	33.961	3.116	5.193
0.02	0.05	1.0001	1001.5	32.521	3.116	5.193
0.03	0.08	1.0002	1001.5	31.678	3.116	5.193
0.04	0.10	1.0003	1001.6	31.081	3.116	5.193
0.05	0.13	1.0004	1001.6	30.617	3.116	5.193
0.06	0.15	1.0004	1001.6	30.239	3.116	5.193
0.07	0.18	1.0005	1001.6	29.918	3.116	5.193
0.08	0.20	1.0006	1001.7	29.641	3.116	5.193
0.09	0.23	1.0007	1001.7	29.396	3.116	5.193
0.10	0.25	1.0007	1001.7	29.178	3.116	5.193
0.2	0.51	1.0015	1002.0	27.738	3.116	5.193
0.3	0.76	1.0022	1002.3	26.896	3.116	5.193
0.4	1.01	1.0029	1002.6	26.298	3.117	5.194
0.5	1.26	1.0036	1002.9	25.834	3.117	5.194
0.6	1.51	1.0044	1003.2	25.456	3.117	5.194
0.7	1.76	1.0051	1003.5	25.136	3.117	5.194
0.8	2.02	1.0058	1003.8	24.858	3.117	5.194
0.9	2.27	1.0065	1004.1	24.614	3.118	5.194
1.0	2.52	1.0073	1004.4	24.395	3.118	5.194
1.5	3.76	1.0109	1005.9	23.553	3.119	5.194
2.0	4.99	1.0145	1007.3	22.956	3.120	5.195
2.5	6.22	1.0182	1008.8	22.492	3.120	5.195
3.0	7.44	1.0218	1010.3	22.114	3.121	5.195
3.5	8.65	1.0255	1011.8	21.794	3.122	5.195
4.0	9.85	1.0292	1013.3	21.517	3.123	5.195
4.5	11.04	1.0329	1014.8	21.273	3.124	5.195
5.0	12.22	1.0366	1016.3	21.054	3.125	5.195
5.5	13.40	1.0403	1017.9	20.857	3.126	5.195
6.0	14.56	1.0440	1019.4	20.676	3.127	5.195
6.5	15.72	1.0477	1020.9	20.510	3.127	5.195
7.0	16.87	1.0514	1022.4	20.357	3.128	5.195
7.5	18.01	1.0551	1024.0	20.214	3.129	5.195
8.0	19.14	1.0589	1025.5	20.080	3.130	5.194
8.5	20.27	1.0626	1027.1	19.955	3.131	5.194
9.0	21.38	1.0664	1028.6	19.837	3.132	5.194
9.5	22.49	1.0701	1030.2	19.725	3.133	5.194
10.0	23.59	1.0739	1031.7	19.619	3.133	5.194
15.0	34.19	1.1116	1047.3	18.782	3.141	5.192
20.0	44.10	1.1491	1062.9	18.189	3.149	5.193
25.0	53.40	1.1862	1078.3	17.729	3.156	5.196
30	62.17	1.2227	1093.5	17.353	3.162	5.199
35	70.46	1.2587	1108.5	17.035	3.169	5.202
40	78.31	1.2942	1123.4	16.759	3.174	5.203
45	85.77	1.3294	1138.2	16.516	3.180	5.203
50	92.85	1.3643	1152.9	16.299	3.185	5.201
55	99.60	1.3991	1167.5	16.102	3.191	5.198
60	106.03	1.4337	1182.1	15.923	3.196	5.195
65	112.18	1.4681	1196.6	15.758	3.200	5.191
70	118.05	1.5024	1211.2	15.606	3.205	5.187
75	123.68	1.5365	1225.6	15.465	3.209	5.183

Table II.17 (*Continued*)

p	ρ	z	h	s	c_v	c_p
80	129.08	1.5703	1240.0	15.332	3.214	5.180
85	134.28	1.6039	1254.3	15.208	3.218	5.178
90	139.29	1.6371	1268.6	15.090	3.222	5.178
95	144.14	1.6700	1282.7	14.979	3.226	5.178
100	148.83	1.7025	1296.7	14.873	3.230	5.179

$T = 200$ K

p	ρ	z	h	s	c_v	c_p
0.01	0.02	1.0001	1053.4	34.227	3.116	5.193
0.02	0.05	1.0001	1053.4	32.787	3.116	5.193
0.03	0.07	1.0002	1053.5	31.945	3.116	5.193
0.04	0.10	1.0003	1053.5	31.347	3.116	5.193
0.05	0.12	1.0003	1053.5	30.884	3.116	5.193
0.06	0.14	1.0004	1053.5	30.505	3.116	5.193
0.07	0.17	1.0005	1053.6	30.185	3.116	5.193
0.08	0.19	1.0006	1053.6	29.907	3.116	5.193
0.09	0.22	1.0006	1053.6	29.663	3.116	5.193
0.10	0.24	1.0007	1053.7	29.444	3.116	5.193
0.2	0.48	1.0014	1054.0	28.004	3.116	5.193
0.3	0.72	1.0021	1054.3	27.162	3.116	5.193
0.4	0.96	1.0028	1054.5	26.564	3.117	5.193
0.5	1.20	1.0034	1054.8	26.101	3.117	5.193
0.6	1.44	1.0041	1055.1	25.722	3.117	5.194
0.7	1.68	1.0048	1055.4	25.402	3.117	5.194
0.8	1.92	1.0055	1055.7	25.125	3.117	5.194
0.9	2.15	1.0062	1056.0	24.880	3.117	5.194
1.0	2.39	1.0069	1056.3	24.661	3.117	5.194
1.5	3.57	1.0103	1057.8	23.819	3.118	5.194
2.0	4.75	1.0138	1059.3	23.222	3.119	5.194
2.5	5.92	1.0172	1060.8	22.759	3.120	5.194
3.0	7.07	1.0207	1062.3	22.380	3.121	5.194
3.5	8.23	1.0242	1063.8	22.060	3.121	5.194
4.0	9.37	1.0277	1065.3	21.783	3.122	5.194
4.5	10.50	1.0311	1066.8	21.539	3.123	5.194
5.0	11.63	1.0346	1068.3	21.321	3.124	5.194
5.5	12.75	1.0381	1069.8	21.123	3.125	5.194
6.0	13.86	1.0416	1071.3	20.943	3.125	5.194
6.5	14.97	1.0452	1072.9	20.777	3.126	5.194
7.0	16.07	1.0487	1074.4	20.623	3.127	5.193
7.5	17.16	1.0522	1075.9	20.480	3.128	5.193
8.0	18.24	1.0558	1077.5	20.347	3.128	5.193
8.5	19.31	1.0593	1079.0	20.221	3.129	5.193
9.0	20.38	1.0628	1080.5	20.103	3.130	5.192
9.5	21.44	1.0664	1082.1	19.991	3.130	5.192
10.0	22.50	1.0700	1083.6	19.885	3.131	5.192
15.0	32.66	1.1057	1099.2	19.048	3.138	5.189
20.0	42.18	1.1413	1114.8	18.455	3.145	5.189
25.0	51.15	1.1765	1130.2	17.995	3.151	5.191
30	59.62	1.2112	1145.5	17.619	3.157	5.194
35	67.65	1.2454	1160.5	17.301	3.162	5.196
40	75.27	1.2792	1175.4	17.026	3.168	5.198
45	82.52	1.3126	1190.2	16.783	3.173	5.198
50	89.43	1.3458	1204.8	16.565	3.177	5.197
55	96.02	1.3788	1219.5	16.369	3.182	5.195
60	102.31	1.4116	1234.0	16.189	3.187	5.192
65	108.33	1.4442	1248.5	16.025	3.191	5.188

Table II.17 (*Continued*)

p	ρ	z	h	s	c_v	c_p
70	114.10	1.4767	1263.0	15.872	3.195	5.184
75	119.63	1.5091	1277.4	15.730	3.199	5.181
80	124.94	1.5412	1291.8	15.598	3.203	5.178
85	130.06	1.5731	1306.1	15.473	3.207	5.176
90	134.99	1.6048	1320.3	15.356	3.210	5.174
95	139.76	1.6361	1334.5	15.244	3.214	5.174
100	144.38	1.6671	1348.5	15.138	3.217	5.175

$T = 210$ K

p	ρ	z	h	s	c_v	c_p
0.01	0.02	1.0001	1105.3	34.480	3.116	5.193
0.02	0.05	1.0001	1105.4	33.040	3.116	5.193
0.03	0.07	1.0002	1105.4	32.198	3.116	5.193
0.04	0.09	1.0003	1105.4	31.601	3.116	5.193
0.05	0.11	1.0003	1105.4	31.137	3.116	5.193
0.06	0.14	1.0004	1105.5	30.758	3.116	5.193
0.07	0.16	1.0005	1105.5	30.438	3.116	5.193
0.08	0.18	1.0005	1105.5	30.161	3.116	5.193
0.09	0.21	1.0006	1105.6	29.916	3.116	5.193
0.10	0.23	1.0007	1105.6	29.697	3.116	5.193
0.2	0.46	1.0013	1105.9	28.257	3.116	5.193
0.3	0.69	1.0020	1106.2	27.415	3.116	5.193
0.4	0.91	1.0026	1106.5	26.818	3.116	5.193
0.5	1.14	1.0033	1106.8	26.354	3.117	5.193
0.6	1.37	1.0039	1107.1	25.976	3.117	5.193
0.7	1.60	1.0046	1107.4	25.655	3.117	5.193
0.8	1.82	1.0052	1107.7	25.378	3.117	5.193
0.9	2.05	1.0059	1108.0	25.133	3.117	5.193
1.0	2.28	1.0065	1108.3	24.915	3.117	5.194
1.5	3.41	1.0098	1109.7	24.073	3.118	5.194
2.0	4.53	1.0131	1111.2	23.475	3.119	5.194
2.5	5.64	1.0164	1112.7	23.012	3.119	5.194
3.0	6.74	1.0197	1114.2	22.634	3.120	5.194
3.5	7.84	1.0230	1115.7	22.314	3.121	5.194
4.0	8.93	1.0263	1117.2	22.037	3.121	5.194
4.5	10.02	1.0296	1118.7	21.792	3.122	5.193
5.0	11.10	1.0329	1120.2	21.574	3.123	5.193
5.5	12.17	1.0362	1121.7	21.376	3.124	5.193
6.0	13.23	1.0396	1123.3	21.196	3.124	5.193
6.5	14.29	1.0429	1124.8	21.030	3.125	5.193
7.0	15.34	1.0462	1126.3	20.877	3.126	5.192
7.5	16.38	1.0496	1127.8	20.734	3.126	5.192
8.0	17.42	1.0529	1129.4	20.600	3.127	5.192
8.5	18.45	1.0563	1130.9	20.475	3.127	5.191
9.0	19.47	1.0597	1132.5	20.356	3.128	5.191
9.5	20.49	1.0630	1134.0	20.244	3.129	5.191
10.0	21.50	1.0664	1135.5	20.138	3.129	5.190
15.0	31.25	1.1003	1151.1	19.301	3.136	5.187
20.0	40.43	1.1341	1166.7	18.708	3.141	5.186
25.0	49.08	1.1677	1182.1	18.248	3.147	5.187
30.	57.27	1.2008	1197.4	17.873	3.152	5.189
35	65.05	1.2334	1212.5	17.555	3.157	5.191
40	72.45	1.2656	1227.4	17.279	3.162	5.193
45	79.51	1.2974	1242.1	17.036	3.166	5.194
50	86.24	1.3290	1256.8	16.819	3.171	5.193

Table II.17 (*Continued*)

p	ρ	z	h	s	c_v	c_p
55	92.68	1.3604	1271.4	16.622	3.175	5.192
60	98.84	1.3916	1285.9	16.443	3.179	5.189
65	104.74	1.4227	1300.4	16.278	3.183	5.186
70	110.40	1.4536	1314.8	16.125	3.186	5.182
75	115.83	1.4843	1329.2	15.983	3.190	5.179
80	121.06	1.5149	1343.6	15.850	3.193	5.176
85	126.09	1.5453	1357.9	15.726	3.197	5.174
90	130.95	1.5755	1372.1	15.608	3.200	5.172
95	135.65	1.6054	1386.2	15.497	3.203	5.172
100	140.20	1.6351	1400.2	15.391	3.207	5.172
			$T=220$ K			
0.01	0.02	1.0001	1157.3	34.722	3.116	5.193
0.02	0.04	1.0001	1157.3	33.282	3.116	5.193
0.03	0.07	1.0002	1157.3	32.440	3.116	5.193
0.04	0.09	1.0002	1157.3	31.842	3.116	5.193
0.05	0.11	1.0003	1157.4	31.379	3.116	5.193
0.06	0.13	1.0004	1157.4	31.000	3.116	5.193
0.07	0.15	1.0004	1157.4	30.680	3.116	5.193
0.08	0.17	1.0005	1157.5	30.402	3.116	5.193
0.09	0.20	1.0006	1157.5	30.158	3.116	5.193
0.10	0.22	1.0006	1157.5	29.939	3.116	5.193
0.2	0.44	1.0012	1157.8	28.499	3.116	5.193
0.3	0.66	1.0019	1158.1	27.657	3.116	5.193
0.4	0.87	1.0025	1158.4	27.059	3.116	5.193
0.5	1.09	1.0031	1158.7	26.596	3.117	5.193
0.6	1.31	1.0037	1159.0	26.217	3.117	5.193
0.7	1.53	1.0044	1159.3	25.897	3.117	5.193
0.8	1.74	1.0050	1159.6	25.620	3.117	5.193
0.9	1.96	1.0056	1159.9	25.375	3.117	5.193
1.0	2.17	1.0062	1160.2	25.156	3.117	5.193
1.5	3.25	1.0094	1161.7	24.314	3.118	5.193
2.0	4.32	1.0125	1163.2	23.717	3.118	5.193
2.5	5.39	1.0156	1164.7	23.254	3.119	5.193
3.0	6.44	1.0187	1166.1	22.875	3.120	5.193
3.5	7.49	1.0219	1167.6	22.555	3.120	5.193
4.0	8.54	1.0250	1169.1	22.278	3.121	5.193
4.5	9.58	1.0282	1170.7	22.034	3.121	5.193
5.0	10.61	1.0313	1172.2	21.816	3.122	5.193
5.5	11.63	1.0345	1173.7	21.618	3.123	5.192
6.0	12.65	1.0377	1175.2	21.438	3.123	5.192
6.5	13.67	1.0408	1176.7	21.272	3.124	5.192
7.0	14.67	1.0440	1178.2	21.118	3.124	5.191
7.5	15.67	1.0472	1179.8	20.975	3.125	5.191
8.0	16.67	1.0504	1181.3	20.842	3.126	5.191
8.5	17.65	1.0536	1182.8	20.716	3.126	5.190
9.0	18.64	1.0568	1184.4	20.598	3.127	5.190
9.5	19.61	1.0600	1185.9	20.486	3.127	5.190
10.0	20.58	1.0632	1187.4	20.380	3.128	5.189
15.0	29.96	1.0954	1203.0	19.542	3.133	5.186
20.0	38.81	1.1277	1218.5	18.949	3.139	5.184
25.0	47.17	1.1597	1234.0	18.489	3.143	5.184
30	55.11	1.1913	1249.2	18.114	3.148	5.185
35	62.65	1.2224	1264.3	17.796	3.153	5.187
40	69.84	1.2532	1279.3	17.521	3.157	5.189

Table II.17 (*Continued*)

p	ρ	z	h	s	c_v	c_p
45	76.71	1.2836	1294.1	17.278	3.161	5.190
50	83.28	1.3138	1308.7	17.060	3.165	5.190
55	89.57	1.3437	1323.3	16.863	3.169	5.189
60	95.59	1.3735	1337.8	16.684	3.172	5.187
65	101.37	1.4031	1352.3	16.519	3.176	5.184
70	106.92	1.4326	1366.7	16.366	3.179	5.181
75	112.26	1.4619	1381.0	16.224	3.182	5.178
80	117.40	1.4911	1395.3	16.091	3.185	5.175
85	122.36	1.5201	1409.6	15.966	3.188	5.172
90	127.14	1.5490	1423.8	15.849	3.191	5.171
95	131.77	1.5776	1437.9	15.737	3.194	5.170
100	136.26	1.6060	1451.9	15.631	3.197	5.170

$T = 230$ K

p	ρ	z	h	s	c_v	c_p
0.01	0.02	1.0001	1209.2	34.953	3.116	5.193
0.02	0.04	1.0001	1209.2	33.513	3.116	5.193
0.03	0.06	1.0002	1209.2	32.671	3.116	5.193
0.04	0.08	1.0002	1209.3	32.073	3.116	5.193
0.05	0.10	1.0003	1209.3	31.610	3.116	5.193
0.06	0.13	1.0004	1209.3	31.231	3.116	5.193
0.07	0.15	1.0004	1209.4	30.911	3.116	5.193
0.08	0.17	1.0005	1209.4	30.633	3.116	5.193
0.09	0.19	1.0005	1209.4	30.389	3.116	5.193
0.10	0.21	1.0006	1209.5	30.170	3.116	5.193
0.2	0.42	1.0012	1209.8	28.730	3.116	5.193
0.3	0.63	1.0018	1210.0	27.888	3.116	5.193
0.4	0.84	1.0024	1210.3	27.290	3.116	5.193
0.5	1.04	1.0030	1210.6	26.827	3.116	5.193
0.6	1.25	1.0036	1210.9	26.448	3.117	5.193
0.7	1.46	1.0042	1211.2	26.128	3.117	5.193
0.8	1.67	1.0048	1211.5	25.851	3.117	5.193
0.9	1.87	1.0054	1211.8	25.606	3.117	5.193
1.0	2.08	1.0060	1212.1	25.387	3.117	5.193
1.5	3.11	1.0089	1213.6	24.545	3.118	5.193
2.0	4.14	1.0119	1215.1	23.948	3.118	5.193
2.5	5.16	1.0149	1216.6	23.485	3.119	5.193
3.0	6.17	1.0179	1218.1	23.106	3.119	5.193
3.5	7.18	1.0209	1219.6	22.786	3.120	5.193
4.0	8.18	1.0239	1221.1	22.509	3.120	5.193
4.5	9.17	1.0269	1222.6	22.265	3.121	5.192
5.0	10.16	1.0299	1224.1	22.046	3.121	5.192
5.5	11.15	1.0329	1225.6	21.849	3.122	5.192
6.0	12.12	1.0359	1227.1	21.668	3.122	5.191
6.5	13.09	1.0389	1228.6	21.502	3.123	5.191
7.0	14.06	1.0420	1230.1	21.349	3.123	5.191
7.5	15.02	1.0450	1231.7	21.206	3.124	5.190
8.0	15.98	1.0480	1233.2	21.072	3.124	5.190
8.5	16.93	1.0511	1234.7	20.947	3.125	5.190
9.0	17.87	1.0541	1236.3	20.828	3.126	5.189
9.5	18.81	1.0572	1237.8	20.717	3.126	5.189
10.0	19.74	1.0603	1239.3	20.610	3.127	5.188
15.0	28.78	1.0910	1254.8	19.773	3.131	5.185
20.0	37.32	1.1218	1270.3	19.179	3.136	5.182
25.0	45.41	1.1524	1285.8	18.720	3.141	5.182
30	53.10	1.1826	1301.1	18.344	3.145	5.182

Table II.17 (*Continued*)

p	ρ	z	h	s	c_v	c_p
35	60.42	1.2124	1316.2	18.027	3.149	5.184
40	67.42	1.2419	1331.1	17.751	3.153	5.186
45	74.10	1.2710	1345.9	17.508	3.156	5.187
50	80.51	1.2999	1360.6	17.291	3.160	5.187
55	86.65	1.3285	1375.2	17.094	3.163	5.186
60	92.55	1.3570	1389.7	16.914	3.166	5.185
65	98.21	1.3853	1404.1	16.749	3.170	5.182
70	103.66	1.4134	1418.5	16.596	3.173	5.180
75	108.90	1.4415	1432.8	16.454	3.176	5.177
80	113.96	1.4694	1447.1	16.321	3.178	5.174
85	118.83	1.4971	1461.3	16.196	3.181	5.172
90	123.55	1.5247	1475.5	16.078	3.184	5.170
95	128.11	1.5522	1489.6	15.967	3.187	5.169
100	132.53	1.5794	1503.6	15.861	3.189	5.168

$T = 240$ K

p	ρ	z	h	s	c_v	c_p
0.01	0.02	1.0001	1261.1	35.174	3.116	5.193
0.02	0.04	1.0001	1261.1	33.734	3.116	5.193
0.03	0.06	1.0002	1261.2	32.892	3.116	5.193
0.04	0.08	1.0002	1261.2	32.294	3.116	5.193
0.05	0.10	1.0003	1261.2	31.831	3.116	5.193
0.06	0.12	1.0003	1261.3	31.452	3.116	5.193
0.07	0.14	1.0004	1261.3	31.132	3.116	5.193
0.08	0.16	1.0005	1261.3	30.854	3.116	5.193
0.09	0.18	1.0005	1261.4	30.610	3.116	5.193
0.10	0.20	1.0006	1261.4	30.391	3.116	5.193
0.2	0.40	1.0011	1261.7	28.951	3.116	5.193
0.3	0.60	1.0017	1262.0	28.109	3.116	5.193
0.4	0.80	1.0023	1262.3	27.511	3.116	5.193
0.5	1.00	1.0028	1262.6	27.048	3.116	5.193
0.6	1.20	1.0034	1262.9	26.669	3.116	5.193
0.7	1.40	1.0040	1263.2	26.349	3.117	5.193
0.8	1.60	1.0046	1263.5	26.072	3.117	5.193
0.9	1.80	1.0051	1263.8	25.827	3.117	5.193
1.0	1.99	1.0057	1264.1	25.608	3.117	5.193
1.5	2.98	1.0085	1265.5	24.766	3.117	5.193
2.0	3.97	1.0114	1267.0	24.169	3.118	5.193
2.5	4.94	1.0143	1268.5	23.706	3.118	5.193
3.0	5.92	1.0171	1270.0	23.327	3.119	5.193
3.5	6.88	1.0200	1271.5	23.007	3.119	5.192
4.0	7.84	1.0228	1273.0	22.730	3.120	5.192
4.5	8.80	1.0257	1274.5	22.486	3.120	5.192
5.0	9.75	1.0286	1276.0	22.267	3.121	5.192
5.5	10.70	1.0315	1277.5	22.070	3.121	5.191
6.0	11.64	1.0343	1279.0	21.889	3.122	5.191
6.5	12.57	1.0372	1280.5	21.723	3.122	5.191
7.0	13.50	1.0401	1282.1	21.570	3.123	5.190
7.5	14.42	1.0430	1283.6	21.427	3.123	5.190
8.0	15.34	1.0459	1285.1	21.293	3.124	5.190
8.5	16.26	1.0488	1286.6	21.168	3.124	5.189
9.0	17.16	1.0517	1288.1	21.049	3.124	5.189
9.5	18.07	1.0547	1289.7	20.937	3.125	5.188
10.0	18.97	1.0576	1291.2	20.831	3.125	5.188
15.0	27.68	1.0869	1306.6	19.993	3.130	5.184
20.0	35.94	1.1164	1322.2	19.400	3.134	5.181
25.0	43.77	1.1456	1337.6	18.940	3.138	5.180

Table II.17 (*Continued*)

p	ρ	z	h	s	c_v	c_p
30	51.23	1.1746	1352.9	18.565	3.142	5.180
35	58.35	1.2033	1368.0	18.247	3.145	5.181
40	65.15	1.2315	1383.0	17.972	3.149	5.183
45	71.67	1.2595	1397.8	17.729	3.152	5.184
50	77.92	1.2871	1412.5	17.512	3.155	5.184
55	83.92	1.3146	1427.0	17.315	3.158	5.184
60	89.69	1.3418	1441.5	17.135	3.161	5.183
65	95.24	1.3689	1455.9	16.970	3.164	5.181
70	100.59	1.3959	1470.3	16.817	3.167	5.179
75	105.74	1.4228	1484.6	16.674	3.170	5.176
80	110.71	1.4495	1498.8	16.541	3.172	5.174
85	115.51	1.4761	1513.0	16.416	3.175	5.171
90	120.15	1.5026	1527.2	16.298	3.177	5.169
95	124.64	1.5289	1541.3	16.187	3.180	5.168
100	129.00	1.5550	1555.3	16.081	3.182	5.167
			$T = 250$ K			
0.01	0.02	1.0001	1313.1	35.386	3.116	5.193
0.02	0.04	1.0001	1313.1	33.946	3.116	5.193
0.03	0.06	1.0002	1313.1	33.104	3.116	5.193
0.04	0.08	1.0002	1313.1	32.506	3.116	5.193
0.05	0.10	1.0003	1313.2	32.043	3.116	5.193
0.06	0.12	1.0003	1313.2	31.664	3.116	5.193
0.07	0.13	1.0004	1313.2	31.344	3.116	5.193
0.08	0.15	1.0004	1313.3	31.066	3.116	5.193
0.09	0.17	1.0005	1313.3	30.822	3.116	5.193
0.10	0.19	1.0005	1313.3	30.603	3.116	5.193
0.2	0.38	1.0011	1313.6	29.163	3.116	5.193
0.3	0.58	1.0016	1313.9	28.321	3.116	5.193
0.4	0.77	1.0022	1314.2	27.723	3.116	5.193
0.5	0.96	1.0027	1314.5	27.260	3.116	5.193
0.6	1.15	1.0033	1314.8	26.881	3.116	5.193
0.7	1.34	1.0038	1315.1	26.561	3.117	5.193
0.8	1.53	1.0044	1315.4	26.284	3.117	5.193
0.9	1.72	1.0049	1315.7	26.039	3.117	5.193
1.0	1.92	1.0055	1316.0	25.820	3.117	5.193
1.5	2.86	1.0082	1317.5	24.978	3.117	5.193
2.0	3.81	1.0109	1319.0	24.381	3.118	5.193
2.5	4.75	1.0137	1320.4	23.918	3.118	5.193
3.0	5.68	1.0164	1321.9	23.539	3.119	5.192
3.5	6.61	1.0191	1323.4	23.219	3.119	5.192
4.0	7.54	1.0219	1324.9	22.942	3.119	5.192
4.5	8.46	1.0246	1326.4	22.698	3.120	5.192
5.0	9.37	1.0274	1327.9	22.479	3.120	5.191
5.5	10.28	1.0301	1329.4	22.282	3.121	5.191
6.0	11.19	1.0329	1330.9	22.101	3.121	5.191
6.5	12.09	1.0356	1332.4	21.935	3.122	5.190
7.0	12.98	1.0384	1334.0	21.782	3.122	5.190
7.5	13.87	1.0412	1335.5	21.639	3.122	5.190
8.0	14.76	1.0440	1337.0	21.505	3.123	5.189
8.5	15.64	1.0467	1338.5	21.379	3.123	5.189
9.0	16.51	1.0495	1340.0	21.261	3.124	5.188
9.5	17.38	1.0523	1341.6	21.149	3.124	5.188
10.0	18.25	1.0551	1343.1	21.043	3.124	5.187
15.0	26.67	1.0832	1358.5	20.205	3.128	5.183
20.0	34.65	1.1114	1374.0	19.611	3.132	5.180

Table II.17 (*Continued*)

p	ρ	z	h	s	c_v	c_p
25.0	42.25	1.1395	1389.4	19.152	3.136	5.178
30	49.49	1.1673	1404.7	18.776	3.139	5.178
35	56.41	1.1948	1419.8	18.459	3.142	5.179
40	63.03	1.2220	1434.8	18.184	3.146	5.180
45	69.39	1.2488	1449.6	17.941	3.149	5.181
50	75.49	1.2754	1464.3	17.723	3.152	5.182
55	81.36	1.3018	1478.8	17.526	3.154	5.182
60	87.01	1.3279	1493.3	17.347	3.157	5.181
65	92.44	1.3539	1507.7	17.181	3.160	5.180
70	97.69	1.3798	1522.0	17.028	3.162	5.178
75	102.75	1.4056	1536.3	16.886	3.165	5.176
80	107.63	1.4312	1550.6	16.752	3.167	5.173
85	112.36	1.4568	1564.7	16.627	3.169	5.171
90	116.93	1.4822	1578.9	16.509	3.172	5.169
95	121.35	1.5074	1592.9	16.398	3.174	5.168
100	125.65	1.5326	1606.9	16.292	3.176	5.167

$T = 260$ K

p	ρ	z	h	s	c_v	c_p
0.01	0.02	1.0001	1365.0	35.589	3.116	5.193
0.02	0.04	1.0001	1365.0	34.150	3.116	5.193
0.03	0.06	1.0002	1365.0	33.307	3.116	5.193
0.04	0.07	1.0002	1365.1	32.710	3.116	5.193
0.05	0.09	1.0003	1365.1	32.246	3.116	5.193
0.06	0.11	1.0003	1365.1	31.867	3.116	5.193
0.07	0.13	1.0004	1365.2	31.547	3.116	5.193
0.08	0.15	1.0004	1365.2	31.270	3.116	5.193
0.09	0.17	1.0005	1365.2	31.025	3.116	5.193
0.10	0.19	1.0005	1365.2	30.806	3.116	5.193
0.2	0.37	1.0010	1365.5	29.367	3.116	5.193
0.3	0.55	1.0016	1365.8	28.524	3.116	5.193
0.4	0.74	1.0021	1366.1	27.927	3.116	5.193
0.5	0.92	1.0026	1366.4	27.463	3.116	5.193
0.6	1.11	1.0031	1366.7	27.085	3.116	5.193
0.7	1.29	1.0037	1367.0	26.765	3.116	5.193
0.8	1.48	1.0042	1367.3	26.487	3.117	5.193
0.9	1.66	1.0047	1367.6	26.243	3.117	5.193
1.0	1.84	1.0052	1367.9	26.024	3.117	5.193
1.5	2.76	1.0079	1369.4	25.182	3.117	5.193
2.0	3.66	1.0105	1370.9	24.584	3.117	5.193
2.5	4.57	1.0131	1372.4	24.121	3.118	5.192
3.0	5.47	1.0157	1373.9	23.743	3.118	5.192
3.5	6.36	1.0184	1375.3	23.423	3.119	5.192
4.0	7.25	1.0210	1376.8	23.146	3.119	5.192
4.5	8.14	1.0236	1378.3	22.901	3.119	5.191
5.0	9.02	1.0263	1379.8	22.683	3.120	5.191
5.5	9.90	1.0289	1381.3	22.485	3.120	5.191
6.0	10.77	1.0315	1382.8	22.305	3.121	5.190
6.5	11.64	1.0342	1384.3	22.139	3.121	5.190
7.0	12.50	1.0368	1385.9	21.985	3.121	5.190
7.5	13.36	1.0395	1387.4	21.842	3.122	5.189
8.0	14.21	1.0422	1388.9	21.709	3.122	5.189
8.5	15.06	1.0448	1390.4	21.583	3.122	5.188
9.0	15.91	1.0475	1391.9	21.465	3.123	5.188
9.5	16.75	1.0502	1393.4	21.353	3.123	5.188
10.0	17.59	1.0528	1395.0	21.246	3.124	5.187

Table II.17 (*Continued*)

p	ρ	z	h	s	c_v	c_p
15.0	25.72	1.0798	1410.3	20.408	3.127	5.183
20.0	33.46	1.1068	1425.8	19.815	3.131	5.179
25.0	40.83	1.1338	1441.2	19.355	3.134	5.177
30	47.86	1.1605	1456.5	18.979	3.137	5.177
35	54.60	1.1870	1471.6	18.662	3.140	5.177
40	61.05	1.2131	1486.6	18.387	3.143	5.178
45	67.25	1.2390	1501.4	18.144	3.146	5.179
50	73.21	1.2646	1516.1	17.926	3.148	5.180
55	78.95	1.2899	1530.7	17.730	3.151	5.180
60	84.47	1.3151	1545.1	17.550	3.153	5.180
65	89.81	1.3401	1559.5	17.384	3.156	5.179
70	94.95	1.3650	1573.8	17.231	3.158	5.177
75	99.92	1.3898	1588.1	17.089	3.160	5.176
80	104.72	1.4144	1602.3	16.955	3.162	5.173
85	109.37	1.4390	1616.5	16.830	3.164	5.171
90	113.87	1.4634	1630.6	16.712	3.166	5.169
95	118.23	1.4877	1644.6	16.601	3.168	5.168
100	122.47	1.5119	1658.6	16.495	3.170	5.166

$T = 270$ K

p	ρ	z	h	s	c_v	c_p
0.01	0.02	1.0000	1416.9	35.785	3.116	5.193
0.02	0.04	1.0001	1416.9	34.346	3.116	5.193
0.03	0.05	1.0002	1417.0	33.503	3.116	5.193
0.04	0.07	1.0002	1417.0	32.906	3.116	5.193
0.05	0.09	1.0003	1417.0	32.442	3.116	5.193
0.06	0.11	1.0003	1417.1	32.063	3.116	5.193
0.07	0.12	1.0004	1417.1	31.743	3.116	5.193
0.08	0.14	1.0004	1417.1	31.466	3.116	5.193
0.09	0.16	1.0005	1417.2	31.221	3.116	5.193
0.10	0.18	1.0005	1417.2	31.002	3.116	5.193
0.2	0.36	1.0010	1417.5	29.563	3.116	5.193
0.3	0.53	1.0015	1417.8	28.720	3.116	5.193
0.4	0.71	1.0020	1418.1	28.123	3.116	5.193
0.5	0.89	1.0025	1418.4	27.659	3.116	5.193
0.6	1.07	1.0030	1418.7	27.281	3.116	5.193
0.7	1.24	1.0035	1419.0	26.961	3.116	5.193
0.8	1.42	1.0040	1419.3	26.683	3.116	5.193
0.9	1.60	1.0045	1419.5	26.439	3.117	5.193
1.0	1.77	1.0050	1419.8	26.220	3.117	5.193
1.5	2.65	1.0076	1421.3	25.378	3.117	5.193
2.0	3.53	1.0101	1422.8	24.780	3.117	5.192
2.5	4.40	1.0126	1424.3	24.317	3.118	5.192
3.0	5.27	1.0151	1425.8	23.939	3.118	5.192
3.5	6.13	1.0176	1427.3	23.619	3.118	5.192
4.0	6.99	1.0202	1428.8	23.342	3.119	5.192
4.5	7.85	1.0227	1430.2	23.097	3.119	5.191
5.0	8.70	1.0252	1431.7	22.879	3.119	5.191
5.5	9.54	1.0278	1433.2	22.681	3.120	5.191
6.0	10.38	1.0303	1434.7	22.501	3.120	5.190
6.5	11.22	1.0329	1436.2	22.335	3.120	5.190
7.0	12.05	1.0354	1437.7	22.181	3.121	5.189
7.5	12.88	1.0379	1439.3	22.038	3.121	5.189
8.0	13.71	1.0405	1440.8	21.904	3.121	5.189
8.5	14.53	1.0431	1442.3	21.779	3.122	5.188
9.0	15.35	1.0456	1443.8	21.660	3.122	5.188
9.5	16.16	1.0482	1445.3	21.548	3.122	5.187

Table II.17 (*Continued*)

p	ρ	z	h	s	c_v	c_p
10.0	16.97	1.0508	1446.8	21.442	3.123	5.187
15.0	24.84	1.0766	1462.1	20.604	3.126	5.182
20.0	32.34	1.1026	1477.5	20.010	3.129	5.179
25.0	39.50	1.1285	1492.9	19.550	3.132	5.176
30	46.34	1.1543	1508.2	19.175	3.135	5.176
35	52.90	1.1798	1523.4	18.857	3.138	5.176
40	59.19	1.2050	1538.4	18.582	3.140	5.177
45	65.24	1.2299	1553.2	18.339	3.143	5.178
50	71.06	1.2546	1567.9	18.122	3.145	5.178
55	76.67	1.2790	1582.5	17.925	3.148	5.179
60	82.09	1.3033	1596.9	17.745	3.150	5.179
65	87.31	1.3273	1611.3	17.580	3.152	5.178
70	92.36	1.3513	1625.6	17.427	3.154	5.177
75	97.24	1.3751	1639.8	17.284	3.156	5.175
80	101.97	1.3989	1654.0	17.151	3.158	5.173
85	106.54	1.4225	1668.2	17.025	3.160	5.171
90	110.97	1.4460	1682.3	16.907	3.162	5.170
95	115.27	1.4694	1696.3	16.796	3.164	5.168
100	119.44	1.4927	1710.3	16.690	3.166	5.167

$T = 280$ K

p	ρ	z	h	s	c_v	c_p
0.01	0.02	1.0000	1468.8	35.974	3.116	5.193
0.02	0.03	1.0001	1468.9	34.534	3.116	5.193
0.03	0.05	1.0001	1468.9	33.692	3.116	5.193
0.04	0.07	1.0002	1468.9	33.095	3.116	5.193
0.05	0.09	1.0002	1469.0	32.631	3.116	5.193
0.06	0.10	1.0003	1469.0	32.252	3.116	5.193
0.07	0.12	1.0003	1469.0	31.932	3.116	5.193
0.08	0.14	1.0004	1469.1	31.655	3.116	5.193
0.09	0.15	1.0004	1469.1	31.410	3.116	5.193
0.10	0.17	1.0005	1469.1	31.191	3.116	5.193
0.2	0.34	1.0010	1469.4	29.751	3.116	5.193
0.3	0.52	1.0015	1469.7	28.909	3.116	5.193
0.4	0.69	1.0019	1470.0	28.312	3.116	5.193
0.5	0.86	1.0024	1470.3	27.848	3.116	5.193
0.6	1.03	1.0029	1470.6	27.470	3.116	5.193
0.7	1.20	1.0034	1470.9	27.149	3.116	5.193
0.8	1.37	1.0039	1471.2	26.872	3.116	5.193
0.9	1.54	1.0044	1471.5	26.627	3.116	5.193
1.0	1.71	1.0048	1471.8	26.409	3.117	5.193
1.5	2.56	1.0073	1473.3	25.567	3.117	5.193
2.0	3.41	1.0097	1474.7	24.969	3.117	5.192
2.5	4.25	1.0121	1476.2	24.506	3.118	5.192
3.0	5.08	1.0146	1477.7	24.128	3.118	5.192
3.5	5.92	1.0170	1479.2	23.808	3.118	5.192
4.0	6.75	1.0194	1480.7	23.531	3.118	5.191
4.5	7.57	1.0218	1482.2	23.286	3.119	5.191
5.0	8.39	1.0243	1483.6	23.068	3.119	5.191
5.5	9.21	1.0267	1485.1	22.870	3.119	5.190
6.0	10.02	1.0292	1486.6	22.689	3.120	5.190
6.5	10.83	1.0316	1488.1	22.523	3.120	5.190
7.0	11.64	1.0341	1489.6	22.370	3.120	5.189
7.5	12.44	1.0365	1491.1	22.227	3.121	5.189
8.0	13.24	1.0390	1492.7	22.093	3.121	5.189
8.5	14.03	1.0414	1494.2	21.967	3.121	5.188
9.0	14.82	1.0439	1495.7	21.849	3.122	5.188

Table II.17 (*Continued*)

p	ρ	z	h	s	c_v	c_p
9.5	15.61	1.0464	1497.2	21.737	3.122	5.187
10.0	16.39	1.0488	1498.7	21.631	3.122	5.187
15.0	24.02	1.0737	1514.0	20.792	3.125	5.182
20.0	31.30	1.0987	1529.3	20.198	3.128	5.178
25.0	38.25	1.1236	1544.7	19.738	3.131	5.176
30	44.91	1.1485	1560.0	19.363	3.133	5.175
35	51.30	1.1730	1575.1	19.045	3.136	5.175
40	57.44	1.1974	1590.1	18.770	3.138	5.175
45	63.34	1.2214	1605.0	18.528	3.140	5.176
50	69.03	1.2452	1619.7	18.310	3.143	5.177
55	74.53	1.2688	1634.2	18.113	3.145	5.178
60	79.83	1.2922	1648.7	17.934	3.147	5.178
65	84.95	1.3155	1663.1	17.768	3.149	5.177
70	89.91	1.3386	1677.4	17.615	3.151	5.176
75	94.71	1.3616	1691.6	17.472	3.153	5.175
80	99.35	1.3844	1705.8	17.339	3.155	5.173
85	103.85	1.4072	1719.9	17.213	3.156	5.172
90	108.21	1.4299	1734.0	17.095	3.158	5.170
95	112.45	1.4525	1748.0	16.984	3.160	5.168
100	116.57	1.4750	1761.9	16.877	3.161	5.167

$T = 290$ K

p	ρ	z	h	s	c_v	c_p
0.01	0.02	1.0000	1520.8	36.156	3.116	5.193
0.02	0.03	1.0001	1520.8	34.717	3.116	5.193
0.03	0.05	1.0001	1520.8	33.874	3.116	5.193
0.04	0.07	1.0002	1520.9	33.277	3.116	5.193
0.05	0.08	1.0002	1520.9	32.813	3.116	5.193
0.06	0.10	1.0003	1520.9	32.435	3.116	5.193
0.07	0.12	1.0003	1521.0	32.114	3.116	5.193
0.08	0.13	1.0004	1521.0	31.837	3.116	5.193
0.09	0.15	1.0004	1521.0	31.592	3.116	5.193
0.10	0.17	1.0005	1521.0	31.373	3.116	5.193
0.2	0.33	1.0009	1521.3	29.934	3.116	5.193
0.3	0.50	1.0014	1521.6	29.091	3.116	5.193
0.4	0.66	1.0019	1521.9	28.494	3.116	5.193
0.5	0.83	1.0023	1522.2	28.030	3.116	5.193
0.6	0.99	1.0028	1522.5	27.652	3.116	5.193
0.7	1.16	1.0033	1522.8	27.332	3.116	5.193
0.8	1.32	1.0037	1523.1	27.054	3.116	5.193
0.9	1.49	1.0042	1523.4	26.810	3.116	5.193
1.0	1.65	1.0047	1523.7	26.591	3.116	5.193
1.5	2.47	1.0070	1525.2	25.749	3.117	5.193
2.0	3.29	1.0093	1526.7	25.151	3.117	5.192
2.5	4.10	1.0117	1528.1	24.688	3.117	5.192
3.0	4.91	1.0140	1529.6	24.310	3.118	5.192
3.5	5.72	1.0164	1531.1	23.990	3.118	5.192
4.0	6.52	1.0187	1532.6	23.713	3.118	5.191
4.5	7.32	1.0210	1534.1	23.468	3.119	5.191
5.0	8.11	1.0234	1535.6	23.250	3.119	5.191
5.5	8.90	1.0257	1537.0	23.052	3.119	5.190
6.0	9.69	1.0281	1538.5	22.872	3.119	5.190
6.5	10.47	1.0305	1540.0	22.706	3.120	5.190
7.0	11.25	1.0328	1541.5	22.552	3.120	5.189
7.5	12.03	1.0352	1543.0	22.409	3.120	5.189
8.0	12.80	1.0375	1544.5	22.275	3.121	5.188

Table II.17 (*Continued*)

p	ρ	z	h	s	c_v	c_p
8.5	13.57	1.0399	1546.0	22.150	3.121	5.188
9.0	14.33	1.0423	1547.5	22.031	3.121	5.188
9.5	15.10	1.0447	1549.1	21.919	3.121	5.187
10.0	15.85	1.0470	1550.6	21.813	3.122	5.187
15.0	23.25	1.0709	1565.8	20.974	3.124	5.182
20.0	30.32	1.0950	1581.1	20.380	3.127	5.178
25.0	37.08	1.1191	1596.5	19.920	3.129	5.176
30	43.57	1.1430	1611.7	19.545	3.132	5.174
35	49.80	1.1668	1626.9	19.227	3.134	5.174
40	55.79	1.1903	1641.9	18.952	3.136	5.174
45	61.56	1.2136	1656.7	18.709	3.138	5.175
50	67.12	1.2366	1671.4	18.492	3.140	5.176
55	72.50	1.2594	1686.0	18.295	3.142	5.176
60	77.69	1.2820	1700.5	18.115	3.144	5.177
65	82.72	1.3044	1714.8	17.950	3.146	5.177
70	87.58	1.3267	1729.1	17.796	3.148	5.176
75	92.30	1.3489	1743.3	17.654	3.150	5.175
80	96.86	1.3710	1757.5	17.520	3.151	5.174
85	101.29	1.3930	1771.6	17.395	3.153	5.172
90	105.59	1.4149	1785.7	17.277	3.155	5.170
95	109.76	1.4367	1799.7	17.165	3.156	5.169
100	103.82	1.4584	1813.6	17.059	3.158	5.167

$T = 300$ K

p	ρ	z	h	s	c_v	c_p
0.01	0.02	1.0000	1572.7	36.333	3.116	5.193
0.02	0.03	1.0001	1572.7	34.893	3.116	5.193
0.03	0.05	1.0001	1572.8	34.050	3.116	5.193
0.04	0.06	1.0002	1572.8	33.453	3.116	5.193
0.05	0.08	1.0002	1572.8	32.989	3.116	5.193
0.06	0.10	1.0003	1572.9	32.611	3.116	5.193
0.07	0.11	1.0003	1572.9	32.290	3.116	5.193
0.08	0.13	1.0004	1572.9	32.013	3.116	5.193
0.09	0.14	1.0004	1572.9	31.768	3.116	5.193
0.10	0.16	1.0005	1573.0	31.550	3.116	5.193
0.2	0.32	1.0009	1573.3	30.110	3.116	5.193
0.3	0.48	1.0014	1573.6	29.268	3.116	5.193
0.4	0.64	1.0018	1573.9	28.670	3.116	5.193
0.5	0.80	1.0023	1574.2	28.207	3.116	5.193
0.6	0.96	1.0027	1574.4	27.828	3.116	5.193
0.7	1.12	1.0032	1574.7	27.508	3.116	5.193
0.8	1.28	1.0036	1575.0	27.230	3.116	5.193
0.9	1.44	1.0041	1575.3	26.986	3.116	5.193
1.0	1.60	1.0045	1575.6	26.767	3.116	5.193
1.5	2.39	1.0068	1577.1	25.925	3.117	5.193
2.0	3.18	1.0090	1578.6	25.328	3.117	5.192
2.5	3.97	1.0113	1580.1	24.864	3.117	5.192
3.0	4.75	1.0135	1581.5	24.486	3.117	5.192
3.5	5.53	1.0158	1583.0	24.166	3.118	5.192
4.0	6.30	1.0180	1584.5	23.889	3.118	5.191
4.5	7.08	1.0203	1586.0	23.644	3.118	5.191
5.0	7.85	1.0226	1587.5	23.426	3.119	5.191
5.5	8.61	1.0248	1589.0	23.228	3.119	5.190
6.0	9.37	1.0271	1590.4	23.047	3.119	5.190
6.5	10.13	1.0294	1591.9	22.881	3.119	5.190
7.0	10.89	1.0317	1593.4	22.728	3.120	5.189
7.5	11.64	1.0339	1594.9	22.585	3.120	5.189

Table II.17 (*Continued*)

p	ρ	z	h	s	c_v	c_p
8.0	12.39	1.0362	1596.4	22.451	3.120	5.188
8.5	13.13	1.0385	1597.9	22.325	3.120	5.188
9.0	13.88	1.0408	1599.4	22.207	3.121	5.188
9.5	14.62	1.0431	1600.9	22.095	3.121	5.187
10.0	15.35	1.0454	1602.4	21.989	3.121	5.187
15.0	22.53	1.0684	1617.6	21.150	3.124	5.182
20.0	29.40	1.0916	1632.9	20.556	3.126	5.178
25.0	35.98	1.1149	1648.2	20.096	3.128	5.175
30	42.30	1.1380	1663.5	19.720	3.130	5.174
35	48.38	1.1609	1678.6	19.402	3.132	5.173
40	54.23	1.1837	1693.6	19.127	3.134	5.173
45	59.87	1.2062	1708.5	18.885	3.136	5.174
50	65.31	1.2285	1723.2	18.667	3.138	5.175
55	70.58	1.2505	1737.8	18.470	3.140	5.175
60	75.67	1.2724	1752.2	18.291	3.142	5.176
65	80.60	1.2941	1766.6	18.125	3.144	5.176
70	85.37	1.3157	1780.9	17.972	3.145	5.175
75	90.00	1.3372	1795.1	17.829	3.147	5.175
80	94.50	1.3585	1809.2	17.696	3.148	5.174
85	98.85	1.3798	1823.3	17.570	3.150	5.172
90	103.09	1.4010	1837.4	17.452	3.151	5.171
95	107.20	1.4220	1851.4	17.340	3.153	5.169
100	111.20	1.4430	1865.3	17.234	3.154	5.168

$$T = 400 \ K$$

p	ρ	z	h	s	c_v	c_p
0.01	0.01	1.0000	2092.0	37.827	3.116	5.193
0.02	0.02	1.0001	2092.0	36.387	3.116	5.193
0.03	0.04	1.0001	2092.1	35.544	3.116	5.193
0.04	0.05	1.0001	2092.1	34.947	3.116	5.193
0.05	0.06	1.0002	2092.1	34.483	3.116	5.193
0.06	0.07	1.0002	2092.2	34.105	3.116	5.193
0.07	0.08	1.0002	2092.2	33.784	3.116	5.193
0.08	0.10	1.0003	2092.2	33.507	3.116	5.193
0.09	0.11	1.0003	2092.3	33.262	3.116	5.193
0.10	0.12	1.0003	2092.3	33.043	3.116	5.193
0.2	0.24	1.0007	2092.6	31.604	3.116	5.193
0.3	0.36	1.0010	2092.9	30.761	3.116	5.193
0.4	0.48	1.0013	2093.2	30.164	3.116	5.193
0.5	0.60	1.0017	2093.4	29.700	3.116	5.193
0.6	0.72	1.0020	2093.7	29.322	3.116	5.193
0.7	0.84	1.0023	2094.0	29.002	3.116	5.193
0.8	0.96	1.0027	2094.3	28.724	3.116	5.193
0.9	1.08	1.0030	2094.6	28.480	3.116	5.193
1.0	1.20	1.0033	2094.9	28.261	3.116	5.193
1.5	1.80	1.0050	2096.3	27.419	3.116	5.192
2.0	2.39	1.0067	2097.8	26.821	3.116	5.192
2.5	2.98	1.0083	2099.3	26.358	3.116	5.192
3.0	3.57	1.0100	2100.7	25.979	3.117	5.192
3.5	4.16	1.0117	2102.2	25.659	3.117	5.192
4.0	4.75	1.0133	2103.6	25.382	3.117	5.191
4.5	5.34	1.0150	2105.1	25.138	3.117	5.191
5.0	5.92	1.0167	2106.5	24.919	3.117	5.191
5.5	6.50	1.0183	2108.0	24.721	3.117	5.191
6.0	7.08	1.0200	2109.4	24.541	3.117	5.190
6.5	7.66	1.0217	2110.9	24.374	3.117	5.190

Table II.17 (*Continued*)

p	ρ	z	h	s	c_v	c_p
7.0	8.23	1.0233	2112.4	24.221	3.118	5.190
7.5	8.81	1.0250	2113.8	24.078	3.118	5.189
8.0	9.38	1.0267	2115.3	23.944	3.118	5.189
8.5	9.95	1.0284	2116.8	23.818	3.118	5.189
9.0	10.52	1.0300	2118.2	23.699	3.118	5.188
9.5	11.08	1.0317	2119.7	23.587	3.118	5.188
10.0	11.65	1.0334	2121.2	23.481	3.118	5.188
15.0	17.19	1.0503	2135.9	22.641	3.119	5.184
20.0	22.55	1.0673	2150.8	22.046	3.120	5.181
25.0	27.75	1.0844	2165.8	21.585	3.122	5.177
30	32.78	1.1016	2180.8	21.208	3.123	5.174
35	37.65	1.1187	2195.8	20.890	3.124	5.172
40	42.39	1.1358	2210.7	20.615	3.125	5.171
45	46.98	1.1527	2225.6	20.372	3.125	5.170
50	51.45	1.1696	2240.3	20.155	3.126	5.170
55	55.80	1.1863	2255.0	19.958	3.127	5.170
60	60.03	1.2029	2269.5	19.779	3.128	5.171
65	64.15	1.2194	2283.9	19.613	3.129	5.171
70	68.18	1.2357	2298.2	19.460	3.130	5.172
75	72.10	1.2520	2312.5	19.318	3.131	5.173
80	75.93	1.2681	2326.6	19.184	3.131	5.173
85	79.66	1.2841	2340.6	19.058	3.132	5.173
90	83.32	1.3001	2354.6	18.940	3.133	5.174
95	86.88	1.3159	2368.5	18.828	3.134	5.174
100	90.37	1.3317	2382.4	18.722	3.134	5.173

$T = 500$ K

p	ρ	z	h	s	c_v	c_p
0.01	0.01	1.0000	2611.3	38.985	3.116	5.193
0.02	0.02	1.0001	2611.4	37.545	3.116	5.193
0.03	0.03	1.0001	2611.4	36.703	3.116	5.193
0.04	0.04	1.0001	2611.4	36.106	3.116	5.193
0.05	0.05	1.0001	2611.4	35.642	3.116	5.193
0.06	0.06	1.0002	2611.5	35.263	3.116	5.193
0.07	0.07	1.0002	2611.5	34.943	3.116	5.193
0.08	0.08	1.0002	2611.5	34.666	3.116	5.193
0.09	0.09	1.0002	2611.6	34.421	3.116	5.193
0.10	0.10	1.0003	2611.6	34.202	3.116	5.193
0.2	0.19	1.0005	2611.9	32.762	3.116	5.193
0.3	0.29	1.0008	2612.2	31.920	3.116	5.193
0.4	0.38	1.0011	2612.5	31.323	3.116	5.193
0.5	0.48	1.0013	2612.7	30.859	3.116	5.193
0.6	0.58	1.0016	2613.0	30.480	3.116	5.193
0.7	0.67	1.0018	2613.3	30.160	3.116	5.193
0.8	0.77	1.0021	2613.6	29.883	3.116	5.193
0.9	0.86	1.0024	2613.9	29.638	3.116	5.193
1.0	0.96	1.0026	2614.2	29.419	3.116	5.193
1.5	1.44	1.0040	2615.6	28.577	3.116	5.193
2.0	1.92	1.0053	2617.0	27.980	3.116	5.192
2.5	2.39	1.0066	2618.5	27.516	3.116	5.192
3.0	2.87	1.0079	2619.9	27.138	3.116	5.192
3.5	3.34	1.0092	2621.3	26.818	3.116	5.192
4.0	3.81	1.0106	2622.8	26.541	3.116	5.192
4.5	4.28	1.0119	2624.2	26.296	3.116	5.191
5.0	4.75	1.0132	2625.6	26.077	3.117	5.191
5.5	5.22	1.0145	2627.1	25.879	3.117	5.191

Table II.17 (*Continued*)

p	ρ	z	h	s	c_v	c_p
6.0	5.69	1.0158	2628.5	25.699	3.117	5.191
6.5	6.15	1.0171	2629.9	25.533	3.117	5.191
7.0	6.62	1.0185	2631.4	25.379	3.117	5.190
7.5	7.08	1.0198	2632.8	25.236	3.117	5.190
8.0	7.54	1.0211	2634.2	25.102	3.117	5.190
8.5	8.00	1.0224	2635.7	24.976	3.117	5.190
9.0	8.46	1.0237	2637.1	24.857	3.117	5.189
9.5	8.92	1.0251	2638.6	24.745	3.117	5.189
10.0	9.38	1.0264	2640.0	24.639	3.117	5.189
15.0	13.89	1.0397	2654.5	23.798	3.118	5.186
20.0	18.29	1.0531	2669.0	23.202	3.118	5.183
25.0	22.57	1.0665	2683.7	22.740	3.119	5.181
30	26.74	1.0801	2698.4	22.363	3.119	5.178
35	30.81	1.0936	2713.2	22.045	3.120	5.176
40	34.78	1.1072	2727.9	21.769	3.121	5.174
45	38.66	1.1207	2742.7	21.526	3.121	5.172
50	42.44	1.1342	2757.4	21.309	3.122	5.171
55	46.14	1.1477	2772.0	21.112	3.122	5.171
60	49.76	1.1611	2786.5	20.933	3.123	5.171
65	53.29	1.1743	2801.0	20.767	3.123	5.171
70	56.75	1.1875	2815.4	20.614	3.124	5.171
75	60.14	1.2007	2829.7	20.472	3.124	5.172
80	63.46	1.2137	2843.8	20.338	3.124	5.172
85	66.72	1.2267	2858.0	20.213	3.125	5.173
90	69.91	1.2396	2872.0	20.095	3.125	5.174
95	73.03	1.2524	2885.9	19.982	3.126	5.174
100	76.10	1.2651	2899.8	19.876	3.126	5.175

$$T = 600 \text{ K}$$

p	ρ	z	h	s	c_v	c_p
0.01	0.01	1.0000	3130.6	39.932	3.116	5.193
0.02	0.02	1.0000	3130.7	38.492	3.116	5.193
0.03	0.02	1.0001	3130.7	37.650	3.116	5.193
0.04	0.03	1.0001	3130.7	37.052	3.116	5.193
0.05	0.04	1.0001	3130.8	36.589	3.116	5.193
0.06	0.05	1.0001	3130.8	36.210	3.116	5.193
0.7	0.06	1.0002	3130.8	35.890	3.116	5.193
0.08	0.06	1.0002	3130.8	35.613	3.116	5.193
0.09	0.07	1.0002	3130.9	35.368	3.116	5.193
0.10	0.08	1.0002	3130.9	35.149	3.116	5.193
0.2	0.16	1.0004	3131.2	33.709	3.116	5.193
0.3	0.24	1.0007	3131.5	32.867	3.116	5.193
0.4	0.32	1.0009	3131.7	32.269	3.116	5.193
0.5	0.40	1.0011	3132.0	31.806	3.116	5.193
0.6	0.48	1.0013	3132.3	31.427	3.116	5.193
0.7	0.56	1.0015	3132.6	31.107	3.116	5.193
0.8	0.64	1.0017	3132.9	30.830	3.116	5.193
0.9	0.72	1.0020	3133.2	30.585	3.116	5.193
1.0	0.80	1.0022	3133.5	30.366	3.116	5.193
1.5	1.20	1.0033	3134.9	29.524	3.116	5.193
2.0	1.60	1.0044	3136.3	28.927	3.116	5.193
2.5	1.99	1.0055	3137.7	28.463	3.116	5.192
3.9	2.39	1.0065	3139.1	28.084	3.116	5.192
3.5	2.79	1.0076	3140.5	27.764	3.116	5.192
4.0	3.18	1.0087	3141.9	27.487	3.116	5.192
4.5	3.58	1.0098	3143.4	27.243	3.116	5.192

Table II.17 (*Continued*)

p	ρ	z	h	s	c_v	c_p
5.0	3.97	1.0109	3144.8	27.024	3.116	5.192
5.5	4.36	1.0120	3146.2	26.826	3.116	5.191
6.0	4.75	1.0131	3147.6	26.645	3.116	5.191
6.5	5.14	1.0142	3149.0	26.479	3.116	5.191
7.0	5.53	1.0153	3150.4	26.325	3.116	5.191
7.5	5.92	1.0164	3151.8	26.182	3.116	5.191
8.0	6.31	1.0174	3153.3	26.048	3.116	5.191
8.5	6.70	1.0185	3154.7	25.922	3.117	5.190
9.0	7.08	1.0196	3156.1	25.804	3.117	5.190
9.5	7.47	1.0207	3157.5	25.691	3.117	5.190
10.0	7.85	1.0218	3158.9	25.585	3.117	5.190
15.0	11.65	1.0328	3173.2	24.744	3.117	5.188
20.0	15.37	1.0438	3187.5	24.147	3.117	5.186
25.0	19.02	1.0548	3201.9	23.685	3.118	5.183
30	22.58	1.0660	3216.4	23.308	3.118	5.181
35	26.07	1.0772	3230.9	22.989	3.118	5.179
40	29.49	1.0884	3245.5	22.713	3.119	5.177
45	32.84	1.0996	3260.1	22.469	3.119	5.176
50	36.12	1.1108	3274.6	22.252	3.119	5.174
55	39.33	1.1220	3289.2	22.055	3.120	5.173
60	42.48	1.1331	3303.7	21.875	3.120	5.172
65	45.58	1.1443	3318.1	21.710	3.120	5.172
70	48.61	1.1553	3332.5	21.557	3.121	5.172
75	51.59	1.1663	3346.8	21.415	3.121	5.172
80	54.52	1.1773	3361.1	21.281	3.121	5.172
85	57.40	1.1882	3375.2	21.156	3.121	5.173
90	60.22	1.1991	3389.3	21.038	3.122	5.173
95	63.00	1.2099	3403.4	20.926	3.122	5.174
100	65.73	1.2206	3417.3	20.820	3.122	5.175

$T = 700$ K

p	ρ	z	h	s	c_v	c_p
0.01	0.01	1.0000	3650.0	40.733	3.116	5.193
0.02	0.01	1.0000	3650.0	39.293	3.116	5.193
0.03	0.02	1.0001	3650.0	38.451	3.116	5.193
0.04	0.03	1.0001	3650.0	37.853	3.116	5.193
0.05	0.03	1.0001	3650.1	37.389	3.116	5.193
0.06	0.04	1.0001	3650.1	37.011	3.116	5.193
0.07	0.05	1.0001	3650.1	36.691	3.116	5.193
0.08	0.06	1.0001	3650.2	36.413	3.116	5.193
0.09	0.06	1.0002	3650.2	36.169	3.116	5.193
0.10	0.07	1.0002	3650.2	35.950	3.116	5.193
0.2	0.14	1.0004	3650.5	34.510	3.116	5.193
0.3	0.21	1.0006	3650.8	33.668	3.116	5.193
0.4	0.27	1.0007	3651.1	33.070	3.116	5.193
0.5	0.34	1.0009	3651.3	32.606	3.116	5.193
0.6	0.41	1.0011	3651.6	32.228	3.116	5.193
0.7	0.48	1.0013	3651.9	31.908	3.116	5.193
0.8	0.55	1.0015	3652.2	31.630	3.116	5.193
0.9	0.62	1.0017	3652.5	31.386	3.116	5.193
1.0	0.69	1.0019	3652.7	31.167	3.116	5.193
1.5	1.03	1.0028	3654.1	30.325	3.116	5.193
2.0	1.37	1.0037	3655.5	29.727	3.116	5.193
2.5	1.71	1.0046	3656.9	29.264	3.116	5.193
3.0	2.05	1.0056	3658.3	28.885	3.116	5.192
3.5	2.39	1.0065	3659.7	28.565	3.116	5.192

Table II.17 (*Continued*)

p	ρ	z	h	s	c_v	c_p
4.0	2.73	1.0074	3661.1	28.287	3.116	5.192
4.5	3.07	1.0084	3662.5	28.043	3.116	5.192
5.0	3.41	1.0093	3663.9	27.824	3.116	5.192
5.5	3.74	1.0102	3665.3	27.626	3.116	5.192
6.0	4.08	1.0111	3666.8	27.445	3.116	5.192
6.5	4.42	1.0121	3668.2	27.279	3.116	5.192
7.0	4.75	1.0130	3669.6	27.125	3.116	5.191
7.5	5.09	1.0139	3671.0	26.982	3.116	5.191
8.0	5.42	1.0149	3672.4	26.848	3.116	5.191
8.5	5.75	1.0158	3673.8	26.722	3.116	5.191
9.0	6.09	1.0167	3675.2	26.604	3.116	5.191
9.5	6.42	1.0176	3676.6	26.491	3.116	5.191
10.0	6.75	1.0186	3678.0	26.385	3.116	5.191
15.0	10.04	1.0279	3692.0	25.544	3.117	5.189
20.0	13.26	1.0372	3706.2	24.947	3.117	5.187
25.0	16.43	1.0466	3720.4	24.484	3.117	5.186
30	19.54	1.0561	3734.6	24.106	3.117	5.184
35	22.59	1.0656	3749.0	23.787	3.117	5.182
40	25.59	1.0751	3763.4	23.511	3.118	5.180
45	28.53	1.0846	3777.8	23.267	3.118	5.179
50	31.43	1.0942	3792.2	23.050	3.118	5.177
55	34.27	1.1037	3806.6	22.853	3.118	5.176
60	37.06	1.1133	3821.1	22.673	3.119	5.175
65	39.81	1.1228	3835.4	22.508	3.119	5.174
70	42.52	1.1323	3849.8	22.355	3.119	5.174
75	45.17	1.1418	3864.1	22.212	3.119	5.174
80	47.79	1.1512	3878.4	22.079	3.119	5.173
85	50.37	1.1606	3892.6	21.953	3.119	5.174
90	52.90	1.1700	3906.7	21.835	3.120	5.174
95	55.40	1.1793	3920.8	21.723	3.120	5.174
100	57.86	1.1886	3934.8	21.617	3.120	5.175

$T = 800$ K

p	ρ	z	h	s	c_v	c_p
0.01	0.01	1.0000	4169.3	41.426	3.116	5.193
0.02	0.01	1.0000	4169.3	39.986	3.116	5.193
0.03	0.02	1.0000	4169.3	39.144	3.116	5.193
0.04	0.02	1.0001	4169.4	38.546	3.116	5.193
0.05	0.03	1.0001	4169.4	38.083	3.116	5.193
0.06	0.04	1.0001	4169.4	37.704	3.116	5.193
0.07	0.04	1.0001	4169.4	37.384	3.116	5.193
0.08	0.05	1.0001	4169.5	37.107	3.116	5.193
0.09	0.05	1.0001	4169.5	36.862	3.116	5.193
0.10	0.06	1.0002	4169.5	36.643	3.116	5.193
0.2	0.12	1.0003	4169.8	35.203	3.116	5.193
0.3	0.18	1.0005	4170.1	34.361	3.116	5.193
0.4	0.24	1.0006	4170.4	33.763	3.116	5.193
0.5	0.30	1.0008	4170.6	33.300	3.116	5.193
0.6	0.36	1.0010	4170.9	32.921	3.116	5.193
0.7	0.42	1.0011	4171.2	32.601	3.116	5.193
0.8	0.48	1.0013	4171.5	32.324	3.116	5.193
0.9	0.54	1.0015	4171.8	32.079	3.116	5.193
1.0	0.60	1.0016	4172.0	31.860	3.116	5.193
1.5	0.90	1.0024	4173.4	31.018	3.116	5.193
2.0	1.20	1.0032	4174.8	30.420	3.116	5.193
2.5	1.50	1.0040	4176.2	29.957	3.116	5.193

Table II.17 (Continued)

p	ρ	z	h	s	c_v	c_p
3.0	1.80	1.0049	4177.6	29.578	3.116	5.193
3.5	2.09	1.0057	4179.0	29.258	3.116	5.192
4.0	2.39	1.0065	4180.4	28.981	3.116	5.192
4.5	2.69	1.0073	4181.8	28.736	3.116	5.192
5.0	2.98	1.0081	4183.2	28.517	3.116	5.192
5.5	3.28	1.0089	4184.5	28.319	3.116	5.192
6.0	3.58	1.0097	4185.9	28.139	3.116	5.192
6.5	3.87	1.0105	4187.3	27.973	3.116	5.192
7.0	4.17	1.0113	4188.7	27.819	3.116	5.192
7.5	4.46	1.0121	4190.1	27.675	3.116	5.192
8.0	4.75	1.0129	4191.5	27.541	3.116	5.192
8.5	5.05	1.0137	4192.9	27.416	3.116	5.191
9.0	5.34	1.0146	4194.3	27.297	3.116	5.191
9.5	5.63	1.0154	4195.7	27.185	3.116	5.191
10.0	5.92	1.0162	4197.0	27.078	3.116	5.191
15.0	8.81	1.0243	4211.0	26.236	3.116	5.190
20.0	11.66	1.0324	4225.0	25.640	3.117	5.189
25.0	14.46	1.0405	4239.0	25.177	3.117	5.187
30	17.21	1.0487	4253.1	24.799	3.117	5.186
35	19.93	1.0570	4267.3	24.479	3.117	5.184
40	22.60	1.0652	4281.5	24.203	3.117	5.183
45	25.22	1.0735	4295.8	23.959	3.117	5.181
50	27.81	1.0818	4310.1	23.741	3.117	5.180
55	30.36	1.0901	4324.4	23.544	3.118	5.179
60	32.87	1.0985	4338.7	23.364	3.118	5.177
65	35.34	1.1068	4353.0	23.199	3.118	5.177
70	37.78	1.1151	4367.3	23.046	3.118	5.176
75	40.17	1.1234	4381.6	22.903	3.118	5.175
80	42.54	1.1317	4395.8	22.770	3.118	5.175
85	44.87	1.1399	4410.0	22.644	3.118	5.175
90	47.17	1.1481	4424.1	22.526	3.118	5.175
95	49.44	1.1563	4438.2	22.414	3.119	5.175
100	51.67	1.1645	4452.3	22.308	3.119	5.175

$$T = 900 \text{ K}$$

p	ρ	z	h	s	c_v	c_p
0.01	0.01	1.0000	4688.6	42.038	3.116	5.193
0.02	0.01	1.0000	4688.6	40.598	3.116	5.193
0.03	0.02	1.0000	4688.6	39.756	3.116	5.193
0.04	0.02	1.0001	4688.7	39.158	3.116	5.193
0.05	0.03	1.0001	4688.7	38.695	3.116	5.193
0.06	0.03	1.0001	4688.7	38.316	3.116	5.193
0.07	0.04	1.0001	4688.7	37.996	3.116	5.193
0.08	0.04	1.0001	4688.8	37.718	3.116	5.193
0.09	0.05	1.0001	4688.8	37.474	3.116	5.193
0.10	0.05	1.0001	4688.8	37.255	3.116	5.193
0.2	0.11	1.0003	4689.1	35.815	3.116	5.193
0.3	0.16	1.0004	4689.4	34.973	3.116	5.193
0.4	0.21	1.0006	4689.7	34.375	3.116	5.193
0.5	0.27	1.0007	4689.9	33.912	3.116	5.193
0.6	0.32	1.0009	4690.2	33.533	3.116	5.193
0.7	0.37	1.0010	4690.5	33.213	3.116	5.193
0.8	0.43	1.0011	4690.8	32.935	3.116	5.193
0.9	0.48	1.0013	4691.0	32.691	3.116	5.193
1.0	0.53	1.0014	4691.3	32.472	3.116	5.193
1.5	0.80	1.0022	4692.7	31.630	3.116	5.193

Table II.17 (*Continued*)

p	ρ	ε	h	s	c_v	c_p
2.0	1.07	1.0029	4694.1	31.032	3.116	5.193
2.5	1.33	1.0036	4695.5	30.569	3.116	5.193
3.0	1.60	1.0043	4696.9	30.190	3.116	5.193
3.5	1.86	1.0050	4698.2	29.870	3.116	5.193
4.0	2.13	1.0057	4699.6	29.592	3.116	5.192
4.5	2.39	1.0064	4701.0	29.348	3.116	5.192
5.0	2.66	1.0072	4702.4	29.129	3.116	5.192
5.5	2.92	1.0079	4703.8	28.931	3.116	5.192
6.0	3.18	1.0086	4705.1	28.750	3.116	5.192
6.5	3.44	1.0093	4706.5	28.584	3.116	5.192
7.0	3.71	1.0100	4707.9	28.430	3.116	5.192
7.5	3.97	1.0107	4709.3	28.287	3.116	5.192
8.0	4.23	1.0115	4710.7	28.153	3.116	5.192
8.5	4.49	1.0122	4712.0	28.027	3.116	5.192
9.0	4.75	1.0129	4713.4	27.908	3.116	5.192
9.5	5.01	1.0136	4714.8	27.796	3.116	5.192
10.0	5.27	1.0143	4716.2	27.690	3.116	5.191
15.0	7.85	1.0215	4730.0	26.848	3.116	5.190
20.0	10.40	1.0287	4743.9	26.251	3.116	5.189
25.0	12.91	1.0359	4757.8	25.788	3.116	5.188
30	15.38	1.0431	4771.7	25.410	3.117	5.187
35	17.82	1.0504	4785.8	25.090	3.117	5.186
40	20.23	1.0576	4799.9	24.813	3.117	5.184
45	22.60	1.0650	4814.0	24.569	3.117	5.183
50	24.94	1.0723	4828.1	24.351	3.117	5.182
55	27.25	1.0796	4842.3	24.154	3.117	5.181
60	29.52	1.0870	4856.5	23.974	3.117	5.180
65	31.77	1.0944	4870.8	23.809	3.117	5.179
70	33.98	1.1017	4885.0	23.655	3.117	5.178
75	36.17	1.1091	4899.2	23.513	3.117	5.177
80	38.33	1.1165	4913.4	23.379	3.118	5.177
85	40.46	1.1238	4927.6	23.254	3.118	5.176
90	42.56	1.1311	4941.7	23.136	3.118	5.176
95	44.64	1.1385	4955.8	23.024	3.118	5.176
100	46.69	1.1457	4969.9	22.918	3.118	5.176

$T = 1000$ K

p	ρ	ε	h	s	c_v	c_p
0.01	0.00	1.0000	5207.9	42.585	3.116	5.193
0.02	0.01	1.0000	5207.9	41.145	3.116	5.193
0.03	0.01	1.0000	5207.9	40.303	3.116	5.193
0.04	0.02	1.0001	5208.0	39.705	3.116	5.193
0.05	0.02	1.0001	5208.0	39.242	3.116	5.193
0.06	0.03	1.0001	5208.0	38.863	3.116	5.193
0.07	0.03	1.0001	5208.1	38.543	3.116	5.193
0.08	0.04	1.0001	5208.1	38.265	3.116	5.193
0.09	0.04	1.0001	5208.1	38.021	3.116	5.193
0.10	0.05	1.0001	5208.1	37.802	3.116	5.193
0.2	0.10	1.0003	5208.4	36.362	3.116	5.193
0.3	0.14	1.0004	5208.7	35.520	3.116	5.193
0.4	0.19	1.0005	5209.0	34.922	3.116	5.193
0.5	0.24	1.0006	5209.2	34.459	3.116	5.193
0.6	0.29	1.0008	5209.5	34.080	3.116	5.193
0.7	0.34	1.0009	5209.8	33.760	3.116	5.193
0.8	0.38	1.0010	5210.1	33.482	3.116	5.193
0.9	0.43	1.0012	5210.3	33.238	3.116	5.193

Table II.17 (*Continued*)

p	ρ	z	h	s	c_v	c_p
1.0	0.48	1.0013	5210.6	33.019	3.116	5.193
1.5	0.72	1.0019	5212.0	32.177	3.116	5.193
2.0	0.96	1.0026	5213.4	31.579	3.116	5.193
2.5	1.20	1.0032	5214.7	31.116	3.116	5.193
3.0	1.44	1.0039	5216.1	30.737	3.116	5.193
3.5	1.68	1.0045	5217.5	30.417	3.116	5.193
4.0	1.92	1.0051	5218.9	30.139	3.116	5.193
4.5	2.15	1.0058	5220.2	29.895	3.116	5.193
5.0	2.39	1.0064	5221.6	29.676	3.116	5.192
5.5	2.63	1.0071	5223.0	29.478	3.116	5.192
6.0	2.87	1.0077	5224.4	29.297	3.116	5.192
6.5	3.10	1.0084	5225.7	29.131	3.116	5.192
7.0	3.34	1.0090	5227.1	28.977	3.116	5.192
7.5	3.58	1.0096	5228.5	28.834	3.116	5.192
8.0	3.81	1.0103	5229.8	28.700	3.116	5.192
8.5	4.05	1.0109	5231.2	28.574	3.116	5.192
9.0	4.28	1.0116	5232.6	28.455	3.116	5.192
9.5	4.52	1.0122	5234.0	28.343	3.116	5.192
10.0	4.75	1.0128	5235.3	28.237	3.116	5.192
15.0	7.08	1.0193	5249.1	27.395	3.116	5.191
20.0	9.39	1.0257	5262.8	26.798	3.116	5.190
25.0	11.66	1.0321	5276.6	26.334	3.116	5.189
30	13.91	1.0386	5290.5	25.956	3.116	5.188
35	16.12	1.0451	5304.4	25.637	3.116	5.187
40	18.31	1.0516	5318.4	25.360	3.117	5.186
45	20.47	1.0582	5332.4	25.116	3.117	5.185
50	22.61	1.0647	5346.4	24.897	3.117	5.184
55	24.71	1.0713	5360.5	24.700	3.117	5.183
60	26.80	1.0779	5374.6	24.520	3.117	5.182
65	28.85	1.0845	5388.7	24.354	3.117	5.181
70	30.88	1.0911	5402.9	24.201	3.117	5.180
75	32.89	1.0977	5417.0	24.058	3.117	5.179
80	34.87	1.1043	5431.1	23.925	3.117	5.179
85	36.83	1.1110	5445.3	23.799	3.117	5.178
90	38.77	1.1175	5459.4	23.681	3.117	5.178
95	40.68	1.1241	5473.5	23.569	3.117	5.177
100	42.58	1.1307	5487.5	23.463	3.117	5.177

$$T = 1100 \text{ K}$$

p	ρ	z	h	s	c_v	c_p
0.01	0.00	1.0000	5727.2	43.080	3.116	5.193
0.02	0.01	1.0000	5727.2	41.640	3.116	5.193
0.03	0.01	1.0000	5727.3	40.798	3.116	5.193
0.04	0.02	1.0000	5727.3	40.200	3.116	5.193
0.05	0.02	1.0001	5727.3	39.737	3.116	5.193
0.06	0.03	1.0001	5727.3	39.358	3.116	5.193
0.07	0.03	1.0001	5727.4	39.038	3.116	5.193
0.08	0.04	1.0001	5727.4	38.760	3.116	5.193
0.09	0.04	1.0001	5727.4	38.516	3.116	5.193
0.10	0.04	1.0001	5727.5	38.297	3.116	5.193
0.2	0.09	1.0002	5727.7	36.857	3.116	5.193
0.3	0.13	1.0003	5728.0	36.015	3.116	5.193
0.4	0.17	1.0005	5728.3	35.417	3.116	5.193
0.5	0.22	1.0006	5728.5	34.954	3.116	5.193
0.6	0.26	1.0007	5728.8	34.575	3.116	5.193
0.7	0.31	1.0008	5729.1	34.255	3.116	5.193

Table II.17 (*Continued*)

p	ρ	z	h	s	c_v	c_p
0.8	0.35	1.0009	5729.4	33.977	3.116	5.193
0.9	0.39	1.0010	5729.6	33.733	3.116	5.193
1.0	0.44	1.0012	5729.9	33.514	3.116	5.193
1.5	0.66	1.0017	5731.3	32.672	3.116	5.193
2.0	0.87	1.0023	5732.7	32.074	3.116	5.193
2.5	1.09	1.0029	5734.0	31.611	3.116	5.193
3.0	1.31	1.0035	5735.4	31.232	3.116	5.193
3.5	1.53	1.0041	5736.8	30.912	3.116	5.193
4.0	1.74	1.0047	5738.1	30.634	3.116	5.193
4.5	1.96	1.0052	5739.5	30.390	3.116	5.193
5.0	2.18	1.0058	5740.9	30.171	3.116	5.193
5.5	2.39	1.0064	5742.2	29.973	3.116	5.192
6.0	2.61	1.0070	5743.6	29.792	3.116	5.192
6.5	2.82	1.0076	5745.0	29.626	3.116	5.192
7.0	3.04	1.0082	5746.3	29.472	3.116	5.192
7.5	3.25	1.0087	5747.7	29.329	3.116	5.192
8.0	3.47	1.0093	5749.1	29.195	3.116	5.192
8.5	3.68	1.0099	5750.4	29.069	3.116	5.192
9.0	3.90	1.0105	5751.8	28.950	3.116	5.192
9.5	4.11	1.0111	5753.2	28.838	3.116	5.192
10.0	4.33	1.0116	5754.5	28.731	3.116	5.192
15.0	6.45	1.0175	5768.2	27.889	3.116	5.191
20.0	8.55	1.0233	5781.9	27.292	3.116	5.191
25.0	10.63	1.0291	5795.6	26.829	3.116	5.190
30	12.69	1.0350	5809.3	26.451	3.116	5.189
35	14.72	1.0409	5823.2	26.131	3.116	5.188
40	16.72	1.0468	5837.0	25.854	3.116	5.187
45	18.71	1.0527	5850.9	25.610	3.116	5.186
50	20.67	1.0586	5864.9	25.392	3.116	5.185
55	22.61	1.0646	5878.8	25.194	3.117	5.184
60	24.53	1.0705	5892.9	25.014	3.117	5.183
65	26.43	1.0765	5906.9	24.848	3.117	5.182
70	28.30	1.0825	5920.9	24.695	3.117	5.182
75	30.16	1.0885	5935.0	24.552	3.117	5.181
80	31.99	1.0945	5949.1	24.418	3.117	5.180
85	33.80	1.1005	5963.2	24.293	3.117	5.180
90	35.60	1.1065	5977.2	24.175	3.117	5.179
95	37.37	1.1124	5991.3	24.063	3.117	5.179
100	39.13	1.1184	6005.3	23.957	3.117	5.178
			$T = 1200$ K			
0.01	0.00	1.0000	6246.5	43.532	3.116	5.193
0.02	0.01	1.0000	6246.5	42.092	3.116	5.193
0.03	0.01	1.0000	6246.6	41.250	3.116	5.193
0.04	0.02	1.0000	6246.6	40.652	3.116	5.193
0.05	0.02	1.0001	6246.6	40.189	3.116	5.193
0.06	0.02	1.0001	6246.7	39.810	3.116	5.193
0.07	0.03	1.0001	6246.7	39.490	3.116	5.193
0.08	0.03	1.0001	6246.7	39.212	3.116	5.193
0.09	0.04	1.0001	6246.7	38.968	3.116	5.193
0.10	0.04	1.0001	6246.8	38.749	3.116	5.193
0.2	0.08	1.0002	6247.0	37.309	3.116	5.193
0.3	0.12	1.0003	6247.3	36.467	3.116	5.193
0.4	0.16	1.0004	6247.6	35.839	3.116	5.193
0.5	0.20	1.0005	6247.9	35.406	3.116	5.193
0.6	0.24	1.0006	6248.1	35.027	3.116	5.193

Table II.17 (Continued)

p	ρ	z	h	s	c_v	c_p
0.7	0.28	1.0007	6248.4	34.707	3.116	5.193
0.8	0.32	1.0009	6248.7	34.429	3.116	5.193
0.9	0.36	1.0010	6248.9	34.185	3.116	5.193
1.0	0.40	1.0011	6249.2	33.966	3.116	5.193
1.5	0.60	1.0016	6250.6	33.123	3.116	5.193
2.0	0.80	1.0021	6251.9	32.526	3.116	5.193
2.5	1.00	1.0027	6253.3	32.062	3.116	5.193
3.0	1.20	1.0032	6254.7	31.684	3.116	5.193
3.5	1.40	1.0037	6256.0	31.364	3.116	5.193
4.0	1.60	1.0043	6257.4	31.086	3.116	5.193
4.5	1.80	1.0048	6258.8	30.842	3.116	5.193
5.0	2.00	1.0053	6260.1	30.623	3.116	5.193
5.5	2.19	1.0059	6261.5	30.425	3.116	5.193
6.0	2.39	1.0064	6262.8	30.244	3.116	5.193
6.5	2.59	1.0069	6264.2	30.078	3.116	5.192
7.0	2.79	1.0075	6265.6	29.924	3.116	5.192
7.5	2.98	1.0080	6266.9	29.781	3.116	5.192
8.0	3.18	1.0085	6268.3	29.647	3.116	5.192
8.5	3.38	1.0091	6269.6	29.521	3.116	5.192
9.0	3.58	1.0096	6271.0	29.402	3.116	5.192
9.5	3.77	1.0101	6272.4	29.290	3.116	5.192
10.0	3.97	1.0107	6273.7	29.183	3.116	5.192
15.0	5.92	1.0160	6287.3	28.341	3.116	5.192
20.0	7.86	1.0213	6300.9	27.744	3.116	5.191
25.0	9.77	1.0266	6314.6	27.281	3.116	5.190
30	11.66	1.0320	6328.3	26.902	3.116	5.190
35	13.54	1.0373	6342.0	26.582	3.116	5.189
40	15.39	1.0427	6355.8	26.305	3.116	5.188
45	17.22	1.0481	6369.6	26.061	3.116	5.187
50	19.04	1.0535	6383.4	25.843	3.116	5.186
55	20.84	1.0590	6397.3	25.645	3.116	5.185
60	22.61	1.0644	6411.2	25.465	3.116	5.185
65	24.37	1.0699	6425.2	25.299	3.116	5.184
70	26.11	1.0753	6439.2	25.146	3.117	5.183
75	27.84	1.0808	6453.2	25.003	3.117	5.182
80	29.54	1.0863	6467.2	24.869	3.117	5.182
85	31.23	1.0918	6481.2	24.744	3.117	5.181
90	32.91	1.0972	6495.2	24.626	3.117	5.180
95	34.56	1.1027	6509.2	24.514	3.117	5.180
100	36.20	1.1082	6523.2	24.408	3.117	5.180
			$T=1300$ K			
0.01	0.00	1.0000	6765.8	43.947	3.116	5.193
0.02	0.01	1.0000	6765.9	42.508	3.116	5.193
0.03	0.01	1.0000	6765.9	41.665	3.116	5.193
0.04	0.01	1.0000	6765.9	41.068	3.116	5.193
0.05	0.02	1.0000	6765.9	40.604	3.116	5.193
0.06	0.02	1.0001	6766.0	40.225	3.116	5.193
0.07	0.03	1.0001	6766.0	39.905	3.116	5.193
0.08	0.03	1.0001	6766.0	39.628	3.116	5.193
0.09	0.03	1.0001	6766.1	39.383	3.116	5.193
0.10	0.04	1.0001	6766.1	39.164	3.116	5.193
0.2	0.07	1.0002	6766.4	37.725	3.116	5.193
0.3	0.11	1.0003	6766.6	36.882	3.116	5.193
0.4	0.15	1.0004	6766.9	36.285	3.116	5.193

Table II.17 (*Continued*)

p	ρ	z	h	s	c_v	c_p
0.5	0.19	1.0005	6767.2	35.821	3.116	5.193
0.6	0.22	1.0006	6767.4	35.442	3.116	5.193
0.7	0.26	1.0007	6767.7	35.122	3.116	5.193
0.8	0.30	1.0008	6768.0	34.845	3.116	5.193
0.9	0.33	1.0009	6768.3	34.600	3.116	5.193
1.0	0.37	1.0010	6768.5	34.381	3.116	5.193
1.5	0.55	1.0015	6769.9	33.539	3.116	5.193
2.0	0.74	1.0020	6771.2	32.942	3.116	5.193
2.5	0.92	1.0025	6772.6	32.478	3.116	5.193
3.0	1.11	1.0029	6774.0	32.099	3.116	5.193
3.5	1.29	1.0034	6775.3	31.779	3.116	5.193
4.0	1.48	1.0039	6776.7	31.502	3.116	5.193
4.5	1.66	1.0044	6778.0	31.257	3.116	5.193
5.0	1.84	1.0049	6779.4	31.038	3.116	5.193
5.5	2.03	1.0054	6780.7	30.840	3.116	5.193
6.0	2.21	1.0059	6782.1	30.660	3.116	5.193
6.5	2.39	1.0064	6783.5	30.493	3.116	5.193
7.0	2.57	1.0069	6784.8	30.340	3.116	5.193
7.5	2.76	1.0074	6786.2	30.196	3.116	5.192
8.0	2.94	1.0079	6787.5	30.062	3.116	5.192
8.5	3.12	1.0083	6788.9	29.936	3.116	5.192
9.0	3.30	1.0088	6790.2	29.818	3.116	5.192
9.5	3.49	1.0093	6791.6	29.705	3.116	5.192
10.0	3.67	1.0098	6792.9	29.599	3.116	5.192
15.0	5.47	1.0147	6806.5	28.757	3.116	5.192
20.0	7.26	1.0196	6820.1	28.159	3.116	5.191
25.0	9.04	1.0245	6833.6	27.696	3.116	5.191
30	10.79	1.0294	6847.3	27.318	3.116	5.190
35	12.53	1.0344	6860.9	26.998	3.116	5.189
40	14.25	1.0393	6874.6	26.721	3.116	5.189
45	15.96	1.0443	6888.3	26.476	3.116	5.188
50	17.65	1.0493	6902.1	26.258	3.116	5.187
55	19.32	1.0543	6915.9	26.060	3.116	5.186
60	20.98	1.0593	6929.8	25.880	3.116	5.186
65	22.62	1.0643	6943.6	25.714	3.116	5.185
70	24.24	1.0693	6957.5	25.561	3.116	5.184
75	25.85	1.0743	6971.5	25.418	3.116	5.184
80	27.45	1.0794	6985.4	25.284	3.116	5.183
85	29.03	1.0844	6999.4	25.159	3.117	5.182
90	30.59	1.0895	7013.3	25.040	3.117	5.182
95	32.14	1.0945	7027.3	24.928	3.117	5.181
100	33.68	1.0996	7041.2	24.822	3.117	5.181
$T=1400$ K						
0.01	0.00	1.0000	7285.1	44.332	3.116	5.193
0.02	0.01	1.0000	7285.2	42.892	3.116	5.193
0.03	0.01	1.0000	7285.2	42.050	3.116	5.193
0.04	0.01	1.0000	7285.2	41.453	3.116	5.193
0.05	0.02	1.0000	7285.3	40.989	3.116	5.193
0.06	0.02	1.0001	7285.3	40.610	3.116	5.193
0.07	0.02	1.0001	7285.3	40.290	3.116	5.193
0.08	0.03	1.0001	7285.3	40.013	3.116	5.193
0.09	0.03	1.0001	7285.4	39.768	3.116	5.193
0.10	0.03	1.0001	7285.4	39.549	3.116	5.193
0.2	0.07	1.0002	7285.7	38.109	3.116	5.193

Table II.17 (*Continued*)

p	ρ	z	h	s	c_v	c_p
0.3	0.10	1.0003	7285.9	37.267	3.116	5.193
0.4	0.14	1.0004	7286.2	36.670	3.116	5.193
0.5	0.17	1.0005	7286.5	36.206	3.116	5.193
0.6	0.21	1.0005	7286.7	35.827	3.116	5.193
0.7	0.24	1.0006	7287.0	35.507	3.116	5.193
0.8	0.27	1.0007	7287.3	35.230	3.116	5.193
0.9	0.31	1.0008	7287.6	34.985	3.116	5.193
1.0	0.34	1.0009	7287.8	34.766	3.116	5.193
1.5	0.52	1.0014	7289.2	33.924	3.116	5.193
2.0	0.69	1.0018	7290.5	33.326	3.116	5.193
2.5	0.86	1.0023	7291.9	32.863	3.116	5.193
3.0	1.03	1.0027	7293.2	32.484	3.116	5.193
3.5	1.20	1.0032	7294.6	32.164	3.116	5.193
4.0	1.37	1.0036	7296.0	31.887	3.116	5.193
4.5	1.54	1.0041	7297.3	31.642	3.116	5.193
5.0	1.71	1.0046	7298.7	31.423	3.116	5.193
5.5	1.88	1.0050	7300.0	31.225	3.116	5.193
6.0	2.05	1.0055	7301.4	31.044	3.116	5.193
6.5	2.22	1.0059	7302.7	30.878	3.116	5.193
7.0	2.39	1.0064	7304.1	30.724	3.116	5.193
7.5	2.56	1.0068	7305.4	30.581	3.116	5.193
8.0	2.73	1.0073	7306.8	30.447	3.116	5.193
8.5	2.90	1.0077	7308.1	30.321	3.116	5.192
9.0	3.07	1.0082	7309.5	30.202	3.116	5.192
9.5	3.24	1.0086	7310.8	30.090	3.116	5.192
10.0	3.41	1.0091	7312.2	29.984	3.116	5.192
15.0	5.09	1.0136	7325.7	29.141	3.116	5.192
20.0	6.75	1.0182	7339.2	28.544	3.116	5.192
25.0	8.41	1.0227	7352.7	28.081	3.116	5.191
30	10.04	1.0273	7366.3	27.702	3.116	5.190
35	11.66	1.0319	7379.9	27.382	3.116	5.190
40	13.27	1.0364	7393.5	27.105	3.116	5.189
45	14.86	1.0410	7407.2	26.861	3.116	5.189
50	16.44	1.0456	7420.9	26.642	3.116	5.188
55	18.01	1.0502	7434.6	26.445	3.116	5.187
60	19.56	1.0549	7448.4	26.264	3.116	5.187
65	21.10	1.0595	7462.2	26.098	3.116	5.186
70	22.62	1.0642	7476.0	25.945	3.116	5.185
75	24.13	1.0688	7489.9	25.802	3.116	5.185
80	25.63	1.0735	7503.7	25.668	3.116	5.184
85	27.11	1.0782	7517.6	25.543	3.116	5.183
90	28.58	1.0828	7531.5	25.424	3.116	5.183
95	30.04	1.0875	7545.5	25.312	3.116	5.182
100	31.48	1.0922	7559.4	25.206	3.116	5.182
$T = 1500$ K						
0.01	0.00	1.0000	7804.5	44.691	3.116	5.193
0.02	0.01	1.0000	7804.5	43.251	3.116	5.193
0.03	0.01	1.0000	7804.5	42.408	3.116	5.193
0.04	0.01	1.0000	7804.5	41.811	3.116	5.193
0.05	0.02	1.0000	7804.6	41.347	3.116	5.193
0.06	0.02	1.0001	7804.6	40.969	3.116	5.193
0.07	0.02	1.0001	7804.6	40.648	3.116	5.193
0.08	0.03	1.0001	7804.6	40.371	3.116	5.193
0.09	0.03	1.0001	7804.7	40.126	3.116	5.193

Table II.17 (*Continued*)

p	ρ	z	h	s	c_v	c_p
0.10	0.03	1.0001	7804.7	39.908	3.116	5.193
0.2	0.06	1.0002	7805.0	38.468	3.116	5.193
0.3	0.10	1.0003	7805.2	37.625	3.116	5.193
0.4	0.13	1.0003	7805.5	37.028	3.116	5.193
0.5	0.16	1.0004	7805.8	36.564	3.116	5.193
0.6	0.19	1.0005	7806.1	36.186	3.116	5.193
0.7	0.22	1.0006	7806.3	35.865	3.116	5.193
0.8	0.26	1.0007	7806.6	35.588	3.116	5.193
0.9	0.29	1.0008	7806.9	35.343	3.116	5.193
1.0	0.32	1.0008	7807.1	35.125	3.116	5.193
1.5	0.48	1.0013	7808.5	34.282	3.116	5.193
2.0	0.64	1.0017	7809.8	33.685	3.116	5.193
2.5	0.80	1.0021	7811.2	33.221	3.116	5.193
3.0	0.96	1.0025	7812.5	32.842	3.116	5.193
3.5	1.12	1.0030	7813.9	32.522	3.116	5.193
4.0	1.28	1.0034	7815.2	32.245	3.116	5.193
4.5	1.44	1.0038	7816.6	32.000	3.116	5.193
5.0	1.60	1.0042	7817.9	31.781	3.116	5.193
5.5	1.76	1.0047	7819.3	31.583	3.116	5.193
6.0	1.92	1.0051	7820.6	31.403	3.116	5.193
6.5	2.07	1.0055	7822.0	31.236	3.116	5.193
7.0	2.23	1.0059	7823.3	31.083	3.116	5.193
7.5	2.39	1.0064	7824.7	30.939	3.116	5.193
8.0	2.55	1.0068	7826.0	30.805	3.116	5.193
8.5	2.71	1.0072	7827.4	30.679	3.116	5.193
9.0	2.87	1.0076	7828.7	30.561	3.116	5.193
9.5	3.02	1.0081	7830.1	30.448	3.116	5.192
10.0	3.18	1.0085	7831.4	30.342	3.116	5.192
15.0	4.75	1.0127	7844.9	29.500	3.116	5.192
20.0	6.31	1.0169	7858.4	28.902	3.116	5.192
25.0	7.86	1.0212	7871.8	28.439	3.116	5.191
30	9.39	1.0254	7885.4	28.060	3.116	5.191
35	10.91	1.0297	7898.9	27.740	3.116	5.190
40	12.42	1.0339	7912.5	27.463	3.116	5.190
45	13.91	1.0382	7926.1	27.219	3.116	5.189
50	15.39	1.0425	7939.7	27.000	3.116	5.189
55	16.86	1.0468	7953.4	26.803	3.116	5.188
60	18.32	1.0511	7967.1	26.622	3.116	5.187
65	19.77	1.0554	7980.8	26.456	3.116	5.187
70	21.20	1.0597	7994.6	26.303	3.116	5.186
75	22.62	1.0641	8008.4	26.160	3.116	5.186
80	24.03	1.0684	8022.2	26.026	3.116	5.185
85	25.43	1.0727	8036.0	25.900	3.116	5.184
90	26.82	1.0771	8049.9	25.782	3.116	5.183
95	28.19	1.0815	8063.8	25.670	3.116	5.183
100	29.56	1.0858	8077.6	25.564	3.116	5.183

Table II.18

p	w	μ	k	f	α/α_0	γ/γ_0
			$T=2.2$ K			
0.01	210.0	−4.77	643.75	0.01	0.034	21.458
0.02	211.4	−4.82	326.73	0.01	0.034	10.866
0.03	212.8	−4.87	221.04	0.01	0.034	7.334
0.04	214.2	−4.93	168.19	0.01	0.034	5.566
0.05	215.5	−4.98	136.46	0.01	0.034	4.505
0.06	216.9	−5.03	115.31	0.01	0.034	3.796
0.07	218.2	−5.08	100.19	0.01	0.033	3.289
0.08	219.5	−5.13	88.84	0.01	0.033	2.909
0.09	220.8	−5.18	80.02	0.01	0.033	2.612
0.10	222.1	−5.24	72.95	0.01	0.033	2.374
0.2	234.0	−5.77	41.04	0.01	0.032	1.294
0.3	244.8	−6.33	30.30	0.01	0.031	0.924
0.4	254.7	−6.92	24.86	0.01	0.030	0.732
0.5	263.8	−7.54	21.55	0.01	0.029	0.614
0.6	272.3	−8.19	19.31	0.01	0.029	0.533
0.7	280.3	−8.88	17.69	0.01	0.028	0.473
0.8	287.8	−9.61	16.46	0.02	0.027	0.428
0.9	295.0	−10.39	15.49	0.02	0.026	0.392
1.0	301.9	−11.22	14.71	0.02	0.026	0.362
1.5	333.4	−16.11	12.35	0.04	0.024	0.274
2.0	362.9	−22.17	11.27	0.08	0.023	0.232
2.5	393.5	−28.21	10.85	0.15	0.023	0.213
3.0	424.9	−31.46	10.77	0.27	0.024	0.205
3.5	452.4	−29.88	10.66	0.50	0.026	0.205
4.0	471.7	−24.84	10.32	0.90	0.028	0.211
			$T=2.3$ K			
0.01	211.0	−3.18	648.71	0.01	0.059	36.812
0.02	212.4	−3.20	329.16	0.01	0.058	18.518
0.03	213.8	−3.22	222.62	0.01	0.058	12.418
0.04	215.1	−3.24	169.34	0.01	0.057	9.367
0.05	216.5	−3.26	137.37	0.01	0.057	7.536
0.06	217.8	−3.28	116.04	0.01	0.056	6.314
0.07	219.1	−3.29	100.80	0.01	0.056	5.441
0.08	220.4	−3.31	89.36	0.01	0.055	4.785
0.09	221.6	−3.33	80.46	0.01	0.055	4.275
0.10	222.9	−3.35	73.34	0.01	0.054	3.866
0.2	234.6	−3.53	41.16	0.01	0.051	2.017
0.3	245.0	−3.70	30.31	0.01	0.047	1.391
0.4	254.6	−3.86	24.80	0.01	0.045	1.073
0.5	263.3	−4.02	21.44	0.01	0.042	0.879
0.6	271.4	−4.17	19.17	0.01	0.040	0.747
0.7	279.0	−4.31	17.51	0.02	0.039	0.652
0.8	286.1	−4.46	16.24	0.02	0.037	0.580
0.9	292.8	−4.59	15.23	0.02	0.036	0.523
1.0	299.1	−4.73	14.41	0.02	0.034	0.477
1.5	326.4	−5.34	11.82	0.05	0.030	0.339
2.0	348.9	−5.85	10.41	0.09	0.027	0.274
2.5	368.3	−6.23	9.50	0.16	0.026	0.240
3.0	385.7	−6.46	8.86	0.29	0.026	0.222
3.5	401.9	−6.53	8.40	0.51	0.027	0.215
4.0	417.2	−6.42	8.06	0.91	0.029	0.215
4.5	432.1	−6.17	7.81	1.59	0.031	0.219

Table II.18 (Continued)

p	w	μ	k	f	α/α_0	γ/γ_0

$T = 2.4$ K

p	w	μ	k	f	α/α_0	γ/γ_0
0.01	211.6	−2.80	650.24	0.01	0.081	50.103
0.02	213.0	−2.82	329.92	0.01	0.080	25.150
0.03	214.3	−2.83	223.13	0.01	0.080	16.831
0.04	215.7	−2.84	169.72	0.01	0.079	12.670
0.05	217.0	−2.86	137.66	0.01	0.078	10.172
0.06	218.3	−2.87	116.28	0.01	0.077	8.507
0.07	219.6	−2.88	101.01	0.01	0.076	7.316
0.08	220.9	−2.90	89.54	0.01	0.076	6.423
0.09	222.1	−2.91	80.62	0.01	0.075	5.728
0.10	223.4	−2.92	73.48	0.01	0.074	5.171
0.2	235.0	−3.05	41.21	0.01	0.068	2.657
0.3	245.4	−3.16	30.33	0.01	0.063	1.809
0.4	254.9	−3.27	24.81	0.01	0.058	1.381
0.5	263.6	−3.37	21.44	0.02	0.055	1.120
0.6	271.6	−3.46	19.15	0.02	0.052	0.945
0.7	279.1	−3.55	17.49	0.02	0.049	0.818
0.8	286.2	−3.64	16.22	0.02	0.047	0.722
0.9	292.8	−3.72	15.21	0.03	0.044	0.647
1.0	299.0	−3.80	14.39	0.03	0.043	0.587
1.5	326.2	−4.15	11.79	0.05	0.036	0.406
2.0	348.4	−4.42	10.37	0.10	0.032	0.319
2.5	367.6	−4.64	9.45	0.17	0.030	0.271
3.0	384.6	−4.78	8.80	0.30	0.029	0.245
3.5	400.2	−4.85	8.33	0.53	0.029	0.231
4.0	414.9	−4.85	7.97	0.91	0.030	0.225
4.5	429.0	−4.77	7.69	1.56	0.031	0.225
5.0	442.9	−4.62	7.49	2.65	0.033	0.228

$T = 2.5$

p	w	μ	k	f	α/α_0	γ/γ_0
0.01	88.5	17.63	1.63	0.01	1.216	1.110
0.02	213.2	−2.68	329.43	0.01	0.101	30.925
0.03	214.6	−2.69	222.81	0.01	0.100	20.677
0.04	215.9	−2.71	169.49	0.01	0.099	15.551
0.05	217.3	−2.72	137.49	0.01	0.098	12.475
0.06	218.6	−2.73	116.14	0.01	0.097	10.423
0.07	219.9	−2.74	100.89	0.01	0.096	8.957
0.08	221.2	−2.76	89.44	0.01	0.095	7.857
0.09	222.4	−2.77	80.53	0.01	0.094	7.001
0.10	223.7	−2.78	73.40	0.01	0.093	6.316
0.2	235.3	−2.89	41.18	0.01	0.084	3.223
0.3	245.7	−2.99	30.31	0.01	0.077	2.183
0.4	255.2	−3.08	24.79	0.02	0.071	1.657
0.5	263.8	−3.16	21.43	0.02	0.066	1.339
0.6	271.9	−3.24	19.14	0.02	0.062	1.125
0.7	279.4	−3.31	17.48	0.02	0.059	0.971
0.8	286.4	−3.38	16.21	0.03	0.056	0.854
0.9	293.0	−3.45	15.20	0.03	0.053	0.763
1.0	299.2	−3.51	14.38	0.03	0.051	0.690
1.5	326.4	−3.78	11.79	0.06	0.042	0.471
2.0	348.7	−3.99	10.37	0.11	0.037	0.364
2.5	367.9	−4.16	9.45	0.18	0.034	0.305
3.0	385.0	−4.28	8.81	0.32	0.032	0.270
3.5	400.6	−4.35	8.33	0.54	0.032	0.250

Table II.18 (*Continued*)

p	w	μ	k	f	α/α_0	γ/γ_0
4.0	415.2	−4.37	7.97	0.92	0.032	0.239
4.5	429.2	−4.34	7.69	1.54	0.032	0.234
5.0	442.8	−4.26	7.48	2.57	0.034	0.233
5.5	456.3	−4.13	7.32	4.24	0.035	0.236
			$T=2.6$ K			
0.01	90.8	16.57	1.64	0.01	1.189	1.097
0.02	213.1	−2.61	327.74	0.01	0.122	36.009
0.03	214.5	−2.62	221.71	0.01	0.121	24.066
0.04	215.9	−2.63	168.68	0.01	0.119	18.092
0.05	217.2	−2.64	136.85	0.01	0.118	14.508
0.06	218.5	−2.66	115.62	0.01	0.116	12.117
0.07	219.8	−2.67	100.44	0.01	0.115	10.409
0.08	221.1	−2.68	89.05	0.01	0.113	9.127
0.09	222.4	−2.69	80.20	0.01	0.112	8.129
0.10	223.6	−2.70	73.10	0.01	0.111	7.331
0.2	235.3	−2.81	41.05	0.01	0.100	3.729
0.3	245.8	−2.91	30.23	0.02	0.091	2.518
0.4	255.3	−2.99	24.74	0.02	0.084	1.908
0.5	264.0	−3.07	21.39	0.02	0.078	1.538
0.6	272.0	−3.14	19.11	0.02	0.073	1.290
0.7	279.5	−3.20	17.46	0.03	0.068	1.111
0.8	286.6	−3.27	16.19	0.03	0.065	0.976
0.9	293.2	−3.32	15.19	0.03	0.061	0.871
1.0	299.4	−3.38	14.37	0.04	0.058	0.786
1.5	326.7	−3.62	11.79	0.07	0.048	0.532
2.0	349.1	−3.80	10.38	0.12	0.041	0.408
2.5	368.5	−3.95	9.47	0.20	0.038	0.338
3.0	385.7	−4.06	8.83	0.34	0.035	0.296
3.5	401.4	−4.13	8.35	0.56	0.034	0.270
4.0	416.1	−4.16	7.99	0.93	0.034	0.254
4.5	430.1	−4.16	7.71	1.53	0.034	0.246
5.0	443.6	−4.11	7.49	2.50	0.035	0.242
5.5	456.9	−4.02	7.33	4.05	0.036	0.241
6.0	470.1	−3.91	7.21	6.52	0.037	0.242
			$T=2.8$ K			
0.01	95.2	14.83	1.65	0.01	1.151	1.078
0.02	211.8	−2.43	320.10	0.02	0.165	44.650
0.03	213.2	−2.45	216.68	0.02	0.162	29.833
0.04	214.6	−2.46	164.94	0.02	0.160	22.423
0.05	216.0	−2.47	133.89	0.02	0.158	17.976
0.06	217.3	−2.49	113.18	0.02	0.156	15.010
0.07	218.7	−2.50	98.38	0.02	0.153	12.891
0.08	220.0	−2.51	87.27	0.02	0.151	11.301
0.09	221.3	−2.53	78.62	0.02	0.149	10.064
0.10	222.6	−2.54	71.70	0.02	0.148	9.074
0.2	234.5	−2.65	40.41	0.02	0.131	4.606
0.3	245.1	−2.75	29.83	0.02	0.119	3.106
0.4	254.7	−2.83	24.46	0.02	0.109	2.350
0.5	263.5	−2.91	21.18	0.03	0.100	1.893
0.6	271.7	−2.97	18.95	0.03	0.093	1.586
0.7	279.2	−3.03	17.32	0.03	0.087	1.365
0.8	286.3	−3.09	16.08	0.04	0.082	1.199

Table II.18 (*Continued*)

p	w	μ	k	f	α/α_0	γ/γ_0
0.9	293.0	−3.14	15.09	0.04	0.078	1.068
1.0	299.4	−3.19	14.29	0.05	0.074	0.964
1.5	326.9	−3.40	11.76	0.08	0.060	0.649
2.0	349.7	−3.55	10.38	0.14	0.051	0.494
2.5	369.4	−3.68	9.49	0.23	0.046	0.405
3.0	386.9	−3.78	8.86	0.37	0.042	0.349
3.5	402.9	−3.86	8.39	0.60	0.040	0.313
4.0	417.8	−3.90	8.04	0.96	0.038	0.289
4.5	431.9	−3.92	7.76	1.52	0.038	0.273
5.0	445.5	−3.92	7.54	2.40	0.038	0.263
5.5	458.8	−3.88	7.37	3.77	0.038	0.257
6.0	471.8	−3.83	7.24	5.87	0.038	0.254
6.5	484.6	−3.75	7.13	9.09	0.039	0.252

$T = 3.0$ K

p	w	μ	k	f	α/α_0	γ/γ_0
0.01	99.1	13.47	1.66	0.01	1.125	1.065
0.02	95.9	14.60	1.65	0.02	1.315	1.154
0.03	209.9	−2.19	207.45	0.02	0.209	34.665
0.04	211.4	−2.21	158.09	0.02	0.206	26.059
0.05	212.8	−2.22	128.45	0.02	0.202	20.894
0.06	214.3	−2.24	108.68	0.02	0.199	17.449
0.07	215.7	−2.25	94.55	0.02	0.196	14.988
0.08	217.0	−2.27	83.95	0.02	0.193	13.141
0.09	218.4	−2.28	75.69	0.02	0.191	11.704
0.10	219.7	−2.30	69.08	0.02	0.188	10.554
0.2	232.1	−2.42	39.18	0.03	0.165	5.364
0.3	243.0	−2.53	29.06	0.03	0.142	3.620
0.4	252.9	−2.61	23.90	0.03	0.135	2.741
0.5	261.9	−2.69	20.75	0.04	0.124	2.209
0.6	270.2	−2.76	18.61	0.04	0.115	1.852
0.7	277.9	−2.82	17.04	0.04	0.107	1.595
0.8	285.1	−2.87	15.84	0.05	0.101	1.402
0.9	291.9	−2.92	14.89	0.05	0.095	1.250
1.0	298.4	−2.97	14.12	0.06	0.090	1.128
1.5	326.5	−3.15	11.67	0.10	0.072	0.761
2.0	349.7	−3.30	10.33	0.16	0.061	0.579
2.5	369.7	−3.41	9.47	0.26	0.054	0.472
3.0	387.6	−3.51	8.86	0.41	0.049	0.404
3.5	403.9	−3.58	8.41	0.64	0.046	0.358
4.0	419.1	−3.63	8.06	0.99	0.044	0.327
4.5	433.4	−3.67	7.79	1.53	0.042	0.305
5.0	447.1	−3.69	7.57	2.34	0.042	0.290
5.5	460.3	−3.69	7.40	3.57	0.041	0.279
6.0	473.3	−3.67	7.20	5.40	0.041	0.272
6.5	486.0	−3.63	7.15	8.14	0.041	0.266
7.0	498.5	−3.58	7.07	12.20	0.041	0.262
7.5	510.9	−3.52	7.01	18.21	0.041	0.260
8.0	523.2	−3.45	6.96	27.07	0.041	0.257

$T = 3.2$ K

p	w	μ	k	f	α/α_0	γ/γ_0
0.01	102.8	12.35	1.66	0.01	1.106	1.056
0.02	100.1	13.20	1.66	0.02	1.254	1.127
0.03	96.9	14.10	1.66	0.03	1.477	1.222
0.04	206.4	−1.90	148.45	0.03	0.261	29.207

Table II.18 (Continued)

p	w	μ	k	f	α/α_0	γ/γ_0
0.05	207.9	−1.92	120.80	0.03	0.256	23.427
0.06	209.4	−1.94	102.36	0.03	0.251	19.572
0.07	210.9	−1.96	89.17	0.03	0.247	16.817
0.08	212.4	−1.98	79.27	0.03	0.243	14.750
0.09	213.8	−2.00	71.57	0.03	0.239	13.141
0.10	215.2	−2.01	65.39	0.03	0.235	11.854
0.2	228.2	−2.16	37.44	0.03	0.204	6.041
0.3	239.6	−2.27	27.96	0.04	0.181	4.086
0.4	249.9	−2.37	23.11	0.04	0.163	3.100
0.5	259.1	−2.45	20.14	0.05	0.149	2.502
0.6	267.7	−2.51	18.12	0.05	0.138	2.101
0.7	275.6	−2.57	16.64	0.06	0.128	1.812
0.8	283.1	−2.63	15.50	0.06	0.120	1.594
0.9	290.1	−2.68	14.60	0.07	0.113	1.423
1.0	296.7	−2.72	13.87	0.07	0.107	1.286
1.5	325.4	−2.90	11.53	0.12	0.085	0.871
2.0	349.1	−3.04	10.26	0.19	0.072	0.663
2.5	369.6	−3.15	9.43	0.29	0.063	0.540
3.0	387.9	−3.23	8.85	0.44	0.057	0.461
3.5	404.5	−3.31	8.41	0.68	0.053	0.407
4.0	420.0	−3.36	8.07	1.03	0.050	0.369
4.5	434.5	−3.41	7.80	1.54	0.048	0.341
5.0	448.3	−3.44	7.59	2.31	0.046	0.321
5.5	461.6	−3.45	7.42	3.43	0.045	0.306
6.0	474.6	−3.46	7.28	5.06	0.044	0.295
6.5	487.2	−3.45	7.17	7.44	0.044	0.286
7.0	499.6	−3.43	7.08	10.89	0.044	0.279
7.5	511.8	−3.41	7.01	15.87	0.043	0.274
8.0	523.9	−3.37	6.96	23.03	0.043	0.269
8.5	535.9	−3.33	6.92	33.31	0.043	0.265
9.0	547.7	−3.28	6.89	48.01	0.042	0.261
			$T=3.4$ K			
0.01	106.4	11.42	1.66	0.01	1.091	1.048
0.02	104.0	12.05	1.66	0.02	1.211	1.108
0.03	101.3	12.72	1.66	0.03	1.378	1.182
0.04	98.2	13.44	1.66	0.03	1.626	1.279
0.05	201.5	−1.60	111.46	0.04	0.323	25.635
0.06	203.1	−1.63	94.64	0.04	0.317	21.429
0.07	204.7	−1.65	82.61	0.04	0.310	18.423
0.08	206.3	−1.67	73.58	0.04	0.304	16.167
0.09	207.8	−1.69	66.54	0.04	0.299	14.411
0.10	209.4	−1.71	60.91	0.04	0.293	13.005
0.2	223.3	−1.89	35.34	0.04	0.249	6.655
0.3	235.3	−2.02	26.63	0.05	0.219	4.515
0.4	246.0	−2.12	22.16	0.05	0.196	3.434
0.5	255.7	−2.21	19.41	0.06	0.178	2.778
0.6	264.5	−2.28	17.53	0.06	0.163	2.337
0.7	272.7	−2.34	16.15	0.07	0.151	2.019
0.8	280.4	−2.40	15.10	0.07	0.141	1.779
0.9	287.6	−2.45	14.25	0.08	0.132	1.590
1.0	294.4	−2.49	13.56	0.09	0.125	1.439
1.5	323.9	−2.67	11.37	0.14	0.099	0.979
2.0	348.2	−2.80	10.16	0.21	0.083	0.747
2.5	269.2	−2.90	9.37	0.32	0.072	0.610

Table II.18 (*Continued*)

p	w	μ	k	f	α/α_0	γ/γ_0
3.0	387.9	−2.99	8.81	0.48	0.065	0.520
3.5	404.9	−3.06	8.40	0.72	0.060	0.458
4.0	420.6	−3.11	8.07	1.07	0.056	0.413
4.5	435.3	−3.16	7.81	1.57	0.053	0.380
5.0	449.3	−3.20	7.60	2.29	0.051	0.355
5.5	462.7	−3.22	7.43	3.33	0.050	0.336
6.0	475.7	−3.24	7.30	4.81	0.049	0.322
6.5	488.3	−3.25	7.18	6.92	0.048	0.310
7.0	500.6	−3.25	7.09	9.91	0.047	0.301
7.5	512.7	−3.24	7.02	14.14	0.046	0.293
8.0	524.7	−3.23	6.96	20.11	0.046	0.286
8.5	536.4	−3.21	6.92	28.48	0.045	0.280
9.0	548.0	−3.19	6.88	40.21	0.045	0.275
9.5	559.5	−3.17	6.86	56.59	0.044	0.269
10.0	570.8	−3.14	6.84	79.42	0.043	0.264

$$T = 3.5 \text{ K}$$

p	w	μ	k	f	α/α_0	γ/γ_0
0.01	108.1	11.00	1.66	0.01	1.085	1.045
0.02	105.9	11.56	1.66	0.02	1.194	1.100
0.03	103.4	12.14	1.66	0.03	1.342	1.167
0.04	100.6	12.75	1.66	0.03	1.550	1.252
0.05	197.7	−1.44	106.28	0.04	0.365	26.615
0.06	199.4	−1.46	90.36	0.04	0.357	22.257
0.07	201.1	−1.49	78.98	0.04	0.349	19.142
0.08	202.8	−1.51	70.43	0.04	0.342	16.803
0.09	204.4	−1.54	63.77	0.04	0.335	14.983
0.10	206.0	−1.56	58.43	0.04	0.328	13.526
0.2	220.4	−1.75	34.19	0.05	0.276	6.939
0.3	232.8	−1.89	25.90	0.05	0.240	4.717
0.4	243.8	−2.00	21.64	0.06	0.214	3.593
0.5	253.7	−2.09	19.01	0.06	0.193	2.911
0.6	262.8	−2.17	17.21	0.07	0.177	2.451
0.7	271.1	−2.23	15.89	0.07	0.164	2.120
0.8	278.9	−2.29	14.87	0.08	0.152	1.869
0.9	286.3	−2.34	14.06	0.09	0.143	1.672
1.0	293.2	−2.39	13.40	0.10	0.135	1.514
1.5	323.1	−2.57	11.27	0.15	0.106	1.033
2.0	347.7	−2.69	10.10	0.23	0.089	0.790
2.5	368.9	−2.79	9.34	0.34	0.077	0.645
3.0	387.8	−2.88	8.79	0.50	0.070	0.550
3.5	405.0	−2.94	8.38	0.74	0.064	0.484
4.0	420.8	−3.00	8.07	1.09	0.60	0.436
4.5	435.7	−3.05	7.81	1.58	0.057	0.400
5.0	449.8	−3.09	7.61	2.29	0.054	0.374
5.5	463.2	−3.12	7.44	3.29	0.052	0.353
6.0	476.2	−3.14	7.30	4.71	0.051	0.336
6.5	488.8	−3.15	7.19	6.71	0.050	0.323
7.0	501.2	−3.16	7.10	9.52	0.049	0.312
7.5	513.2	−3.16	7.02	13.45	0.048	0.303
8.0	525.1	−3.15	6.96	18.94	0.048	0.296
8.5	536.7	−3.14	6.92	26.57	0.047	0.289
9.0	548.3	−3.13	6.88	37.17	0.046	0.283
9.5	559.6	−3.12	6.85	51.82	0.045	0.277
10.0	570.8	−3.10	6.83	72.05	0.045	0.271

Table II.18 (*Continued*)

p	w	μ	k	j	α/α_0	γ/γ_0

$T = 3.6$ K

p	w	μ	k	j	α/α_0	γ/γ_0
0.01	109.7	10.61	1.67	0.01	1.079	1.042
0.02	107.7	11.10	1.66	0.02	1.179	1.093
0.03	105.4	11.60	1.66	0.03	1.311	1.154
0.04	102.8	12.13	1.66	0.03	1.490	1.229
0.05	99.8	12.70	1.66	0.04	1.750	1.324
0.06	195.4	−1.29	85.83	0.04	0.403	23.012
0.07	197.2	−1.32	75.14	0.04	0.394	19.800
0.08	198.9	−1.35	67.10	0.05	0.385	17.389
0.09	200.7	−1.38	60.84	0.05	0.376	15.511
0.10	202.3	−1.40	55.81	0.05	0.368	14.008
0.2	217.4	−1.61	32.98	0.05	0.305	7.208
0.3	230.2	−1.77	25.14	0.06	0.263	4.910
0.4	241.5	−1.89	21.10	0.06	0.233	3.746
0.5	251.6	−1.98	18.60	0.07	0.210	3.039
0.6	260.9	−2.06	16.88	0.07	0.192	2.562
0.7	269.4	−2.13	15.62	0.08	0.177	2.218
0.8	277.4	−2.19	14.64	0.09	0.164	1.957
0.9	284.8	−2.24	13.85	0.10	0.154	1.753
1.0	291.8	−2.29	13.22	0.10	0.145	1.588
1.5	322.2	−2.47	11.18	0.16	0.113	1.087
2.0	347.1	−2.60	10.04	0.24	0.095	0.832
2.5	368.6	−2.69	9.30	0.36	0.083	0.680
3.0	387.7	−2.77	8.77	0.53	0.074	0.580
3.5	405.0	−2.84	8.37	0.77	0.068	0.510
4.0	421.0	−2.90	8.06	1.11	0.063	0.459
4.5	436.0	−2.94	7.81	1.60	0.060	0.421
5.0	450.2	−2.98	7.61	2.29	0.057	0.392
5.5	463.7	−3.01	7.44	3.26	0.055	0.369
6.0	476.7	−3.04	7.31	4.63	0.053	0.351
6.5	489.4	−3.05	7.19	6.53	0.052	0.337
7.0	501.7	−3.07	7.10	9.17	0.051	0.325
7.5	513.7	−3.07	7.03	12.85	0.050	0.315
8.0	525.5	−3.07	6.96	17.92	0.049	0.306
8.5	537.1	−3.07	6.92	24.92	0.049	0.298
9.0	548.5	−3.06	6.88	34.55	0.048	0.291
9.5	559.8	−3.05	6.84	47.75	0.047	0.285
10.0	570.9	−3.04	6.82	65.82	0.046	0.279

$T = 3.8$ K

p	w	μ	k	j	α/α_0	γ/γ_0
0.01	113.0	9.91	1.67	0.01	1.069	1.037
0.02	111.1	10.29	1.66	0.02	1.154	1.081
0.03	109.1	10.67	1.66	0.03	1.261	1.132
0.04	106.9	11.07	1.66	0.04	1.399	1.192
0.05	104.5	11.49	1.66	0.04	1.583	1.265
0.06	101.7	11.93	1.66	0.05	1.815	1.357
0.07	188.3	−0.96	66.84	0.05	0.511	20.909
0.08	190.2	−1.00	59.93	0.05	0.496	18.383
0.09	192.2	−1.04	54.53	0.06	0.482	16.416
0.10	194.0	−1.07	50.20	0.06	0.469	14.839
0.2	210.6	−1.34	30.40	0.06	0.376	7.694
0.3	224.5	−1.52	23.53	0.07	0.317	5.268
0.4	236.5	−1.66	19.96	0.07	0.277	4.035
0.5	247.2	−1.77	17.73	0.08	0.247	3.283
0.6	256.9	−1.86	16.18	0.09	0.224	2.775

Table II.18 (Continued)

p	w	μ	k	f	α/α_0	ν/ν_0
0.7	265.8	−1.94	15.04	0.10	0.206	2.407
0.8	274.1	−2.00	14.16	0.10	0.191	2.128
0.9	281.8	−2.06	13.45	0.11	0.178	1.909
1.0	289.0	−2.11	12.86	0.12	0.167	1.733
1.5	320.3	−2.29	10.98	0.18	0.129	1.193
2.0	345.9	−2.42	9.92	0.27	0.108	0.917
2.5	367.9	−2.52	9.22	0.39	0.093	0.751
3.0	387.4	−2.59	8.72	0.57	0.083	0.641
3.5	405.1	−2.66	8.34	0.81	0.076	0.563
4.0	421.4	−2.71	8.04	1.16	0.071	0.506
4.5	436.6	−2.76	7.80	1.64	0.067	0.464
5.0	450.9	−2.80	7.61	2.30	0.063	0.430
5.5	464.6	−2.83	7.45	3.22	0.061	0.404
6.0	477.7	−2.85	7.31	4.49	0.059	0.383
6.5	490.3	−2.88	7.20	6.23	0.057	0.366
7.0	502.6	−2.89	7.11	8.60	0.056	0.351
7.5	514.6	−2.90	7.03	11.85	0.054	0.339
8.0	526.3	−2.91	6.97	16.26	0.053	0.329
8.5	537.8	−2.92	6.92	22.24	0.052	0.319
9.0	549.1	−2.92	6.87	30.32	0.051	0.311
9.5	560.2	−2.92	6.84	41.24	0.050	0.303
10.0	571.1	−2.92	6.81	55.93	0.049	0.296

$T = 4.0$ K

p	w	μ	k	f	α/α_0	ν/ν_0
0.01	116.1	9.29	1.67	0.01	1.061	1.033
0.02	114.5	9.58	1.67	0.02	1.135	1.071
0.03	112.7	9.88	1.66	0.03	1.223	1.115
0.04	110.8	10.19	1.66	0.04	1.333	1.165
0.05	108.7	10.50	1.66	0.04	1.472	1.224
0.06	106.4	10.83	1.67	0.05	1.656	1.294
0.07	103.8	11.18	1.67	0.06	1.909	1.379
0.08	100.8	11.57	1.67	0.06	2.288	1.487
0.09	182.1	−0.64	47.60	0.07	0.640	17.070
0.10	184.3	−0.69	44.05	0.07	0.617	15.454
0.2	203.0	−1.05	27.64	0.07	0.468	8.100
0.3	218.1	−1.28	21.82	0.08	0.384	5.584
0.4	231.0	−1.44	18.75	0.09	0.330	4.296
0.5	242.3	−1.57	16.80	0.09	0.291	3.508
0.6	252.5	−1.67	15.45	0.10	0.262	2.974
0.7	261.9	−1.75	14.44	0.11	0.238	2.586
0.8	270.5	−1.83	13.65	0.12	0.220	2.291
0.9	278.5	−1.89	13.01	0.13	0.204	2.059
1.0	286.0	−1.94	12.48	0.14	0.191	1.872
1.5	318.3	−2.14	10.76	0.21	0.147	1.296
2.0	344.5	−2.27	9.78	0.30	0.121	1.000
2.5	367.1	−2.36	9.13	0.43	0.105	0.821
3.0	387.0	−2.44	8.66	0.61	0.093	0.701
3.5	405.0	−2.50	8.30	0.86	0.085	0.617
4.0	421.6	−2.55	8.02	1.21	0.079	0.554
4.5	437.0	−2.60	7.79	1.68	0.074	0.507
5.0	451.5	−2.63	7.60	2.32	0.070	0.470
5.5	465.3	−2.67	7.45	3.20	0.067	0.440
6.0	478.5	−2.69	7.32	4.39	0.064	0.416
6.5	491.2	−2.71	7.21	5.99	0.062	0.396
7.0	503.5	−2.73	7.11	8.16	0.060	0.379

Table II.18 (*Continued*)

p	w	μ	k	f	α/α_0	ν/ν_0
7.5	515.5	−2.75	7.03	11.06	0.059	0.365
8.0	527.1	−2.76	6.97	14.96	0.057	0.353
8.5	538.5	−2.77	6.91	20.16	0.056	0.342
9.0	549.7	−2.77	6.87	27.09	0.055	0.332
9.5	560.7	−2.78	6.83	36.30	0.054	0.323
10.0	571.5	−2.78	6.80	48.53	0.053	0.315

$T = 4.2$ K

0.01	119.2	8.74	1.67	0.01	1.055	1.030
0.02	117.7	8.97	1.67	0.02	1.118	1.063
0.03	116.1	9.21	1.67	0.03	1.194	1.101
0.04	114.4	9.45	1.67	0.04	1.284	1.143
0.05	112.6	9.69	1.67	0.04	1.394	1.192
0.06	110.6	9.94	1.67	0.05	1.531	1.248
0.07	108.5	10.20	1.67	0.06	1.707	1.313
0.08	106.1	10.47	1.67	0.07	1.945	1.392
0.09	103.3	10.77	1.67	0.07	2.284	1.488
0.10	172.5	−0.22	37.23	0.08	0.861	15.771
0.2	194.3	−0.73	24.67	0.08	0.596	8.407
0.3	211.0	−1.02	20.01	0.09	0.470	5.848
0.4	224.9	−1.22	17.47	0.10	0.394	4.526
0.5	237.1	−1.37	15.84	0.11	0.343	3.712
0.6	247.9	−1.49	14.68	0.12	0.305	3.157
0.7	257.6	−1.58	13.80	0.13	0.276	2.753
0.8	266.7	−1.66	13.11	0.14	0.252	2.444
0.9	275.0	−1.73	12.55	0.15	0.233	2.201
1.0	282.8	−1.79	12.08	0.16	0.217	2.004
1.5	316.1	−2.00	10.53	0.23	0.165	1.397
2.0	343.1	−2.13	9.64	0.33	0.136	1.082
2.5	366.1	−2.23	9.03	0.47	0.117	0.890
3.0	386.4	−2.30	8.60	0.66	0.104	0.762
3.5	404.8	−2.36	8.26	0.91	0.094	0.670
4.0	421.6	−2.41	7.99	1.26	0.087	0.602
4.5	437.3	−2.46	7.77	1.72	0.081	0.550
5.0	452.0	−2.49	7.59	2.35	0.077	0.509
5.5	465.9	−2.52	7.44	3.19	0.073	0.476
6.0	479.2	−2.55	7.31	4.32	0.070	0.449
6.5	492.0	−2.57	7.20	5.81	0.068	0.427
7.0	504.3	−2.59	7.11	7.80	0.065	0.408
7.5	516.2	−2.60	7.03	10.44	0.064	0.392
8.0	527.8	−2.62	6.97	13.93	0.062	0.378
8.5	539.2	−2.63	6.91	18.52	0.061	0.366
9.0	550.2	−2.63	6.86	24.56	0.059	0.355
9.5	561.1	−2.64	6.82	32.48	0.058	0.345
10.0	571.7	−2.65	6.79	42.85	0.057	0.335

$T = 4.4$ K

0.01	122.1	8.24	1.67	0.01	1.049	1.027
0.02	120.8	8.43	1.67	0.02	1.105	1.057
0.03	119.3	8.62	1.67	0.03	1.170	1.089
0.04	117.8	8.81	1.67	0.04	1.246	1.126
0.05	116.2	9.00	1.67	0.04	1.335	1.167
0.06	114.5	9.19	1.67	0.05	1.443	1.213
0.07	112.7	9.39	1.67	0.06	1.575	1.266

Table II.18 (*Continued*)

p	w	μ	k	f	α/α_0	ν/ν_0
0.08	110.7	9.60	1.67	0.07	1.741	1.326
0.09	108.5	9.81	1.67	0.07	1.957	1.397
0.10	106.1	10.03	1.68	0.08	2.253	1.482
0.2	184.1	−0.34	21.46	0.10	0.787	8.585
0.3	203.1	−0.74	18.08	0.11	0.584	6.049
0.4	218.3	−0.99	16.13	0.11	0.474	4.718
0.5	231.3	−1.17	14.82	0.12	0.404	3.890
0.6	242.8	−1.31	13.87	0.13	0.355	3.321
0.7	253.1	−1.41	13.14	0.14	0.318	2.905
0.8	262.6	−1.50	12.55	0.16	0.289	2.586
0.9	371.3	−1.58	12.07	0.17	0.266	2.334
1.0	279.4	−1.64	11.66	0.18	0.247	2.129
1.5	313.8	−1.87	10.29	0.26	0.185	1.493
2.0	341.4	−2.01	9.48	0.36	0.151	1.162
2.5	365.0	−2.10	8.93	0.51	0.129	0.958
3.0	385.7	−2.18	8.52	0.70	0.114	0.821
3.5	404.4	−2.24	8.20	0.96	0.104	0.723
4.0	421.5	−2.29	7.95	1.31	0.095	0.650
4.5	437.4	−2.33	7.74	1.77	0.089	0.593
5.0	452.2	−2.36	7.57	2.38	0.084	0.549
5.5	466.3	−2.39	7.43	3.19	0.080	0.513
6.0	479.7	−2.42	7.30	4.27	0.076	0.483
6.5	492.5	−2.44	7.20	5.67	0.073	0.458
7.0	504.9	−2.46	7.11	7.52	0.071	0.437
7.5	516.8	−2.47	7.03	9.94	0.069	0.419
8.0	528.4	−2.49	6.96	13.10	0.067	0.404
8.5	539.7	−2.50	6.90	17.21	0.065	0.390
9.0	550.7	−2.51	6.85	22.54	0.063	0.378
9.5	561.5	−2.51	6.81	29.46	0.062	0.366
10.0	572.0	−2.52	6.78	38.42	0.061	0.356
15.0	667.9	−2.62	6.60	490.05	0.046	0.270

$T = 4.5$ K

p	w	μ	k	f	α/α_0	ν/ν_0
0.01	123.6	8.01	1.67	0.01	1.046	1.026
0.02	122.3	8.18	1.67	0.02	1.099	1.054
0.03	120.9	8.35	1.67	0.03	1.160	1.085
0.04	119.5	8.52	1.67	0.04	1.229	1.119
0.05	118.0	8.69	1.67	0.05	1.311	1.157
0.06	116.4	8.86	1.67	0.05	1.408	1.199
0.07	114.7	9.04	1.67	0.06	1.524	1.247
0.08	112.9	9.22	1.67	0.07	1.667	1.301
0.09	110.9	9.40	1.67	0.07	1.847	1.363
0.10	108.7	9.59	1.68	0.08	2.083	1.436
0.2	178.3	−0.11	19.75	1.10	0.924	8.611
0.3	198.7	−0.58	17.08	0.11	0.656	6.122
0.4	214.8	−0.87	15.43	0.12	0.522	4.798
0.5	228.3	−1.07	14.30	0.13	0.440	3.968
0.6	240.1	−1.21	13.45	0.14	0.384	3.395
0.7	250.7	−1.33	12.80	0.15	0.342	2.975
0.8	260.4	−1.43	12.26	0.17	0.310	2.652
0.9	269.3	−1.51	11.82	0.18	0.284	2.396
1.0	277.6	−1.57	11.45	0.19	0.263	2.188
1.5	312.5	−1.81	10.17	0.27	0.195	1.540
2.0	340.5	−1.95	9.40	0.38	0.159	1.201
2.5	364.4	−2.05	8.87	0.53	0.136	0.992

Table II.18 (*Continued*)

p	w	μ	k	f	α/α_0	γ/γ_0
3.0	385.3	−2.12	8.48	0.72	0.120	0.851
3.5	404.1	−2.18	8.17	0.98	0.108	0.749
4.0	421.4	−2.23	7.93	1.33	0.100	0.673
4.5	437.4	−2.27	7.73	1.79	0.093	0.615
5.0	452.3	−2.31	7.56	2.40	0.087	0.568
5.5	466.4	−2.33	7.42	3.20	0.083	0.531
6.0	479.9	−2.36	7.30	4.25	0.079	0.500
6.5	492.8	−2.38	7.19	5.62	0.076	0.474
7.0	505.1	−2.40	7.10	7.40	0.073	0.452
7.5	517.1	−2.41	7.02	9.73	0.071	0.433
8.0	528.6	−2.42	6.96	12.74	0.069	0.417
8.5	539.9	−2.42	6.90	16.65	0.067	0.402
9.0	550.9	−2.44	6.85	21.69	0.066	0.389
9.5	561.6	−2.45	6.81	28.19	0.064	0.377
10.0	572.1	−2.46	6.77	36.55	0.063	0.367
15.0	667.3	−2.55	6.58	441.95	0.048	0.279

$T = 4.6$ K

p	w	μ	k	f	α/α_0	γ/γ_0
0.01	125.0	7.80	1.67	0.01	1.044	1.024
0.02	123.8	7.95	1.67	0.02	1.094	1.051
0.03	122.5	8.10	1.67	0.03	1.150	1.080
0.04	121.1	8.25	1.67	0.04	1.215	1.112
0.05	119.7	8.41	1.67	0.05	1.290	1.147
0.06	118.2	8.56	1.67	0.05	1.377	1.186
0.07	116.6	8.71	1.67	0.06	1.480	1.230
0.08	114.9	8.87	1.67	0.07	1.605	1.278
0.09	113.1	9.03	1.67	0.08	1.758	1.334
0.10	111.1	9.20	1.68	0.08	1.952	1.398
0.2	171.9	0.15	17.96	0.11	1.110	8.583
0.3	194.1	−0.41	16.04	0.12	0.743	6.174
0.4	211.0	−0.74	14.72	0.13	0.577	4.866
0.5	225.1	−0.96	13.76	0.14	0.480	4.038
0.6	237.3	−1.12	13.03	0.15	0.415	3.464
0.7	248.3	−1.25	12.45	0.16	0.368	3.041
0.8	258.2	−1.35	11.97	0.18	0.331	2.715
0.9	267.3	−1.43	11.57	0.19	0.303	2.456
1.0	275.7	−1.50	11.23	0.20	0.279	2.245
1.5	311.2	−1.75	10.04	0.29	0.206	1.585
2.0	339.6	−1.89	9.32	0.40	0.167	1.239
2.5	363.7	−1.99	8.81	0.55	0.143	1.025
3.0	384.8	−2.07	8.44	0.75	0.126	0.879
3.5	403.8	−2.13	8.14	1.01	0.113	0.775
4.0	421.2	−2.18	7.91	1.36	0.104	0.697
4.5	437.3	−2.22	7.71	1.82	0.097	0.636
5.0	452.3	−2.25	7.55	2.42	0.091	0.588
5.5	466.5	−2.28	7.41	3.21	0.086	0.549
6.0	480.0	−2.30	7.29	4.23	0.082	0.517
6.5	492.9	−2.32	7.18	5.57	0.079	0.490
7.0	505.3	−2.34	7.10	7.30	0.076	0.467
7.5	517.3	−2.35	7.02	9.53	0.074	0.447
8.0	528.8	−2.37	6.95	12.42	0.072	0.430
8.5	540.1	−2.38	6.89	16.14	0.070	0.415
9.0	551.0	−2.39	6.84	20.92	0.068	0.401
9.5	561.7	−2.39	6.80	27.04	0.066	0.389
10.0	572.2	−2.40	6.76	34.88	0.065	0.377

Table II.18 (*Continued*)

p	w	μ	k	f	α/α_0	γ/γ_0
15.0	666.7	−2.49	6.56	400.68	0.050	0.287

$T = 4.8$ K

p	w	μ	k	f	α/α_0	γ/γ_0
0.01	127.8	7.39	1.67	0.01	1.040	1.022
0.02	126.7	7.52	1.67	0.02	1.084	1.146
0.03	125.5	7.64	1.67	0.03	1.134	1.072
0.04	124.3	7.76	1.67	0.04	1.190	1.100
0.05	123.0	7.89	1.67	0.05	1.254	1.131
0.06	121.7	8.01	1.67	0.05	1.326	1.164
0.07	120.3	8.14	1.67	0.06	1.410	1.201
0.08	118.8	8.26	1.67	0.07	1.508	1.242
0.09	117.3	8.38	1.68	0.08	1.623	1.287
0.10	115.6	8.51	1.68	0.08	1.763	1.337
0.2	156.3	0.87	13.99	0.12	1.811	8.290
0.3	183.8	−0.02	13.87	0.13	0.983	6.198
0.4	202.9	−0.46	13.25	0.15	0.715	4.960
0.5	218.3	−0.73	12.66	0.16	0.575	4.152
0.6	231.5	−0.93	12.16	0.17	0.487	3.583
0.7	243.0	−1.08	11.73	0.18	0.425	3.159
0.8	253.5	−1.19	11.36	0.20	0.380	2.830
0.9	263.0	−1.29	11.05	0.21	0.344	2.566
1.0	271.8	−1.37	10.77	0.23	0.316	2.351
1.5	308.5	−1.63	9.78	0.31	0.229	1.673
2.0	337.6	−1.79	9.14	0.43	0.184	1.313
2.5	362.2	−1.89	8.69	0.59	0.156	1.089
3.0	383.8	−1.97	8.34	0.79	0.137	0.936
3.5	403.1	−2.03	8.07	1.06	0.124	0.826
4.0	420.7	−2.08	7.85	1.41	0.113	0.743
4.5	437.0	−2.12	7.67	1.87	0.105	0.678
5.0	452.3	−2.15	7.51	2.46	0.098	0.626
5.5	466.6	−2.17	7.38	3.23	0.093	0.584
6.0	480.2	−2.20	7.27	4.21	0.089	0.550
6.5	493.1	−2.22	7.17	5.49	0.085	0.521
7.0	505.6	−2.23	7.08	7.12	0.082	0.496
7.5	517.6	−2.25	7.00	9.20	0.079	0.474
8.0	529.1	−2.26	6.94	11.87	0.077	0.456
8.5	540.3	−2.27	6.88	15.26	0.074	0.439
9.0	551.2	−2.28	6.83	19.58	0.072	0.424
9.5	561.8	−2.28	6.78	25.07	0.071	0.411
10.0	572.2	−2.29	6.74	32.01	0.069	0.399
15.0	665.5	−2.37	6.53	334.17	0.053	0.305

$T = 5$ K

p	w	μ	k	f	α/α_0	γ/γ_0
0.01	130.6	7.03	1.67	0.01	1.036	1.020
0.02	129.5	7.13	1.67	0.02	1.076	1.042
0.03	128.5	7.23	1.67	0.03	1.120	1.065
0.04	127.3	7.33	1.67	0.04	1.170	1.090
0.05	126.2	7.43	1.67	0.05	1.224	1.117
0.06	125.0	7.53	1.67	0.05	1.286	1.147
0.07	123.8	7.63	1.67	0.06	1.355	1.178
0.08	122.5	7.73	1.67	0.07	1.435	1.213
0.09	121.1	7.83	1.68	0.08	1.526	1.250
0.10	119.7	7.93	1.68	0.08	1.632	1.292
0.2	132.3	2.33	8.90	0.14	5.190	7.293

Table II.18 (*Continued*)

p	w	μ	k	f	α/α_0	ν/ν_0
0.3	171.6	0.48	11.54	0.15	1.390	6.084
0.4	193.9	−0.14	11.70	0.16	0.907	4.989
0.5	210.9	−0.49	11.51	0.17	0.696	4.227
0.6	225.1	−0.73	11.25	0.19	0.574	3.674
0.7	237.4	−0.90	10.99	0.20	0.494	3.256
0.8	248.4	−1.03	10.74	0.22	0.436	2.928
0.9	258.4	−1.14	10.51	0.23	0.392	2.664
1.0	267.6	−1.23	10.30	0.25	0.357	2.447
1.5	305.6	−1.52	9.50	0.34	0.254	1.755
2.0	335.5	−1.69	8.95	0.47	0.203	1.383
2.5	360.6	−1.80	8.55	0.63	0.171	1.150
3.0	382.6	−1.88	8.24	0.84	0.150	0.991
3.5	402.2	−1.94	7.99	1.11	0.134	0.875
4.0	420.1	−1.98	7.79	1.47	0.122	0.788
4.5	436.6	−2.02	7.62	1.92	0.113	0.719
5.0	452.0	−2.05	7.47	2.51	0.106	0.664
5.5	466.4	−2.08	7.35	3.25	0.100	0.620
6.0	480.1	−2.10	7.24	4.21	0.095	0.583
6.5	493.2	−2.12	7.14	5.42	0.091	0.551
7.0	505.7	−2.14	7.06	6.97	0.087	0.525
7.5	517.7	−2.15	6.98	8.93	0.084	0.501
8.0	529.2	−2.16	6.92	11.41	0.082	0.481
8.5	540.4	−2.17	6.86	14.54	0.079	0.463
9.0	551.3	−2.18	6.81	18.48	0.077	0.447
9.5	561.9	−2.19	6.76	23.44	0.075	0.433
10.0	572.2	−2.19	6.72	29.66	0.073	0.420
15.0	664.3	−2.26	6.49	283.54	0.057	0.322

$$T = 5.2 \text{ K}$$

p	w	μ	k	f	α/α_0	ν/ν_0
0.01	133.2	6.69	1.67	0.01	1.033	1.019
0.02	132.3	6.78	1.67	0.02	1.069	1.038
0.03	131.3	6.86	1.67	0.03	1.109	1.059
0.04	130.3	6.95	1.67	0.04	1.152	1.082
0.05	129.3	7.03	1.67	0.05	1.200	1.106
0.06	128.2	7.11	1.67	0.06	1.253	1.132
0.07	127.1	7.19	1.67	0.06	1.312	1.159
0.08	125.9	7.27	1.67	0.07	1.378	1.189
0.09	124.7	7.35	1.68	0.08	1.453	1.221
0.10	123.5	7.43	1.68	0.09	1.537	1.256
0.2	106.9	8.01	1.81	0.14	4.607	1.949
0.3	156.8	1.19	8.99	0.17	2.239	5.756
0.4	183.7	0.25	10.09	0.18	1.194	4.935
0.5	202.8	−0.21	10.33	0.19	0.857	4.255
0.6	218.2	−0.51	10.32	0.21	0.684	3.735
0.7	231.4	−0.71	10.22	0.22	0.576	3.331
0.8	243.0	−0.87	10.09	0.24	0.501	3.010
0.9	253.5	−0.99	9.95	0.25	0.446	2.748
1.0	263.1	−1.10	9.81	0.27	0.403	2.531
1.5	302.5	−1.42	9.21	0.37	0.281	1.831
2.0	333.1	−1.59	8.75	0.50	0.222	1.450
2.5	358.8	−1.71	8.41	0.67	0.186	1.210
3.0	381.1	−1.79	8.13	0.89	0.162	1.044
3.5	401.1	−1.85	7.91	1.17	0.145	0.923
4.0	419.3	−1.90	7.72	1.52	0.132	0.831
4.5	436.0	−1.94	7.56	1.98	0.122	0.759

Table II.18 (Continued)

p	w	μ	k	f	α/α_0	ν/ν_0
5.0	451.5	−1.97	7.43	2.55	0.114	0.702
5.5	466.1	−1.99	7.31	3.28	0.107	0.654
6.0	479.9	−2.02	7.20	4.21	0.102	0.615
6.5	493.1	−2.03	7.11	5.38	0.097	0.581
7.0	505.6	−2.05	7.03	6.85	0.093	0.553
7.5	517.6	−2.06	6.96	8.70	0.090	0.528
8.0	529.2	−2.07	6.89	11.02	0.087	0.507
8.5	540.4	−2.08	6.84	13.93	0.084	0.488
9.0	551.2	−2.09	6.79	17.56	0.082	0.470
9.5	561.8	−2.09	6.74	22.08	0.079	0.455
10.0	572.0	−2.10	6.70	27.71	0.077	0.441
15.0	663.2	−2.15	6.45	244.24	0.061	0.338

$T = 5.5$ K

p	w	μ	k	f	α/α_0	ν/ν_0
0.01	137.2	6.24	1.67	0.01	1.029	1.016
0.02	136.3	6.31	1.67	0.02	1.061	1.034
0.03	135.5	6.37	1.67	0.03	1.094	1.052
0.04	134.6	6.44	1.67	0.04	1.131	1.071
0.05	133.7	6.50	1.67	0.05	1.171	1.092
0.06	132.8	6.56	1.67	0.06	1.214	1.114
0.07	131.8	6.62	1.67	0.06	1.262	1.137
0.08	130.8	6.68	1.68	0.07	1.314	1.161
0.09	129.8	6.74	1.68	0.08	1.371	1.187
0.10	128.8	6.80	1.68	0.09	1.434	1.215
0.2	116.5	7.22	1.76	0.15	2.858	1.661
0.3	124.7	3.44	4.44	0.19	9.440	4.281
0.4	165.9	1.03	7.55	0.20	2.009	4.639
0.5	189.3	0.28	8.49	0.22	1.221	4.189
0.6	207.0	−0.13	8.88	0.24	0.906	3.758
0.7	221.6	−0.41	9.04	0.25	0.733	3.396
0.8	234.4	−0.61	9.09	0.27	0.621	3.096
0.9	245.7	−0.77	9.09	0.29	0.543	2.845
1.0	256.0	−0.89	9.06	0.31	0.485	2.633
1.5	297.4	−1.26	8.75	0.42	0.326	1.934
2.0	329.3	−1.46	8.44	0.56	0.253	1.543
2.5	355.8	−1.58	8.17	0.74	0.210	1.293
3.0	378.7	−1.67	7.95	0.96	0.182	1.120
3.5	399.2	−1.73	7.76	1.25	0.162	0.992
4.0	417.7	−1.78	7.60	1.61	0.147	0.895
4.5	434.7	−1.82	7.46	2.06	0.135	0.818
5.0	450.5	−1.85	7.34	2.63	0.126	0.756
5.5	465.3	−1.88	7.24	3.34	0.118	0.705
6.0	479.3	−1.90	7.14	4.23	0.112	0.662
6.5	492.6	−1.92	7.06	5.34	0.107	0.626
7.0	505.2	−1.93	6.98	6.72	0.102	0.595
7.5	517.3	−1.94	6.91	8.44	0.098	0.568
8.0	528.9	−1.95	6.85	10.56	0.095	0.544
8.5	540.1	−1.96	6.80	13.19	0.092	0.523
9.0	550.9	−1.97	6.75	16.43	0.089	0.504
9.5	561.4	−1.97	6.70	20.43	0.086	0.488
10.0	571.6	−1.98	6.66	25.35	0.084	0.472
15.0	661.6	−2.02	6.40	200.07	0.066	0.363
20.0	737.5	−2.13	6.28	1395.44	0.050	0.277

Table II.18 (*Continued*)

p	w	μ	k	f	α/α_0	ν/ν_0
			$T = 6$ K			
0.01	143.4	5.61	1.67	0.01	1.024	1.014
0.02	142.8	5.65	1.67	0.02	1.049	1.028
0.03	142.0	5.69	1.67	0.03	1.076	1.043
0.04	141.3	5.73	1.67	0.04	1.105	1.058
0.05	140.6	5.77	1.67	0.05	1.135	1.074
0.06	139.9	5.81	1.67	0.06	1.168	1.091
0.07	139.1	5.85	1.68	0.07	1.202	1.109
0.08	138.4	5.89	1.68	0.07	1.240	1.127
0.09	137.6	5.92	1.68	0.08	1.279	1.147
0.10	136.8	5.96	1.68	0.09	1.322	1.167
0.2	128.2	6.21	1.74	0.16	2.028	1.445
0.3	120.0	5.77	2.00	0.21	4.258	2.022
0.4	134.0	3.16	3.58	0.24	5.604	3.264
0.5	163.8	1.43	5.46	0.27	2.509	3.690
0.6	186.1	0.64	6.47	0.29	1.550	3.572
0.7	203.7	0.18	7.05	0.31	1.137	3.353
0.8	218.5	−0.13	7.40	0.33	0.909	3.128
0.9	231.3	−0.35	7.62	0.35	0.764	2.921
1.0	242.9	−0.52	7.76	0.37	0.662	2.736
1.5	288.1	−1.01	7.95	0.50	0.414	2.073
2.0	322.0	−1.25	7.88	0.65	0.312	1.679
2.5	349.9	−1.39	7.75	0.84	0.255	1.419
3.0	373.9	−1.49	7.62	1.08	0.218	1.236
3.5	395.2	−1.56	7.49	1.38	0.193	1.099
4.0	414.3	−1.61	7.38	1.75	0.174	0.993
4.5	431.9	−1.65	7.27	2.20	0.159	0.910
5.0	448.1	−1.68	7.18	2.76	0.147	0.841
5.5	463.3	−1.71	7.09	3.45	0.138	0.785
6.0	477.5	−1.73	7.01	4.29	0.130	0.737
6.5	491.0	−1.75	6.94	5.32	0.123	0.697
7.0	503.8	−1.76	6.88	6.59	0.118	0.662
7.5	516.1	−1.77	6.82	8.13	0.113	0.631
8.0	527.8	−1.78	6.76	10.01	0.108	0.604
8.5	539.0	−1.79	6.71	12.29	0.104	0.581
9.0	549.9	−1.80	6.67	15.06	0.101	0.559
9.5	560.4	−1.80	6.62	18.41	0.098	0.540
10.0	570.5	−1.81	6.58	22.48	0.095	0.523
15.0	659.0	−1.83	6.31	151.32	0.075	0.402
20.0	732.5	−1.90	6.17	906.42	0.058	0.316
25.0	797.9	−2.05	6.10	5003.29	0.042	0.232
			$T = 6,5$ K			
0.01	149.4	5.08	1.67	0.01	1.020	1.011
0.02	148.9	5.11	1.67	0.02	1.041	1.023
0.03	148.3	5.14	1.67	0.03	1.063	1.036
0.04	147.7	5.16	1.67	0.04	1.086	1.048
0.05	147.1	5.19	1.67	0.05	1.110	1.062
0.06	146.5	5.21	1.68	0.06	1.135	1.075
0.07	145.9	5.24	1.68	0.07	1.162	1.089
0.08	145.3	5.26	1.68	0.07	1.190	1.104
0.09	144.7	5.29	1.68	0.08	1.220	1.119
0.10	144.1	5.31	1.68	0.09	1.251	1.135
0.2	137.7	5.46	1.73	0.17	1.696	1.333

Table II.18 (*Continued*)

p	w	μ	k	f	α/α_0	γ/γ_0
0.3	131.7	5.30	1.86	0.23	2.584	1.647
0.4	131.3	4.32	2.29	0.27	4.055	2.171
0.5	145.3	2.73	3.28	0.31	3.913	2.777
0.6	166.2	1.59	4.38	0.33	2.608	3.069
0.7	185.4	0.89	5.20	0.36	1.795	3.092
0.8	201.8	0.43	5.78	0.38	1.351	3.003
0.9	216.2	0.11	6.18	0.41	1.085	2.876
1.0	228.9	−0.12	6.48	0.44	0.911	2.741
1.5	277.9	−0.76	7.13	0.58	0.524	2.168
2.0	314.0	−1.05	7.28	0.75	0.381	1.788
2.5	343.3	−1.22	7.29	0.95	0.305	1.527
3.0	368.3	−1.33	7.25	1.21	0.258	1.338
3.5	390.4	−1.41	7.19	1.51	0.226	1.195
4.0	410.2	−1.46	7.12	1.89	0.202	1.084
4.5	428.3	−1.51	7.06	2.34	00184	0.994
5.0	445.0	−1.54	6.99	2.90	0.170	0.921
5.5	460.5	−1.57	6.93	3.57	0.158	0.860
6.0	475.1	−1.59	6.87	4.38	0.148	0.808
6.5	488.8	−1.61	6.81	5.35	0.140	0.764
7.0	501.8	−1.63	6.75	6.53	0.134	0.725
7.5	514.2	−1.64	6.70	7.94	0.128	0.692
8.0	526.1	−1.65	6.66	9.64	0.122	0.662
8.5	537.4	−1.66	6.61	11.67	0.118	0.636
9.0	548.3	−1.66	6.57	14.10	0.114	0.612
9.5	558.9	−1.67	6.53	17.01	0.110	0.591
10.0	569.0	−1.67	6.50	20.47	0.107	0.571
15.0	656.7	−1.69	6.22	120.54	0.084	0.439
20.0	728.3	−1.73	6.07	635.12	0.067	0.352
25.0	791.3	−1.82	5.98	3095.29	0.050	0.271

$$T = 7 \text{ K}$$

p	w	μ	k	f	α/α_0	γ/γ_0
0.01	155.2	4.64	1.67	0.01	1.017	1.010
0.02	154.7	4.66	1.67	0.02	1.034	1.020
0.03	154.2	4.67	1.67	0.03	1.052	1.030
0.04	153.8	4.69	1.67	0.04	1.071	1.041
0.05	153.3	4.71	1.67	0.05	1.091	1.052
0.06	152.8	4.73	1.68	0.06	1.111	1.063
0.07	152.3	4.74	1.68	0.07	1.133	1.075
0.08	151.8	4.76	1.68	0.08	1.155	1.087
0.09	151.3	4.77	1.68	0.08	1.178	1.099
0.10	150.8	4.79	1.69	0.09	1.203	1.112
0.2	145.9	4.87	1.73	0.17	1.516	1.263
0.3	141.4	4.78	1.81	0.24	2.022	1.474
0.4	139.5	4.31	2.02	0.29	2.768	1.771
0.5	144.1	3.37	2.45	0.34	3.326	2.142
0.6	156.1	2.37	3.11	0.38	3.085	2.474
0.7	171.7	1.59	3.81	0.41	2.455	2.666
0.8	187.3	1.02	4.41	0.44	1.898	2.728
0.9	201.9	0.61	4.90	0.47	1.506	2.708
1.0	215.2	0.30	5.29	0.50	1.237	2.646
1.5	267.2	−0.51	6.31	0.66	0.658	2.217
2.0	305.3	−0.86	6.68	0.84	0.462	1.869
2.5	336.0	−1.06	6.82	1.06	0.363	1.615
3.0	362.1	−1.18	6.87	1.33	0.303	1.426
3.5	385.0	−1.27	6.87	1.65	0.262	1.280

Table II.18 (*Continued*)

p	x	μ	k	f	a/a_0	γ/γ_0
4.0	405.5	−1.33	6.85	2.03	0.233	1.165
4.5	424.1	−1.38	6.82	2.49	0.211	1.072
5.0	441.2	−1.42	6.78	3.04	0.194	0.995
5.5	457.1	−1.45	6.74	3.69	0.180	0.930
6.0	472.0	−1.48	6.70	4.48	0.168	0.875
6.5	486.0	−1.49	6.66	5.41	0.159	0.827
7.0	499.3	−1.51	6.62	6.52	0.150	0.786
7.5	511.9	−1.52	6.58	7.83	0.143	0.749
8.0	523.9	−1.53	6.54	9.39	0.137	0.717
8.5	535.4	−1.54	6.50	11.24	0.132	0.689
9.0	546.4	−1.55	6.47	13.42	0.127	0.663
9.5	557.0	−1.56	6.43	15.99	0.123	0.639
10.0	567.2	−1.56	6.40	19.03	0.119	0.618
15.0	654.5	−1.58	6.14	99.90	0.093	0.475
20.0	724.8	−1.60	5.98	471.75	0.075	0.385
25.0	785.9	−1.66	5.87	2067.31	0.059	0.307
30.0	841.9	−1.79	5.81	8565.10	0.043	0.229
			$T = 7.5$ K			
0.01	160.7	4.26	1.67	0.01	1.014	1.008
0.02	160.3	4.27	1.67	0.02	1.029	1.017
0.03	159.9	4.29	1.67	0.03	1.045	1.026
0.04	159.5	4.30	1.67	0.04	1.060	1.035
0.05	159.1	4.31	1.67	0.05	1.077	1.045
0.06	158.7	4.32	1.68	0.06	1.094	1.054
0.07	158.3	4.33	1.68	0.07	1.111	1.064
0.08	157.9	4.34	1.68	0.08	1.129	1.074
0.09	157.5	4.35	1.68	0.09	1.148	1.084
0.10	157.2	4.36	1.69	0.09	1.167	1.095
0.2	153.3	4.40	1.72	0.18	1.403	1.215
0.3	149.8	4.33	1.79	0.25	1.738	1.371
0.4	148.0	4.05	1.92	0.31	2.186	1.571
0.5	149.6	3.49	2.16	0.36	2.626	1.812
0.6	156.0	2.75	2.53	0.41	2.785	2.060
0.7	166.3	2.07	2.99	0.45	2.600	2.262
0.8	178.5	1.51	3.47	0.49	2.250	2.394
0.9	191.3	1.07	3.91	0.52	1.889	2.457
1.0	203.9	0.72	4.31	0.56	1.584	2.468
1.5	256.7	−0.26	5.54	0.73	0.818	2.219
2.0	296.5	−0.67	6.08	0.93	0.555	1.921
2.5	328.4	−0.90	6.34	1.17	0.427	1.684
3.0	355.5	−1.05	6.47	1.45	0.352	1.500
3.5	379.2	−1.15	6.54	1.78	0.302	1.355
4.0	400.3	−1.22	6.56	2.17	0.266	1.238
4.5	419.5	−1.27	6.57	2.63	0.240	1.142
5.0	437.1	−1.31	6.56	3.18	0.219	1.062
5.5	453.4	−1.35	6.54	3.83	0.203	0.995
6.0	468.6	−1.37	6.52	4.59	0.189	0.937
6.5	482.9	−1.39	6.50	5.49	0.178	0.887
7.0	496.4	−1.41	6.47	6.54	0.168	0.843
7.5	509.2	−1.43	6.44	7.78	0.160	0.804
8.0	521.4	−1.44	6.41	9.23	0.152	0.770
8.5	533.0	−1.45	6.38	10.93	0.146	0.739
9.0	544.1	−1.46	6.35	12.92	0.140	0.711
9.5	554.8	−1.46	6.33	15.24	0.135	0.686

Table II.18 (*Continued*)

p	w	μ	k	f	α/α_0	γ/γ_0
10.0	565.2	−1.47	6.30	17.95	0.131	0.653
15.0	652.4	−1.49	6.06	85.39	0.101	0.509
20.0	721.8	−1.50	5.89	366.80	0.083	0.416
25.0	781.3	−1.54	5.78	1466.50	0.067	0.339
30	835.3	−1.62	5.70	5554.50	0.051	0.265
35	886.6	−1.77	5.67	20164.05	0.035	0.190

$T = 8$ K

p	w	μ	k	f	α/α_0	γ/γ_0
0.01	166.1	3.94	1.67	0.01	1.012	1.007
0.02	165.8	3.94	1.67	0.02	1.025	1.015
0.03	165.4	3.95	1.67	0.03	1.038	1.023
0.04	165.1	3.96	1.67	0.04	1.052	1.031
0.05	164.8	3.96	1.68	0.05	1.066	1.039
0.06	164.4	3.97	1.68	0.06	1.080	1.047
0.07	164.1	3.98	1.68	0.07	1.094	1.055
0.08	163.8	3.98	1.68	0.08	1.109	1.064
0.09	163.4	2.99	1.68	0.09	1.125	1.072
0.10	163.1	3.99	1.69	0.10	1.141	1.081
0.2	160.0	4.01	1.72	0.18	1.326	1.181
0.3	157.4	3.95	1.78	0.26	1.568	1.302
0.4	155.9	3.75	1.87	0.32	1.869	1.450
0.5	156.5	3.38	2.02	0.38	2.183	1.621
0.6	160.1	2.87	2.23	0.44	2.395	1.803
0.7	166.9	2.32	2.56	0.48	2.418	1.971
0.8	175.8	1.82	2.90	0.53	2.282	2.106
0.9	185.9	1.40	3.25	0.57	2.065	2.200
1.0	196.6	1.05	3.59	0.61	1.826	2.254
1.5	247.1	−0.01	4.84	0.81	0.994	2.179
2.0	287.7	−0.49	5.51	1.03	0.660	1.946
2.5	320.7	−0.76	5.87	1.28	0.499	1.733
3.0	348.7	−0.92	6.07	1.57	0.406	1.559
3.5	373.1	−1.03	6.20	1.91	0.345	1.417
4.0	394.9	−1.11	6.27	2.31	0.302	1.301
4.5	414.6	−1.17	6.31	2.77	0.271	1.205
5.0	432.6	−1.22	6.33	3.32	0.246	1.124
5.5	449.3	−1.25	6.34	3.96	0.227	1.054
6.0	464.8	−1.28	6.33	4.71	0.211	0.994
6.5	479.4	−1.31	6.33	5.58	0.198	0.942
7.0	493.2	−1.33	6.31	6.59	0.186	0.896
7.5	506.2	−1.34	6.30	7.77	0.177	0.856
8.0	518.6	−1.36	6.28	9.13	0.168	0.820
8.5	530.4	−1.37	6.26	10.72	0.161	0.787
9.0	541.7	−1.38	6.24	12.56	0.155	0.758
9.5	552.5	−1.38	6.22	14.68	0.149	0.731
10.0	562.9	−1.39	6.19	17.14	0.144	0.707
15.0	650.4	−1.41	5.99	74.79	0.110	0.542
20.0	719.1	−1.42	5.82	295.78	0.091	0.445
25.0	777.6	−1.44	5.70	1091.68	0.075	0.369
30	829.9	−1.50	5.61	3824.05	0.059	0.298
35	879.0	−1.61	5.56	12855.00	0.044	0.227
40	927.2	−1.78	5.55	41798.56	0.029	0.154

$T = 8.5$ K

p	w	μ	k	f	α/α_0	γ/γ_0
0.01	171.3	3.65	1.67	0.01	1.011	1.007
0.02	171.0	3.66	1.67	0.02	1.022	1.013
0.03	170.7	3.66	1.67	0.03	1.033	1.020

Table II.18 (*Continued*)

p	w	μ	k	f	α/α_0	γ/γ_0
0.04	170.4	3.66	1.67	0.04	1.045	1.027
0.05	170.2	3.67	1.68	0.05	1.057	1.034
0.06	169.9	3.67	1.68	0.06	1.069	1.041
0.07	169.6	3.67	1.68	0.07	1.081	1.048
0.08	169.3	3.68	1.68	0.08	1.094	1.056
0.09	169.1	3.68	1.68	0.09	1.107	1.063
0.10	168.8	3.68	1.69	0.10	1.120	1.071
0.2	166.3	3.68	1.72	0.18	1.270	1.155
0.3	164.3	3.62	1.77	0.26	1.455	1.253
0.4	163.2	3.47	1.84	0.34	1.674	1.368
0.5	163.4	3.21	1.95	0.40	1.903	1.498
0.6	165.8	2.84	2.11	0.46	2.091	1.636
0.7	170.4	2.41	2.32	0.51	2.181	1.771
0.8	176.9	1.99	2.57	0.56	2.160	1.889
0.9	184.8	1.61	2.83	0.61	2.058	1.984
1.0	193.4	1.29	3.10	0.66	1.911	2.054
1.5	239.2	0.22	4.25	0.88	1.166	2.105
2.0	279.5	−0.32	4.97	1.11	0.774	1.945
2.5	313.1	−0.61	5.41	1.38	0.577	1.764
3.0	341.8	−0.80	5.69	1.68	0.464	1.604
3.5	266.9	−0.92	5.86	2.03	0.392	1.469
4.0	389.2	−1.01	5.97	2.44	0.341	1.356
4.5	409.4	−1.08	6.05	2.91	0.304	1.260
5.0	427.9	−1.13	6.10	3.46	0.275	1.179
5.5	445.0	−1.17	6.13	4.09	0.253	1.109
6.0	460.9	−1.20	6.14	4.83	0.234	1.048
6.5	475.8	−1.23	6.15	5.67	0.219	0.994
7.0	489.8	−1.25	6.15	6.65	0.206	0.947
7.5	503.0	−1.27	6.15	7.78	0.195	0.905
8.0	515.6	−1.28	6.14	9.08	0.185	0.867
8.5	527.6	−1.29	6.13	10.57	0.177	0.833
9.0	539.1	−1.30	6.12	12.28	0.169	0.802
9.5	550.0	−1.31	6.10	14.25	0.163	0.774
10.0	560.5	−1.32	6.09	16.51	0.157	0.749
15.0	648.5	−1.35	5.91	66.80	0.119	0.574
20.0	716.9	−1.36	5.75	245.66	0.098	0.473
25.0	774.4	−1.37	5.62	845.06	0.082	0.396
30	825.5	−1.41	5.53	2763.74	0.067	0.327
35	872.8	−1.49	5.47	8684.52	0.052	0.260
40	918.5	−1.61	5.44	26417.97	0.037	0.192

$T = 9$ K

p	w	μ	k	f	α/α_0	γ/γ_0
0.01	176.3	3.40	1.67	0.01	1.010	1.006
0.02	176.0	3.40	1.67	0.02	1.019	1.012
0.03	175.8	3.41	1.67	0.03	1.029	1.018
0.04	175.6	3.41	1.67	0.04	1.039	1.024
0.05	175.4	3.41	1.68	0.05	1.049	1.030
0.06	175.1	3.41	1.68	0.06	1.060	1.036
0.07	174.9	3.41	1.68	0.07	1.070	1.043
0.08	174.7	3.41	1.68	0.08	1.081	1.049
0.09	174.5	3.41	1.68	0.09	1.092	1.056
0.10	174.3	3.41	1.69	0.10	1.104	1.062
0.2	172.3	3.40	1.72	0.19	1.228	1.134
0.3	170.7	3.34	1.76	0.27	1.375	1.216
0.4	169.9	3.22	1.82	0.34	1.542	1.309

Table II.18 (*Continued*)

P	w	μ	k	f	α/α_0	ν/ν_0
0.5	170.1	3.02	1.91	0.41	1.716	1.412
0.6	171.8	2.75	2.02	0.48	1.870	1.521
0.7	175.1	2.41	2.18	0.54	1.972	1.630
0.8	180.0	2.07	2.36	0.60	2.004	1.730
0.9	186.2	1.74	2.56	0.65	1.971	1.816
1.0	193.3	1.44	2.77	0.70	1.892	1.887
1.5	233.3	0.41	3.76	0.94	1.307	2.012
2.0	272.2	−0.15	4.50	1.20	0.890	1.921
2.5	305.9	−0.48	4.99	1.48	0.661	1.776
3.0	335.1	−0.68	5.31	1.79	0.528	1.635
3.5	360.7	−0.82	5.53	2.15	0.442	1.510
4.0	383.6	−0.92	5.68	2.57	0.382	1.401
4.5	404.2	−0.99	5.79	3.05	0.339	1.308
5.0	423.1	−1.05	5.86	3.60	0.306	1.228
5.5	440.6	−1.09	5.92	4.23	0.280	1.158
6.0	456.8	−1.13	5.95	4.95	0.258	1.096
6.5	472.0	−1.16	5.98	5.78	0.241	1.042
7.0	486.3	−1.18	5.99	6.73	0.226	0.994
7.5	499.7	−1.20	6.00	7.81	0.214	0.950
8.0	512.5	−1.22	6.00	9.05	0.203	0.912
8.5	524.7	−1.23	6.00	10.47	0.193	0.876
9.0	536.3	−1.24	5.99	12.08	0.185	0.845
9.5	547.4	−1.25	5.99	13.92	0.177	0.815
10.0	558.1	−1.26	5.98	16.02	0.171	0.789
15.0	646.7	−1.30	5.83	60.62	0.129	0.606
20.0	714.9	−1.30	5.68	209.01	0.106	0.499
25.0	771.8	−1.31	5.56	675.52	0.089	0.422
30	821.9	−1.34	5.46	2078.90	0.074	0.355
35	867.8	−1.39	5.39	6154.10	0.060	0.291
40	911.6	−1.49	5.34	17650.25	0.046	0.226
45	954.8	−1.63	5.33	49308.14	0.031	0.161
50	999.4	−1.84	5.37		0.018	0.094

$T = 9.5$ K

P	w	μ	k	f	α/α_0	ν/ν_0
0.01	181.2	3.18	1.67	0.01	1.008	1.005
0.02	181.0	3.18	1.67	0.02	1.017	1.011
0.03	180.8	3.18	1.67	0.03	1.026	1.016
0.04	180.6	3.18	1.67	0.04	1.034	1.021
0.05	180.4	3.18	1.68	0.05	1.043	1.027
0.06	180.2	3.18	1.68	0.06	1.053	1.032
0.07	180.0	3.18	1.68	0.07	1.062	1.038
0.08	179.8	3.18	1.68	0.08	1.071	1.044
0.09	179.7	3.18	1.68	0.09	1.081	1.049
0.10	179.5	3.18	1.69	0.10	1.090	1.055
0.2	177.9	3.15	1.71	0.19	1.195	1.118
0.3	176.7	3.09	1.75	0.27	1.315	1.188
0.4	176.2	2.99	1.80	0.35	1.447	1.265
0.5	176.4	2.83	1.88	0.43	1.584	1.349
0.6	177.7	2.62	1.97	0.50	1.710	1.438
0.7	180.3	2.36	2.09	0.56	1.805	1.527
0.8	184.2	2.08	2.23	0.62	1.857	1.611
0.9	189.2	1.79	2.38	0.68	1.861	1.688
1.0	195.0	1.53	2.55	0.74	1.826	1.753
1.5	229.8	0.57	3.39	1.01	1.402	1.913
2.0	266.1	−0.00	4.09	1.28	0.999	1.880

Table II.18 (*Continued*)

p	w	μ	k	f	α/α_0	γ/γ_0
2.5	299.3	−0.35	4.60	1.57	0.747	1.773
3.0	328.7	−0.57	4.96	1.90	0.594	1.653
3.5	354.6	−0.73	5.21	2.27	0.495	1.539
4.0	377.9	−0.83	5.40	2.69	0.426	1.438
4.5	399.0	−0.91	5.53	3.18	0.376	1.349
5.0	418.3	−0.97	5.63	3.73	0.338	1.270
5.5	436.1	−1.02	5.70	4.35	0.308	1.201
6.0	452.7	−1.06	5.76	5.07	0.284	1.140
6.5	468.1	−1.09	5.80	5.88	0.264	1.086
7.0	482.7	−1.12	5.83	6.81	0.247	1.037
7.5	496.4	−1.14	5.85	7.86	0.233	0.993
8.0	509.4	−1.16	5.86	9.05	0.221	0.954
8.5	521.7	−1.17	5.87	10.40	0.210	0.918
9.0	533.5	−1.18	5.87	11.93	0.201	0.885
9.5	544.8	−1.20	5.87	13.67	0.192	0.855
10.0	555.6	−1.20	5.87	15.63	0.185	0.827
15.0	644.9	−1.25	5.76	55.72	0.138	0.636
20.0	713.0	−1.26	5.62	181.41	0.114	0.525
25.0	769.5	−1.26	5.50	554.60	0.096	0.446
30	818.9	−1.28	5.40	1616.66	0.082	0.380
35	863.8	−1.32	5.32	4537.93	0.067	0.318
40	906.1	−1.39	5.27	12350.70	0.054	0.258
45	947.3	−1.50	5.24	32760.61	0.040	0.196
50	988.6	−1.66	5.25	85028.13	0.026	0.133

$$T = 10 \text{ K}$$

p	w	μ	k	f	α/α_0	γ/γ_0
0.01	185.9	2.98	1.67	0.01	1.008	1.005
0.02	185.7	2.98	1.67	0.02	1.015	1.009
0.03	185.6	2.98	1.67	0.03	1.023	1.014
0.04	185.4	2.98	1.67	0.04	1.031	1.019
0.05	185.3	2.98	1.68	0.05	1.038	1.024
0.06	185.1	2.97	1.68	0.06	1.046	1.029
0.07	185.0	2.97	1.68	0.07	1.055	1.034
0.08	184.8	2.97	1.68	0.08	1.063	1.039
0.09	184.7	2.97	1.68	0.09	1.071	1.044
0.10	184.5	2.97	1.69	0.10	1.079	1.049
0.2	183.3	2.94	1.71	0.19	1.169	1.104
0.3	182.4	2.88	1.75	0.28	1.269	1.165
0.4	182.1	2.79	1.79	0.36	1.377	1.231
0.5	182.4	2.66	1.85	0.44	1.487	1.301
0.6	183.6	2.49	1.93	0.51	1.591	1.375
0.7	185.7	2.28	2.02	0.58	1.676	1.449
0.8	188.9	2.04	2.13	0.64	1.732	1.521
0.9	192.9	1.80	2.26	0.71	1.754	1.588
1.0	197.3	1.57	2.40	0.77	1.746	1.647
1.5	228.1	0.68	3.11	1.06	1.452	1.819
2.0	261.5	0.13	3.74	1.35	1.092	1.828
2.5	293.6	−0.23	4.25	1.66	0.831	1.756
3.0	322.7	−0.47	4.63	2.00	0.661	1.659
3.5	348.9	−0.63	4.91	2.38	0.549	1.559
4.0	372.4	−0.75	5.12	2.82	0.471	1.466
4.5	393.9	−0.84	5.28	3.30	0.415	1.382
5.0	413.5	−0.90	5.40	3.86	0.372	1.307
5.5	431.6	−0.96	5.50	4.48	0.338	1.240
6.0	448.5	−1.00	5.57	5.19	0.311	1.180

Table II.18 (*Continued*)

p	w	μ	k	j	α/α_0	γ/γ_0
6.5	464.2	−1.03	5.62	5.99	0.288	1.125
7.0	479.0	−1.06	5.66	6.89	0.269	1.077
7.5	492.9	−1.08	5.69	7.91	0.254	1.033
8.0	506.1	−1.10	5.72	9.06	0.240	0.993
8.5	518.7	−1.12	5.73	10.36	0.228	0.956
9.0	530.6	−1.13	5.74	11.82	0.217	0.923
9.5	542.1	−1.14	5.75	13.47	0.208	0.892
10.0	553.0	−1.15	5.76	15.32	0.200	0.864
15.0	643.1	−1.21	5.69	51.78	0.148	0.666
20.0	711.4	−1.22	5.56	160.12	0.121	0.550
25.0	767.6	−1.22	5.44	465.63	0.103	0.469
30	816.4	−1.23	5.34	1292.85	0.088	0.403
35	860.5	−1.26	5.26	3460.04	0.075	0.344
40	901.8	−1.31	5.20	8985.31	0.061	0.287
45	941.4	−1.40	5.16	22753.92	0.048	0.229
50	980.5	−1.52	5.15	56403.63	0.035	0.170
55	1020.3	−1.70	5.17		0.022	0.110

$T = 11$ K

p	w	μ	k	j	α/α_0	γ/γ_0
0.01	195.0	2.64	1.67	0.01	1.006	1.004
0.02	194.9	2.64	1.67	0.02	1.012	1.008
0.03	194.8	2.63	1.67	0.03	1.018	1.012
0.04	194.7	2.63	1.67	0.04	1.024	1.016
0.05	194.6	2.63	1.68	0.05	1.031	1.020
0.06	194.5	2.63	1.68	0.06	1.037	1.024
0.07	194.4	2.62	1.68	0.07	1.043	1.028
0.08	194.3	2.62	1.68	0.08	1.050	1.032
0.09	194.2	2.62	1.68	0.09	1.056	1.036
0.10	194.2	2.61	1.69	0.10	1.063	1.040
0.2	193.5	2.57	1.71	0.19	1.131	1.084
0.3	193.1	2.52	1.74	0.28	1.203	1.131
0.4	193.1	2.44	1.78	0.37	1.279	1.180
0.5	193.5	2.35	1.82	0.45	1.355	1.233
0.6	194.6	2.22	1.88	0.53	1.428	1.287
0.7	196.3	2.08	1.94	0.61	1.492	1.341
0.8	198.6	1.92	2.02	0.68	1.543	1.395
0.9	201.6	1.74	2.11	0.75	1.576	1.446
1.0	205.2	1.57	2.20	0.82	1.591	1.493
1.5	228.9	0.82	2.72	1.15	1.465	1.658
2.0	256.6	0.32	3.22	1.48	1.214	1.713
2.5	285.1	−0.03	3.67	1.82	0.975	1.696
3.0	312.7	−0.29	4.06	2.19	0.792	1.642
3.5	338.5	−0.47	4.37	2.59	0.661	1.572
4.0	362.2	−0.60	4.61	3.04	0.566	1.499
4.5	384.1	−0.70	4.81	3.54	0.496	1.428
5.0	404.2	−0.77	4.97	4.10	0.442	1.362
5.5	422.8	−0.83	5.09	4.72	0.400	1.301
6.0	440.2	−0.88	5.19	5.42	0.367	1.244
6.5	456.4	−0.92	5.27	6.20	0.339	1.193
7.0	471.7	−0.95	5.34	7.07	0.316	1.146
7.5	486.0	−0.98	5.39	8.04	0.296	1.102
8.0	499.6	−1.00	5.44	9.13	0.280	1.063
8.5	512.6	−1.02	5.47	10.34	0.265	1.026
9.0	524.9	−1.04	5.50	11.69	0.252	0.992
9.5	536.6	−1.05	5.52	13.20	0.241	0.961

Table II.18 (*Continued*)

p	w	μ	k	f	α/α_0	γ/γ_0
10.0	547.8	−1.07	5.53	14.87	0.231	0.931
15.0	639.7	−1.13	5.54	45.88	0.169	0.723
20.0	708.4	−1.15	5.45	129.84	0.137	0.599
25.0	764.3	−1.15	5.34	346.48	0.117	0.513
30	812.5	−1.16	5.24	884.72	0.101	0.447
35	855.6	−1.17	5.16	2181.46	0.088	0.390
40	895.5	−1.20	5.09	5226.57	0.075	0.337
45	933.3	−1.25	5.05	12224.20	0.063	0.286
50	969.8	−1.33	5.02	28009.48	0.051	0.235
55	1005.6	−1.44	5.01	63051.75	0.030	0.183
60	1041.4	−1.60	5.01	—	0.027	0.130

$T = 12$ K

p	w	μ	k	f	α/α_0	γ/γ_0
0.01	203.8	2.36	1.67	0.01	1.005	1.003
0.02	203.7	2.35	1.67	0.02	1.010	1.007
0.03	203.6	2.35	1.67	0.03	1.015	1.010
0.04	203.6	2.35	1.67	0.04	1.020	1.013
0.05	203.5	2.34	1.68	0.05	1.025	1.017
0.06	203.5	2.34	1.68	0.06	1.030	1.020
0.07	203.4	2.33	1.68	0.07	1.035	1.023
0.08	203.3	2.33	1.68	0.08	1.040	1.027
0.09	203.3	2.33	1.68	0.09	1.045	1.030
0.10	203.2	2.32	1.68	0.10	1.051	1.034
0.2	202.9	2.28	1.71	0.19	1.104	1.069
0.3	202.9	2.23	1.73	0.29	1.159	1.107
0.4	203.1	2.16	1.77	0.38	1.215	1.146
0.5	203.8	2.08	1.80	0.46	1.272	1.187
0.6	204.8	1.99	1.85	0.55	1.325	1.228
0.7	206.3	1.88	1.90	0.63	1.374	1.270
0.8	208.3	1.76	1.95	0.71	1.415	1.312
0.9	210.7	1.63	2.02	0.79	1.446	1.352
1.0	213.6	1.50	2.09	0.86	1.467	1.390
1.5	232.8	0.88	2.48	1.23	1.427	1.536
2.0	256.1	0.43	2.88	1.59	1.262	1.607
2.5	280.8	0.11	3.26	1.96	1.073	1.621
3.0	305.9	−0.14	3.60	2.35	0.903	1.600
3.5	330.4	−0.32	3.91	2.77	0.765	1.557
4.0	353.6	−0.47	4.17	3.24	0.659	1.505
4.5	375.3	−0.57	4.38	3.75	0.578	1.450
5.0	395.6	−0.66	4.56	4.32	0.515	1.395
5.5	414.6	−0.73	4.71	4.94	0.466	1.342
6.0	432.3	−0.78	4.84	5.63	0.425	1.292
6.5	448.9	−0.82	4.94	6.40	0.392	1.245
7.0	464.5	−0.86	5.03	7.24	0.365	1.200
7.5	479.3	−0.89	5.10	8.18	0.342	1.159
8.0	493.3	−0.92	5.16	9.22	0.322	1.121
8.5	506.5	−0.94	5.21	10.37	0.304	1.085
9.0	519.1	−0.96	5.25	11.63	0.289	1.052
9.5	531.1	−0.97	5.29	13.03	0.276	1.021
10.0	542.6	−0.99	5.31	14.57	0.264	0.991
15.0	636.3	−1.07	5.40	41.72	0.190	0.777
20.0	705.8	−1.09	5.34	109.72	0.154	0.646
25.0	761.8	−1.10	5.25	272.67	0.131	0.555
30	809.6	−1.10	5.15	649.55	0.114	0.487
35	852.2	−1.11	5.08	1496.42	0.100	0.430

Table II.18 (*Continued*)

p	w	μ	k	f	α/α_0	γ/γ_0
40	891.5	−1.13	5.01	3354.06	0.088	0.381
45	928.5	−1.16	4.97	7346.16	0.077	0.335
50	963.9	−1.20	4.93	15775.40	0.065	0.290
55	998.2	−1.27	4.91	33302.67	0.055	0.246
60	1031.6	−1.37	4.90	69260.88	0.044	0.201
65	1064.4	−1.51	4.90	—	0.033	0.155
70	1097.1	−1.72	4.91	—	0.023	0.108
75	1130.6	−2.02	4.94	—	0.012	0.061

$T = 13$ K

p	w	μ	k	f	α/α_0	γ/γ_0
0.01	212.1	2.12	1.67	0.01	1.004	1.003
0.02	212.1	2.11	1.67	0.02	1.008	1.006
0.03	212.0	2.11	1.67	0.03	1.012	1.008
0.04	212.0	2.11	1.67	0.04	1.016	1.011
0.05	212.0	2.10	1.67	0.05	1.021	1.014
0.6	212.0	2.10	1.68	0.06	1.025	1.017
0.7	211.9	2.09	1.68	0.07	1.029	1.020
0.08	211.9	2.09	1.68	0.08	1.033	1.023
0.09	211.9	2.08	1.68	0.09	1.037	1.026
0.10	211.9	2.08	1.68	0.10	1.041	1.028
0.2	211.9	2.04	1.70	0.19	1.084	1.058
0.3	212.1	1.99	1.73	0.29	1.128	1.089
0.4	212.5	1.93	1.76	0.38	1.171	1.121
0.5	213.3	1.86	1.79	0.47	1.214	1.154
0.6	214.4	1.79	1.82	0.56	1.255	1.187
0.7	215.8	1.70	1.87	0.64	1.293	1.221
0.8	217.5	1.61	1.91	0.73	1.326	1.254
0.9	219.6	1.51	1.96	0.81	1.353	1.285
1.0	222.1	1.40	2.01	0.89	1.373	1.317
1.5	238.3	0.90	2.32	1.29	1.374	1.444
2.0	258.3	0.50	2.65	1.67	1.268	1.517
2.5	279.9	0.20	2.96	2.07	1.127	1.547
3.0	302.3	−0.03	3.26	2.49	0.982	1.546
3.5	324.8	−0.21	3.54	2.93	0.853	1.526
4.0	346.9	−0.35	3.79	3.42	0.745	1.493
4.5	368.1	−0.47	4.01	3.94	0.658	1.453
5.0	388.2	−0.56	4.20	4.51	0.588	1.410
5.5	407.1	−0.63	4.37	5.14	0.531	1.366
6.0	425.0	−0.69	4.51	5.83	0.485	1.323
6.5	441.9	−0.74	4.63	6.58	0.447	1.281
7.0	457.8	−0.78	4.73	7.41	0.415	1.242
7.5	472.8	−0.81	4.82	8.32	0.388	1.204
8.0	487.1	−0.84	4.90	9.32	0.365	1.168
8.5	500.6	−0.86	4.96	10.41	0.345	1.134
9.0	513.5	−0.88	5.01	11.62	0.327	1.102
9.5	525.8	−0.90	5.06	12.93	0.312	1.072
10.0	537.6	−0.92	5.10	14.37	0.298	1.044
15.0	633.1	−1.01	5.26	38.66	0.213	0.828
20.0	703.4	−1.04	5.24	95.62	0.171	0.691
25.0	759.6	−1.06	5.16	223.81	0.145	0.596
30	807.3	−1.06	5.08	502.89	0.127	0.524
35	849.8	−1.06	5.00	1094.16	0.112	0.467
40	888.8	−1.07	4.95	2318.67	0.100	0.419
45	925.7	−1.09	4.90	4805.85	0.089	0.376
50	961.0	−1.12	4.88	9773.86	0.079	0.337

Table II.18 (*Continued*)

p	w	μ	k	f	α/α_0	γ/γ_0
55	995.1	−1.16	4.86	19553.06	0.069	0.299
60	1028.1	−1.22	4.85	38556.39	0.060	0.262
65	1060.0	−1.31	4.85	75065.06	0.050	0.224
70	1091.0	−1.43	4.85	—	0.041	0.186
75	1120.9	−1.62	4.85	—	0.032	0.147
80	1149.9	−1.88	4.85	—	0.023	0.107
85	1178.6	−2.31	4.85	—	0.014	0.066

$T = 14$ K

p	w	μ	k	f	α/α_0	γ/γ_0
0.01	220.1	1.91	1.67	0.01	1.003	1.002
0.02	220.1	1.91	1.67	0.02	1.007	1.005
0.03	220.1	1.90	1.67	0.03	1.010	1.007
0.04	220.1	1.90	1.67	0.04	1.014	1.010
0.05	220.1	1.90	1.67	0.05	1.017	1.012
0.06	220.1	1.89	1.68	0.06	1.021	1.015
0.07	220.1	1.89	1.68	0.07	1.024	1.017
0.08	220.1	1.88	1.68	0.08	1.028	1.019
0.09	220.1	1.88	1.68	0.09	1.031	1.022
0.10	220.1	1.87	1.68	0.10	1.034	1.024
0.2	220.3	1.83	1.70	0.20	1.069	1.050
0.3	220.7	1.78	1.72	0.29	1.104	1.076
0.4	221.4	1.73	1.75	0.38	1.139	1.102
0.5	222.2	1.67	1.78	0.48	1.173	1.129
0.6	223.4	1.61	1.81	0.57	1.205	1.157
0.7	224.7	1.54	1.84	0.66	1.235	1.184
0.8	226.4	1.46	1.88	0.74	1.261	1.211
0.9	228.3	1.38	1.92	0.83	1.284	1.238
1.0	230.5	1.30	1.96	0.92	1.302	1.264
1.5	244.7	0.88	2.22	1.33	1.322	1.374
2.0	262.2	0.53	2.49	1.75	1.255	1.444
2.5	281.3	0.26	2.75	2.17	1.150	1.480
3.0	301.2	0.05	3.01	2.61	1.033	1.492
3.5	321.7	−0.12	3.25	3.07	0.919	1.486
4.0	342.3	−0.26	3.48	3.57	0.817	1.468
4.5	362.5	−0.37	3.70	4.11	0.729	1.441
5.0	382.0	−0.47	3.89	4.69	0.655	1.409
5.5	400.7	−0.54	4.06	5.32	0.594	1.375
6.0	418.6	−0.61	4.21	6.01	0.544	1.340
6.5	435.5	−0.66	4.34	6.75	0.502	1.305
7.0	451.5	−0.70	4.46	7.57	0.466	1.270
7.5	466.8	−0.74	4.56	8.45	0.435	1.237
8.0	481.3	−0.77	4.65	9.42	0.409	1.204
8.5	495.1	−0.80	4.72	10.48	0.386	1.173
9.0	508.2	−0.82	4.79	11.63	0.366	1.143
9.5	520.8	−0.84	4.84	12.88	0.349	1.115
10.0	532.8	−0.86	4.89	14.24	0.333	1.088
15.0	630.1	−0.96	5.12	36.34	0.237	0.875
20.0	701.2	−1.00	5.13	85.31	0.189	0.734
25.0	757.7	−1.02	5.08	189.75	0.160	0.635
30	805.5	−1.02	5.00	405.63	0.140	0.560
35	847.9	−1.03	4.94	840.53	0.124	0.501
40	887.0	−1.03	4.89	1697.99	0.111	0.453
45	924.0	−1.04	4.85	3357.77	0.101	0.413
50	959.6	−1.05	4.83	6519.84	0.091	0.377
55	994.3	−1.08	4.83	12460.71	0.082	0.344

Table II.18 (Continued)

p	w	μ	k	f	α/α_0	γ/γ_0
60	1028.0	−1.12	4.83	23485.96	0.074	0.313
65	1060.9	−1.17	4.83	43724.84	0.066	0.283
70	1092.7	−1.25	4.84	80513.19	0.058	0.253
75	1123.3	−1.35	4.85	—	0.050	0.223
80	1152.4	−1.51	4.86	—	0.043	0.192
85	1180.0	−1.74	4.86	—	0.035	0.160
90	1205.8	−2.10	4.85	—	0.028	0.127
95	1229.7	−2.73	4.84	—	0.020	0.093
100	1251.5	−4.08	4.81	—	0.012	0.057

$T = 15$ K

p	w	μ	k	f	α/α_0	γ/γ_0
0.01	227.9	1.73	1.67	0.01	1.003	1.002
0.02	227.9	1.73	1.67	0.02	1.006	1.004
0.03	227.9	1.73	1.67	0.03	1.009	1.006
0.04	227.9	1.72	1.67	0.04	1.012	1.008
0.05	228.0	1.72	1.67	0.05	1.014	1.011
0.06	228.0	1.71	1.68	0.06	1.017	1.013
0.07	228.0	1.71	1.68	0.07	1.020	1.015
0.08	228.0	1.70	1.68	0.08	1.023	1.017
0.09	228.0	1.70	1.68	0.09	1.026	1.019
0.10	228.1	1.70	1.68	0.10	1.029	1.021
0.2	228.4	1.65	1.70	0.20	1.058	1.043
0.3	229.0	1.61	1.72	0.29	1.087	1.065
0.4	229.8	1.56	1.74	0.39	1.115	1.088
0.5	230.7	1.51	1.77	0.48	1.142	1.111
0.6	231.9	1.45	1.80	0.57	1.168	1.134
0.7	233.2	1.40	1.83	0.67	1.192	1.157
0.8	234.8	1.33	1.86	0.76	1.213	1.179
0.9	236.6	1.27	1.89	0.85	1.232	1.202
1.0	238.6	1.20	1.93	0.94	1.248	1.224
1.5	251.4	0.85	2.14	1.37	1.275	1.319
2.0	267.1	0.54	2.37	1.81	1.233	1.385
2.5	284.2	0.30	2.59	2.25	1.155	1.423
3.0	302.2	0.10	2.81	2.71	1.062	1.442
3.5	320.8	−0.05	3.03	3.19	0.965	1.445
4.0	339.7	−0.19	3.24	3.70	0.873	1.437
4.5	358.7	−0.30	3.44	4.25	0.789	1.421
5.0	377.4	−0.39	3.62	4.84	0.716	1.399
5.5	395.6	−0.47	3.79	5.48	0.653	1.374
6.0	413.1	−0.53	3.94	6.16	0.600	1.346
6.5	429.9	−0.59	4.08	6.90	0.554	1.317
7.0	446.0	−0.63	4.20	7.71	0.515	1.288
7.5	461.3	−0.67	4.31	8.58	0.482	1.259
8.0	475.9	−0.71	4.41	9.52	0.453	1.231
8.5	489.9	−0.74	4.49	10.54	0.428	1.203
9.0	503.2	−0.76	4.57	11.65	0.406	1.176
9.5	516.0	−0.78	4.64	12.85	0.386	1.149
10.0	528.2	−0.80	4.69	14.14	0.368	1.124
15.0	627.2	−0.91	4.99	34.52	0.261	0.918
20.0	699.2	−0.96	5.04	77.51	0.208	0.775
25.0	756.1	−0.98	5.00	165.01	0.175	0.672
30	803.9	−0.99	4.93	337.90	0.153	0.594
35	846.3	−0.99	4.88	671.33	0.136	0.534
40	885.5	−1.00	4.83	1301.35	0.122	0.485
45	922.9	−1.00	4.81	2471.15	0.111	0.445

Table II.18 (*Continued*)

p	w	μ	k	f	α/α_0	γ/γ_0
50	959.1	−1.01	4.80	4610.57	0.102	0.412
55	994.7	−1.02	4.80	8471.90	0.094	0.383
60	1029.7	−1.04	4.82	15359.82	0.086	0.357
65	1064.2	−1.07	4.84	27519.39	0.079	0.333
70	1098.0	−1.12	4.87	48784.72	0.073	0.310
75	1131.0	−1.18	4.90	85660.19	0.067	0.288
80	1163.1	−1.27	4.93	—	0.061	0.266
85	1194.1	−1.39	4.96	—	0.055	0.243
90	1223.9	−1.58	4.99	—	0.049	0.220
95	1252.7	−1.86	5.01	—	0.044	0.196
100	1281.0	−2.33	5.03	—	0.038	0.170

$T = 16$ K

p	w	μ	k	f	α/α_0	γ/γ_0
0.01	235.4	1.58	1.67	0.01	1.002	1.002
0.02	235.4	1.57	1.67	0.02	1.005	1.004
0.03	235.4	1.57	1.67	0.03	1.007	1.006
0.04	235.5	1.57	1.67	0.04	1.010	1.007
0.05	235.5	1.56	1.67	0.05	1.012	1.009
0.06	235.5	1.56	1.68	0.06	1.015	1.011
0.07	235.6	1.55	1.68	0.07	1.017	1.013
0.08	235.6	1.55	1.68	0.08	1.020	1.015
0.09	235.7	1.55	1.68	0.09	1.022	1.017
0.10	235.7	1.54	1.68	0.10	1.025	1.019
0.2	236.2	1.50	1.70	0.20	1.049	1.038
0.3	236.9	1.46	1.72	0.29	1.073	1.057
0.4	237.8	1.41	1.74	0.39	1.096	1.076
0.5	238.8	1.37	1.76	0.49	1.118	1.096
0.6	240.0	1.32	1.79	0.58	1.139	1.116
0.7	241.3	1.27	1.81	0.67	1.159	1.135
0.8	242.9	1.21	1.84	0.77	1.177	1.155
0.9	244.6	1.16	1.87	0.86	1.192	1.174
1.0	246.5	1.10	1.90	0.95	1.206	1.192
1.5	258.2	0.80	2.08	1.40	1.235	1.275
2.0	272.5	0.53	2.28	1.85	1.209	1.336
2.5	288.1	0.31	2.47	2.32	1.150	1.375
3.0	304.5	0.14	2.67	2.79	1.076	1.397
3.5	321.4	−0.01	2.86	3.29	0.995	1.405
4.0	338.9	−0.13	3.04	3.82	0.914	1.404
4.5	356.6	−0.24	3.22	4.38	0.837	1.396
5.0	374.3	−0.33	3.39	4.97	0.767	1.382
5.5	391.7	−0.41	3.56	5.61	0.705	1.364
6.0	408.7	−0.47	3.71	6.30	0.650	1.344
6.5	425.3	−0.53	3.85	7.04	0.603	1.321
7.0	441.2	−0.58	3.97	7.83	0.562	1.297
7.5	456.4	−0.62	4.09	8.69	0.527	1.273
8.0	471.1	−0.65	4.19	9.61	0.496	1.248
8.5	485.1	−0.68	4.28	10.61	0.468	1.224
9.0	498.6	−0.71	4.36	11.68	0.444	1.200
9.5	511.5	−0.73	4.44	12.83	0.423	1.176
10.0	523.9	−0.75	4.50	14.08	0.404	1.153
15.0	624.5	−0.87	4.86	33.07	0.286	0.957
20.0	697.5	−0.92	4.94	71.44	0.227	0.814
25.0	754.7	−0.94	4.92	146.41	0.191	0.708
30	802.6	−0.96	4.87	288.83	0.166	0.628
35	845.0	−0.97	4.82	553.19	0.148	0.564

Table II.18 (*Continued*)

p	w	μ	k	f	α/α_0	γ/γ_0
40	884.3	−0.97	4.78	1034.46	0.133	0.515
45	922.0	−0.97	4.76	1896.15	0.121	0.475
50	958.9	−0.97	4.76	3416.91	0.112	0.442
55	995.5	−0.98	4.78	6067.27	0.104	0.416
60	1032.0	−0.99	4.81	10635.07	0.097	0.393
65	1068.4	−1.00	4.85	18429.94	0.091	0.374
70	1104.6	−1.03	4.90	31613.43	0.086	0.358
75	1140.6	−1.06	4.96	53731.26	0.081	0.342
80	1176.4	−1.11	5.02	90564.50	0.077	0.328
85	1211.9	−1.18	5.09	—	0.073	0.314
90	1247.3	−1.27	5.16	—	0.069	0.300
95	1283.1	−1.40	5.24	—	0.065	0.285
100	1320.5	−1.60	5.33	—	0.062	0.270

$T = 17$ K

p	w	μ	k	f	α/α_0	γ/γ_0
0.01	242.6	1.44	1.67	0.01	1.002	1.002
0.02	242.7	1.44	1.67	0.02	1.004	1.003
0.03	242.7	1.43	1.67	0.03	1.006	1.005
0.04	242.8	1.43	1.67	0.04	1.008	1.007
0.05	242.8	1.42	1.67	0.05	1.011	1.008
0.06	242.9	1.42	1.68	0.06	1.013	1.010
0.07	242.9	1.42	1.68	0.07	1.015	1.012
0.08	243.0	1.41	1.68	0.08	1.017	1.013
0.09	243.0	1.41	1.68	0.09	1.019	1.015
0.10	243.1	1.41	1.68	0.10	1.021	1.016
0.2	243.7	1.37	1.70	0.20	1.042	1.033
0.3	244.5	1.33	1.71	0.30	1.062	1.050
0.4	245.4	1.29	1.73	0.39	1.081	1.067
0.5	246.5	1.24	1.75	0.49	1.099	1.084
0.6	247.7	1.20	1.78	0.58	1.117	1.101
0.7	249.1	1.15	1.80	0.68	1.133	1.118
0.8	250.6	1.11	1.83	0.77	1.148	1.135
0.9	252.3	1.06	1.85	0.87	1.161	1.151
1.0	254.1	1.01	1.88	0.96	1.173	1.167
1.5	265.0	0.75	2.04	1.43	1.201	1.240
2.0	278.2	0.52	2.21	1.89	1.185	1.296
2.5	292.6	0.32	2.38	2.37	1.140	1.334
3.0	307.7	0.16	2.55	2.86	1.080	1.357
3.5	323.3	0.02	2.72	3.38	1.013	1.369
4.0	339.4	−0.09	2.88	3.91	0.943	1.373
4.5	355.8	−0.19	3.05	4.48	0.874	1.370
5.0	372.5	−0.28	3.20	5.09	0.809	1.362
5.5	389.1	−0.35	3.36	5.73	0.749	1.350
6.0	405.5	−0.42	3.50	6.42	0.696	1.335
6.5	421.5	−0.48	3.64	7.16	0.648	1.318
7.0	437.2	−0.52	3.77	7.94	0.606	1.299
7.5	452.3	−0.57	3.88	8.79	0.569	1.279
8.0	466.9	−0.60	3.99	9.70	0.536	1.258
8.5	480.9	−0.63	4.09	10.67	0.507	1.237
9.0	494.4	−0.66	4.17	11.71	0.482	1.217
9.5	507.4	−0.68	4.25	12.83	0.459	1.196
10.0	519.9	−0.70	4.32	14.02	0.438	1.175
15.0	622.0	−0.83	4.74	31.88	0.310	0.991
20.0	695.9	−0.88	4.85	66.61	0.247	0.851
25.0	753.5	−0.91	4.84	132.04	0.208	0.743

Table II.18 (*Continued*)

p	w	μ	k	f	α/α_0	γ/γ_0
30	801.4	−0.93	4.80	252.08	0.180	0.660
35	843.8	−0.94	4.76	467.54	0.160	0.594
40	883.2	−0.94	4.73	847.11	0.144	0.543
45	921.2	−0.95	4.72	1505.27	0.131	0.502
50	958.7	−0.95	4.72	2630.92	0.121	0.470
55	996.3	−0.95	4.75	4533.21	0.113	0.445
60	1034.2	−0.95	4.80	7714.04	0.107	0.425
65	1072.5	−0.96	4.86	12982.92	0.101	0.409
70	1111.1	−0.97	4.93	21636.83	0.097	0.397
75	1150.2	−0.98	5.02	35741.88	0.093	0.387
80	1189.6	−1.01	5.11	58570.61	0.090	0.379
85	1229.5	−1.04	5.21	95283.19	0.088	0.372
90	1270.2	−1.08	5.32	—	0.085	0.366
95	1312.4	−1.15	5.45	—	0.084	0.360
100	1357.2	−1.23	5.61	—	0.082	0.354

$T = 18$ K

p	w	μ	k	f	α/α_0	γ/γ_0
0.01	249.7	1.32	1.67	0.01	1.002	1.001
0.02	249.7	1.31	1.67	0.02	1.004	1.003
0.03	249.8	1.31	1.67	0.03	1.005	1.004
0.04	249.9	1.31	1.67	0.04	1.007	1.006
0.05	249.9	1.30	1.67	0.05	1.009	1.007
0.06	250.0	1.30	1.67	0.06	1.011	1.009
0.07	250.0	1.30	1.68	0.07	1.013	1.010
0.08	250.1	1.29	1.68	0.08	1.015	1.012
0.09	250.2	1.29	1.68	0.09	1.016	1.013
0.10	250.2	1.28	1.68	0.10	1.018	1.015
0.2	251.0	1.25	1.70	0.20	1.036	1.029
0.3	251.8	1.21	1.71	0.30	1.053	1.044
0.4	252.8	1.17	1.73	0.39	1.069	1.059
0.5	253.9	1.13	1.75	0.49	1.085	1.074
0.6	255.2	1.09	1.77	0.59	1.099	1.089
0.7	256.6	1.05	1.79	0.68	1.113	1.104
0.8	258.1	1.01	1.81	0.78	1.125	1.119
0.9	259.7	0.97	1.84	0.88	1.136	1.133
1.0	261.4	0.92	1.86	0.97	1.146	1.147
1.5	271.8	0.70	2.00	1.45	1.172	1.212
2.0	284.1	0.49	2.15	1.93	1.163	1.263
2.5	297.6	0.32	2.31	2.42	1.128	1.299
3.0	311.6	0.17	2.46	2.92	1.080	1.323
3.5	326.1	0.04	2.61	3.45	1.023	1.337
4.0	341.0	−0.06	2.76	4.00	0.963	1.344
4.5	356.3	−0.16	2.90	4.58	0.902	1.345
5.0	371.9	−0.24	3.05	5.19	0.843	1.341
5.5	387.6	−0.31	3.19	5.84	0.787	1.333
6.0	403.3	−0.38	3.33	6.53	0.735	1.323
6.5	418.8	−0.43	3.46	7.26	0.688	1.310
7.0	434.0	−0.48	3.58	8.04	0.646	1.295
7.5	448.9	−0.52	3.70	8.88	0.608	1.280
8.0	463.3	−0.56	3.80	9.77	0.574	1.263
8.5	477.2	−0.59	3.90	10.72	0.544	1.245
9.0	490.7	−0.62	4.00	11.74	0.517	1.227
9.5	503.7	−0.64	4.08	12.82	0.493	1.209
10.0	516.3	−0.66	4.16	13.98	0.471	1.191
15.0	619.6	−0.79	4.61	30.90	0.335	1.021

Table II.18 (Continued)

p	w	μ	k	f	α/α_0	ν/ν_0
20.0	694.5	−0.85	4.76	62.68	0.267	0.884
25.0	752.5	−0.88	4.77	120.66	0.224	0.777
30	800.5	−0.90	4.74	223.81	0.195	0.691
35	842.8	−0.91	4.70	403.47	0.172	0.624
40	882.2	−0.92	4.68	710.89	0.155	0.570
45	920.3	−0.92	4.67	1228.95	0.141	0.528
50	958.3	−0.92	4.69	2090.64	0.131	0.495
55	996.8	−0.92	4.73	3507.57	0.122	0.470
60	1036.0	−0.92	4.79	5814.12	0.115	0.452
65	1076.1	−0.92	4.86	9535.42	0.110	0.439
70	1117.0	−0.92	4.96	15491.11	0.106	0.430
75	1158.7	−0.93	5.06	24953.63	0.103	0.424
80	1201.4	−0.94	5.18	39888.51	0.101	0.421
85	1245.1	−0.95	5.32	63317.84	0.100	0.419
90	1290.3	−0.97	5.47	99866.13	0.099	0.419
95	1337.5	−0.99	5.63	—	0.098	0.420
100	1387.9	−1.03	5.83	—	0.098	0.422

$T = 19$ K

p	w	μ	k	f	α/α_0	ν/ν_0
0.01	256.5	1.21	1.67	0.01	1.002	1.001
0.02	256.6	1.20	1.67	0.02	1.003	1.003
0.03	256.7	1.20	1.67	0.03	1.005	1.004
0.04	256.7	1.20	1.67	0.04	1.006	1.005
0.05	256.8	1.19	1.67	0.05	1.008	1.007
0.06	256.9	1.19	1.67	0.06	1.009	1.008
0.07	256.9	1.19	1.68	0.07	1.011	1.009
0.08	257.0	1.18	1.68	0.08	1.013	1.011
0.09	257.1	1.18	1.68	0.09	1.014	1.012
0.10	257.2	1.18	1.68	0.10	1.016	1.013
0.2	258.0	1.14	1.69	0.20	1.031	1.026
0.3	258.9	1.11	1.71	0.30	1.045	1.040
0.4	259.9	1.07	1.73	0.40	1.059	1.053
0.5	261.1	1.03	1.74	0.49	1.072	1.066
0.6	262.4	1.00	1.76	0.59	1.085	1.079
0.7	263.8	0.96	1.78	0.69	1.096	1.092
0.8	265.2	0.92	1.80	0.79	1.107	1.105
0.9	266.8	0.89	1.83	0.88	1.116	1.118
1.0	268.5	0.85	1.85	0.98	1.124	1.131
1.5	278.4	0.65	1.97	1.47	1.148	1.188
2.0	290.1	0.47	2.11	1.95	1.142	1.235
2.5	302.7	0.31	2.25	2.45	1.116	1.269
3.0	315.9	0.17	2.38	2.97	1.076	1.293
3.5	329.4	0.06	2.52	3.51	1.028	1.308
4.0	343.4	−0.04	2.65	4.07	0.976	1.317
4.5	357.7	−0.13	2.78	4.65	0.923	1.320
5.0	372.3	−0.21	2.92	5.27	0.869	1.320
5.5	387.1	−0.28	3.05	5.92	0.817	1.315
6.0	402.1	−0.34	3.17	6.61	0.768	1.309
6.5	417.0	−0.39	3.30	7.35	0.723	1.299
7.0	431.7	−0.44	3.42	8.12	0.681	1.289
7.5	446.2	−0.48	3.53	8.95	0.643	1.276
8.0	460.4	−0.52	3.64	9.83	0.609	1.263
8.5	474.2	−0.55	3.74	10.77	0.578	1.248
9.0	487.5	−0.58	3.83	11.76	0.551	1.233
9.5	500.5	−0.61	3.92	12.82	0.526	1.218

Table II.18 (*Continued*)

p	w	μ	k	f	α/α_0	γ/γ_0
10.0	513.1	−0.63	4.00	13.95	0.503	1.202
15.0	617.5	−0.76	4.50	30.06	0.360	1.047
20.0	693.3	−0.82	4.67	59.43	0.287	0.916
25.0	751.7	−0.85	4.70	111.48	0.242	0.808
30	799.8	−0.87	4.68	201.54	0.209	0.722
35	841.9	−0.89	4.65	354.27	0.185	0.652
40	881.2	−0.90	4.63	608.86	0.166	0.596
45	919.4	−0.90	4.62	1027.10	0.151	0.552
50	957.8	−0.90	4.65	1705.61	0.140	0.519
55	997.0	−0.90	4.70	2794.37	0.130	0.494
60	1037.3	−0.90	4.77	4524.75	0.123	0.476
65	1078.9	−0.90	4.86	7251.54	0.118	0.464
70	1121.8	−0.89	4.97	11515.32	0.114	0.457
75	1165.9	−0.89	5.10	18138.66	0.111	0.454
80	1211.3	−0.89	5.24	28360.34	0.110	0.455
85	1258.2	−0.89	5.40	44046.14	0.109	0.457
90	1307.0	−0.89	5.58	67989.69	0.109	0.462
95	1358.1	−0.90	5.78	—	0.109	0.468
100	1412.4	−0.91	6.01	—	0.110	0.475

$T = 20$ K

p	w	μ	k	f	α/α_0	γ/γ_0
0.01	263.2	1.11	1.67	0.01	1.001	1.001
0.02	263.3	1.11	1.67	0.02	1.003	1.002
0.03	263.4	1.10	1.67	0.03	1.004	1.004
0.04	263.4	1.10	1.67	0.04	1.006	1.005
0.05	263.5	1.10	1.67	0.05	1.007	1.006
0.06	263.6	1.09	1.67	0.06	1.008	1.007
0.07	263.7	1.09	1.68	0.07	1.010	1.008
0.08	263.7	1.09	1.68	0.08	1.011	1.009
0.09	263.8	1.08	1.68	0.09	1.012	1.011
0.10	263.9	1.08	1.68	0.10	1.014	1.012
0.2	264.8	1.05	1.69	0.20	1.027	1.024
0.3	265.8	1.01	1.71	0.30	1.039	1.036
0.4	266.8	0.98	1.72	0.40	1.051	1.047
0.5	268.0	0.95	1.74	0.50	1.062	1.059
0.6	269.3	0.91	1.76	0.59	1.073	1.071
0.7	270.7	0.88	1.78	0.69	1.083	1.083
0.8	272.2	0.85	1.80	0.79	1.091	1.094
0.9	273.8	0.81	1.82	0.89	1.099	1.106
1.0	275.4	0.78	1.84	0.99	1.106	1.117
1.5	285.0	0.60	1.95	1.48	1.127	1.168
2.0	296.0	0.44	2.07	1.98	1.124	1.210
2.5	308.0	0.29	2.20	2.49	1.103	1.244
3.0	320.5	0.17	2.32	3.01	1.070	1.267
3.5	333.3	0.06	2.44	3.56	1.030	1.283
4.0	346.4	−0.03	2.57	4.12	0.985	1.293
4.5	359.8	−0.11	2.69	4.72	0.938	1.298
5.0	373.5	−0.18	2.81	5.34	0.889	1.299
5.5	387.5	−0.25	2.93	6.00	0.842	1.298
6.0	401.7	−0.31	3.04	6.69	0.796	1.293
6.5	416.0	−0.36	3.16	7.42	0.753	1.287
7.0	430.2	−0.41	3.28	8.20	0.712	1.279
7.5	444.2	−0.45	3.38	9.02	0.675	1.270
8.0	458.1	−0.49	3.49	9.88	0.641	1.259
8.5	471.6	−0.52	3.59	10.81	0.610	1.247

Table II.18 (*Continued*)

p	w	μ	k	f	α/α_0	γ/γ_0
9.0	484.9	−0.55	3.68	11.78	0.582	1.235
9.5	497.7	−0.57	3.77	12.82	0.556	1.222
10.0	510.3	−0.60	3.85	13.92	0.533	1.209
15.0	615.4	−0.73	4.38	29.35	0.383	1.069
20.0	692.2	−0.79	4.59	56.70	0.307	0.94
25.0	751.2	−0.82	4.64	103.93	0.259	0.838
30	799.3	−0.85	4.62	183.65	0.225	0.751
35	841.2	−0.87	4.59	315.62	0.198	0.680
40	880.2	−0.88	4.57	530.49	0.178	0.622
45	918.4	−0.88	4.58	875.47	0.162	0.576
50	957.0	−0.89	4.61	1424.5	0.149	0.541
55	996.7	−0.89	4.66	2281.5	0.139	0.516
60	1038.0	−0.88	4.74	3617.4	0.131	0.498
65	1080.9	−0.88	4.85	5678.5	0.125	0.487
70	1125.4	−0.87	4.97	8835.3	0.121	0.481
75	1171.6	−0.86	5.12	13639.7	0.118	0.480
80	1219.3	−0.86	5.28	20905.5	0.117	0.482
85	1268.8	−0.85	5.46	31837.3	0.116	0.488
90	1320.3	−0.84	5.66	48203.3	0.117	0.496
95	1374.3	−0.84	5.88	72591.8	0.117	0.505
100	1431.5	−0.84	6.13	—	0.119	0.517

$T = 25 \text{ K}$

p	w	μ	k	f	α/α_0	γ/γ_0
0.01	294.3	0.74	1.67	0.01	1.001	1.001
0.02	294.4	0.73	1.67	0.02	1.001	1.002
0.03	294.5	0.73	1.67	0.03	1.002	1.002
0.04	294.6	0.73	1.67	0.04	1.003	1.003
0.05	294.7	0.73	1.67	0.05	1.004	1.004
0.06	294.8	0.72	1.67	0.06	1.004	1.005
0.07	294.9	0.72	1.67	0.07	1.005	1.005
0.08	295.0	0.72	1.68	0.08	1.006	1.006
0.09	295.2	0.72	1.68	0.09	1.006	1.007
0.10	295.3	0.71	1.68	0.10	1.007	1.008
0.2	296.4	0.69	1.69	0.20	1.014	1.015
0.3	297.6	0.66	1.70	0.30	1.020	1.023
0.4	298.9	0.64	1.72	0.40	1.026	1.030
0.5	300.2	0.61	1.73	0.50	1.031	1.037
0.6	301.6	0.59	1.74	0.60	1.036	1.045
0.7	303.1	0.56	1.76	0.70	1.041	1.052
0.8	304.6	0.54	1.77	0.80	1.045	1.059
0.9	306.2	0.52	1.79	0.91	1.048	1.065
1.0	307.8	0.49	1.80	1.01	1.051	1.072
1.5	316.5	0.38	1.89	1.52	1.060	1.104
2.0	326.1	0.29	1.97	2.05	1.060	1.132
2.5	336.2	0.20	2.06	2.58	1.052	1.156
3.0	346.6	0.13	2.15	3.14	1.039	1.177
3.5	357.4	0.06	2.24	3.71	1.020	1.193
4.0	368.3	−0.01	2.33	4.30	0.998	1.207
4.5	379.6	−0.07	2.42	4.92	0.971	1.217
5.0	391.0	−0.12	2.51	5.56	0.942	1.225
5.5	402.5	−0.17	2.60	6.22	0.911	1.229
6.0	414.2	−0.22	2.69	6.91	0.878	1.232
6.5	426.0	−0.27	2.77	7.64	0.846	1.233
7.0	437.8	−0.31	2.86	8.40	0.814	1.231
7.5	449.6	−0.35	2.94	9.19	0.783	1.229

Table II.18 (*Continued*)

p	w	μ	k	f	α/α_0	γ/γ_0
8.0	461.3	−0.39	3.03	10.02	0.753	1.225
8.5	472.9	−0.42	3.10	10.88	0.725	1.220
9.0	484.3	−0.45	3.18	11.79	0.698	1.214
9.5	495.6	−0.48	3.25	12.75	0.673	1.207
10.0	506.6	−0.50	3.32	13.75	0.649	1.200
15.0	605.4	−0.64	3.85	26.83	0.484	1.111
20.0	684.9	−0.70	4.15	47.65	0.393	1.023
25.0	750.3	−0.73	4.33	80.20	0.336	0.946
30	805.8	−0.75	4.43	130.19	0.297	0.879
35	854.1	−0.75	4.49	205.76	0.267	0.821
40	897.2	−0.76	4.53	318.50	0.244	0.771
45	936.2	−0.76	4.55	484.68	0.225	0.727
50	972.2	−0.77	4.56	727.16	0.209	0.688
55	1005.8	−0.77	4.57	1077.73	0.196	0.653
60	1037.5	−0.77	4.57	1580.49	0.184	0.623
65	1067.8	−0.77	4.57	2296.28	0.174	0.595
70	1096.9	−0.78	4.58	3308.62	0.165	0.570
75	1125.0	−0.78	4.59	4731.79	0.157	0.548
80	1152.3	−0.78	4.59	6721.45	0.149	0.527
85	1179.0	−0.79	4.61	9489.05	0.143	0.508
90	1205.2	−0.79	4.62	13320.70	0.137	0.491
95	1231.0	−0.80	4.63	18602.45	0.131	0.475
100	1255.4	−0.81	4.65	25853.50	0.126	0.460

$$T = 30 \text{ K}$$

p	w	μ	k	f	α/α_0	γ/γ_0
0.01	322.4	0.50	1.67	0.01	1.000	1.001
0.02	322.5	0.50	1.67	0.02	1.001	1.001
0.03	322.6	0.50	1.67	0.03	1.001	1.002
0.04	322.7	0.50	1.67	0.04	1.002	1.002
0.05	322.9	0.49	1.67	0.05	1.002	1.003
0.06	323.0	0.49	1.67	0.06	1.002	1.003
0.07	323.1	0.49	1.67	0.07	1.003	1.004
0.08	323.2	0.49	1.67	0.08	1.003	1.004
0.09	323.3	0.49	1.68	0.09	1.004	1.005
0.10	323.4	0.48	1.68	0.10	1.004	1.005
0.2	324.6	0.46	1.69	0.20	1.008	1.011
0.3	325.9	0.44	1.70	0.30	1.011	1.016
0.4	327.2	0.42	1.71	0.40	1.014	1.021
0.5	328.5	0.40	1.72	0.50	1.017	1.026
0.6	329.9	0.39	1.73	0.61	1.020	1.032
0.7	331.3	0.37	1.74	0.71	1.022	1.037
0.8	332.7	0.35	1.75	0.81	1.024	1.041
0.9	334.2	0.33	1.76	0.91	1.025	1.046
1.0	335.7	0.31	1.78	1.02	1.026	1.051
1.5	343.6	0.23	1.84	1.54	1.029	1.073
2.0	352.1	0.16	1.90	2.07	1.027	1.093
2.5	360.9	0.09	1.97	2.62	1.020	1.111
3.0	369.8	0.04	2.03	3.19	1.010	1.126
3.5	378.9	−0.01	2.09	3.78	0.998	1.139
4.0	388.1	−0.05	2.16	4.38	0.984	1.150
4.5	397.4	−0.09	2.22	5.00	0.967	1.160
5.0	406.7	−0.13	2.28	5.65	0.949	1.167
5.5	416.2	−0.16	2.35	6.31	0.929	1.173
6.0	425.7	−0.20	2.41	7.01	0.909	1.178
6.5	435.3	−0.23	2.47	7.72	0.887	1.181

Table II.18 (*Continued*)

p	w	μ	k	f	α/α_0	γ/γ_0
7.0	445.0	−0.26	2.53	8.47	0.864	1.184
7.5	454.7	−0.29	2.60	9.24	0.841	1.185
8.0	464.5	−0.32	2.66	10.04	0.819	1.185
8.5	474.2	−0.35	2.72	10.88	0.796	1.184
9.0	484.0	−0.37	2.78	11.74	0.774	1.182
9.5	493.7	−0.39	2.83	12.65	0.752	1.179
10.0	503.4	−0.42	2.89	13.58	0.731	1.176
15.0	594.3	−0.56	3.37	25.30	0.556	1.125
20.0	671.3	−0.62	3.69	42.64	0.465	1.060
25.0	736.1	−0.65	3.90	67.98	0.400	0.997
30	791.4	−0.66	4.03	104.51	0.354	0.939
35	839.7	−0.67	4.12	155.46	0.320	0.888
40	882.7	−0.67	4.17	229.52	0.294	0.842
45	921.7	−0.67	4.21	331.22	0.272	0.800
50	957.6	−0.67	4.24	471.49	0.254	0.764
55	991.1	−0.67	4.26	663.46	0.238	0.731
60	1022.7	−0.67	4.28	924.30	0.225	0.702
65	1053.0	−0.66	4.29	1276.45	0.214	0.675
70	1082.1	−0.66	4.31	1749.13	0.203	0.652
75	1110.3	−0.66	4.33	2380.23	0.194	0.630
80	1137.9	−0.66	4.34	3218.77	0.186	0.611
85	1164.9	−0.65	4.37	4328.02	0.179	0.594
90	1191.5	−0.65	4.39	5789.34	0.172	0.578
95	1217.7	−0.65	4.41	7707.13	0.166	0.564
100	1243.6	−0.65	4.44	10215.05	0.161	0.551

$$T = 35 \text{ K}$$

p	w	μ	k	f	α/α_0	γ/γ_0
0.01	348.2	0.32	1.67	0.01	1.000	1.000
0.02	348.3	0.31	1.67	0.02	1.000	1.001
0.03	348.4	0.31	1.67	0.03	1.001	1.001
0.04	348.6	0.31	1.67	0.04	1.001	1.002
0.05	348.7	0.31	1.67	0.05	1.001	1.002
0.06	348.8	0.31	1.67	0.06	1.001	1.002
0.07	348.9	0.31	1.67	0.07	1.002	1.003
0.08	349.0	0.30	1.67	0.08	1.002	1.003
0.09	349.2	0.30	1.67	0.09	1.002	1.004
0.10	349.3	0.30	1.67	0.10	1.002	1.004
0.2	350.5	0.29	1.68	0.20	1.004	1.008
0.3	351.7	0.27	1.69	0.30	1.006	1.012
0.4	353.0	0.25	1.70	0.40	1.007	1.016
0.5	354.2	0.24	1.71	0.50	1.009	1.019
0.6	355.5	0.22	1.72	0.61	1.010	1.023
0.7	356.9	0.21	1.73	0.71	1.010	1.027
0.8	358.2	0.19	1.74	0.81	1.011	1.030
0.9	359.6	0.18	1.75	0.92	1.012	1.034
1.0	361.0	0.17	1.76	1.02	1.012	1.037
1.5	368.3	0.10	1.80	1.55	1.011	1.053
2.0	375.9	0.04	1.85	2.09	1.006	1.068
2.5	383.7	−0.00	1.90	2.64	0.999	1.081
3.0	391.7	−0.04	1.95	3.21	0.991	1.092
3.5	399.7	−0.08	2.00	3.80	0.980	1.102
4.0	407.8	−0.11	2.05	4.40	0.969	1.111
4.5	415.9	−0.14	2.10	5.03	0.956	1.119
5.0	424.0	−0.17	2.15	5.67	0.943	1.126
5.5	432.1	−0.19	2.19	6.34	0.929	1.131

Table II.18 (*Continued*)

p	w	μ	k	f	α/α_0	γ/γ_0
6.0	440.2	−0.22	2.24	7.02	0.914	1.136
6.5	448.4	−0.24	2.29	7.73	0.898	1.140
7.0	456.6	−0.26	2.33	8.46	0.882	1.142
7.5	464.8	−0.28	2.38	9.22	0.865	1.145
8.0	473.1	−0.30	2.42	10.00	0.848	1.146
8.5	481.4	−0.32	2.47	10.81	0.831	1.147
9.0	489.8	−0.34	2.51	11.64	0.814	1.147
9.5	498.2	−0.36	2.56	12.50	0.796	1.147
10.0	506.5	−0.38	2.60	13.40	0.779	1.146
15.0	588.2	−0.51	3.01	24.21	0.628	1.117
20.0	661.4	−0.57	3.32	39.37	0.524	1.071
25.0	724.5	−0.60	3.54	60.47	0.454	1.021
30	779.2	−0.61	3.69	89.51	0.404	0.972
35	827.1	−0.62	3.79	129.03	0.366	0.927
40	869.7	−0.62	3.86	182.29	0.336	0.885
45	908.3	−0.62	3.91	253.40	0.312	0.847
50	943.8	−0.62	3.95	347.63	0.292	0.812
55	976.9	−0.61	3.98	471.58	0.275	0.781
60	1008.2	−0.61	4.00	633.59	0.260	0.752
65	1038.1	−0.61	4.03	844.17	0.247	0.727
70	1066.9	−0.60	4.05	1116.43	0.235	0.703
75	1094.9	−0.60	4.07	1466.79	0.225	0.683
80	1122.4	−0.60	4.10	1915.68	0.216	0.664
85	1149.3	−0.59	4.12	2488.58	0.207	0.647
90	1175.9	−0.59	4.15	3217.04	0.200	0.632
95	1202.3	−0.58	4.18	4140.17	0.193	0.619
100	1228.4	−0.58	4.22	5306.29	0.187	0.607
			$T = 40$ K			
0.01	372.2	0.16	1.67	0.01	1.000	1.000
0.02	372.4	0.16	1.67	0.02	1.000	1.001
0.03	372.5	0.16	1.67	0.03	1.000	1.001
0.04	372.6	0.16	1.67	0.04	1.000	1.001
0.05	372.7	0.16	1.67	0.05	1.000	1.002
0.06	372.8	0.16	1.67	0.06	1.001	1.002
0.07	372.9	0.16	1.67	0.07	1.001	1.002
0.08	373.1	0.15	1.67	0.08	1.001	1.002
0.09	373.2	0.15	1.67	0.09	1.001	1.003
0.10	373.3	0.15	1.67	0.10	1.001	1.003
0.2	374.5	0.14	1.68	0.20	1.002	1.006
0.3	375.6	0.13	1.69	0.30	1.002	1.009
0.4	376.9	0.11	1.70	0.40	1.003	1.012
0.5	378.1	0.10	1.70	0.51	1.003	1.014
0.6	379.3	0.09	1.71	0.61	1.003	1.017
0.7	380.6	0.08	1.72	0.71	1.003	1.020
0.8	381.9	0.07	1.72	0.81	1.003	1.022
0.9	383.1	0.05	1.73	0.92	1.003	1.025
1.0	384.5	0.04	1.74	1.02	1.003	1.028
1.5	391.2	−0.01	1.78	1.55	0.999	1.040
2.0	398.2	−0.05	1.82	2.09	0.993	1.050
2.5	405.3	−0.09	1.86	2.65	0.986	1.060
3.0	412.6	−0.12	1.90	3.22	0.977	1.069
3.5	419.9	−0.15	1.94	3.80	0.968	1.077
4.0	427.2	−0.18	1.98	4.41	0.957	1.084
4.5	434.5	−0.20	2.02	5.03	0.947	1.090

Table II.18 (*Continued*)

p	w	μ	k	f	α/α_0	γ/γ_0
5.0	441.8	−0.22	2.05	5.67	0.936	1.095
5.5	449.1	−0.23	2.09	6.32	0.924	1.100
6.0	456.3	−0.25	2.13	7.00	0.912	1.104
6.5	463.5	−0.27	2.17	7.70	0.900	1.107
7.0	470.8	−0.28	2.20	8.41	0.887	1.110
7.5	478.0	−0.30	2.24	9.15	0.874	1.112
8.0	485.3	−0.31	2.27	9.91	0.861	1.114
8.5	492.6	−0.33	2.31	10.70	0.847	1.115
9.0	499.9	−0.34	2.35	11.50	0.834	1.116
9.5	507.3	−0.35	2.38	12.34	0.820	1.117
10.0	514.6	−0.37	2.42	13.19	0.806	1.117
15.0	587.9	−0.48	2.76	23.33	0.673	1.101
20.0	656.6	−0.54	3.04	37.00	0.572	1.069
25.0	717.7	−0.57	3.26	55.30	0.499	1.029
30	771.4	−0.59	3.41	79.61	0.446	0.989
35	818.7	−0.59	3.53	111.60	0.405	0.949
40	861.0	−0.59	3.61	153.32	0.374	0.912
45	899.1	−0.59	3.67	207.32	0.347	0.877
50	934.2	−0.59	3.71	276.70	0.325	0.845
55	966.8	−0.59	3.75	365.27	0.307	0.815
60	997.5	−0.58	3.78	477.70	0.290	0.787
65	1026.9	−0.58	3.81	619.68	0.276	0.762
70	1055.2	−0.58	3.83	798.13	0.263	0.739
75	1082.7	−0.57	3.86	1021.45	0.251	0.719
80	1109.7	−0.57	3.89	1299.84	0.241	0.700
85	1136.4	−0.56	3.92	1645.62	0.232	0.683
90	1162.7	−0.56	3.95	2073.69	0.223	0.668
95	1189.0	−0.56	3.98	2602.02	0.216	0.655
100	1215.1	−0.55	4.02	3252.24	0.209	0.643

$$T = 45 \text{ K}$$

p	w	μ	k	f	α/α_0	γ/γ_0
0.01	394.8	0.04	1.67	0.01	1.000	1.000
0.02	394.9	0.04	1.67	0.02	1.000	1.000
0.03	395.0	0.04	1.67	0.03	1.000	1.001
0.04	395.2	0.04	1.67	0.04	1.000	1.001
0.05	395.3	0.04	1.67	0.05	1.000	1.001
0.06	395.4	0.03	1.67	0.06	1.000	1.001
0.07	395.5	0.03	1.67	0.07	1.000	1.002
0.08	395.6	0.03	1.67	0.08	1.000	1.002
0.09	395.7	0.03	1.67	0.09	1.000	1.002
0.10	395.8	0.03	1.67	0.10	1.000	1.002
0.2	397.0	0.02	1.68	0.20	1.000	1.004
0.3	398.1	0.01	1.68	0.30	1.000	1.007
0.4	399.3	−0.00	1.69	0.40	1.000	1.009
0.5	400.4	−0.01	1.70	0.51	1.000	1.011
0.6	401.6	−0.02	1.70	0.61	0.999	1.013
0.7	402.8	−0.03	1.71	0.71	0.999	1.015
0.8	404.0	−0.04	1.72	0.81	0.998	1.017
0.9	405.2	−0.05	1.72	0.92	0.998	1.019
1.0	406.5	−0.06	1.73	1.02	0.997	1.021
1.5	412.8	−0.10	1.76	1.55	0.992	1.030
2.0	419.3	−0.13	1.80	2.09	0.985	1.038
2.5	425.9	−0.17	1.83	2.65	0.977	1.045
3.0	432.6	−0.19	1.86	3.21	0.969	1.052
3.5	439.4	−0.22	1.90	3.80	0.959	1.058

Table II.18 (*Continued*)

p	w	μ	k	f	α/α_0	γ/γ_0
4.0	446.1	−0.23	1.93	4.40	0.950	1.063
4.5	452.9	−0.25	1.96	5.01	0.940	1.068
5.0	459.6	−0.27	1.99	5.64	0.930	1.072
5.5	466.2	−0.28	2.02	6.29	0.920	1.076
6.0	472.9	−0.29	2.05	6.96	0.909	1.080
6.5	479.5	−0.30	2.08	7.64	0.899	1.082
7.0	486.1	−0.32	2.11	8.35	0.888	1.085
7.5	492.7	−0.33	2.14	9.07	0.877	1.087
8.0	499.3	−0.34	2.17	9.81	0.866	1.089
8.5	505.9	−0.35	2.20	10.57	0.855	1.090
9.0	512.4	−0.36	2.23	11.35	0.844	1.091
9.5	519.0	−0.37	2.26	12.16	0.832	1.092
10.0	525.6	−0.38	2.29	12.99	0.821	1.093
15.0	592.0	−0.46	2.58	22.59	0.707	1.085
20.0	656.2	−0.52	2.83	35.15	0.610	1.061
25.0	714.9	−0.56	3.04	51.47	0.537	1.030
30	767.4	−0.57	3.20	72.54	0.482	0.997
35	814.2	−0.58	3.32	99.52	0.440	0.963
40	856.0	−0.58	3.41	133.83	0.407	0.930
45	893.8	−0.58	3.47	177.13	0.379	0.898
50	928.4	−0.58	3.52	231.45	0.356	0.868
55	960.4	−0.57	3.56	299.17	0.336	0.840
60	990.6	−0.57	3.60	383.17	0.318	0.813
65	1019.3	−0.57	3.62	486.88	0.303	0.789
70	1046.9	−0.56	3.65	614.35	0.289	0.766
75	1073.8	−0.56	3.68	770.42	0.276	0.746
80	1100.2	−0.56	3.71	960.82	0.265	0.727
85	1126.3	−0.55	3.74	1192.33	0.254	0.710
90	1152.2	−0.55	3.77	1472.97	0.245	0.695
95	1178.0	−0.54	3.80	1812.23	0.236	0.681
100	1203.8	−0.54	3.84	2221.28	0.228	0.669

$$T = 50 \text{ K}$$

p	w	μ	k	f	α/α_0	γ/γ_0
0.01	416.2	−0.06	1.67	0.01	1.000	1.000
0.02	416.3	−0.06	1.67	0.02	1.000	1.000
0.03	416.4	−0.06	1.67	0.03	1.000	1.001
0.04	416.5	−0.06	1.67	0.04	1.000	1.001
0.05	416.6	−0.07	1.67	0.05	1.000	1.001
0.06	416.7	−0.07	1.67	0.06	1.000	1.001
0.07	416.8	−0.07	1.67	0.07	1.000	1.001
0.08	416.9	−0.07	1.67	0.08	1.000	1.001
0.09	417.0	−0.07	1.67	0.09	1.000	1.002
0.10	417.1	−0.07	1.67	0.10	1.000	1.002
0.2	418.2	−0.08	1.68	0.20	0.999	1.003
0.3	419.3	−0.09	1.68	0.30	0.999	1.005
0.4	420.5	−0.09	1.69	0.40	0.998	1.007
0.5	421.6	−0.10	1.69	0.51	0.997	1.008
0.6	422.7	−0.11	1.70	0.61	0.997	1.010
0.7	423.8	−0.12	1.70	0.71	0.996	1.011
0.8	425.0	−0.12	1.71	0.81	0.995	1.013
0.9	426.2	−0.13	1.72	0.92	0.994	1.014
1.0	427.3	−0.14	1.72	1.02	0.993	1.016
1.5	433.3	−0.17	1.75	1.55	0.987	1.023
2.0	439.4	−0.20	1.78	2.09	0.980	1.029
2.5	445.6	−0.23	1.81	2.64	0.972	1.035

Table II.18 (*Continued*)

p	w	μ	k	f	α/α_0	γ/γ_0
3.0	451.9	−0.25	1.83	3.21	0.963	1.040
3.5	458.2	−0.27	1.86	3.79	0.954	1.044
4.0	464.5	−0.29	1.89	4.38	0.945	1.049
4.5	470.8	−0.30	1.92	4.99	0.936	1.052
5.0	477.1	−0.31	1.95	5.61	0.926	1.056
5.5	483.3	−0.32	1.97	6.25	0.917	1.059
6.0	489.5	−0.33	2.00	6.91	0.907	1.061
6.5	495.7	−0.34	2.03	7.58	0.898	1.064
7.0	501.8	−0.35	2.05	8.27	0.888	1.066
7.5	507.9	−0.36	2.08	8.98	0.879	1.068
8.0	514.0	−0.37	2.10	9.70	0.869	1.069
8.5	520.1	−0.37	2.13	10.44	0.860	1.070
9.0	526.1	−0.38	2.15	11.21	0.850	1.071
9.5	532.2	−0.39	2.18	11.99	0.840	1.072
10.0	538.2	−0.39	2.20	12.79	0.830	1.073
15.0	599.0	−0.46	2.45	21.96	0.731	1.069
20.0	658.9	−0.51	2.67	33.65	0.642	1.052
25.0	715.2	−0.55	2.87	48.49	0.570	1.028
30	766.4	−0.56	3.03	67.20	0.514	1.000
35	812.5	−0.57	3.15	90.65	0.471	0.971
40	853.9	−0.57	3.24	119.84	0.436	0.942
45	891.4	−0.57	3.31	155.95	0.408	0.913
50	925.6	−0.57	3.37	200.35	0.384	0.886
55	957.2	−0.57	3.41	254.68	0.363	0.859
60	986.8	−0.56	3.45	320.81	0.344	0.834
65	1014.9	−0.56	3.48	400.97	0.328	0.811
70	1041.8	−0.56	3.50	497.75	0.313	0.789
75	1067.9	−0.55	3.53	614.15	0.299	0.768
80	1093.6	−0.55	3.56	753.69	0.287	0.749
85	1118.9	−0.55	3.59	920.47	0.276	0.732
90	1144.1	−0.54	3.62	1119.23	0.265	0.717
95	1169.3	−0.54	3.65	1355.51	0.256	0.702
100	1194.6	−0.54	3.69	1635.70	0.247	0.690

$T = 55$ K

p	w	μ	k	f	α/α_0	γ/γ_0
0.01	436.5	−0.15	1.67	0.01	1.000	1.000
0.02	436.6	−0.15	1.67	0.02	1.000	1.000
0.03	436.7	−0.15	1.67	0.03	1.000	1.000
0.04	436.8	−0.15	1.67	0.04	1.000	1.001
0.05	436.9	−0.15	1.67	0.05	1.000	1.001
0.06	437.0	−0.15	1.67	0.06	1.000	1.001
0.07	437.1	−0.15	1.67	0.07	1.000	1.001
0.08	437.2	−0.15	1.67	0.08	0.999	1.001
0.09	437.3	−0.15	1.67	0.09	0.999	1.001
0.10	437.4	−0.15	1.67	0.10	0.999	1.001
0.2	438.5	−0.16	1.68	0.20	0.999	1.003
0.3	439.5	−0.16	1.68	0.30	0.998	1.004
0.4	440.6	−0.17	1.69	0.40	0.997	1.005
0.5	441.7	−0.18	1.69	0.51	0.996	1.006
0.6	442.8	−0.18	1.70	0.61	0.995	1.008
0.7	443.9	−0.19	1.70	0.71	0.994	1.009
0.8	445.0	−0.20	1.70	0.81	0.993	1.010
0.9	446.1	−0.20	1.71	0.92	0.992	1.011
1.0	447.2	−0.21	1.71	1.02	0.991	1.012
1.5	452.8	−0.24	1.74	1.55	0.984	1.018

Table II.18 (*Continued*)

p	w	μ	k	f	α/α_0	γ/γ_0
2.0	458.6	−0.26	1.76	2.09	0.977	1.022
2.5	464.4	−0.28	1.79	2.64	0.969	1.027
3.0	470.4	−0.30	1.81	3.20	0.960	1.031
3.5	476.3	−0.32	1.84	3.77	0.951	1.034
4.0	482.3	−0.33	1.86	4.36	0.942	1.038
4.5	488.2	−0.34	1.89	4.97	0.933	1.041
5.0	494.1	−0.35	1.91	5.58	0.924	1.043
5.5	500.0	−0.36	1.94	6.21	0.915	1.046
6.0	505.9	−0.37	1.96	6.86	0.907	1.048
6.5	511.7	−0.38	1.98	7.52	0.898	1.050
7.0	517.5	−0.38	2.01	8.20	0.889	1.051
7.5	523.3	−0.39	2.03	8.89	0.880	1.053
8.0	529.0	−0.39	2.05	9.60	0.872	1.054
8.5	531.7	−0.40	2.07	10.32	0.863	1.055
9.0	540.3	−0.40	2.09	11.06	0.854	1.056
9.5	546.0	−0.41	2.12	11.82	0.846	1.057
10.0	551.6	−0.41	2.14	12.60	0.837	1.057
15.0	608.0	−0.47	2.35	21.41	0.749	1.055
20.0	664.1	−0.51	2.55	32.41	0.667	1.043
25.0	717.9	−0.54	2.73	46.09	0.598	1.024
30	767.7	−0.56	2.89	63.02	0.543	1.001
35	813.0	−0.57	3.01	83.85	0.499	0.976
40	854.0	−0.57	3.11	109.33	0.463	0.951
45	891.2	−0.57	3.18	140.32	0.434	0.925
50	925.2	−0.57	3.24	177.81	0.409	0.900
55	956.5	−0.56	3.29	222.96	0.388	0.875
60	985.6	−0.56	3.32	277.08	0.369	0.852
65	1013.1	−0.56	3.35	341.69	0.352	0.829
70	1039.3	−0.56	3.38	418.54	0.336	0.808
75	1064.7	−0.55	3.41	509.63	0.322	0.788
80	1089.5	−0.55	3.43	617.26	0.309	0.769
85	1113.9	−0.55	3.46	744.08	0.297	0.752
90	1138.3	−0.54	3.49	893.11	0.286	0.736
95	1162.7	−0.54	3.52	1067.82	0.275	0.721
100	1187.2	−0.54	3.56	1272.17	0.265	0.708
			$T = 60$ K			
0.01	455.9	−0.21	1.67	0.01	1.000	1.000
0.02	456.0	−0.21	1.67	0.02	1.000	1.000
0.03	456.1	−0.21	1.67	0.03	1.000	1.000
0.04	456.2	−0.21	1.67	0.04	1.000	1.000
0.05	456.3	−0.22	1.67	0.05	1.000	1.001
0.06	456.4	−0.22	1.67	0.06	0.999	1.001
0.07	456.5	−0.22	1.67	0.07	0.999	1.001
0.08	456.6	−0.22	1.67	0.08	0.999	1.001
0.09	456.7	−0.22	1.67	0.09	0.999	1.001
0.10	456.8	−0.22	1.67	0.10	0.999	1.001
0.2	457.8	−0.22	1.68	0.20	0.998	1.002
0.3	458.8	−0.23	1.68	0.30	0.997	1.003
0.4	459.9	−0.23	1.68	0.40	0.996	1.004
0.5	460.9	−0.24	1.69	0.50	0.995	1.005
0.6	461.9	−0.24	1.69	0.61	0.994	1.006
0.7	463.0	−0.25	1.70	0.71	0.993	1.007
0.8	464.0	−0.26	1.70	0.81	0.992	1.008
0.9	465.1	−0.26	1.70	0.92	0.990	1.009

Table II.18 (*Continued*)

p	w	μ	k	f	α/α_0	γ/γ_0
1.0	466.2	−0.27	1.71	1.02	0.989	1.010
1.5	471.5	−0.29	1.73	1.55	0.982	1.014
2.0	477.0	−0.31	1.75	2.08	0.975	1.017
2.5	482.6	−0.33	1.78	2.63	0.967	1.021
3.0	488.2	−0.34	1.80	3.19	0.958	1.024
3.5	493.8	−0.36	1.82	3.76	0.950	1.027
4.0	499.5	−0.37	1.84	4.34	0.941	1.029
4.5	501.1	−0.38	1.86	4.94	0.932	1.032
5.0	510.8	−0.39	1.89	5.55	0.924	1.034
5.5	516.4	−0.40	1.91	6.17	0.915	1.036
6.0	522.0	−0.40	1.93	6.81	0.907	1.037
6.5	527.5	−0.41	1.95	7.46	0.898	1.039
7.0	533.0	−0.41	1.97	8.12	0.890	1.040
7.5	538.5	−0.42	1.99	8.80	0.882	1.041
8.0	543.9	−0.42	2.01	9.50	0.874	1.042
8.5	549.3	−0.43	2.03	10.21	0.866	1.043
9.0	554.7	−0.43	2.05	10.93	0.858	1.044
9.5	560.0	−0.43	2.07	11.67	0.850	1.045
10.0	565.4	−0.44	2.09	12.42	0.842	1.045
15.0	618.3	−0.47	2.27	20.92	0.764	1.044
20.0	671.0	−0.51	2.46	31.36	0.689	1.034
25.0	722.3	−0.54	2.63	44.12	0.622	1.019
30	770.6	−0.55	2.77	59.65	0.568	1.000
35	815.1	−0.56	2.89	78.47	0.524	0.979
40	855.8	−0.57	2.99	101.15	0.488	0.957
45	892.7	−0.57	3.07	128.35	0.458	0.934
50	926.5	−0.57	3.13	160.80	0.433	0.911
55	957.5	−0.56	3.18	199.36	0.411	0.888
60	986.3	−0.56	3.22	244.98	0.391	0.866
65	1013.3	−0.56	3.25	298.75	0.374	0.845
70	1039.0	−0.55	3.28	361.91	0.358	0.825
75	1063.6	−0.55	3.30	435.86	0.344	0.805
80	1087.6	−0.55	3.33	522.18	0.330	0.787
85	1111.2	−0.55	3.35	622.67	0.317	0.770
90	1134.7	−0.54	3.38	739.38	0.306	0.753
95	1158.2	−0.54	3.41	874.61	0.294	0.738
100	1181.8	−0.54	3.44	1030.96	0.284	0.725

$T = 65$ K

p	w	μ	k	f	α/α_0	γ/γ_0
0.01	474.5	−0.27	1.67	0.01	1.000	1.000
0.02	474.6	−0.27	1.67	0.02	1.000	1.000
0.03	474.7	−0.27	1.67	0.03	1.000	1.000
0.04	474.8	−0.27	1.67	0.04	1.000	1.000
0.05	474.9	−0.27	1.67	0.05	0.999	1.000
0.06	475.0	−0.27	1.67	0.06	0.999	1.000
0.07	475.1	−0.27	1.67	0.07	0.999	1.001
0.08	475.2	−0.27	1.67	0.08	0.999	1.001
0.09	475.3	−0.27	1.67	0.09	0.999	1.001
0.10	475.4	−0.27	1.67	0.10	0.999	1.001
0.2	476.3	−0.28	1.67	0.20	0.998	1.002
0.3	477.3	−0.28	1.68	0.30	0.997	1.002
0.4	478.3	−0.29	1.68	0.40	0.996	1.003
0.5	479.3	−0.29	1.69	0.50	0.994	1.004
0.6	480.3	−0.30	1.69	0.61	0.993	1.005
0.7	481.3	−0.30	1.69	0.71	0.992	1.005

Table II.18 (*Continued*)

p	w	μ	k	f	α/α_0	γ/γ_0
0.8	482.3	−0.30	1.70	0.81	0.991	1.006
0.9	483.4	−0.31	1.70	0.92	0.989	1.007
1.0	484.4	−0.31	1.71	1.02	0.988	1.007
1.5	489.5	−0.33	1.72	1.54	0.981	1.011
2.0	494.8	−0.35	1.74	2.08	0.974	1.014
2.5	500.1	−0.37	1.76	2.62	0.966	1.016
3.0	505.4	−0.38	1.78	3.18	0.958	1.019
3.5	510.8	−0.39	1.80	3.75	0.949	1.021
4.0	516.2	−0.40	1.82	4.33	0.941	1.023
4.5	521.6	−0.41	1.84	4.92	0.932	1.025
5.0	527.0	−0.42	1.86	5.52	0.924	1.026
5.5	532.3	−0.43	1.88	6.13	0.916	1.028
6.0	537.7	−0.43	1.90	6.76	0.908	1.029
6.5	543.0	−0.44	1.92	7.40	0.900	1.030
7.0	548.3	−0.44	1.94	8.06	0.892	1.031
7.5	553.5	−0.44	1.96	8.72	0.884	1.032
8.0	558.7	−0.45	1.98	9.40	0.877	1.033
8.5	563.9	−0.45	1.99	10.10	0.869	1.034
9.0	569.0	−0.45	2.01	10.81	0.862	1.034
9.5	574.1	−0.45	2.03	11.53	0.854	1.035
10.0	579.2	−0.46	2.05	12.27	0.847	1.035
15.0	629.3	−0.48	2.21	20.49	0.775	1.035
20.0	679.1	−0.51	2.38	30.45	0.706	1.027
25.0	728.1	−0.54	2.54	42.46	0.644	1.015
30	774.9	−0.55	2.68	56.88	0.591	0.999
35	818.5	−0.56	2.80	74.11	0.547	0.981
40	858.7	−0.57	2.90	94.62	0.510	0.961
45	895.5	−0.57	2.98	118.90	0.480	0.941
50	929.1	−0.57	3.04	147.54	0.454	0.920
55	960.0	−0.56	3.09	181.19	0.432	0.900
60	988.6	−0.56	3.13	220.55	0.413	0.879
65	1015.2	−0.56	3.16	266.44	0.395	0.859
70	1040.4	−0.55	3.19	319.77	0.379	0.840
75	1064.5	−0.55	3.22	381.55	0.364	0.821
80	1087.7	−0.55	3.24	452.93	0.351	0.803
85	1110.5	−0.55	3.26	535.19	0.338	0.786
90	1133.1	−0.54	3.29	629.75	0.325	0.770
95	1155.6	−0.54	3.31	738.23	0.314	0.755
100	1178.3	−0.54	3.34	862.43	0.303	0.741
			$T = 70$K			
0.01	492.4	−0.32	1.67	0.01	1.000	1.000
0.02	492.5	−0.32	1.67	0.02	1.000	1.000
0.03	492.6	−0.32	1.67	0.03	1.000	1.000
0.04	492.7	−0.32	1.67	0.04	1.000	1.000
0.05	492.8	−0.32	1.67	0.05	0.999	1.000
0.06	492.9	−0.32	1.67	0.06	0.999	1.000
0.07	493.0	−0.32	1.67	0.07	0.999	1.000
0.08	493.0	−0.32	1.67	0.08	0.999	1.001
0.09	493.1	−0.32	1.67	0.09	0.999	1.001
0.10	493.2	−0.32	1.67	0.10	0.999	1.001
0.2	494.2	−0.32	1.67	0.20	0.998	1.001
0.3	495.1	−0.33	1.68	0.30	0.997	1.002
0.4	496.1	−0.33	1.68	0.40	0.995	1.003
0.5	497.1	−0.33	1.68	0.50	0.994	1.003

Table II.18 (*Continued*)

p	w	μ	k	f	α/α_0	γ/γ_0
0.6	498.0	−0.34	1.69	0.61	0.993	1.004
0.7	499.0	−0.34	1.69	0.71	0.992	1.004
0.8	500.0	−0.34	1.69	0.81	0.990	1.005
0.9	501.0	−0.35	1.70	0.91	0.989	1.005
1.0	501.9	−0.35	1.70	1.02	0.988	1.006
1.5	506.9	−0.37	1.72	1.54	0.981	1.008
2.0	511.9	−0.38	1.74	2.07	0.973	1.011
2.5	517.0	−0.40	1.76	2.62	0.965	1.013
3.0	522.1	−0.41	1.77	3.17	0.957	1.015
3.5	527.3	−0.42	1.79	3.73	0.949	1.016
4.0	532.4	−0.43	1.81	4.31	0.941	1.018
4.5	537.6	−0.44	1.83	4.89	0.933	1.019
5.0	542.8	−0.45	1.85	5.49	0.925	1.021
5.5	547.9	−0.45	1.86	6.10	0.917	1.022
6.0	553.0	−0.46	1.88	6.72	0.910	1.023
6.5	558.1	−0.46	1.90	7.35	0.902	1.024
7.0	563.2	−0.46	1.92	7.99	0.894	1.025
7.5	568.2	−0.47	1.93	8.65	0.887	1.025
8.0	573.2	−0.47	1.95	9.32	0.880	1.026
8.5	578.2	−0.47	1.97	10.00	0.872	1.027
9.0	583.1	−0.47	1.98	10.69	0.865	1.027
9.5	588.0	−0.47	2.00	11.40	0.858	1.027
10.0	592.9	−0.48	2.01	12.12	0.851	1.028
15.0	640.8	−0.49	2.17	20.11	0.785	1.027
20.0	688.1	−0.52	2.32	29.67	0.722	1.021
25.0	734.9	−0.54	2.46	41.06	0.622	1.011
30	780.2	−0.55	2.59	54.56	0.611	0.997
35	822.9	−0.56	2.71	70.51	0.567	0.981
40	862.6	−0.57	2.81	89.28	0.531	0.964
45	899.1	−0.57	2.89	111.28	0.500	0.946
50	932.7	−0.57	2.96	136.95	0.475	0.928
55	963.5	−0.56	3.01	166.82	0.452	0.909
60	991.9	−0.56	3.06	201.42	0.433	0.890
65	1018.4	−0.56	3.09	241.38	0.415	0.872
70	1043.3	−0.55	3.12	287.40	0.399	0.853
75	1066.8	−0.55	3.14	340.22	0.384	0.836
80	1089.5	−0.55	3.16	400.71	0.370	0.818
85	1111.5	−0.55	3.18	469.80	0.357	0.802
90	1133.2	−0.54	3.21	548.54	0.345	0.786
95	1154.8	−0.54	3.23	638.09	0.333	0.771
100	1176.5	−0.54	3.26	739.75	0.321	0.757
			$T = 75$ K			
0.01	509.7	−0.35	1.67	0.01	1.000	1.000
0.02	509.7	−0.35	1.67	0.02	1.000	1.000
0.03	509.8	−0.35	1.67	0.03	1.000	1.000
0.04	509.9	−0.35	1.67	0.04	1.000	1.000
0.05	510.0	−0.36	1.67	0.05	0.999	1.000
0.06	510.1	−0.36	1.67	0.06	0.999	1.000
0.07	510.2	−0.36	1.67	0.07	0.999	1.000
0.08	510.3	−0.36	1.67	0.08	0.999	1.000
0.09	510.4	−0.36	1.67	0.09	0.999	1.000
0.10	510.5	−0.36	1.67	0.10	0.999	1.001
0.2	511.4	−0.36	1.67	0.20	0.998	1.001
0.3	512.3	−0.36	1.68	0.30	0.996	1.002

Table II.18 (*Continued*)

p	w	μ	k	$\hat{f}$	α/α_0	γ/γ_0
0.4	513.3	−0.37	1.68	0.40	0.995	1.002
0.5	514.2	−0.37	1.68	0.50	0.994	1.002
0.6	515.1	−0.37	1.69	0.61	0.993	1.003
0.7	516.1	−0.38	1.69	0.71	0.991	1.003
0.8	517.0	−0.38	1.69	0.81	0.990	1.004
0.9	518.0	−0.38	1.70	0.91	0.989	1.004
1.0	518.9	−0.38	1.70	1.02	0.987	1.005
1.5	523.7	−0.40	1.72	1.54	0.980	1.007
2.0	528.5	−0.41	1.73	2.07	0.973	1.009
2.5	533.4	−0.43	1.75	2.61	0.966	1.010
3.0	538.3	−0.44	1.77	3.16	0.958	1.012
3.5	543.2	−0.45	1.78	3.72	0.950	1.013
4.0	548.2	−0.45	1.80	4.29	0.942	1.014
4.5	553.2	−0.46	1.82	4.87	0.934	1.015
5.0	558.1	−0.47	1.83	5.46	0.927	1.016
5.5	563.1	−0.47	1.85	6.06	0.919	1.017
6.0	568.0	−0.48	1.86	6.68	0.912	1.018
6.5	572.9	−0.48	1.88	7.30	0.904	1.019
7.0	577.8	−0.48	1.90	7.94	0.897	1.019
7.5	582.7	−0.49	1.91	8.58	0.890	1.020
8.0	587.5	−0.49	1.93	9.24	0.883	1.020
8.5	592.3	−0.49	1.94	9.91	0.876	1.021
9.0	597.1	−0.49	1.96	10.59	0.869	1.021
9.5	601.8	−0.49	1.97	11.29	0.862	1.021
10.0	606.5	−0.49	1.99	11.99	0.856	1.022
15.0	652.5	−0.50	2.13	19.77	0.794	1.021
20.0	697.7	−0.52	2.27	28.99	0.735	1.016
25.0	742.6	−0.54	2.40	39.84	0.679	1.007
30	786.3	−0.55	2.53	52.58	0.629	0.995
35	828.0	−0.56	2.64	67.48	0.586	0.982
40	867.2	−0.57	2.74	84.85	0.550	0.967
45	903.4	−0.57	2.82	105.00	0.519	0.951
50	936.9	−0.57	2.89	128.32	0.493	0.934
55	967.7	−0.57	2.94	155.20	0.471	0.917
60	996.1	−0.56	2.99	186.09	0.451	0.900
65	1022.5	−0.56	3.02	221.46	0.434	0.883
70	1047.2	−0.55	3.05	261.87	0.418	0.866
75	1070.4	−0.55	3.08	307.90	0.403	0.849
80	1092.6	−0.55	3.10	360.18	0.389	0.832
85	1114.0	−0.55	3.12	419.45	0.376	0.816
90	1134.9	−0.54	3.14	486.49	0.363	0.801
95	1155.7	−0.54	3.16	562.16	0.351	0.786
100	1176.4	−0.54	3.18	647.42	0.340	0.772
			$T = 80$ K			
0.01	526.4	−0.39	1.67	0.01	1.000	1.000
0.02	526.5	−0.39	1.67	0.02	1.000	1.000
0.03	526.5	−0.39	1.67	0.03	1.000	1.000
0.04	526.6	−0.39	1.67	0.04	1.000	1.000
0.05	526.7	−0.39	1.67	0.05	0.999	1.000
0.06	526.8	−0.39	1.67	0.06	0.999	1.000
0.07	526.9	−0.39	1.67	0.07	0.999	1.000
0.08	527.0	−0.39	1.67	0.08	0.999	1.000
0.09	527.1	−0.39	1.67	0.09	0.999	1.000
0.10	527.2	−0.39	1.67	0.10	0.999	1.000

Table II.18 (*Continued*)

p	w	μ	k	f	α/α_0	γ/γ_0
0.2	528.1	−0.39	1.67	0.20	0.998	1.001
0.3	529.0	−0.39	1.68	0.30	0.996	1.001
0.4	529.9	−0.40	1.68	0.40	0.995	1.002
0.5	530.8	−0.40	1.68	0.50	0.994	1.002
0.6	531.7	−0.40	1.68	0.61	0.993	1.002
0.7	532.6	−0.40	1.69	0.71	0.991	1.003
0.8	533.5	−0.41	1.69	0.81	0.990	1.003
0.9	534.4	−0.41	1.69	0.91	0.989	1.003
1.0	535.3	−0.41	1.70	1.02	0.987	1.004
1.5	535.9	−0.43	1.71	1.54	0.980	1.005
2.0	544.6	−0.44	1.73	2.07	0.973	1.007
2.5	549.3	−0.45	1.74	2.61	0.966	1.008
3.0	554.0	−0.46	1.76	3.15	0.959	1.009
3.5	558.8	−0.47	1.77	3.71	0.951	1.010
4.0	563.5	−0.47	1.79	4.28	0.943	1.011
4.5	568.3	−0.48	1.80	4.85	0.936	1.012
5.0	573.1	−0.49	1.82	5.44	0.929	1.013
5.5	577.9	−0.49	1.83	6.03	0.921	1.014
6.0	582.7	−0.50	1.85	6.64	0.914	1.014
6.5	587.4	−0.50	1.86	7.26	0.907	1.015
7.0	592.1	0.50	1.88	7.88	0.900	1.015
7.5	596.8	−0.50	1.89	8.52	0.893	1.016
8.0	601.5	−0.51	1.91	9.17	0.886	1.016
8.5	606.2	−0.51	1.92	9.83	0.879	1.016
9.0	610.8	−0.51	1.94	10.50	0.873	1.016
9.5	615.4	−0.51	1.95	11.18	0.866	1.017
10.0	619.9	−0.51	1.96	11.88	0.860	1.017
15.0	664.3	−0.51	2.10	19.47	0.801	1.016
20.0	707.7	−0.53	2.22	28.39	0.746	1.011
25.0	750.8	−0.54	2.35	38.79	0.694	1.004
30	793.1	−0.55	2.47	50.89	0.646	0.994
35	833.8	−0.56	2.58	64.91	0.604	0.982
40	872.3	−0.57	2.67	81.11	0.568	0.969
45	908.3	−0.57	2.76	99.76	0.537	0.955
50	941.7	−0.57	2.83	121.16	0.511	0.940
55	972.5	−0.57	2.88	145.64	0.488	0.924
60	1000.9	−0.56	2.93	173.56	0.468	0.908
65	1027.3	−0.56	2.97	205.31	0.451	0.893
70	1051.9	−0.55	3.00	241.31	0.435	0.877
75	1074.9	−0.55	3.02	282.03	0.420	0.861
80	1096.8	−0.55	3.04	327.99	0.407	0.845
85	1117.7	−0.54	3.06	379.72	0.394	0.830
90	1138.0	−0.54	3.08	437.85	0.381	0.815
95	1158.0	−0.54	3.10	503.03	0.369	0.800
100	1177.8	−0.54	3.12	575.99	0.358	0.786
			$T = 85$ K			
0.01	542.6	−0.41	1.67	0.01	1.000	1.000
0.02	542.6	−0.41	1.67	0.02	1.000	1.000
0.03	542.7	−0.41	1.67	0.03	1.000	1.000
0.04	542.8	−0.41	1.67	0.04	1.000	1.000
0.05	542.9	−0.41	1.67	0.05	0.999	1.000
0.06	543.0	−0.41	1.67	0.06	0.999	1.000
0.07	543.1	−0.41	1.67	0.07	0.999	1.000
0.08	543.2	−0.41	1.67	0.08	0.999	1.000

Table II.18 (*Continued*)

p	w	μ	k	f	α/α_0	γ/γ_0
0.09	543.3	−0.42	1.67	0.09	0.999	1.000
0.10	543.3	−0.42	1.67	0.10	0.999	1.000
0.2	544.2	−0.42	1.67	0.20	0.998	1.001
0.3	545.1	−0.42	1.67	0.30	0.996	1.001
0.4	546.0	−0.42	1.68	0.40	0.995	1.001
0.5	546.8	−0.42	1.68	0.50	0.994	1.002
0.6	547.7	−0.43	1.68	0.61	0.993	1.002
0.7	548.6	−0.43	1.69	0.71	0.991	1.002
0.8	549.5	−0.43	1.69	0.81	0.990	1.002
0.9	550.4	−0.43	1.69	0.91	0.989	1.003
1.0	551.2	−0.44	1.69	1.02	0.987	1.003
1.5	555.7	−0.45	1.71	1.54	0.981	1.004
2.0	560.2	−0.46	1.72	2.06	0.974	1.005
2.5	564.7	−0.47	1.74	2.60	0.967	1.006
3.0	569.3	−0.48	1.75	3.15	0.959	1.007
3.5	573.9	−0.48	1.77	3.70	0.952	1.008
4.0	578.5	−0.49	1.78	4.26	0.945	1.009
4.5	583.1	−0.50	1.79	4.83	0.938	1.010
5.0	587.8	−0.50	1.81	5.42	0.930	1.010
5.5	592.4	−0.51	1.82	6.01	0.923	1.011
6.0	597.0	−0.51	1.84	6.61	0.916	1.011
6.5	601.6	−0.51	1.85	7.22	0.909	1.011
7.0	606.2	−0.52	1.86	7.84	0.903	1.012
7.5	610.7	−0.52	1.88	8.47	0.896	1.012
8.0	615.2	−0.52	1.89	9.11	0.889	1.012
8.5	619.8	−0.52	1.90	9.76	0.883	1.013
9.0	624.2	−0.52	1.92	10.42	0.877	1.013
9.5	628.7	−0.52	1.93	11.09	0.870	1.013
10.0	633.1	−0.52	1.94	11.77	0.864	1.013
15.0	676.2	−0.52	2.07	19.20	0.808	1.012
20.0	718.0	−0.53	2.19	27.85	0.756	1.008
25.0	759.5	−0.54	2.30	37.86	0.707	1.001
30	800.4	−0.55	2.41	49.41	0.661	0.992
35	840.1	−0.56	2.52	62.69	0.620	0.982
40	878.0	−0.57	2.61	77.91	0.584	0.970
45	913.6	−0.57	2.70	95.31	0.554	0.958
50	946.8	−0.57	2.77	115.13	0.527	0.944
55	977.6	−0.57	2.83	137.64	0.505	0.930
60	1006.2	−0.56	2.88	163.15	0.485	0.916
65	1032.6	−0.56	2.92	191.97	0.467	0.901
70	1057.1	−0.56	2.95	224.44	0.451	0.886
75	1080.1	−0.55	2.97	260.95	0.437	0.872
80	1101.7	−0.55	3.00	301.89	0.423	0.857
85	1122.3	−0.54	3.01	347.71	0.410	0.842
90	1142.2	−0.54	3.03	398.89	0.398	0.828
95	1161.5	−0.54	3.05	455.94	0.387	0.814
100	1180.6	−0.54	3.06	519.44	0.375	0.800

$T = 90$ K

p	w	μ	k	f	α/α_0	γ/γ_0
0.01	558.3	−0.44	1.67	0.01	1.000	1.000
0.02	558.4	−0.44	1.67	0.02	1.000	1.000
0.03	558.5	−0.44	1.67	0.03	1.000	1.000
0.04	558.5	−0.44	1.67	0.04	1.000	1.000
0.05	558.6	−0.44	1.67	0.05	0.999	1.000
0.06	558.7	−0.44	1.67	0.06	0.999	1.000

Table II.18 (Continued)

p	w	μ	k	f	α/α₀	γ/γ₀
0.07	558.8	−0.44	1.67	0.07	0.999	1.000
0.08	558.9	−0.44	1.67	0.08	0.999	1.000
0.09	559.0	−0.44	1.67	0.09	0.999	1.000
0.10	559.0	−0.44	1.67	0.10	0.999	1.000
0.2	559.9	−0.44	1.67	0.20	0.998	1.001
0.3	560.7	−0.44	1.67	0.30	0.996	1.001
0.4	561.6	−0.44	1.68	0.40	0.995	1.001
0.5	562.4	−0.45	1.68	0.50	0.994	1.001
0.6	563.3	−0.45	1.68	0.61	0.993	1.001
0.7	564.1	−0.45	1.68	0.71	0.991	1.002
0.8	565.0	−0.45	1.69	0.81	0.990	1.002
0.9	565.9	−0.45	1.69	0.91	0.989	1.002
1.0	566.7	−0.46	1.69	1.01	0.987	1.002
1.5	571.0	−0.47	1.71	1.53	0.981	1.003
2.0	575.4	−0.48	1.72	2.06	0.974	1.004
2.5	579.8	−0.48	1.73	2.60	0.967	1.005
3.0	584.2	−0.49	1.75	3.14	0.960	1.006
3.5	588.6	−0.50	1.76	3.69	0.953	1.006
4.0	593.1	−0.51	1.77	4.25	0.946	1.007
4.5	597.6	−0.51	1.79	4.82	0.939	1.007
5.0	602.0	−0.52	1.80	5.39	0.932	1.008
5.5	606.5	−0.52	1.81	5.98	0.926	1.008
6.0	611.0	−0.52	1.83	6.57	0.919	1.009
6.5	615.5	−0.53	1.84	7.18	0.912	1.009
7.0	619.9	−0.53	1.85	7.79	0.906	1.009
7.5	624.3	−0.53	1.86	8.41	0.899	1.009
8.0	628.7	−0.53	1.88	9.05	0.893	1.009
8.5	633.1	−0.53	1.89	9.69	0.887	1.010
9.0	637.5	−0.53	1.90	10.34	0.880	1.010
9.5	641.8	−0.53	1.91	11.00	0.874	1.010
10.0	646.1	−0.53	1.93	11.67	0.869	1.010
15.0	687.9	−0.53	2.04	18.96	0.814	1.008
20.0	728.4	−0.54	2.15	27.38	0.765	1.005
25.0	768.5	−0.54	2.26	37.05	0.718	0.999
30	808.1	−0.55	2.37	48.12	0.675	0.991
35	846.8	−0.56	2.47	60.76	0.635	0.982
40	884.0	−0.57	2.56	75.16	0.600	0.972
45	919.3	−0.57	2.64	91.49	0.569	0.960
50	952.3	−0.57	2.72	109.99	0.543	0.948
55	983.1	−0.57	2.78	130.87	0.520	0.936
60	1011.7	−0.56	2.83	154.38	0.500	0.922
65	1038.2	−0.56	2.87	180.79	0.482	0.909
70	1062.8	−0.56	2.90	210.39	0.466	0.895
75	1085.8	−0.55	2.93	243.47	0.452	0.881
80	1107.3	−0.55	2.95	280.38	0.439	0.868
85	1127.7	−0.54	2.97	321.46	0.426	0.854
90	1147.2	−0.54	2.99	367.11	0.414	0.840
95	1166.1	−0.54	3.00	417.73	0.403	0.826
100	1184.5	−0.54	3.02	473.78	0.392	0.813

$T = 95$ K

p	w	μ	k	f	α/α₀	γ/γ₀
0.01	573.6	−0.46	1.67	0.01	1.000	1.000
0.02	573.7	−0.46	1.67	0.02	1.000	1.000
0.03	573.7	−0.46	1.67	0.03	1.000	1.000
0.04	573.8	−0.46	1.67	0.04	1.000	1.000

Table II.18 (*Continued*)

p	w	μ	k	f	α/α_0	γ/γ_0
0.05	573.9	−0.46	1.67	0.05	0.999	1.000
0.06	574.0	−0.46	1.67	0.06	0.999	1.000
0.07	574.1	−0.46	1.67	0.07	0.999	1.000
0.08	574.2	−0.46	1.67	0.08	0.999	1.000
0.09	574.2	−0.46	1.67	0.09	0.999	1.000
0.10	574.3	−0.46	1.67	0.10	0.999	1.000
0.2	575.1	−0.46	1.67	0.20	0.998	1.000
0.3	576.0	−0.46	1.67	0.30	0.996	1.001
0.4	576.8	−0.46	1.68	0.40	0.995	1.001
0.5	577.6	−0.46	1.68	0.50	0.994	1.001
0.6	578.5	−0.47	1.68	0.61	0.993	1.001
0.7	579.3	−0.47	1.68	0.71	0.991	1.001
0.8	580.1	−0.47	1.69	0.81	0.990	1.002
0.9	580.9	−0.47	1.69	0.91	0.989	1.002
1.0	581.8	−0.47	1.69	1.01	0.988	1.002
1.5	586.0	−0.48	1.70	1.53	0.981	1.003
2.0	590.2	−0.49	1.72	2.06	0.975	1.003
2.5	594.4	−0.50	1.73	2.59	0.968	1.004
3.0	598.7	−0.51	1.74	3.13	0.961	1.005
3.5	603.0	−0.51	1.75	3.68	0.955	1.005
4.0	607.4	−0.52	1.77	4.24	0.948	1.005
4.5	611.7	−0.52	1.78	4.80	0.941	1.006
5.0	616.0	−0.53	1.79	5.37	0.935	1.006
5.5	620.4	−0.53	1.80	5.96	0.928	1.006
6.0	624.7	−0.53	1.82	6.55	0.921	1.007
6.5	629.0	−0.54	1.83	7.14	0.915	1.007
7.0	633.4	−0.54	1.84	7.75	0.909	1.007
7.5	637.7	−0.54	1.85	8.37	0.902	1.007
8.0	641.9	−0.54	1.86	8.99	0.896	1.007
8.5	646.2	−0.54	1.88	9.63	0.890	1.007
9.0	650.5	−0.54	1.89	10.27	0.884	1.007
9.5	654.7	−0.55	1.90	10.92	0.878	1.007
10.0	658.9	−0.55	1.91	11.58	0.873	1.007
15.0	699.7	−0.54	2.02	18.74	0.820	1.006
20.0	739.0	−0.54	2.13	26.95	0.773	1.002
25.0	777.8	−0.55	2.23	36.32	0.729	0.997
30	816.2	−0.56	2.33	46.98	0.687	0.990
35	853.9	−0.56	2.42	59.07	0.648	0.982
40	890.4	−0.57	2.51	72.75	0.614	0.973
45	925.2	−0.57	2.60	88.19	0.584	0.963
50	958.0	−0.57	2.67	105.56	0.557	0.952
55	988.8	−0.57	2.73	125.06	0.534	0.940
60	1017.4	−0.57	2.78	146.91	0.514	0.928
65	1044.0	−0.56	2.82	171.31	0.497	0.916
70	1068.8	−0.56	2.86	198.52	0.481	0.903
75	1091.8	−0.55	2.89	228.79	0.467	0.890
80	1113.4	−0.55	2.91	262.39	0.453	0.877
85	1133.7	−0.54	2.93	299.61	0.441	0.864
90	1152.9	−0.54	2.95	340.77	0.429	0.851
95	1171.4	−0.54	2.96	386.21	0.418	0.838
100	1189.4	−0.53	2.98	436.29	0.407	0.826

$T = 100$ K

p	w	μ	k	f	α/α_0	γ/γ_0
0.01	588.5	−0.47	1.67	0.01	1.000	1.000
0.02	588.6	−0.47	1.67	0.02	1.000	1.000

Table II.18 (*Continued*)

p	w	μ	k	f	α/α_0	γ/γ_0
0.03	588.6	−0.47	1.67	0.03	1.000	1.000
0.04	588.7	−0.47	1.67	0.04	1.000	1.000
0.05	588.8	−0.47	1.67	0.05	0.999	1.000
0.06	588.9	−0.47	1.67	0.06	0.999	1.000
0.07	589.0	−0.47	1.67	0.07	0.999	1.000
0.08	589.0	−0.47	1.67	0.08	0.999	1.000
0.09	589.1	−0.47	1.67	0.09	0.999	1.000
0.10	589.2	−0.47	1.67	0.10	0.999	1.000
0.2	590.0	−0.47	1.67	0.20	0.998	1.000
0.3	590.8	−0.48	1.67	0.30	0.996	1.000
0.4	591.6	−0.48	1.68	0.40	0.995	1.001
0.5	592.4	−0.48	1.68	0.50	0.994	1.001
0.6	593.2	−0.48	1.68	0.60	0.993	1.001
0.7	594.0	−0.48	1.68	0.71	0.992	1.001
0.8	594.8	−0.48	1.69	0.81	0.990	1.001
0.9	595.7	−0.49	1.69	0.91	0.989	1.001
1.0	596.5	−0.49	1.69	1.01	0.988	1.002
1.5	600.5	−0.49	1.70	1.53	0.982	1.002
2.0	604.6	−0.50	1.71	2.06	0.975	1.003
2.5	608.8	−0.51	1.72	2.59	0.969	1.003
3.0	612.9	−0.52	1.74	3.13	0.963	1.004
3.5	617.1	−0.52	1.75	3.67	0.956	1.004
4.0	621.3	−0.53	1.76	4.23	0.950	1.004
4.5	625.5	−0.53	1.77	4.79	0.943	1.005
5.0	629.7	−0.54	1.78	5.36	0.937	1.005
5.5	633.9	−0.54	1.80	5.93	0.930	1.005
6.0	638.2	−0.54	1.81	6.52	0.924	1.005
6.5	642.4	−0.55	1.82	7.11	0.918	1.005
7.0	646.6	−0.55	1.83	7.71	0.911	1.005
7.5	650.8	−0.55	1.84	8.32	0.905	1.005
8.0	654.9	−0.55	1.85	8.94	0.899	1.005
8.5	659.1	−0.55	1.86	9.57	0.894	1.005
9.0	663.2	−0.55	1.88	10.21	0.888	1.005
9.5	667.3	−0.55	1.89	10.85	0.882	1.005
10.0	671.4	−0.55	1.90	11.50	0.877	1.005
15.0	711.3	−0.55	2.00	18.54	0.826	1.003
20.0	749.5	−0.55	2.10	26.57	0.781	1.000
25.0	787.2	−0.55	2.20	35.67	0.739	0.995
30	824.5	−0.56	2.29	45.97	0.698	0.989
35	861.3	−0.56	2.39	57.58	0.661	0.982
40	897.0	−0.57	2.47	70.64	0.627	0.974
45	931.3	−0.57	2.55	85.30	0.597	0.964
50	963.9	−0.57	2.62	101.71	0.571	0.955
55	994.6	−0.57	2.69	120.04	0.548	0.944
60	1023.3	−0.57	2.74	140.47	0.528	0.933
65	1050.0	−0.56	2.78	163.18	0.510	0.922
70	1074.9	−0.56	2.82	188.39	0.494	0.910
75	1098.1	−0.55	2.85	216.30	0.480	0.989
80	1119.7	−0.55	2.88	247.15	0.467	0.886
85	1140.0	−0.54	2.90	281.18	0.455	0.873
90	1159.2	−0.54	2.91	318.65	0.444	0.862
95	1177.4	−0.54	2.93	359.84	0.433	0.849
100	1195.0	−0.53	2.94	405.05	0.422	0.837

Table II.18 (*Continued*)

p	w	μ	k	f	α/α_0	γ/γ_0
			$T=110$ K			
0.01	617.2	−0.50	1.67	0.01	1.000	1.000
0.02	617.3	−0.50	1.67	0.02	1.000	1.000
0.03	617.3	−0.50	1.67	0.03	1.000	1.000
0.04	617.4	−0.50	1.67	0.04	1.000	1.000
0.05	617.5	−0.50	1.67	0.05	0.999	1.000
0.06	617.6	−0.50	1.67	0.06	0.999	1.000
0.07	617.6	−0.50	1.67	0.07	0.999	1.000
0.08	617.7	−0.50	1.67	0.08	0.999	1.000
0.09	617.8	−0.50	1.67	0.09	0.999	1.000
0.10	617.9	−0.50	1.67	0.10	0.999	1.000
0.2	618.6	−0.50	1.67	0.20	0.998	1.000
0.3	619.4	−0.50	1.67	0.30	0.997	1.000
0.4	620.2	−0.50	1.67	0.40	0.995	1.000
0.5	620.9	−0.50	1.68	0.50	0.994	1.001
0.6	621.7	−0.50	1.68	0.60	0.993	1.001
0.7	622.5	−0.51	1.68	0.71	0.992	1.001
0.8	623.2	−0.51	1.68	0.81	0.991	1.001
0.9	624.0	−0.51	1.69	0.91	0.990	1.001
1.0	624.8	−0.51	1.69	1.01	0.989	1.001
1.5	628.7	−0.52	1.70	1.53	0.983	1.001
2.0	632.5	−0.52	1.71	2.05	0.977	1.002
2.5	636.5	−0.53	1.72	2.58	0.971	1.002
3.0	640.4	−0.53	1.73	3.11	0.965	1.002
3.5	644.3	−0.54	1.74	3.66	0.959	1.002
4.0	648.3	−0.54	1.75	4.21	0.953	1.002
4.5	652.3	−0.55	1.76	4.76	0.947	1.003
5.0	656.3	−0.55	1.77	5.33	0.941	1.003
5.5	660.3	−0.56	1.78	5.90	0.935	1.003
6.0	664.3	−0.56	1.79	6.47	0.929	1.003
6.5	668.3	−0.56	1.80	7.06	0.923	1.003
7.0	672.3	−0.56	1.81	7.65	0.917	1.003
7.5	676.2	−0.56	1.82	8.25	0.911	1.003
8.0	680.2	−0.57	1.83	8.86	0.906	1.003
8.5	684.2	−0.57	1.84	9.47	0.900	1.002
9.0	688.1	−0.57	1.85	10.09	0.895	1.002
9.5	692.0	−0.57	1.87	10.72	0.889	1.002
10.0	696.0	−0.57	1.88	11.36	0.884	1.002
15.0	734.1	−0.56	1.97	18.20	0.836	1.000
20.0	770.6	−0.56	2.06	25.90	0.794	0.997
25.0	806.3	−0.56	2.15	34.56	0.756	0.993
30	841.7	−0.56	2.23	44.25	0.719	0.988
35	876.7	−0.56	2.32	55.06	0.684	0.982
40	911.0	−0.57	2.40	67.11	0.652	0.975
45	944.3	−0.57	2.48	80.50	0.622	0.968
50	976.3	−0.57	2.55	95.35	0.596	0.959
55	1006.6	−0.57	2.61	111.79	0.573	0.951
60	1035.3	−0.57	2.66	129.95	0.553	0.942
65	1062.2	−0.56	2.71	149.98	0.535	0.932
70	1087.4	−0.56	2.75	172.03	0.519	0.922
75	1110.9	−0.55	2.79	196.26	0.505	0.912
80	1132.8	−0.55	2.81	222.82	0.492	0.901
85	1153.3	−0.54	2.84	251.90	0.480	0.891
90	1172.5	−0.54	2.86	283.69	0.469	0.880

Table II.18 (*Continued*)

p	w	μ	k	f	α/α_0	γ/γ_0
95	1190.6	−0.54	2.87	318.39	0.459	0.869
100	1207.9	−0.53	2.88	356.20	0.449	0.858

$T = 120$ K

p	w	μ	k	f	α/α_0	γ/γ_0
0.01	644.6	−0.52	1.67	0.01	1.000	1.000
0.02	644.7	−0.52	1.67	0.02	1.000	1.000
0.03	644.8	−0.52	1.67	0.03	1.000	1.000
0.04	644.8	−0.52	1.67	0.04	1.000	1.000
0.05	644.9	−0.52	1.67	0.05	0.999	1.000
0.06	645.0	−0.52	1.67	0.06	0.999	1.000
0.07	645.1	−0.52	1.67	0.07	0.999	1.000
0.08	645.1	−0.52	1.67	0.08	0.999	1.000
0.09	645.2	−0.52	1.67	0.09	0.999	1.000
0.10	645.3	−0.52	1.67	0.10	0.999	1.000
0.2	646.0	−0.52	1.67	0.20	0.998	1.000
0.3	646.7	−0.52	1.67	0.30	0.997	1.000
0.4	647.5	−0.52	1.67	0.40	0.996	1.000
0.5	648.2	−0.52	1.68	0.50	0.995	1.000
0.6	648.9	−0.52	1.68	0.60	0.994	1.000
0.7	649.7	−0.52	1.68	0.71	0.992	1.000
0.8	650.4	−0.52	1.68	0.81	0.991	1.000
0.9	651.1	−0.53	1.68	0.91	0.990	1.001
1.0	651.9	−0.53	1.69	1.01	0.989	1.001
1.5	655.6	−0.53	1.69	1.53	0.984	1.001
2.0	659.3	−0.54	1.70	2.05	0.978	1.001
2.5	663.0	−0.54	1.71	2.57	0.972	1.001
3.0	666.7	−0.55	1.72	3.11	0.967	1.001
3.5	670.5	−0.55	1.73	3.64	0.961	1.001
4.0	674.3	−0.56	1.74	4.19	0.956	1.001
4.5	678.0	−0.56	1.75	4.74	0.950	1.001
5.0	681.8	−0.56	1.76	5.30	0.944	1.001
5.5	685.6	−0.57	1.77	5.86	0.939	1.001
6.0	689.4	−0.57	1.78	6.43	0.933	1.001
6.5	693.3	−0.57	1.79	7.01	0.928	1.001
7.0	697.1	−0.57	1.80	7.59	0.922	1.001
7.5	700.9	−0.57	1.81	8.19	0.917	1.001
8.0	704.7	−0.58	1.82	8.78	0.912	1.001
8.5	708.4	−0.58	1.83	9.39	0.906	1.001
9.0	712.2	−0.58	1.84	10.00	0.901	1.000
9.5	716.0	−0.58	1.85	10.62	0.896	1.000
10.0	719.7	−0.58	1.86	11.24	0.891	1.000
15.0	756.4	−0.58	1.94	17.91	0.846	0.998
20.0	791.5	−0.57	2.03	25.35	0.806	0.995
25.0	825.6	−0.56	2.11	33.64	0.770	0.991
30	859.3	−0.56	2.19	42.84	0.736	0.987
35	892.7	−0.56	2.26	53.02	0.703	0.982
40	925.6	−0.57	2.34	64.27	0.673	0.976
45	957.9	−0.57	2.41	76.67	0.645	0.970
50	989.1	−0.57	2.48	90.32	0.619	0.963
55	1019.1	−0.57	2.54	105.31	0.596	0.956
60	1047.6	−0.57	2.60	121.76	0.576	0.948
65	1074.6	−0.57	2.65	139.76	0.558	0.940
70	1100.0	−0.56	2.69	159.45	0.542	0.932
75	1123.8	−0.56	2.73	180.93	0.527	0.923
80	1146.1	−0.55	2.76	204.33	0.515	0.914

Table II.18 (*Continued*)

p	w	μ	k	f	α/α_0	γ/γ_0
85	1166.9	−0.55	2.78	229.79	0.503	0.904
90	1186.4	−0.54	2.80	257.45	0.492	0.895
95	1204.7	−0.54	2.82	287.45	0.482	0.885
100	1222.0	−0.53	2.84	319.95	0.473	0.875

$T = 130$ K

p	w	μ	k	f	α/α_0	γ/γ_0
0.01	670.9	−0.53	1.67	0.01	1.000	1.000
0.02	671.0	−0.53	1.67	0.02	1.000	1.000
0.03	671.1	−0.53	1.67	0.03	1.000	1.000
0.04	671.2	−0.53	1.67	0.04	1.000	1.000
0.05	671.2	−0.53	1.67	0.05	0.999	1.000
0.06	671.3	−0.53	1.67	0.06	0.999	1.000
0.07	671.4	−0.53	1.67	0.07	0.999	1.000
0.08	671.4	−0.53	1.67	0.08	0.999	1.000
0.09	671.5	−0.53	1.67	0.09	0.999	1.000
0.10	671.6	−0.53	1.67	0.10	0.999	1.000
0.2	672.3	−0.53	1.67	0.20	0.998	1.000
0.3	673.0	−0.53	1.67	0.30	0.997	1.000
0.4	673.7	−0.53	1.67	0.40	0.996	1.000
0.5	674.4	−0.54	1.68	0.50	0.995	1.000
0.6	675.1	−0.54	1.68	0.60	0.994	1.000
0.7	675.8	−0.54	1.68	0.71	0.993	1.000
0.8	676.5	−0.54	1.68	0.81	0.992	1.000
0.9	677.2	−0.54	1.68	0.91	0.991	1.000
1.0	677.9	−0.54	1.68	1.01	0.990	1.000
1.5	681.4	−0.54	1.69	1.52	0.985	1.000
2.0	685.0	−0.55	1.70	2.04	0.979	1.000
2.5	688.5	−0.55	1.71	2.57	0.974	1.000
3.0	692.1	−0.56	1.72	3.10	0.969	1.000
3.5	695.7	−0.56	1.73	3.63	0.964	1.000
4.0	699.3	−0.56	1.74	4.17	0.958	1.000
4.5	702.9	−0.57	1.74	4.72	0.953	1.000
5.0	706.5	−0.57	1.75	5.28	0.948	1.000
5.5	710.1	−0.57	1.76	5.83	0.943	1.000
6.0	713.8	−0.58	1.77	6.40	0.937	1.000
6.5	717.4	−0.58	1.78	6.97	0.932	1.000
7.0	721.1	−0.58	1.79	7.55	0.927	1.000
7.5	724.7	−0.58	1.80	8.13	0.922	1.000
8.0	728.3	−0.58	1.81	8.72	0.917	1.000
8.5	732.0	−0.58	1.81	9.32	0.912	0.999
9.0	735.6	−0.59	1.82	9.92	0.907	0.999
9.5	739.2	−0.59	1.83	10.53	0.902	0.999
10.0	742.8	−0.59	1.84	11.14	0.898	0.999
15.0	778.1	−0.58	1.92	17.66	0.854	0.996
20.0	812.0	−0.57	2.00	24.89	0.817	0.994
25.0	844.9	−0.57	2.07	32.88	0.783	0.990
30	877.1	−0.57	2.15	41.67	0.751	0.987
35	909.1	−0.57	2.22	51.34	0.720	0.982
40	940.8	−0.57	2.29	61.95	0.692	0.977
45	971.9	−0.57	2.36	73.56	0.665	0.972
50	1002.3	−0.57	2.42	86.25	0.640	0.966
55	1031.8	−0.57	2.48	100.11	0.617	0.960
60	1060.0	−0.57	2.54	115.21	0.597	0.954
65	1087.0	−0.57	2.59	131.64	0.579	0.947
70	1112.5	−0.56	2.63	149.49	0.562	0.939

Table II.18 (*Continued*)

p	w	μ	k	f	α/α₀	γ/γ₀
75	1136.6	−0.56	2.67	168.87	0.548	0.932
80	1159.2	−0.55	2.70	189.86	0.535	0.924
85	1180.5	−0.55	2.73	212.57	0.523	0.916
90	1200.4	−0.54	2.76	237.11	0.512	0.908
95	1219.0	−0.54	2.78	263.59	0.503	0.899
100	1236.5	−0.53	2.79	292.14	0.494	0.890

$T = 140$ K

p	w	μ	k	f	α/α₀	γ/γ₀
0.01	696.3	−0.54	1.67	0.01	1.000	1.000
0.02	696.3	−0.54	1.67	0.02	1.000	1.000
0.03	696.4	−0.54	1.67	0.03	1.000	1.000
0.04	696.5	−0.54	1.67	0.04	1.000	1.000
0.05	696.5	−0.54	1.67	0.05	1.000	1.000
0.06	696.6	−0.54	1.67	0.06	0.999	1.000
0.07	696.7	−0.54	1.67	0.07	0.999	1.000
0.08	696.7	−0.54	1.67	0.08	0.999	1.000
0.09	696.8	−0.54	1.67	0.09	0.999	1.000
0.10	696.9	−0.54	1.67	0.10	0.999	1.000
0.2	697.5	−0.54	1.67	0.20	0.998	1.000
0.3	698.2	−0.54	1.67	0.30	0.997	1.000
0.4	698.9	−0.54	1.67	0.40	0.996	1.000
0.5	699.6	−0.55	1.67	0.50	0.995	1.000
0.6	700.2	−0.55	1.68	0.60	0.994	1.000
0.7	700.9	−0.55	1.68	0.70	0.993	1.000
0.8	701.6	−0.55	1.68	0.81	0.992	1.000
0.9	702.3	−0.55	1.68	0.91	0.991	1.000
1.0	702.9	−0.55	1.68	1.01	0.990	1.000
1.5	706.3	−0.55	1.69	1.52	0.985	1.000
2.0	709.7	−0.56	1.70	2.04	0.981	1.000
2.5	713.1	−0.56	1.71	2.56	0.976	1.000
3.0	716.6	−0.56	1.71	3.09	0.971	1.000
3.5	720.0	−0.57	1.72	3.62	0.966	1.000
4.0	723.4	−0.57	1.73	4.16	0.961	1.000
4.5	726.9	−0.57	1.74	4.71	0.956	1.000
5.0	730.4	−0.58	1.75	5.26	0.951	1.000
5.5	733.8	−0.58	1.75	5.81	0.946	1.000
6.0	737.3	−0.58	1.76	6.37	0.941	0.999
6.5	740.8	−0.58	1.77	6.94	0.936	0.999
7.0	744.3	−0.58	1.78	7.51	0.931	0.999
7.5	747.8	−0.59	1.79	8.08	0.927	0.999
8.0	751.3	−0.59	1.80	8.67	0.922	0.999
8.5	754.8	−0.59	1.80	9.26	0.917	0.998
9.0	758.3	−0.59	1.81	9.85	0.913	0.998
9.5	761.7	−0.59	1.82	10.45	0.908	0.998
10.0	765.2	−0.59	1.83	11.06	0.904	0.998
15.0	799.3	−0.59	1.90	17.45	0.862	0.996
20.0	832.1	−0.58	1.98	24.50	0.826	0.993
25.0	863.9	−0.57	2.05	32.23	0.794	0.990
30	895.0	−0.57	2.11	40.69	0.764	0.986
35	925.8	−0.57	2.18	49.93	0.735	0.983
40	956.2	−0.57	2.24	60.01	0.708	0.978
45	986.4	−0.57	2.31	70.98	0.682	0.974
50	1015.9	−0.57	2.37	82.90	0.658	0.969
55	1044.8	−0.57	2.43	95.84	0.636	0.964
60	1072.6	−0.57	2.48	109.86	0.616	0.958

Table II.18 (*Continued*)

p	w	μ	k	f	α/α_0	γ/γ_0
65	1099.4	−0.57	2.53	125.03	0.598	0.952
70	1124.9	−0.56	2.58	141.44	0.581	0.946
75	1149.2	−0.56	2.62	159.15	0.567	0.939
80	1172.1	−0.55	2.66	178.25	0.553	0.932
85	1193.8	−0.55	2.69	198.82	0.542	0.925
90	1214.1	−0.54	2.71	220.94	0.531	0.918
95	1233.1	−0.54	2.74	244.71	0.521	0.911
100	1251.0	−0.53	2.75	270.21	0.512	0.903

$T = 150$ K

p	w	μ	k	f	α/α_0	γ/γ_0
0.01	720.7	−0.55	1.67	0.01	1.000	1.000
0.02	720.8	−0.55	1.67	0.02	1.000	1.000
0.03	720.8	−0.55	1.67	0.03	1.000	1.000
0.04	720.9	−0.55	1.67	0.04	1.000	1.000
0.05	721.0	−0.55	1.67	0.05	1.000	1.000
0.06	721.0	−0.55	1.67	0.06	0.999	1.000
0.07	721.1	−0.55	1.67	0.07	0.999	1.000
0.08	721.2	−0.55	1.67	0.08	0.999	1.000
0.09	721.2	−0.55	1.67	0.09	0.999	1.000
0.10	721.3	−0.55	1.67	0.10	0.999	1.000
0.2	721.9	−0.55	1.67	0.20	0.998	1.000
0.3	722.6	−0.55	1.67	0.30	0.997	1.000
0.4	723.2	−0.55	1.67	0.40	0.996	1.000
0.5	723.9	−0.55	1.67	0.50	0.995	1.000
0.6	724.5	−0.55	1.68	0.60	0.994	1.000
0.7	725.2	−0.55	1.68	0.70	0.994	1.000
0.8	725.8	−0.55	1.68	0.81	0.993	1.000
0.9	726.5	−0.55	1.68	0.91	0.992	1.000
1.0	727.1	−0.56	1.68	1.01	0.991	1.000
1.5	730.4	−0.56	1.69	1.52	0.986	1.000
2.0	733.7	−0.56	1.70	2.04	0.982	1.000
2.5	736.9	−0.56	1.70	2.56	0.977	1.000
3.0	740.2	−0.57	1.71	3.08	0.972	1.000
3.5	743.5	−0.57	1.72	3.62	0.968	1.000
4.0	746.8	−0.57	1.73	4.15	0.963	1.000
4.5	750.2	−0.58	1.73	4.69	0.958	0.999
5.0	753.5	−0.58	1.74	5.24	0.954	0.999
5.5	756.8	−0.58	1.75	5.79	0.949	0.999
6.0	760.2	−0.58	1.76	6.34	0.945	0.999
6.5	763.5	−0.59	1.76	6.91	0.940	0.999
7.0	766.9	−0.59	1.77	7.47	0.936	0.999
7.5	770.3	−0.59	1.78	8.04	0.931	0.998
8.0	773.6	−0.59	1.79	8.62	0.927	0.998
8.5	777.0	−0.59	1.79	9.20	0.922	0.998
9.0	780.3	−0.59	1.80	9.79	0.918	0.998
9.5	783.7	−0.59	1.81	10.38	0.913	0.998
10.0	787.0	−0.59	1.82	10.98	0.909	0.997
15.0	820.0	−0.59	1.89	17.27	0.869	0.995
20.0	851.9	−0.59	1.96	24.16	0.835	0.992
25.0	882.7	−0.58	2.02	31.67	0.804	0.989
30	912.8	−0.57	2.08	39.85	0.776	0.986
35	942.5	−0.57	2.15	48.74	0.749	0.983
40	971.9	−0.57	2.21	58.37	0.723	0.979
45	1001.1	−0.57	2.27	68.81	0.698	0.975
50	1029.8	−0.57	2.32	80.09	0.675	0.971

Table II.18 (*Continued*)

p	w	μ	k	$\hat{f}$	α/α_0	γ/γ_0
55	1057.9	−0.57	2.38	92.27	0.653	0.966
60	1085.4	−0.57	2.43	105.41	0.634	0.962
65	1111.9	−0.56	2.48	119.57	0.615	0.956
70	1137.3	−0.56	2.53	134.81	0.599	0.951
75	1161.7	−0.56	2.57	151.18	0.584	0.945
80	1184.8	−0.55	2.61	168.76	0.571	0.939
85	1206.7	−0.55	2.64	187.62	0.559	0.933
90	1227.4	−0.54	2.67	207.81	0.548	0.927
95	1246.9	−0.54	2.70	229.42	0.538	0.920
100	1265.2	−0.53	2.72	252.52	0.529	0.914

$T = 160$ K

p	w	μ	k	$\hat{f}$	α/α_0	γ/γ_0
0.01	744.3	−0.56	1.67	0.01	1.000	1.000
0.02	744.4	−0.56	1.67	0.02	1.000	1.000
0.03	744.5	−0.56	1.67	0.03	1.000	1.000
0.04	744.5	−0.56	1.67	0.04	1.000	1.000
0.05	744.6	−0.56	1.67	0.05	1.000	1.000
0.06	744.6	−0.56	1.67	0.06	0.999	1.000
0.07	744.7	−0.56	1.67	0.07	0.999	1.000
0.08	744.8	−0.56	1.67	0.08	0.999	1.000
0.09	744.8	−0.56	1.67	0.09	0.999	1.000
0.10	744.9	−0.56	1.67	0.10	0.999	1.000
0.2	745.5	−0.56	1.67	0.20	0.998	1.000
0.3	746.2	−0.56	1.67	0.30	0.997	1.000
0.4	746.8	−0.56	1.67	0.40	0.997	1.000
0.5	747.4	−0.56	1.67	0.50	0.996	1.000
0.6	748.0	−0.56	1.67	0.60	0.995	1.000
0.7	748.7	−0.56	1.68	0.70	0.994	1.000
0.8	749.3	−0.56	1.68	0.81	0.993	1.000
0.9	749.9	−0.56	1.68	0.91	0.992	1.000
1.0	750.5	−0.56	1.68	1.01	0.991	1.000
1.5	753.7	−0.56	1.69	1.52	0.987	1.000
2.0	756.8	−0.57	1.69	2.03	0.983	1.000
2.5	760.0	−0.57	1.70	2.55	0.978	1.000
3.0	763.2	−0.57	1.71	3.08	0.974	1.000
3.5	766.4	−0.57	1.71	3.61	0.970	0.999
4.0	769.6	−0.58	1.72	4.14	0.965	0.999
4.5	772.8	−0.58	1.73	4.68	0.961	0.999
5.0	776.0	−0.58	1.74	5.22	0.956	0.999
5.5	779.2	−0.58	1.74	5.77	0.952	0.999
6.0	782.4	−0.59	1.75	6.32	0.948	0.999
6.5	785.6	−0.59	1.76	6.88	0.943	0.998
7.0	788.9	−0.59	1.76	7.44	0.939	0.998
7.5	792.1	−0.59	1.77	8.01	0.935	0.998
8.0	795.3	−0.59	1.78	8.58	0.931	0.998
8.5	798.6	−0.59	1.78	9.16	0.926	0.998
9.0	801.8	−0.60	1.79	9.74	0.922	0.997
9.5	805.1	−0.60	1.80	10.32	0.918	0.997
10.0	808.3	−0.60	1.81	10.91	0.914	0.997
15.0	840.3	−0.60	1.87	17.12	0.876	0.995
20.0	871.3	−0.59	1.94	23.86	0.843	0.992
25	901.3	−0.58	2.00	31.19	0.813	0.989
30	930.5	−0.57	2.06	39.13	0.786	0.987
35	959.3	−0.57	2.12	47.71	0.760	0.984

Table II.18 (Continued)

p	w	μ	k	f	α/α₀	γ/γ₀
40	987.7	−0.57	2.17	56.97	0.736	0.980
45	1015.9	−0.57	2.23	66.95	0.712	0.977
50	1043.8	−0.57	2.29	77.70	0.690	0.973
55	1071.3	−0.57	2.34	89.26	0.669	0.969
60	1098.2	−0.57	2.39	101.67	0.650	0.965
65	1124.4	−0.56	2.44	114.98	0.632	0.960
70	1149.6	−0.56	2.49	129.25	0.615	0.955
75	1173.9	−0.56	2.53	144.53	0.600	0.950
80	1197.2	−0.56	2.57	160.87	0.587	0.945
85	1219.3	−0.55	2.60	178.33	0.575	0.940
90	1240.4	−0.55	2.63	196.96	0.563	0.934
95	1260.2	−0.54	2.66	216.83	0.553	0.929
100	1279.0	−0.53	2.68	237.99	0.544	0.923

$T = 170$ K

p	w	μ	k	f	α/α₀	γ/γ₀
0.01	767.2	−0.56	1.67	0.01	1.000	1.000
0.02	767.3	−0.56	1.67	0.02	1.000	1.000
0.03	767.4	−0.56	1.67	0.03	1.000	1.000
0.04	767.4	−0.56	1.67	0.04	1.000	1.000
0.05	767.5	−0.56	1.67	0.05	1.000	1.000
0.06	767.5	−0.56	1.67	0.06	1.000	1.000
0.07	767.6	−0.56	1.67	0.07	0.999	1.000
0.08	767.7	−0.56	1.67	0.08	0.999	1.000
0.09	767.7	−0.56	1.67	0.09	0.999	1.000
0.10	767.8	−0.56	1.67	0.10	0.999	1.000
0.2	768.4	−0.56	1.67	0.20	0.998	1.000
0.3	769.0	−0.56	1.67	0.30	0.998	1.000
0.4	769.6	−0.56	1.67	0.40	0.997	1.000
0.5	770.2	−0.56	1.67	0.50	0.996	1.000
0.6	770.8	−0.56	1.67	0.60	0.995	1.000
0.7	771.4	−0.56	1.68	0.70	0.994	1.000
0.8	772.0	−0.56	1.68	0.81	0.993	1.000
0.9	772.6	−0.56	1.68	0.91	0.993	1.000
1.0	773.3	−0.56	1.68	1.01	0.992	1.000
1.5	776.3	−0.57	1.69	1.52	0.988	1.000
2.0	779.3	−0.57	1.69	2.03	0.984	1.000
2.5	782.4	−0.57	1.70	2.55	0.979	1.000
3.0	785.5	−0.57	1.71	3.07	0.975	0.999
3.5	788.5	−0.58	1.71	3.60	0.971	0.999
4.0	791.6	−0.58	1.72	4.13	0.967	0.999
4.5	794.7	−0.58	1.72	4.67	0.963	0.999
5.0	797.8	−0.58	1.73	5.21	0.959	0.999
5.5	800.9	−0.58	1.74	5.75	0.955	0.999
6.0	804.0	−0.59	1.74	6.30	0.951	0.998
6.5	807.1	−0.59	1.75	6.86	0.947	0.998
7.0	810.3	−0.59	1.76	7.41	0.942	0.998
7.5	813.4	−0.59	1.76	7.98	0.938	0.998
8.0	816.5	−0.59	1.77	8.54	0.934	0.998
8.5	819.7	−0.59	1.78	9.11	0.930	0.997
9.0	822.8	−0.60	1.78	9.69	0.927	0.997
9.5	825.9	−0.60	1.79	10.27	0.923	0.997
10.0	829.1	−0.60	1.70	10.86	0.919	0.997
15.0	860.1	−0.60	1.86	16.98	0.882	0.994
20.0	890.3	−0.59	1.82	23.61	0.850	0.992
25.0	919.5	−0.58	1.989	30.77	0.822	0.989

Table II.18 (*Continued*)

p	w	μ	k	f	α/α_0	γ/γ_0
30	948.0	−0.58	2.04	38.50	0.795	0.987
35	975.9	−0.57	2.09	46.81	0.771	0.984
40	1003.5	−0.57	2.15	55.76	0.748	0.981
45	1030.9	−0.57	2.20	65.36	0.725	0.978
50	1058.0	−0.56	2.25	75.65	0.704	0.975
55	1084.8	−0.56	2.30	86.68	0.684	0.971
60	1111.2	−0.56	2.35	98.47	0.665	0.967
65	1136.9	−0.56	2.40	111.08	0.647	0.963
70	1161.9	−0.56	2.45	124.54	0.630	0.959
75	1186.1	−0.56	2.49	138.91	0.615	0.955
80	1209.4	−0.56	2.53	154.21	0.602	0.950
85	1231.7	−0.55	2.56	170.51	0.589	0.946
90	1252.9	−0.55	2.59	187.86	0.578	0.941
95	1273.1	−0.54	2.62	206.29	0.568	0.935
100	1292.3	−0.54	2.65	225.87	0.559	0.930

$T = 180$ K

p	w	μ	k	f	α/α_0	γ/γ_0
0.01	789.5	−0.56	1.67	0.01	1.000	1.000
0.02	789.5	−0.56	1.67	0.02	1.000	1.000
0.03	789.6	−0.56	1.67	0.03	1.000	1.000
0.04	789.6	−0.56	1.67	0.04	1.000	1.000
0.05	789.7	−0.56	1.67	0.05	1.000	1.000
0.06	789.8	−0.56	1.67	0.06	1.000	1.000
0.07	789.8	−0.56	1.67	0.07	0.999	1.000
0.08	789.9	−0.56	1.67	0.08	0.999	1.000
0.09	789.9	−0.56	1.67	0.09	0.999	1.000
0.10	790.0	−0.56	1.67	0.10	0.999	1.000
0.2	790.6	−0.56	1.67	0.20	0.998	1.000
0.3	791.2	−0.56	1.67	0.30	0.998	1.000
0.4	791.8	−0.57	1.67	0.40	0.997	1.000
0.5	792.4	−0.57	1.67	0.50	0.996	1.000
0.6	793.0	−0.57	1.67	0.60	0.995	1.000
0.7	793.5	−0.57	1.68	0.70	0.995	1.000
0.8	794.1	−0.57	1.68	0.80	0.994	1.000
0.9	794.7	−0.57	1.68	0.91	0.993	1.000
1.0	795.3	−0.57	1.68	1.01	0.992	1.000
1.5	798.3	−0.57	1.68	1.52	0.988	1.000
2.0	801.2	−0.57	1.69	2.03	0.984	1.000
2.5	804.2	−0.57	1.70	2.55	0.980	0.999
3.0	807.1	−0.57	1.70	3.07	0.977	0.999
3.5	810.1	−0.58	1.71	3.60	0.973	0.999
4.0	813.1	−0.58	1.72	4.12	0.969	0.999
4.5	816.1	−0.58	1.72	4.66	0.965	0.999
5.0	819.1	−0.58	1.73	5.20	0.961	0.999
5.5	822.1	−0.59	1.73	5.74	0.957	0.998
6.0	825.1	−0.59	1.74	6.28	0.953	0.998

Table II.18 (*Continued*)

p	w	μ	k	f	α/α_0	γ/γ_0
6.5	828.1	−0.59	1.75	6.83	0.949	0.998
7.0	831.1	−0.59	1.75	7.39	0.946	0.998
7.5	834.2	−0.59	1.76	7.95	0.942	0.998
8.0	837.2	−0.59	1.76	8.51	0.938	0.997
8.5	840.2	−0.59	1.77	9.08	0.934	0.997
9.0	843.3	−0.60	1.78	9.65	0.930	0.997
9.5	846.3	−0.60	1.78	10.22	0.927	0.997
10.0	849.3	−0.60	1.79	10.80	0.923	0.997
15.0	879.5	−0.60	1.85	16.86	0.888	0.994
20.0	908.9	−0.60	1.91	23.38	0.857	0.992
25.0	937.5	−0.59	1.96	30.40	0.829	0.990
30	965.2	−0.58	2.02	37.94	0.804	0.987
35	992.5	−0.57	2.07	46.03	0.781	0.985
40	1019.3	−0.57	2.12	54.70	0.758	0.982
45	1046.0	−0.57	2.17	63.97	0.737	0.979
50	1072.4	−0.56	2.22	73.87	0.716	0.976
55	1098.5	−0.56	2.27	84.44	0.697	0.973
60	1124.2	−0.56	2.32	95.71	0.678	0.970
65	1149.5	−0.56	2.36	107.72	0.661	0.966
70	1174.2	−0.56	2.41	120.50	0.645	0.962
75	1198.2	−0.56	2.45	134.09	0.630	0.959
80	1221.5	−0.55	2.49	148.53	0.616	0.954
85	1243.8	−0.55	2.53	163.86	0.603	0.950
90	1265.2	−0.55	2.56	180.12	0.592	0.946
95	1285.7	−0.54	2.59	197.36	0.582	0.941
100	1305.1	−0.54	2.62	215.61	0.572	0.937
			$T=190$ K			
0.01	811.1	−0.57	1.67	0.01	1.000	1.000
0.02	811.2	−0.57	1.67	0.02	1.000	1.000
0.03	811.2	−0.57	1.67	0.03	1.000	1.000
0.04	811.3	−0.57	1.67	0.04	1.000	1.000
0.05	811.3	−0.57	1.67	0.05	1.000	1.000
0.06	811.4	−0.57	1.67	0.06	1.000	1.000
0.07	811.4	−0.57	1.67	0.07	0.999	1.000
0.08	811.5	−0.57	1.67	0.08	0.999	1.000
0.09	811.6	−0.57	1.67	0.09	0.999	1.000
0.10	811.6	−0.57	1.67	0.10	0.999	1.000
0.2	812.2	−0.57	1.67	0.20	0.999	1.000
0.3	812.8	−0.57	1.67	0.30	0.998	1.000
0.4	813.3	−0.57	1.67	0.40	0.997	1.000
0.5	813.9	−0.57	1.67	0.50	0.996	1.000
0.6	814.5	−0.57	1.67	0.60	0.996	1.000
0.7	815.1	−0.57	1.67	0.70	0.995	1.000
0.8	815.6	−0.57	1.68	0.80	0.994	1.000
0.9	816.2	−0.57	1.68	0.91	0.993	1.000
1.0	816.8	−0.57	1.68	1.01	0.993	1.000
1.5	819.6	−0.57	1.68	1.52	0.989	1.000
2.0	822.5	−0.57	1.69	2.03	0.985	1.000
2.5	825.4	−0.57	1.70	2.55	0.981	0.999
3.0	828.2	−0.58	1.70	3.07	0.978	0.999
3.5	831.1	−0.58	1.71	3.59	0.974	0.999
4.0	834.0	−0.58	1.71	4.12	0.970	0.999
4.5	836.9	−0.58	1.72	4.65	0.967	0.999
5.0	839.8	−0.58	1.72	5.19	0.963	0.999

Table II.18 (*Continued*)

p	w	μ	k	f	α/α_0	γ/γ_0
5.5	842.7	−0.59	1.73	5.72	0.959	0.998
6.0	845.7	−0.59	1.74	6.27	0.955	0.998
6.5	848.6	−0.59	1.74	6.82	0.952	0.998
7.0	851.5	−0.59	1.75	7.37	0.948	0.998
7.5	854.4	−0.59	1.75	7.92	0.945	0.998
8.0	857.4	−0.59	1.76	8.48	0.941	0.997
8.5	860.3	−0.59	1.76	9.04	0.937	0.997
9.0	863.3	−0.60	1.77	9.61	0.934	0.997
9.5	866.2	−0.60	1.78	10.18	0.930	0.997
10.0	869.2	−0.60	1.78	10.76	0.927	0.997
15.0	898.5	−0.60	1.84	16.75	0.893	0.994
20.0	927.2	−0.60	1.90	23.18	0.863	0.992
25.0	955.1	−0.59	1.95	30.07	0.836	0.990
30	982.3	−0.58	2.00	37.45	0.812	0.988
35	1008.9	−0.57	2.05	45.35	0.789	0.985
40	1035.1	−0.57	2.10	53.77	0.768	0.983
45	1061.0	−0.57	2.15	62.75	0.747	0.980
50	1086.7	−0.56	2.19	72.32	0.728	0.977
55	1112.2	−0.56	2.24	82.49	0.709	0.975
60	1137.4	−0.56	2.29	93.31	0.691	0.972
65	1162.2	−0.56	2.33	104.80	0.674	0.968
70	1186.6	−0.56	2.37	116.99	0.658	0.965
75	1210.3	−0.56	2.42	129.92	0.643	0.962
80	1233.4	−0.55	2.45	143.62	0.629	0.958
85	1255.7	−0.55	2.49	158.13	0.617	0.954
90	1277.2	−0.55	2.52	173.47	0.605	0.951
95	1297.8	−0.54	2.56	189.69	0.594	0.946
100	1317.6	−0.54	2.58	206.83	0.585	0.942
			$T=200$ K			
0.01	832.2	−0.57	1.67	0.01	1.000	1.000
0.02	832.2	−0.57	1.67	0.02	1.000	1.000
0.03	832.3	−0.57	1.67	0.03	1.000	1.000
0.04	832.3	−0.57	1.67	0.04	1.000	1.000
0.05	832.4	−0.57	1.67	0.05	1.000	1.000
0.06	832.4	−0.57	1.67	0.06	1.000	1.000
0.07	832.5	−0.57	1.67	0.07	1.000	1.000
0.08	832.6	−0.57	1.67	0.08	0.999	1.000
0.09	832.6	−0.57	1.67	0.09	0.999	1.000
0.10	832.7	−0.57	1.67	0.10	0.999	1.000
0.2	833.2	−0.57	1.67	0.20	0.999	1.000
0.3	833.8	−0.57	1.67	0.30	0.998	1.000
0.4	834.3	−0.57	1.67	0.40	0.997	1.000
0.5	834.9	−0.57	1.67	0.50	0.996	1.000
0.6	835.5	−0.57	1.67	0.60	0.996	1.000
0.7	836.0	−0.57	1.67	0.70	0.995	1.000
0.8	836.6	−0.57	1.68	0.80	0.994	1.000
0.9	837.1	−0.57	1.68	0.91	0.994	1.000
1.0	837.7	−0.57	1.68	1.01	0.993	1.000
1.5	840.5	−0.57	1.68	1.52	0.989	1.000
2.0	843.3	−0.57	1.69	2.03	0.986	0.999
2.5	846.0	−0.57	1.69	2.54	0.982	0.999
3.0	848.8	−0.58	1.70	3.06	0.979	0.999
3.5	851.6	−0.58	1.70	3.59	0.975	0.999
4.0	854.4	−0.58	1.71	4.11	0.972	0.999

Table II.18 (Continued)

p	w	μ	k	f	α/α_0	γ/γ_0
4.5	857.3	—0.58	1.72	4.64	0.968	0.999
5.0	860.1	—0.58	1.72	5.18	0.965	0.999
5.5	862.9	—0.58	1.73	5.71	0.961	0.998
6.0	865.7	—0.59	1.73	6.25	0.958	0.998
6.5	868.6	—0.59	1.74	6.80	0.954	0.998
7.0	871.4	—0.59	1.74	7.35	0.951	0.998
7.5	874.3	—0.59	1.75	7.90	0.947	0.998
8.0	877.1	—0.59	1.75	8.46	0.944	0.997
8.5	880.0	—0.59	1.76	9.02	0.940	0.997
9.0	882.8	—0.59	1.77	9.58	0.937	0.997
9.5	885.7	—0.60	1.77	10.15	0.934	0.997
10.0	888.6	—0.60	1.78	10.72	0.930	0.997
15.0	917.1	—0.60	1.83	16.65	0.898	0.994
20.0	945.2	—6.60	1.88	23.00	0.869	0.992
25.0	972.5	—0.59	1.93	29.78	0.843	0.990
30	999.1	—0.58	1.98	37.02	0.819	0.988
35	1025.1	—0.58	2.03	44.73	0.798	0.986
40	1050.7	—0.57	2.08	52.95	0.777	0.983
45	1076.0	—0.57	2.12	61.67	0.757	0.981
50	1101.1	—0.56	2.17	70.94	0.738	0.979
55	1126.0	—0.56	2.21	80.78	0.720	0.976
60	1150.6	—0.56	2.26	91.20	0.702	0.973
65	1174.9	—0.56	2.30	102.24	0.686	0.971
70	1198.9	—0.56	2.34	113.93	0.670	0.968
75	1222.4	—0.56	2.38	126.29	0.655	0.965
80	1245.3	—0.55	2.42	139.35	0.642	0.961
85	1267.5	—0.55	2.46	153.14	0.629	0.958
90	1289.0	—0.55	2.49	167.70	0.617	0.954
95	1309.7	—0.54	2.52	183.05	0.607	0.951
100	1329.6	—0.54	2.55	199.24	0.597	0.947
			$T=210$ K			
0.01	852.7	—0.57	1.67	0.01	1.000	1.000
0.02	852.8	—0.57	1.67	0.02	1.000	1.000
0.03	852.8	—0.57	1.67	0.03	1.000	1.000
0.04	852.9	—0.57	1.67	0.04	1.000	1.000
0.05	852.9	—0.57	1.67	0.05	1.000	1.000
0.06	853.0	—0.57	1.67	0.06	1.000	1.000
0.07	853.0	—0.57	1.67	0.07	1.000	1.000
0.08	853.1	—0.57	1.67	0.08	0.999	1.000
0.09	853.2	—0.57	1.67	0.09	0.999	1.000
0.10	853.2	—0.57	1.67	0.10	0.999	1.000
0.2	853.7	—0.57	1.67	0.20	0.999	1.000
0.3	854.3	—0.57	1.67	0.30	0.998	1.000
0.4	854.8	—0.57	1.67	0.40	0.997	1.000
0.5	855.4	—0.57	1.67	0.50	0.997	1.000
0.6	855.9	—0.57	1.67	0.60	0.996	1.000
0.7	856.5	—0.57	1.67	0.70	0.995	1.000
0.8	857.0	—0.57	1.67	0.80	0.995	1.000
0.9	857.5	—0.57	1.68	0.91	0.994	1.000
1.0	858.1	—0.57	1.68	1.01	0.993	1.000
1.5	860.8	—0.57	1.68	1.51	0.990	1.000
2.0	863.5	—0.57	1.69	2.03	0.987	0.999
2.5	866.2	—0.57	1.69	2.54	0.983	0.999
3.0	868.9	—0.58	1.70	3.06	0.980	0.999

Table II.18 (*Continued*)

p	w	μ	k	f	α/α_0	γ/γ_0
3.5	871.7	−0.58	1.70	3.58	0.976	0.999
4.0	874.4	−0.58	1.71	4.11	0.973	0.999
4.5	877.1	−0.58	1.71	4.63	0.970	0.999
5.0	879.9	−0.58	1.72	5.17	0.966	0.999
5.5	882.6	−0.58	1.72	5.70	0.963	0.998
6.0	885.4	−0.59	1.73	6.24	0.960	0.998
6.5	888.1	−0.59	1.73	6.78	0.956	0.998
7.0	890.9	−0.59	1.74	7.33	0.953	0.998
7.5	893.7	−0.59	1.74	7.88	0.950	0.998
8.0	896.5	−0.59	1.75	8.43	0.947	0.997
8.5	899.2	−0.59	1.75	8.99	0.943	0.997
9.0	902.0	−0.59	1.76	9.55	0.940	0.997
9.5	904.8	−0.59	1.77	10.11	0.937	0.997
10.0	907.6	−0.60	1.77	10.68	0.934	0.997
15.0	935.4	−0.60	1.82	16.56	0.903	0.995
20.0	962.8	−0.60	1.87	22.84	0.874	0.993
25.0	989.6	−0.59	1.92	29.52	0.849	0.990
30	1015.7	−0.58	1.97	36.63	0.826	0.988
35	1041.2	−0.58	2.01	44.19	0.805	0.986
40	1066.2	−0.57	2.06	52.21	0.785	0.984
45	1090.9	−0.57	2.10	60.71	0.766	0.982
50	1115.4	−0.56	2.15	69.72	0.748	0.980
55	1139.8	−0.56	2.19	79.25	0.730	0.977
60	1163.9	−0.56	2.23	89.33	0.713	0.975
65	1187.7	−0.56	2.27	99.98	0.697	0.972
70	1211.3	−0.56	2.31	111.22	0.682	0.970
75	1234.4	−0.55	2.35	123.08	0.667	0.967
80	1257.1	−0.55	2.39	135.59	0.653	0.964
85	1279.2	−0.55	2.43	148.77	0.641	0.961
90	1300.6	−0.55	2.46	162.65	0.629	0.958
95	1321.4	−0.54	2.49	177.25	0.618	0.955
100	1341.4	−0.54	2.52	192.61	0.608	0.951
		$T=220$ K				
0.01	872.8	−0.57	1.67	0.01	1.000	1.000
0.02	872.8	−0.57	1.67	0.02	1.000	1.000
0.03	872.9	−0.57	1.67	0.03	1.000	1.000
0.04	872.9	−0.57	1.67	0.04	1.000	1.000
0.05	873.0	−0.57	1.67	0.05	1.000	1.000
0.06	873.0	−0.57	1.67	0.06	1.000	1.000
0.07	873.1	−0.57	1.67	0.07	1.000	1.000
0.08	873.2	−0.57	1.67	0.08	0.999	1.000
0.09	873.2	−0.57	1.67	0.09	0.999	1.000
0.10	873.3	−0.57	1.67	0.10	0.999	1.000
0.2	873.8	−0.57	1.67	0.20	0.999	1.000
0.3	874.3	−0.57	1.67	0.30	0.998	1.000
0.4	874.8	−0.57	1.67	0.40	0.997	1.000
0.5	875.4	−0.57	1.67	0.50	0.997	1.000
0.6	875.9	−0.57	1.67	0.60	0.996	1.000
0.7	876.4	−0.57	1.67	0.70	0.995	1.000
0.8	877.0	−0.57	1.67	0.80	0.995	1.000
0.9	877.5	−0.57	1.68	0.91	0.994	1.000
1.0	878.0	−0.57	1.68	1.01	0.994	1.000
1.5	880.7	−0.57	1.68	1.51	0.990	1.000
2.0	883.3	−0.57	1.69	2.03	0.987	0.999

Table II.18 (*Continued*)

p	w	μ	k	f	a/a_0	γ/γ_0
2.5	885.9	−0.57	1.69	2.54	0.984	0.999
3.0	888.6	−0.58	1.70	3.06	0.981	0.999
3.5	891.3	−0.58	1.70	3.58	0.978	0.999
4.0	893.9	−0.58	1.71	4.10	0.974	0.999
4.5	896.6	−0.58	1.71	4.63	0.971	0.999
5.0	899.2	−0.58	1.72	5.16	0.968	0.999
5.5	901.9	−0.58	1.72	5.69	0.965	0.998
6.0	904.6	−0.58	1.73	6.23	0.962	0.998
6.5	907.3	−0.59	1.73	6.77	0.958	0.998
7.0	910.0	−0.59	1.74	7.31	0.955	0.998
7.5	912.7	−0.59	1.74	7.86	0.952	0.998
8.0	915.4	−0.59	1.75	8.41	0.949	0.997
8.5	918.1	−0.59	1.75	8.96	0.946	0.997
9.0	920.8	−0.59	1.76	9.52	0.943	0.997
9.5	923.5	−0.59	1.76	10.08	0.940	0.997
10.0	926.2	−0.59	1.77	10.65	0.936	0.997
15.0	953.4	−0.60	1.82	16.49	0.907	0.995
20.0	980.2	−0.60	1.86	22.69	0.880	0.993
25.0	1006.4	−0.59	1.91	29.29	0.855	0.991
30	1032.0	−0.59	1.96	36.28	0.833	0.989
35	1057.0	−0.58	2.00	43.70	0.812	0.987
40	1081.6	−0.57	2.04	51.55	0.793	0.985
45	1105.8	−0.57	2.08	59.85	0.774	0.983
50	1129.8	−0.56	2.13	68.63	0.757	0.981
55	1153.5	−0.56	2.17	77.89	0.740	0.979
60	1177.1	−0.56	2.21	87.67	0.723	0.976
65	1200.5	−0.56	2.25	97.97	0.707	0.974
70	1223.7	−0.56	2.29	108.82	0.692	0.972
75	1246.5	−0.55	2.33	120.25	0.678	0.969
80	1268.9	−0.55	2.36	132.27	0.665	0.967
85	1290.8	−0.55	2.40	144.91	0.652	0.964
90	1312.1	−0.55	2.43	158.19	0.640	0.961
95	1332.9	−0.54	2.46	172.14	0.629	0.958
100	1353.0	−0.54	2.49	186.79	0.619	0.955

$T = 230$ K

p	w	μ	k	f	a/a_0	γ/γ_0
0.01	892.4	−0.57	1.67	0.01	1.000	1.000
0.02	892.4	−0.57	1.67	0.02	1.000	1.000
0.03	892.5	−0.57	1.67	0.03	1.000	1.000
0.04	892.6	−0.57	1.67	0.04	1.000	1.000
0.05	892.6	−0.57	1.67	0.05	1.000	1.000
0.06	892.7	−0.57	1.67	0.06	1.000	1.000
0.07	892.7	−0.57	1.67	0.07	1.000	1.000
0.08	892.8	−0.57	1.67	0.08	1.000	1.000
0.09	892.8	−0.57	1.67	0.09	0.999	1.000
0.10	892.9	−0.57	1.67	0.10	0.999	1.000
0.2	893.4	−0.57	1.67	0.20	0.999	1.000
0.3	893.9	−0.57	1.67	0.30	0.998	1.000
0.4	894.4	−0.57	1.67	0.40	0.998	1.000
0.5	894.9	−0.57	1.67	0.50	0.997	1.000
0.6	895.4	−0.57	1.67	0.60	0.996	1.000
0.7	896.0	−0.57	1.67	0.70	0.996	1.000
0.8	896.5	−0.57	1.67	0.80	0.995	1.000
0.9	897.0	−0.57	1.68	0.90	0.994	1.000
1.0	897.5	−0.57	1.68	1.01	0.994	1.000

Table II.18 (*Continued*)

p	w	μ	k	f	α/α_0	γ/γ_0
1.5	900.1	−0.57	1.68	1.51	0.991	1.000
2.0	902.7	−0.57	1.69	2.02	0.988	0.999
2.5	905.2	−0.57	1.69	2.54	0.985	0.999
3.0	907.8	−0.58	1.69	3.05	0.982	0.999
3.5	910.4	−0.58	1.70	3.57	0.979	0.999
4.0	913.0	−0.58	1.70	4.10	0.975	0.999
4.5	915.6	−0.58	1.71	4.62	0.972	0.999
5.0	918.2	−0.58	1.71	5.15	0.969	0.999
5.5	920.8	−0.58	1.72	5.68	0.966	0.998
6.0	923.4	−0.58	1.72	6.22	0.963	0.998
6.5	926.1	−0.59	1.73	6.76	0.960	0.998
7.0	928.7	−0.59	1.73	7.30	0.957	0.998
7.5	931.3	−0.59	1.74	7.84	0.954	0.998
8.0	933.9	−0.59	1.74	8.39	0.951	0.997
8.5	936.6	−0.59	1.75	8.94	0.948	0.997
9.0	939.2	−0.59	1.75	9.50	0.945	0.997
9.5	941.9	−0.59	1.76	10.06	0.942	0.997
10.0	944.5	−0.59	1.76	10.62	0.939	0.997
15.0	971.0	−0.60	1.81	16.41	0.911	0.995
20.0	997.3	−0.60	1.86	22.56	0.884	0.993
25.0	1023.0	−0.59	1.90	29.07	0.860	0.991
30	1048.1	−0.59	1.94	35.97	0.838	0.989
35	1072.7	−0.58	1.99	43.26	0.818	0.987
40	1096.8	−0.57	2.03	50.95	0.800	0.986
45	1120.5	−0.57	2.07	59.08	0.782	0.984
50	1144.0	−0.56	2.11	67.65	0.765	0.982
55	1167.3	−0.56	2.15	76.67	0.748	0.980
60	1190.4	−0.56	2.19	86.17	0.732	0.978
65	1213.4	−0.56	2.22	96.17	0.717	0.976
70	1236.1	−0.55	2.26	106.67	0.702	0.973
75	1258.5	−0.55	2.30	117.71	0.688	0.971
80	1280.6	−0.55	2.34	129.31	0.675	0.969
85	1302.3	−0.55	2.37	141.47	0.662	0.966
90	1323.5	−0.55	2.40	154.23	0.651	0.964
95	1344.2	−0.54	2.44	167.61	0.640	0.961
100	1364.3	−0.54	2.47	181.62	0.629	0.958
$T = 240$ K						
0.01	911.6	−0.57	1.67	0.01	1.000	1.000
0.02	911.6	−0.57	1.67	0.02	1.000	1.000
0.03	911.7	−0.57	1.67	0.03	1.000	1.000
0.04	911.7	−0.57	1.67	0.04	1.000	1.000
0.05	911.8	−0.57	1.67	0.05	1.000	1.000
0.06	911.8	−0.57	1.67	0.06	1.000	1.000
0.07	911.9	−0.57	1.67	0.07	1.000	1.000
0.08	911.9	−0.57	1.67	0.08	1.000	1.000
0.09	912.0	−0.57	1.67	0.09	0.999	1.000
0.10	912.0	−0.57	1.67	0.10	0.999	1.000
0.2	912.5	−0.57	1.67	0.20	0.999	1.000
0.3	913.1	−0.57	1.67	0.30	0.998	1.000
0.4	913.6	−0.57	1.67	0.40	0.998	1.000
0.5	914.1	−0.57	1.67	0.50	0.997	1.000
0.6	914.6	−0.57	1.67	0.60	0.996	1.000
0.7	915.1	−0.57	1.67	0.70	0.996	1.000
0.8	915.6	−0.57	1.67	0.80	0.995	1.000

Table II.18 (*Continued*)

p	w	μ	k	f	α/α_0	γ/γ_0
0.9	916.1	−0.57	1.67	0.90	0.995	1.000
1.0	916.6	−0.57	1.68	1.01	0.994	1.000
1.5	919.1	−0.57	1.68	1.51	0.991	1.000
2.0	921.6	−0.57	1.68	2.02	0.988	0.999
2.5	924.1	−0.57	1.69	2.54	0.985	0.999
3.0	926.7	−0.58	1.69	3.05	0.982	0.999
3.5	929.2	−0.58	1.70	3.57	0.979	0.999
4.0	931.7	−0.58	1.70	4.09	0.976	0.999
4.5	934.3	−0.58	1.71	4.62	0.974	0.999
5.0	936.8	−0.58	1.71	5.14	0.971	0.999
5.5	939.3	−0.58	1.72	5.68	0.968	0.998
6.0	941.9	−0.58	1.72	6.21	0.965	0.998
6.5	944.5	−0.58	1.73	6.75	0.962	0.998
7.0	947.0	−0.59	1.73	7.29	0.959	0.998
7.5	949.6	−0.59	1.73	7.83	0.956	0.998
8.0	952.2	−0.59	1.74	8.37	0.953	0.998
8.5	954.7	−0.59	1.74	8.92	0.950	0.997
9.0	957.3	−0.59	1.75	9.48	0.947	0.997
9.5	959.9	−0.59	1.75	10.03	0.945	0.997
10.0	962.5	−0.59	1.76	10.59	0.942	0.997
15.0	988.4	−0.60	1.80	16.35	0.914	0.995
20.0	1014.1	−0.60	1.85	22.44	0.889	0.993
25.0	1039.3	−0.59	1.89	28.88	0.865	0.991
30	1064.0	−0.59	1.93	35.68	0.844	0.990
35	1088.2	−0.58	1.97	42.85	0.824	0.988
40	1111.8	−0.57	2.01	50.42	0.806	0.986
45	1135.1	−0.57	2.65	58.38	0.789	0.984
50	1158.2	−0.56	2.09	66.76	0.772	0.983
55	1181.0	−0.56	2.13	75.57	0.756	0.981
60	1203.7	−0.56	2.17	84.83	0.741	0.979
65	1226.2	−0.56	2.20	94.55	0.726	0.977
70	1248.5	−0.55	2.24	104.75	0.712	0.975
75	1270.6	−0.55	2.28	115.44	0.698	0.973
80	1292.4	−0.55	2.31	126.65	0.685	0.971
85	1313.8	−0.55	2.35	138.40	0.672	0.968
90	1334.9	−0.55	2.38	150.69	0.661	0.966
95	1355.5	−0.54	2.41	163.56	0.650	0.964
100	1375.5	−0.54	2.44	177.02	0.639	0.961
			$T=250$ K			
0.01	930.4	−0.57	1.67	0.01	1.000	1.000
0.02	930.4	−0.57	1.67	0.02	1.000	1.000
0.03	930.5	−0.57	1.67	0.03	1.000	1.000
0.04	930.5	−0.57	1.67	0.04	1.000	1.000
0.05	930.6	−0.57	1.67	0.05	1.000	1.000
0.06	930.6	−0.57	1.67	0.06	1.000	1.000
0.07	930.7	−0.57	1.67	0.07	1.000	1.000
0.08	930.7	−0.57	1.67	0.08	1.000	1.000
0.09	930.8	−0.57	1.67	0.09	0.999	1.000
0.10	930.8	−0.57	1.67	0.10	0.999	1.000
0.2	931.3	−0.57	1.67	0.20	0.999	1.000
0.3	931.8	−0.57	1.67	0.30	0.998	1.000
0.4	932.3	−0.57	1.67	0.40	0.998	1.000
0.5	932.8	−0.57	1.67	0.50	0.997	1.000
0.6	933.3	−0.57	1.67	0.60	0.997	1.000

Table II.18 (*Continued*)

p	w	μ	k	f	α/α_0	γ/γ_0
0.7	933.8	−0.57	1.67	0.70	0.996	1.000
0.8	934.3	−0.57	1.67	0.80	0.995	1.000
0.9	934.8	−0.57	1.67	0.90	0.995	1.000
1.0	935.3	−0.57	1.68	1.01	0.994	1.000
1.5	937.7	−0.57	1.68	1.51	0.991	1.000
2.0	940.2	−0.57	1.68	2.02	0.989	0.999
2.5	942.7	−0.57	1.69	2.53	0.986	0.999
3.0	945.1	−0.57	1.69	3.05	0.983	0.999
3.5	947.6	−0.58	1.70	3.57	0.980	0.999
4.0	950.1	−0.58	1.70	4.09	0.977	0.999
4.5	952.6	−0.58	1.71	4.61	0.975	0.999
5.0	955.0	−0.58	1.71	5.14	0.972	0.999
5.5	957.5	−0.58	1.71	5.67	0.969	0.998
6.0	960.0	−0.58	1.72	6.20	0.966	0.998
6.5	962.5	−0.58	1.72	6.74	0.963	0.998
7.0	965.0	−0.58	1.73	7.27	0.961	0.998
7.5	967.5	−0.58	1.73	7.81	0.958	0.998
8.0	970.0	−0.59	1.74	8.36	0.955	0.998
8.5	972.5	−0.59	1.74	8.90	0.952	0.997
9.0	975.1	−0.59	1.74	9.45	0.950	0.997
9.5	977.6	−0.59	1.75	10.01	0.947	0.997
10.0	980.1	−0.59	1.75	10.56	0.944	0.997
15.0	1005.4	−0.60	1.80	16.29	0.918	0.995
20.0	1030.6	−0.60	1.84	22.33	0.893	0.994
25.0	1055.4	−0.59	1.88	28.70	0.870	0.992
30	1079.7	−0.59	1.92	35.42	0.849	0.990
35	1103.4	−0.58	1.96	42.49	0.830	0.988
40	1126.7	−0.57	2.00	49.93	0.812	0.987
45	1149.6	−0.57	2.04	57.74	0.795	0.985
50	1172.3	−0.56	2.07	65.95	0.779	0.983
55	1194.7	−0.56	2.11	74.57	0.764	0.982
60	1216.9	−0.56	2.15	83.61	0.749	0.980
65	1239.0	−0.55	2.18	93.08	0.734	0.978
70	1260.9	−0.55	2.22	103.00	0.720	0.976
75	1282.7	−0.55	2.25	113.39	0.707	0.974
80	1304.1	−0.55	2.29	124.26	0.694	0.972
85	1325.3	−0.55	2.32	135.63	0.682	0.970
90	1346.2	−0.55	2.35	147.51	0.670	0.968
95	1366.6	−0.54	2.39	159.93	0.659	0.966
100	1386.6	−0.54	2.42	172.89	0.649	0.964
			$T = 260$ K			
0.01	948.8	−0.57	1.67	0.01	1.000	1.000
0.02	948.9	−0.57	1.67	0.02	1.000	1.000
0.03	948.9	−0.57	1.67	0.03	1.000	1.000
0.04	949.0	−0.57	1.67	0.04	1.000	1.000
0.05	949.0	−0.57	1.67	0.05	1.000	1.000
0.06	949.0	−0.57	1.67	0.06	1.000	1.000
0.07	949.1	−0.57	1.67	0.07	1.000	1.000
0.08	949.1	−0.57	1.67	0.08	1.000	1.000
0.09	949.2	−0.57	1.67	0.09	1.000	1.000
0.10	949.2	−0.57	1.67	0.10	0.999	1.000
0.2	949.7	−0.57	1.67	0.20	0.999	1.000
0.3	950.2	−0.57	1.67	0.30	0.998	1.000
0.4	950.7	−0.57	1.67	0.40	0.998	1.000

Table II.18 (*Continued*)

p	w	μ	k	j	α/α_0	γ/γ_0
0.5	951.2	−0.57	1.67	0.50	0.997	1.000
0.6	951.7	−0.57	1.67	0.60	0.997	1.000
0.7	952.1	−0.57	1.67	0.70	0.996	1.000
0.8	952.6	−0.57	1.67	0.80	0.996	1.000
0.9	953.1	−0.57	1.67	0.90	0.995	1.000
1.0	953.6	−0.57	1.67	1.01	0.995	1.000
1.5	956.0	−0.57	1.68	1.51	0.992	1.000
2.0	958.4	−0.57	1.68	2.02	0.989	0.999
2.5	960.8	−0.57	1.69	2.53	0.986	0.999
3.0	963.2	−0.57	1.69	3.05	0.984	0.999
3.5	965.7	−0.57	1.70	3.56	0.981	0.999
4.0	968.1	−0.58	1.70	4.08	0.978	0.999
4.5	970.5	−0.58	1.70	4.61	0.976	0.999
5.0	972.9	−0.58	1.71	5.13	0.973	0.999
5.5	975.4	−0.58	1.71	5.66	0.970	0.998
6.0	977.8	−0.58	1.72	6.19	0.968	0.998
6.5	980.2	−0.58	1.72	6.73	0.965	0.998
7.0	982.7	−0.58	1.72	7.26	0.962	0.998
7.5	985.1	−0.58	1.73	7.80	0.960	0.998
8.0	987.6	−0.58	1.73	8.34	0.957	0.998
8.5	990.1	−0.59	1.74	8.89	0.954	0.998
9.0	992.5	−0.59	1.74	9.44	0.952	0.997
9.5	995.0	−0.59	1.75	9.99	0.949	0.997
10.0	997.4	−0.59	1.75	10.54	0.946	0.997
15.0	1022.2	−0.59	1.79	16.23	0.921	0.995
20.0	1046.9	−0.60	1.83	22.23	0.897	0.994
25.0	1071.3	−0.59	1.87	28.54	0.875	0.992
30	1095.2	−0.59	1.91	35.18	0.854	0.991
35	1118.5	−0.58	1.95	42.15	0.836	0.989
40	1141.5	−0.58	1.99	49.48	0.818	0.987
45	1164.0	−0.57	2.02	57.16	0.802	0.986
50	1186.3	−0.56	2.06	65.22	0.786	0.984
55	1208.3	−0.56	2.10	73.66	0.771	0.983
60	1230.1	−0.56	2.13	82.50	0.756	0.981
65	1251.8	−0.55	2.17	91.75	0.742	0.979
70	1273.4	−0.55	2.20	101.42	0.729	0.977
75	1294.7	−0.55	2.23	111.53	0.716	0.976
80	1315.9	−0.55	2.27	122.09	0.703	0.974
85	1336.8	−0.55	2.30	133.12	0.691	0.972
90	1357.4	−0.54	2.33	144.64	0.679	0.970
95	1377.7	−0.54	2.36	156.65	0.668	0.968
100	1397.6	−0.54	2.39	169.17	0.658	0.966
			$T=270$ K			
0.01	966.9	−0.57	1.67	0.01	1.000	1.000
0.02	966.9	−0.57	1.67	0.02	1.000	1.000
0.03	967.0	−0.57	1.67	0.03	1.000	1.000
0.04	967.0	−0.57	1.67	0.04	1.000	1.000
0.05	967.1	−0.57	1.67	0.05	1.000	1.000
0.06	967.1	−0.57	1.67	0.06	1.000	1.000
0.07	967.2	−0.57	1.67	0.07	1.000	1.000
0.08	967.2	−0.57	1.67	0.08	1.000	1.000
0.09	967.3	−0.57	1.67	0.09	1.000	1.000
0.10	967.3	−0.57	1.67	0.10	0.999	1.000
0.2	967.8	−0.57	1.67	0.20	0.999	1.000

p	w	μ	k	$\hat{f}$	α/α_0	γ/γ_0
0.3	968.3	—0.57	1.67	0.30	0.998	1.000
0.4	968.7	—0.57	1.67	0.40	0.998	1.000
0.5	969.2	—0.57	1.67	0.50	0.997	1.000
0.6	969.7	—0.57	1.67	0.60	0.997	1.000
0.7	970.1	—0.57	1.67	0.70	0.996	1.000
0.8	970.6	—0.57	1.67	0.80	0.996	1.000
0.9	971.1	—0.57	1.67	0.90	0.995	1.000
1.0	971.6	—0.57	1.67	1.01	0.995	1.000
1.5	973.9	—0.57	1.68	1.51	0.992	1.000
2.0	976.3	—0.57	1.68	2.02	0.990	1.000
2.5	978.7	—0.57	1.69	2.53	0.987	0.999
3.0	981.0	—0.57	1.69	3.05	0.984	0.999
3.5	983.4	—0.57	1.69	3.56	0.982	0.999
4.0	985.8	—0.57	1.70	4.08	0.979	0.999
4.5	988.1	—0.58	1.70	4.60	0.977	0.999
5.0	990.5	—0.58	1.71	5.13	0.974	0.999
5.5	992.9	—5.58	1.71	5.65	0.971	0.999
6.0	995.3	—0.58	1.71	6.18	0.969	0.998
6.5	997.7	—0.58	1.72	6.72	0.966	0.998
7.0	1000.1	—0.58	1.72	7.25	0.964	0.998
7.5	1002.5	—0.58	1.73	7.79	0.961	0.998
8.0	1004.9	—0.58	1.73	8.33	0.959	0.998
8.5	1007.3	—0.58	1.73	8.87	0.956	0.998
9.0	1009.7	—0.58	1.74	9.42	0.953	0.997
9.5	1012.1	—0.59	1.74	9.97	0.951	0.997
10.0	1014.5	—0.59	1.75	10.52	0.948	0.997
15.0	1038.8	—0.59	1.79	16.18	0.924	0.996
20.0	1063.0	—0.60	1.83	22.14	0.900	0.994
25.0	1086.9	—0.59	1.87	28.39	0.879	0.992
30	1110.4	—0.59	1.90	34.96	0.859	0.991
35	1133.5	—0.58	1.94	41.85	0.841	0.989
40	1156.0	—0.58	1.98	49.07	0.823	0.988
45	1178.2	—0.57	2.01	56.63	0.807	0.986
50	1200.1	—0.56	2.05	64.55	0.792	0.985
55	1221.8	—0.56	2.08	72.83	0.778	0.983
60	1243.3	—0.56	2.11	81.49	0.763	0.982
65	1264.6	—0.55	2.15	90.53	0.750	0.980
70	1285.8	—0.55	2.18	99.98	0.736	0.979
75	1306.8	—0.55	2.21	109.84	0.724	0.977
80	1327.6	—0.55	2.25	120.12	0.711	0.975
85	1348.3	—0.55	2.28	130.85	0.699	0.974
90	1368.6	—0.54	2.31	142.03	0.688	0.972
95	1388.7	—0.54	2.34	153.67	0.677	0.970
100	1408.4	—0.54	2.37	165.80	0.667	0.968
			$T=280$ K			
0.01	984.6	—0.57	1.67	0.01	1.000	1.000
0.02	984.7	—0.57	1.67	0.02	1.000	1.000
0.03	984.7	—0.57	1.67	0.03	1.000	1.000
0.04	984.8	—0.57	1.67	0.04	1.000	1.000
0.05	984.8	—0.57	1.67	0.05	1.000	1.000
0.06	984.9	—0.57	1.67	0.06	1.000	1.000
0.07	984.9	—0.57	1.67	0.07	1.000	1.000
0.08	984.9	—0.57	1.67	0.08	1.000	1.000
0.09	985.0	—0.57	1.67	0.09	1.000	1.000

Table II.18 (*Continued*)

p	w	μ	k	f	α/α_0	γ/γ_0
0.10	985.0	−0.57	1.67	0.10	0.999	1.000
0.2	985.5	−0.57	1.67	0.20	0.999	1.000
0.3	986.0	−0.57	1.67	0.30	0.998	1.000
0.4	986.4	−0.57	1.67	0.40	0.998	1.000
0.5	986.9	−0.57	1.67	0.50	0.997	1.000
0.6	987.4	−0.57	1.67	0.60	0.997	1.000
0.7	987.8	−0.57	1.67	0.70	0.996	1.000
0.8	988.3	−0.57	1.67	0.80	0.996	1.000
0.9	988.7	−0.57	1.67	0.90	0.995	1.000
1.0	989.2	−0.57	1.67	1.00	0.995	1.000
1.5	991.5	−0.57	1.68	1.51	0.992	1.000
2.0	993.8	−0.57	1.68	2.02	0.990	1.000
2.5	996.2	−0.57	1.69	2.53	0.987	0.999
3.0	998.5	−0.57	1.69	3.04	0.985	0.999
3.5	1000.8	−0.57	1.69	3.56	0.982	0.999
4.0	1003.1	−0.57	1.70	4.08	0.980	0.999
4.5	1005.5	−0.57	1.70	4.60	0.977	0.999
5.0	1007.8	−0.58	1.70	5.12	0.975	0.999
5.5	1010.1	−0.58	1.71	5.65	0.972	0.999
6.0	1012.5	−0.58	1.71	6.18	0.970	0.998
6.5	1014.8	−0.58	1.72	6.71	0.967	0.998
7.0	1017.1	−0.58	1.72	7.24	0.965	0.998
7.5	1019.5	−0.58	1.72	7.78	0.963	0.998
8.0	1021.8	−0.58	1.73	8.32	0.960	0.998
8.5	1024.2	−0.58	1.73	8.86	0.958	0.998
9.0	1026.6	−0.58	1.74	9.40	0.955	0.998
9.5	1028.9	−0.58	1.74	9.95	0.953	0.997
10.0	1031.3	−0.58	1.74	10.50	0.950	0.997
15.0	1055.1	−0.59	1.78	16.14	0.926	0.996
20.0	1078.8	−0.59	1.82	22.05	0.904	0.994
25.0	1102.4	−0.59	1.86	28.26	0.883	0.993
30	1125.5	−0.59	1.90	34.76	0.863	0.991
35	1148.2	−0.58	1.93	41.56	0.845	0.990
40	1170.4	−0.58	1.97	48.69	0.829	0.988
45	1192.3	−0.57	2.00	56.14	0.813	0.987
50	1213.9	−0.57	2.03	63.93	0.798	0.986
55	1235.2	−0.56	2.07	72.07	0.784	0.984
60	1256.3	−0.56	2.10	80.56	0.770	0.983
65	1277.3	−0.55	2.13	89.42	0.757	0.981
70	1298.1	−0.55	2.16	98.66	0.744	0.980
75	1318.8	−0.55	2.20	108.29	0.731	0.978
80	1339.4	−0.55	2.23	118.32	0.719	0.977
85	1359.7	−0.55	2.26	128.77	0.707	0.975
90	1379.8	−0.54	2.29	139.65	0.696	0.973
95	1399.7	−0.54	2.32	150.96	0.685	0.972
100	1419.3	−0.54	2.35	162.73	0.675	0.970

$T = 290$ K

p	w	μ	k	f	α/α_0	γ/γ_0
0.01	1002.0	−0.57	1.67	0.01	1.000	1.000
0.02	1002.1	−0.57	1.67	0.02	1.000	1.000
0.03	1002.1	−0.57	1.67	0.03	1.000	1.000
0.04	1002.2	−0.57	1.67	0.04	1.000	1.000
0.05	1002.2	−0.57	1.67	0.05	1.000	1.000
0.06	1002.3	−0.57	1.67	0.06	1.000	1.000
0.07	1002.3	−0.57	1.67	0.07	1.000	1.000

Table II.18 (*Continued*)

p	w	μ	k	j	α/α_0	γ/γ_0
0.08	1002.4	−0.57	1.67	0.08	1.000	1.000
0.09	1002.4	−0.57	1.67	0.09	1.000	1.000
0.10	1002.5	−0.57	1.67	0.10	1.000	1.000
0.2	1002.9	−0.57	1.67	0.20	0.999	1.000
0.3	1003.4	−0.57	1.67	0.30	0.999	1.000
0.4	1003.8	−0.57	1.67	0.40	0.998	1.000
0.5	1004.3	−0.57	1.67	0.50	0.998	1.000
0.6	1004.7	−0.57	1.67	0.60	0.997	1.000
0.7	1005.2	−0.57	1.67	0.70	0.997	1.000
0.8	1005.6	−0.57	1.67	0.80	0.996	1.000
0.9	1006.1	−0.57	1.67	0.90	0.996	1.000
1.0	1006.6	−0.57	1.67	1.00	0.995	1.000
1.5	1008.8	−0.57	1.68	1.51	0.993	1.000
2.0	1011.1	−0.57	1.68	2.02	0.990	1.000
2.5	1013.4	−0.57	1.69	2.53	0.988	0.999
3.0	1015.6	−0.57	1.69	3.04	0.985	0.999
3.5	1017.9	−0.57	1.69	3.56	0.983	0.999
4.0	1020.2	−0.57	1.70	4.08	0.981	0.999
4.5	1022.5	−0.57	1.70	4.60	0.978	0.999
5.0	1024.8	−0.57	1.70	5.12	0.976	0.999
5.5	1027.1	−0.57	1.71	5.64	0.973	0.999
6.0	1029.2	−0.58	1.71	6.17	0.971	0.998
6.5	1031.6	−0.58	1.71	6.70	0.969	0.998
7.0	1033.9	−0.58	1.72	7.23	0.966	0.998
7.5	1036.2	−0.58	1.72	7.77	0.964	0.998
8.0	1038.5	−0.58	1.73	8.31	0.962	0.998
8.5	1040.9	−0.58	1.73	8.84	0.959	0.998
9.0	1043.2	−0.58	1.73	9.39	0.957	0.998
9.5	1045.5	−0.58	1.74	9.93	0.954	0.998
10.0	1047.8	−0.58	1.74	10.48	0.952	0.997
15.0	1071.1	−0.59	1.78	16.09	0.929	0.996
20.0	1094.5	−0.59	1.82	21.97	0.907	0.994
25.0	1117.6	−0.59	1.85	28.13	0.886	0.993
30	1140.4	−0.59	1.89	34.57	0.867	0.992
35	1162.8	−0.58	1.92	41.30	0.850	0.990
40	1184.7	−0.58	1.96	48.34	0.833	0.989
45	1206.3	−0.57	1.99	55.69	0.818	0.988
50	1227.5	−0.57	2.02	63.36	0.803	0.986
55	1248.5	−0.56	2.05	71.36	0.790	0.985
60	1269.3	−0.56	2.09	79.71	0.776	0.983
65	1290.0	−0.55	2.12	88.40	0.763	0.982
70	1310.5	−0.55	2.15	97.45	0.750	0.981
75	1330.9	−0.55	2.18	106.87	0.738	0.979
80	1351.1	−0.55	2.21	116.67	0.725	0.978
85	1371.2	−0.54	2.24	126.87	0.715	0.976
90	1391.0	−0.54	2.27	137.47	0.704	0.975
95	1410.7	−0.54	2.30	148.49	0.693	0.973
100	1430.1	−0.54	2.33	159.93	0.683	0.971

$T = 300$ K

p	w	μ	k	j	α/α_0	γ/γ_0
0.01	1019.2	−0.57	1.67	0.01	1.000	1.000
0.02	1019.2	−0.57	1.67	0.02	1.000	1.000
0.03	1019.3	−0.57	1.67	0.03	1.000	1.000
0.04	1019.3	−0.57	1.67	0.04	1.000	1.000
0.05	1019.4	−0.57	1.67	0.05	1.000	1.000

Table II.18 (*Continued*)

p	w	μ	k	f	α/α_0	γ/γ_0
0.06	1019.4	−0.57	1.67	0.06	1.000	1.000
0.07	1019.4	−0.57	1.67	0.07	1.000	1.000
0.08	1019.5	−0.57	1.67	0.08	1.000	1.000
0.09	1019.5	−0.57	1.67	0.09	1.000	1.000
0.10	1019.6	−0.57	1.67	0.10	1.000	1.000
0.2	1020.0	−0.57	1.67	0.20	0.999	1.000
0.3	1020.5	−0.57	1.67	0.30	0.999	1.000
0.4	1020.9	−0.57	1.67	0.40	0.998	1.000
0.5	1021.4	−0.57	1.67	0.50	0.998	1.000
0.6	1021.8	−0.57	1.67	0.60	0.997	1.000
0.7	1022.3	−0.57	1.67	0.70	0.997	1.000
0.8	1022.7	−0.57	1.67	0.80	0.996	1.000
0.9	1023.2	−0.57	1.67	0.90	0.996	1.000
1.0	1023.6	−0.57	1.67	1.00	0.995	1.000
1.5	1025.8	−0.57	1.68	1.51	0.993	1.000
2.0	1028.1	−0.57	1.68	2.02	0.991	1.000
2.5	1030.3	−0.57	1.68	2.53	0.988	0.999
3.0	1032.5	−0.57	1.69	3.04	0.986	0.999
3.5	1034.8	−0.57	1.69	3.56	0.984	0.999
4.0	1037.0	−0.57	1.70	4.07	0.981	0.999
4.5	1039.2	−0.57	1.70	4.59	0.979	0.999
5.0	1041.5	−0.57	1.70	5.11	0.977	0.999
5.5	1043.7	−0.57	1.71	5.64	0.974	0.999
6.0	1046.0	−0.57	1.71	6.16	0.972	0.999
6.5	1048.2	−0.58	1.71	6.69	0.970	0.998
7.0	1050.5	−0.58	1.72	7.22	0.967	0.998
7.5	1052.7	−0.58	1.72	7.76	0.965	0.998
8.0	1055.0	−0.58	1.72	8.29	0.963	0.998
8.5	1057.3	−0.58	1.73	8.83	0.961	0.998
9.0	1059.5	−0.58	1.73	9.37	0.958	0.998
9.5	1061.8	−0.58	1.73	9.92	0.956	0.998
10.0	1064.1	−0.58	1.74	10.46	0.954	0.997
15.0	1086.9	−0.59	1.77	16.05	0.931	0.996
20.0	1109.9	−0.59	1.81	21.90	0.910	0.995
25.0	1132.6	−0.59	1.85	28.01	0.890	0.993
30	1155.1	−0.59	1.88	34.39	0.871	0.992
35	1177.1	−0.58	1.92	41.06	0.854	0.991
40	1198.8	−0.58	1.95	48.02	0.838	0.989
45	1220.1	−0.57	1.98	55.27	0.823	0.988
50	1241.1	−0.57	2.01	62.84	0.809	0.987
55	1261.8	−0.56	2.04	70.71	0.795	0.985
60	1282.3	−0.56	2.07	78.92	0.782	0.984
65	1302.6	−0.55	2.10	87.45	0.769	0.983
70	1322.8	−0.55	2.13	96.33	0.757	0.982
75	1342.9	−0.55	2.16	105.56	0.745	0.980
80	1362.8	−0.55	2.19	115.16	0.733	0.979
85	1382.6	−0.54	2.22	125.12	0.722	0.977
90	1402.2	−0.54	2.25	135.47	0.711	0.976
95	1421.6	−0.54	2.28	146.21	0.701	0.974
100	1440.8	−0.54	2.31	157.36	0.690	0.973
			$T = 400$ K			
0.01	1176.8	−0.56	1.67	0.01	1.000	1.000
0.02	1176.9	−0.56	1.67	0.02	1.000	1.000
0.03	1176.9	−0.56	1.67	0.03	1.000	1.000

Table II.18 (*Continued*)

p	w	μ	k	f	α/α_0	γ/γ_0
0.04	1176.9	−0.56	1.67	0.04	1.000	1.000
0.05	1177.0	−0.56	1.67	0.05	1.000	1.000
0.06	1177.0	−0.56	1.67	0.06	1.000	1.000
0.07	1177.1	−0.56	1.67	0.07	1.000	1.000
0.08	1177.1	−0.56	1.67	0.08	1.000	1.000
0.09	1177.1	−0.56	1.67	0.09	1.000	1.000
0.10	1177.2	−0.56	1.67	0.10	1.000	1.000
0.2	1177.6	−0.56	1.67	0.20	0.999	1.000
0.3	1177.9	−0.56	1.67	0.30	0.999	1.000
0.4	1178.3	−0.56	1.67	0.40	0.999	1.000
0.5	1178.7	−0.56	1.67	0.50	0.998	1.000
0.6	1179.1	−0.56	1.67	0.60	0.998	1.000
0.7	1179.5	−0.56	1.67	0.70	0.998	1.000
0.8	1179.9	−0.56	1.67	0.80	0.997	1.000
0.9	1180.2	−0.56	1.67	0.90	0.997	1.000
1.0	1180.6	−0.56	1.67	1.00	0.997	1.000
1.5	1182.5	−0.56	1.67	1.51	0.995	1.000
2.0	1184.4	−0.56	1.68	2.01	0.993	1.000
2.5	1186.4	−0.56	1.68	2.52	0.991	1.000
3.0	1188.3	−0.56	1.68	3.03	0.990	0.999
3.5	1190.2	−0.56	1.69	3.54	0.988	0.999
4.0	1192.1	−0.56	1.69	4.05	0.986	0.999
4.5	1194.0	−0.56	1.69	4.57	0.984	0.999
5.0	1195.9	−0.56	1.69	5.08	0.983	0.999
5.5	1197.8	−0.56	1.70	5.60	0.981	0.999
6.0	1199.7	−0.56	1.70	6.12	0.979	0.999
6.5	1201.7	−0.56	1.70	6.64	0.978	0.999
7.0	1203.6	−0.56	1.70	7.17	0.976	0.999
7.5	1205.5	−0.56	1.71	7.69	0.974	0.999
8.0	1207.4	−0.56	1.71	8.22	0.973	0.999
8.5	1209.3	−0.56	1.71	8.74	0.971	0.999
9.0	1211.3	−0.57	1.71	9.27	0.969	0.998
9.5	1213.2	−0.57	1.72	9.81	0.967	0.998
10.0	1215.1	−0.57	1.72	10.34	0.966	0.998
15.0	1234.6	−0.57	1.75	15.77	0.949	0.997
20.0	1254.2	−0.58	1.77	21.38	0.933	0.996
25.0	1273.9	−0.58	1.80	27.18	0.917	0.995
30	1293.5	−0.58	1.83	33.18	0.902	0.995
35	1312.9	−0.58	1.85	39.37	0.887	0.994
40	1332.1	−0.58	1.88	45.76	0.874	0.993
45	1351.1	−0.57	1.91	52.36	0.861	0.992
50	1369.8	−0.57	1.93	59.18	0.849	0.991
55	1388.3	−0.56	1.96	66.21	0.837	0.990
60	1406.5	−0.56	1.98	73.46	0.826	0.989
65	1424.6	−0.56	2.00	80.94	0.816	0.989
70	1442.4	−0.55	2.03	88.64	0.805	0.988
75	1460.2	−0.55	2.05	96.59	0.796	0.987
80	1477.7	−0.54	2.07	104.77	0.786	0.986
85	1495.2	−0.54	2.10	113.19	0.777	0.985
90	1512.6	−0.54	2.12	121.87	0.768	0.984
95	1529.8	−0.54	2.14	130.80	0.759	0.984
100	1547.0	−0.53	2.16	139.99	0.750	0.983
			$T = 500$ K			
0.01	1315.7	−0.55	1.67	0.01	1.000	1.000
0.02	1315.8	−0.55	1.67	0.02	1.000	1.000

Table II.18 (*Continued*)

p	w	μ	k	f	α/α_0	γ/γ_0
0.03	1315.8	—0.55	1.67	0.03	1.000	1.000
0.04	1315.8	—0.55	1.67	0.04	1.000	1.000
0.05	1315.9	—0.55	1.67	0.05	1.000	1.000
0.06	1315.9	—0.55	1.67	0.06	1.000	1.000
0.07	1315.9	—0.55	1.67	0.07	1.000	1.000
0.08	1316.0	—0.55	1.67	0.08	1.000	1.000
0.09	1316.0	—0.55	1.67	0.09	1.000	1.000
0.10	1316.0	—0.55	1.67	0.10	1.000	1.000
0.2	1316.4	—0.55	1.67	0.20	0.999	1.000
0.3	1316.7	—0.55	1.67	0.30	0.999	1.000
0.4	1317.1	—0.55	1.67	0.40	0.999	1.000
0.5	1317.4	—0.55	1.67	0.50	0.999	1.000
0.6	1317.7	—0.55	1.67	0.60	0.998	1.000
0.7	1318.1	—0.55	1.67	0.70	0.998	1.000
0.8	1318.4	—0.55	1.67	0.80	0.998	1.000
0.9	1318.8	—0.55	1.67	0.90	0.998	1.000
1.0	1319.1	—0.55	1.67	1.00	0.997	1.000
1.5	1320.8	—0.55	1.67	1.51	0.996	1.000
2.0	1322.5	—0.55	1.68	2.01	0.995	1.000
2.5	1324.2	—0.55	1.68	2.52	0.993	1.000
3.0	1325.9	—0.55	1.68	3.02	0.992	1.000
3.5	1327.6	—0.55	1.68	3.53	0.990	1.000
4.0	1329.3	—0.55	1.68	4.04	0.989	1.000
4.5	1331.0	—0.55	1.69	4.55	0.988	0.999
5.0	1332.7	—0.55	1.69	5.07	0.986	0.999
5.5	1334.4	—0.55	1.69	5.58	0.985	0.999
6.0	1336.1	—0.55	1.69	6.10	0.984	0.999
6.5	1337.7	—0.55	1.69	6.61	0.982	0.999
7.0	1339.4	—0.55	1.70	7.13	0.981	0.999
7.5	1341.1	—0.55	1.70	7.65	0.980	0.999
8.0	1342.8	—0.55	1.70	8.17	0.978	0.999
8.5	1344.6	—0.55	1.70	8.69	0.977	0.999
9.0	1346.3	—0.55	1.70	9.22	0.976	0.999
9.5	1348.0	—0.55	1.71	9.74	0.974	0.999
10.0	1349.7	—0.56	1.71	10.27	0.973	0.999
15.0	1366.8	—0.56	1.73	15.61	0.960	0.998
20.0	1384.1	—0.56	1.75	21.09	0.947	0.997
25.0	1401.6	—0.57	1.77	26.71	0.934	0.997
30	1419.0	—0.57	1.80	32.48	0.921	0.996
35	1436.5	—0.57	1.82	38.40	0.909	0.995
40	1453.8	—0.57	1.84	44.48	0.897	0.995
45	1471.1	—0.57	1.86	50.72	0.886	0.994
50	1488.2	—0.57	1.88	57.12	0.876	0.994
55	1505.1	—0.56	1.90	63.68	0.865	0.993
60	1521.8	—0.56	1.92	70.40	0.856	0.992
65	1538.3	—0.56	1.94	77.30	0.846	0.992
70	1554.7	—0.55	1.96	84.37	0.837	0.991
75	1570.9	—0.55	1.98	91.62	0.829	0.991
80	1587.0	—0.55	2.00	99.04	0.820	0.990
85	1602.9	—0.54	2.02	106.64	0.812	0.990
90	1618.7	—0.54	2.04	114.43	0.804	0.989
95	1634.5	—0.54	2.05	122.40	0.797	0.988
100	1650.1	—0.54	2.07	130.57	0.789	0.988

Table II.18 (*Continued*)

p	w	μ	k	f	α/α_0	γ/γ_0
\multicolumn{7}{c}{$T=600$ K}						
0.01	1441.3	−0.55	1.67	0.01	1.000	1.000
0.02	1441.3	−0.55	1.67	0.02	1.000	1.000
0.03	1441.4	−0.55	1.67	0.03	1.000	1.000
0.04	1441.4	−0.55	1.67	0.04	1.000	1.000
0.05	1441.4	−0.55	1.67	0.05	1.000	1.000
0.06	1441.5	−0.55	1.67	0.06	1.000	1.000
0.07	1441.5	−0.55	1.67	0.07	1.000	1.000
0.08	1441.5	−0.55	1.67	0.08	1.000	1.000
0.09	1441.5	−0.55	1.67	0.09	1.000	1.000
0.10	1441.6	−0.55	1.67	0.10	1.000	1.000
0.2	1441.9	−0.55	1.67	0.20	1.000	1.000
0.3	1442.2	−0.55	1.67	0.30	0.999	1.000
0.4	1442.5	−0.55	1.67	0.40	0.999	1.000
0.5	1442.8	−0.55	1.67	0.50	0.999	1.000
0.6	1443.1	−0.55	1.67	0.60	0.999	1.000
0.7	1443.4	−0.55	1.67	0.70	0.998	1.000
0.8	1443.7	−0.55	1.67	0.80	0.998	1.000
0.9	1444.0	−0.55	1.67	0.90	0.998	1.000
1.0	1444.4	−0.55	1.67	1.00	0.998	1.000
1.5	1445.9	−0.55	1.67	1.50	0.997	1.000
2.0	1447.4	−0.55	1.67	2.01	0.995	1.000
2.5	1449.0	−0.55	1.68	2.51	0.994	1.000
3.0	1450.5	−0.54	1.68	3.02	0.993	1.000
3.5	1452.1	−0.54	1.68	3.53	0.992	1.000
4.0	1453.6	−0.54	1.68	4.04	0.991	1.000
4.5	1455.1	−0.54	1.68	4.54	0.990	1.000
5.0	1456.7	−0.55	1.68	5.05	0.989	1.000
5.5	1458.2	−0.55	1.69	5.57	0.988	0.999
6.0	1459.8	−0.55	1.69	6.08	0.987	0.999
6.5	1461.3	−0.55	1.69	6.59	0.985	0.999
7.0	1462.8	−0.55	1.69	7.11	0.984	0.999
7.5	1464.4	−0.55	1.69	7.62	0.983	0.999
8.0	1465.9	−0.55	1.69	8.14	0.982	0.999
8.5	1467.5	−0.55	1.70	8.66	0.981	0.999
9.0	1469.0	−0.55	1.70	9.18	0.980	0.999
9.5	1470.5	−0.55	1.70	9.70	0.979	0.999
10.0	1472.1	−0.55	1.70	10.22	0.978	0.999
15.0	1487.6	−0.55	1.72	15.50	0.967	0.999
20.0	1503.2	−0.55	1.74	20.89	0.956	0.998
25.0	1519.0	−0.56	1.76	26.40	0.945	0.998
30	1534.8	−0.56	1.77	32.03	0.934	0.997
35	1550.7	−0.56	1.79	37.79	0.924	0.997
40	1566.6	−0.56	1.81	43.66	0.914	0.996
45	1582.4	−0.56	1.83	49.67	0.904	0.996
50	1598.1	−0.56	1.84	55.80	0.895	0.995
55	1613.7	−0.56	1.86	62.07	0.886	0.995
60	1629.3	−0.56	1.88	68.46	0.877	0.994
65	1644.7	−0.56	1.90	75.00	0.869	0.994
70	1659.9	−0.55	1.91	81.67	0.860	0.993
75	1675.0	−0.55	1.93	88.48	0.853	0.993
80	1690.0	−0.55	1.95	95.43	0.845	0.993
85	1704.9	−0.55	1.96	102.52	0.838	0.992
90	1719.6	−0.54	1.98	109.76	0.831	0.992
95	1734.2	−0.54	1.99	117.14	0.824	0.991

Table II.18 (*Continued*)

p	w	μ	k	f	α/α_0	γ/γ_0
100	1748.8	−0.54	2.01	124.68	0.817	0.991

$T = 700$ K

p	w	μ	k	f	α/α_0	γ/γ_0
0.01	1556.8	−0.54	1.67	0.01	1.000	1.000
0.02	1556.8	−0.54	1.67	0.02	1.000	1.000
0.03	1556.8	−0.54	1.67	0.03	1.000	1.000
0.04	1556.9	−0.54	1.67	0.04	1.000	1.000
0.05	1556.9	−0.54	1.67	0.05	1.000	1.000
0.06	1556.9	−0.54	1.67	0.06	1.000	1.000
0.07	1556.9	−0.54	1.67	0.07	1.000	1.000
0.08	1557.0	−0.54	1.67	0.08	1.000	1.000
0.09	1557.0	−0.54	1.67	0.09	1.000	1.000
0.10	1557.0	−0.54	1.67	0.10	1.000	1.000
0.2	1557.3	−0.54	1.67	0.20	1.000	1.000
0.3	1557.6	−0.54	1.67	0.30	0.999	1.000
0.4	1557.9	−0.54	1.67	0.40	0.999	1.000
0.5	1558.2	−0.54	1.67	0.50	0.999	1.000
0.6	1558.5	−0.54	1.67	0.60	0.999	1.000
0.7	1558.7	−0.54	1.67	0.70	0.999	1.000
0.8	1559.0	−0.54	1.67	0.80	0.998	1.000
0.9	1559.3	−0.54	1.67	0.90	0.998	1.000
1.0	1559.6	−0.54	1.67	1.00	0.998	1.000
1.5	1561.0	−0.54	1.67	1.50	0.997	1.000
2.0	1562.4	−0.54	1.67	2.01	0.996	1.000
2.5	1563.9	−0.54	1.67	2.51	0.995	1.000
3.0	1565.3	−0.54	1.68	3.02	0.994	1.000
3.5	1566.7	−0.54	1.68	3.52	0.993	1.000
4.0	1568.1	−0.54	1.68	4.03	0.992	1.000
4.5	1569.5	−0.54	1.68	4.54	0.991	1.000
5.0	1571.0	−0.54	1.68	5.05	0.990	1.000
5.5	1572.4	−0.54	1.68	5.56	0.990	1.000
6.0	1573.8	−0.54	1.68	6.07	0.989	1.000
6.5	1575.2	−0.54	1.69	6.58	0.988	1.000
7.0	1576.6	−0.54	1.69	7.09	0.987	0.999
7.5	1578.0	−0.54	1.69	7.61	0.986	0.999
8.0	1579.5	−0.54	1.69	8.12	0.985	0.999
8.5	1580.9	−0.54	1.69	8.64	0.984	0.999
9.0	1582.3	−0.54	1.69	9.15	0.983	0.999
9.5	1583.7	−0.54	1.70	9.67	0.982	0.999
10.0	1585.2	−0.54	1.70	10.19	0.981	0.999
15.0	1599.4	−0.54	1.71	15.42	0.972	0.999
20.0	1613.8	−0.55	1.73	20.76	0.962	0.999
25.0	1628.2	−0.55	1.74	26.19	0.953	0.998
30	1642.8	−0.55	1.76	31.72	0.944	0.998
35	1657.4	−0.55	1.77	37.36	0.935	0.997
40	1672.0	−0.56	1.79	43.10	0.926	0.997
45	1686.7	−0.56	1.80	48.94	0.918	0.997
50	1701.3	−0.56	1.82	54.89	0.909	0.996
55	1715.8	−0.56	1.83	60.95	0.901	0.996
60	1730.3	−0.56	1.85	67.13	0.893	0.996
65	1744.8	−0.56	1.86	73.41	0.886	0.995
70	1759.1	−0.55	1.88	79.81	0.878	0.995
75	1773.3	−0.55	1.89	86.32	0.871	0.995
80	1787.4	−0.55	1.91	92.95	0.864	0.994
85	1801.4	−0.55	1.92	99.70	0.857	0.994

Table II.18 (*Continued*)

p	w	μ	k	f	α/α_0	γ/γ_0
90	1815.3	—0.55	1.94	106.56	0.851	0.994
95	1829.1	—0.54	1.95	113.55	0.844	0.993
100	1842.8	—0.54	1.96	120.66	0.838	0.993

$T = 800$ K

p	w	μ	k	f	α/α_0	γ/γ_0
0.01	1664.3	—0.54	1.67	0.01	1.000	1.000
0.02	1664.3	—0.54	1.67	0.02	1.000	1.000
0.03	1664.3	—0.54	1.67	0.03	1.000	1.000
0.04	1664.3	—0.54	1.67	0.04	1.000	1.000
0.05	1664.4	—0.54	1.67	0.05	1.000	1.000
0.06	1664.4	—0.54	1.67	0.06	1.000	1.000
0.07	1664.4	—0.54	1.67	0.07	1.000	1.000
0.08	1664.4	—0.54	1.67	0.08	1.000	1.000
0.09	1664.5	—0.54	1.67	0.09	1.000	1.000
0.10	1664.5	—0.54	1.67	0.10	1.000	1.000
0.2	1664.8	—0.54	1.67	0.20	1.000	1.000
0.3	1665.0	—0.54	1.67	0.30	0.999	1.000
0.4	1665.3	—0.54	1.67	0.40	0.999	1.000
0.5	1665.6	—0.54	1.67	0.50	0.999	1.000
0.6	1665.8	—0.54	1.67	0.60	0.999	1.000
0.7	1666.1	—0.54	1.67	0.70	0.999	1.000
0.8	1666.4	—0.54	1.67	0.80	0.999	1.000
0.9	1666.6	—0.54	1.67	0.90	0.998	1.000
1.0	1666.9	—0.54	1.67	1.00	0.998	1.000
1.5	1668.2	—0.54	1.67	1.50	0.997	1.000
2.0	1669.5	—0.54	1.67	2.01	0.997	1.000
2.5	1670.9	—0.54	1.67	2.51	0.996	1.000
3.0	1672.2	—0.54	1.67	3.01	0.995	1.000
3.5	1673.5	—0.54	1.68	3.52	0.994	1.000
4.0	1674.8	—0.54	1.68	4.03	0.993	1.000
4.5	1676.2	—0.54	1.68	4.53	0.993	1.000
5.0	1677.5	—0.54	1.68	5.04	0.992	1.000
5.5	1678.8	—0.54	1.68	5.55	0.991	1.000
6.0	1680.1	—0.54	1.68	6.06	0.990	1.000
6.5	1681.4	—0.54	1.68	6.57	0.989	1.000
7.0	1682.8	—0.54	1.68	7.08	0.988	1.000
7.5	1684.1	—0.54	1.69	7.59	0.988	1.000
8.0	1685.4	—0.54	1.69	8.10	0.987	1.000
8.5	1686.7	—0.54	1.69	8.62	0.986	1.000
9.0	1688.1	—0.54	1.69	9.13	0.985	0.999
9.5	1689.4	—0.54	1.69	9.65	0.984	0.999
10.0	1690.7	—0.54	1.69	10.16	0.984	0.999
15.0	1704.0	—0.54	1.71	15.37	0.975	0.999
20.0	1717.3	—0.54	1.72	20.66	0.967	0.999
25.0	1730.8	—0.54	1.73	26.03	0.959	0.999
30	1744.3	—0.55	1.75	31.49	0.951	0.998
35	1757.9	—0.55	1.76	37.04	0.943	0.998
40	1771.5	—0.55	1.77	42.68	0.936	0.998
45	1785.1	—0.55	1.79	48.41	0.928	0.997
50	1798.8	—0.55	1.80	54.23	0.920	0.997
55	1812.4	—0.55	1.81	60.14	0.913	0.997
60	1826.1	—0.55	1.83	66.15	0.906	0.996
65	1839.6	—0.55	1.84	72.25	0.899	0.996
70	1853.1	—0.55	1.85	78.45	0.892	0.996
75	1866.6	—0.55	1.87	84.75	0.885	0.996

Table II.18 (Continued)

p	w	μ	k	f	α/α_0	γ/γ_0
80	1879.9	−0.55	1.88	91.15	0.879	0.995
85	1893.2	−0.55	1.89	97.64	0.873	0.995
90	1906.4	−0.55	1.90	104.24	0.867	0.995
95	1919.5	−0.54	1.92	110.94	0.861	0.995
100	1932.5	−0.54	1.93	117.75	0.855	0.994

$T = 900$ K

p	w	μ	k	f	α/α_0	γ/γ_0
0.01	1765.2	−0.53	1.67	0.01	1.000	1.000
0.02	1765.2	−0.53	1.67	0.02	1.000	1.000
0.03	1765.3	−0.53	1.67	0.03	1.000	1.000
0.04	1765.3	−0.53	1.67	0.04	1.000	1.000
0.05	1765.3	−0.53	1.67	0.05	1.000	1.000
0.06	1765.3	−0.53	1.67	0.06	1.000	1.000
0.07	1765.4	−0.53	1.67	0.07	1.000	1.000
0.08	1765.4	−0.53	1.67	0.08	1.000	1.000
0.09	1765.4	−0.53	1.67	0.09	1.000	1.000
0.10	1765.4	−0.53	1.67	0.10	1.000	1.000
0.2	1765.7	−0.53	1.67	0.20	1.000	1.000
0.3	1765.9	−0.53	1.67	0.30	1.000	1.000
0.4	1766.2	−0.53	1.67	0.40	0.999	1.000
0.5	1766.4	−0.53	1.67	0.50	0.999	1.000
0.6	1766.7	−0.53	1.67	0.60	0.999	1.000
0.7	1766.9	−0.53	1.67	0.70	0.999	1.000
0.8	1767.2	−0.53	1.67	0.80	0.999	1.000
0.9	1767.4	−0.53	1.67	0.90	0.999	1.000
1.0	1767.7	−0.53	1.67	1.00	0.999	1.000
1.5	1768.9	−0.53	1.67	1.50	0.998	1.000
2.0	1770.2	−0.53	1.67	2.01	0.997	1.000
2.5	1771.4	−0.53	1.67	2.51	0.996	1.000
3.0	1772.7	−0.53	1.67	3.01	0.996	1.000
3.5	1773.9	−0.53	1.67	3.52	0.995	1.000
4.0	1775.2	−0.53	1.68	4.02	0.994	1.000
4.5	1776.4	−0.53	1.68	4.53	0.993	1.000
5.0	1777.6	−0.53	1.68	5.04	0.993	1.000
5.5	1778.9	−0.53	1.68	5.54	0.992	1.000
6.0	1780.1	−0.53	1.68	6.05	0.991	1.000
6.5	1781.4	−0.53	1.68	6.56	0.990	1.000
7.0	1782.6	−0.53	1.68	7.07	0.990	1.000
7.5	1783.9	−0.53	1.68	7.58	0.989	1.000
8.0	1785.1	−0.53	1.69	8.09	0.988	1.000
8.5	1786.4	−0.53	1.69	8.60	0.988	1.000
9.0	1787.6	−0.53	1.69	9.12	0.987	1.000
9.5	1788.8	−0.53	1.69	9.63	0.986	1.000
10.0	1790.1	−0.53	1.69	10.14	0.985	1.000
15.0	1802.6	−0.53	1.70	15.33	0.978	0.999
20.0	1815.1	−0.54	1.71	20.58	0.971	0.999
25.0	1827.7	−0.54	1.72	25.91	0.964	0.999
30	1840.3	−0.54	1.74	31.32	0.957	0.999
35	1853.1	−0.54	1.75	36.80	0.950	0.998
40	1865.8	−0.54	1.76	42.36	0.943	0.998
45	1878.7	−0.55	1.77	48.00	0.936	0.998
50	1891.5	−0.55	1.78	53.72	0.929	0.998
55	1904.4	−0.55	1.80	59.52	0.923	0.997
60	1917.2	−0.55	1.81	65.41	0.916	0.997
65	1930.0	−0.55	1.82	71.37	0.910	0.997

Продолжение табл. II.18

p	w	μ	k	f	α/α_0	γ/γ_0
70	1942.8	—0.55	1.83	77.43	0.903	0.997
75	1955.6	—0.55	1.84	83.56	0.897	0.996
80	1968.3	—0.55	1.86	89.78	0.891	0.996
85	1980.9	—0.55	1.87	96.09	0.885	0.996
90	1993.5	—0.55	1.88	102.49	0.880	0.996
95	2006.0	—0.54	1.89	108.97	0.874	0.996
100	2018.4	—0.54	1.90	115.55	0.869	0.995

$T = 1000$ K

p	w	μ	k	f	α/α_0	γ/γ_0
0.01	1860.7	—0.53	1.67	0.01	1.000	1.000
0.02	1860.7	—0.53	1.67	0.02	1.000	1.000
0.03	1860.7	—0.53	1.67	0.03	1.000	1.000
0.04	1860.8	—0.53	1.67	0.04	1.000	1.000
0.05	1860.8	—0.53	1.67	0.05	1.000	1.000
0.06	1860.8	—0.53	1.67	0.06	1.000	1.000
0.07	1860.8	—0.53	1.67	0.07	1.000	1.000
0.08	1860.9	—0.53	1.67	0.08	1.000	1.000
0.09	1860.9	—0.53	1.67	0.09	1.000	1.000
0.10	1860.9	—0.53	1.67	0.10	1.000	1.000
0.2	1861.1	—0.53	1.67	0.20	1.000	1.000
0.3	1861.4	—0.53	1.67	0.30	1.000	1.000
0.4	1861.6	—0.53	1.67	0.40	0.999	1.000
0.5	1861.8	—0.53	1.67	0.50	0.999	1.000
0.6	1862.1	—0.53	1.67	0.60	0.999	1.000
0.7	1862.3	—0.53	1.67	0.70	0.999	1.000
0.8	1862.6	—0.53	1.67	0.80	0.999	1.000
0.9	1862.8	—0.53	1.67	0.90	0.999	1.000
1.0	1863.0	—0.53	1.67	1.00	0.999	1.000
1.5	1864.2	—0.53	1.67	1.50	0.998	1.000
2.0	1865.4	—0.53	1.67	2.01	0.997	1.000
2.5	1866.6	—0.53	1.67	2.51	0.997	1.000
3.0	1867.8	—0.53	1.67	3.01	0.996	1.000
3.5	1868.9	—0.53	1.67	3.52	0.995	1.000
4.0	1870.1	—0.53	1.68	4.02	0.995	1.000
4.5	1871.3	—0.53	1.68	4.53	0.994	1.000
5.0	1872.5	—0.53	1.68	5.03	0.993	1.000
5.5	1873.6	—0.53	1.68	5.54	0.993	1.000
6.0	1874.8	—0.53	1.68	6.05	0.992	1.000
6.5	1876.0	—0.53	1.68	6.55	0.991	1.000
7.0	1877.2	—0.53	1.68	7.06	0.991	1.000
7.5	1878.4	—0.53	1.68	7.57	0.990	1.000
8.0	1879.5	—0.53	1.68	8.08	0.990	1.000
8.5	1880.7	—0.53	1.68	8.59	0.989	1.000
9.0	1881.9	—0.53	1.69	9.10	0.988	1.000
9.5	1883.1	—0.53	1.69	9.62	0.988	1.000
10.0	1884.2	—0.53	1.69	10.13	0.987	1.000
15.0	1896.0	—0.53	1.70	15.29	0.981	0.999
20.0	1907.9	—0.53	1.71	20.52	0.974	0.999
25.0	1919.8	—0.53	1.72	25.82	0.968	0.999
30	1931.7	—0.54	1.73	31.18	0.961	0.999
35	1943.7	—0.54	1.74	36.61	0.955	0.999
40	1955.8	—0.54	1.75	42.11	0.949	0.998
45	1967.9	—0.54	1.76	47.68	0.943	0.998
50	1980.0	—0.54	1.77	53.32	0.936	0.998
55	1992.2	—0.54	1.78	59.04	0.930	0.998

Table II.18 (*Continued*)

p	w	μ	k	f	α/α_0	γ/γ_0
60	2004.4	−0.54	1.79	64.82	0.924	0.998
65	2016.5	−0.55	1.81	70.68	0.918	0.997
70	2028.7	−0.55	1.82	76.62	0.913	0.997
75	2040.8	−0.55	1.83	82.63	0.907	0.997
80	2052.9	−0.55	1.84	88.71	0.901	0.997
85	2065.0	−0.55	1.85	94.87	0.896	0.997
90	2077.0	−0.54	1.86	101.11	0.891	0.996
95	2088.9	−0.54	1.87	107.43	0.885	0.996
100	2100.9	−0.54	1.88	113.83	0.880	0.996

$T = 1100$ K

p	w	μ	k	f	α/α_0	γ/γ_0
0.01	1951.5	−0.53	1.67	0.01	1.000	1.000
0.02	1951.5	−0.53	1.67	0.02	1.000	1.000
0.03	1951.6	−0.53	1.67	0.03	1.000	1.000
0.04	1951.6	−0.53	1.67	0.04	1.000	1.000
0.05	1951.6	−0.53	1.67	0.05	1.000	1.000
0.06	1951.6	−0.53	1.67	0.06	1.000	1.000
0.07	1951.6	−0.53	1.67	0.07	1.000	1.000
0.08	1951.7	−0.53	1.67	0.08	1.000	1.000
0.09	1951.7	−0.53	1.67	0.09	1.000	1.000
0.10	1951.7	−0.53	1.67	0.10	1.000	1.000
0.2	1951.9	−0.53	1.67	0.20	1.000	1.000
0.3	1952.2	−0.53	1.67	0.30	1.000	1.000
0.4	1952.4	−0.53	1.67	0.40	1.000	1.000
0.5	1952.6	−0.53	1.67	0.50	0.999	1.000
0.6	1952.8	−0.53	1.67	0.60	0.999	1.000
0.7	1953.1	−0.53	1.67	0.70	0.999	1.000
0.8	1953.3	−0.53	1.67	0.80	0.999	1.000
0.9	1953.5	−0.53	1.67	0.90	0.999	1.000
1.0	1953.7	−0.53	1.67	1.00	0.999	1.000
1.5	1954.9	−0.53	1.67	1.50	0.998	1.000
2.0	1956.0	−0.53	1.67	2.00	0.998	1.000
2.5	1957.1	−0.53	1.67	2.51	0.997	1.000
3.0	1958.2	−0.53	1.67	3.01	0.996	1.000
3.5	1959.4	−0.53	1.67	3.51	0.996	1.000
4.0	1960.5	−0.53	1.67	4.02	0.995	1.000
4.5	1961.6	−0.53	1.68	4.52	0.995	1.000
5.0	1962.7	−0.53	1.68	5.03	0.994	1.000
5.5	1963.8	−0.53	1.68	5.54	0.993	1.000
6.0	1965.0	−0.53	1.68	6.04	0.993	1.000
6.5	1966.1	−0.53	1.68	6.55	0.992	1.000
7.0	1967.2	−0.53	1.68	7.06	0.992	1.000
7.5	1968.3	−0.53	1.68	7.57	0.991	1.000
8.0	1969.4	−0.53	1.68	8.07	0.991	1.000
8.5	1970.6	−0.53	1.68	8.58	0.990	1.000
9.0	1971.7	−0.53	1.68	9.09	0.989	1.000
9.5	1972.8	−0.53	1.68	9.61	0.989	1.000
10.0	1973.9	−0.53	1.69	10.12	0.988	1.000
15.0	1985.1	−0.53	1.70	15.26	0.982	1.000
20.0	1996.4	−0.53	1.70	20.47	0.977	0.999
25.0	2007.7	−0.53	1.71	25.74	0.971	0.999
30	2019.0	−0.53	1.72	31.07	0.965	0.999
35	2030.4	−0.53	1.73	36.46	0.959	0.999
40	2041.9	−0.54	1.74	41.91	0.954	0.999
45	2053.4	−0.54	1.75	47.42	0.948	0.998

Table II.18 (*Continued*)

p	w	μ	k	f	α/α_0	γ/γ_0
50	2064.9	−0.54	1.76	53.00	0.942	0.998
55	2076.5	−0.54	1.77	58.65	0.937	0.998
60	2088.0	−0.54	1.78	64.35	0.931	0.998
65	2099.6	−0.54	1.79	70.13	0.926	0.998
70	2111.2	−0.54	1.80	75.97	0.920	0.998
75	2122.8	−0.54	1.81	81.88	0.915	0.998
80	2134.4	−0.54	1.82	87.85	0.910	0.997
85	2145.9	−0.54	1.83	93.90	0.905	0.997
90	2157.4	−0.54	1.84	100.01	0.900	0.997
95	2168.9	−0.54	1.85	106.19	0.895	0.997
100	2180.3	−0.54	1.86	112.44	0.890	0.997

$T = 1200$ K

p	w	μ	k	f	α/α_0	γ/γ_0
0.01	2038.3	−0.53	1.67	0.01	1.000	1.000
0.02	2038.3	−0.53	1.67	0.02	1.000	1.000
0.03	2038.3	−0.53	1.67	0.03	1.000	1.000
0.04	2038.3	−0.53	1.67	0.04	1.000	1.000
0.05	2038.4	−0.53	1.67	0.05	1.000	1.000
0.06	2038.4	−0.53	1.67	0.06	1.000	1.000
0.07	2038.4	−0.53	1.67	0.07	1.000	1.000
0.08	2038.4	−0.53	1.67	0.08	1.000	1.000
0.09	2038.5	−0.53	1.67	0.09	1.000	1.000
0.10	2038.5	−0.53	1.67	0.10	1.000	1.000
0.2	2038.7	−0.53	1.67	0.20	1.000	1.000
0.3	2038.9	−0.53	1.67	0.30	1.000	1.000
0.4	2039.1	−0.53	1.67	0.40	1.000	1.000
0.5	2039.3	−0.53	1.67	0.50	0.999	1.000
0.6	2039.5	−0.53	1.67	0.60	0.999	1.000
0.7	2039.8	−0.53	1.67	0.70	0.999	1.000
0.8	2040.0	−0.53	1.67	0.80	0.999	1.000
0.9	2040.2	−0.53	1.67	0.90	0.999	1.000
1.0	2040.4	−0.53	1.67	1.00	0.999	1.000
1.5	2041.5	−0.53	1.67	1.50	0.998	1.000
2.0	2042.6	−0.52	1.67	2.00	0.998	1.000
2.5	2043.6	−0.52	1.67	2.51	0.997	1.000
3.0	2044.7	−0.52	1.67	3.01	0.997	1.000
3.5	2045.8	−0.52	1.67	3.51	0.996	1.000
4.0	2046.9	−0.52	1.67	4.02	0.996	1.000
4.5	2047.9	−0.52	1.67	4.52	0.995	1.000
5.0	2049.0	−0.52	1.68	5.03	0.995	1.000
5.5	2050.1	−0.52	1.68	5.53	0.994	1.000
6.0	2051.1	−0.52	1.68	6.04	0.993	1.000
6.5	2052.2	−0.52	1.68	6.55	0.993	1.000
7.0	2053.3	−0.52	1.68	7.05	0.992	1.000
7.5	2054.4	−0.52	1.68	7.56	0.992	1.000
8.0	2055.4	−0.52	1.68	8.07	0.991	1.000
8.5	2056.5	−0.52	1.68	8.58	0.991	1.000
9.0	2057.6	−0.52	1.68	9.09	0.990	1.000
9.5	2058.6	−0.52	1.68	9.60	0.990	1.000
10.0	2059.7	−0.52	1.68	10.11	0.989	1.000
15.0	2070.4	−0.52	1.69	15.24	0.984	1.000
20.0	2081.2	−0.53	1.70	20.43	0.979	0.999
25.0	2092.0	−0.53	1.71	25.67	0.973	0.999
30	2102.8	−0.53	1.72	30.97	0.968	0.999
35	2113.7	−0.53	1.73	36.33	0.963	0.999

Table II.18 (*Continued*)

p	w	μ	k	f	α/α_0	γ/γ_0
40	2124.6	—0.53	1.74	41.74	0.958	0.999
45	2135.5	—0.53	1.75	47.21	0.952	0.999
50	2146.5	—0.53	1.75	52.74	0.947	0.999
55	2157.6	—0.54	1.76	58.32	0.942	0.998
60	2168.6	—0.54	1.77	63.97	0.937	0.998
65	2179.7	—0.54	1.78	69.67	0.932	0.998
70	2190.8	—0.54	1.79	75.44	0.927	0.998
75	2201.8	—0.54	1.80	81.26	0.922	0.998
80	2212.9	—0.54	1.81	87.15	0.917	0.998
85	2224.0	—0.54	1.82	93.09	0.912	0.998
90	2235.0	—0.54	1.83	99.10	0.908	0.997
95	2246.1	—0.54	1.84	105.18	0.903	0.997
100	2257.0	—0.54	1.84	111.31	0.899	0.997
			$T=1300$ K			
0.01	2121.5	—0.52	1.67	0.01	1.000	1.000
0.02	2121.5	—0.52	1.67	0.02	1.000	1.000
0.03	2121.5	—0.52	1.67	0.03	1.000	1.000
0.04	2121.6	—0.52	1.67	0.04	1.000	1.000
0.05	2121.6	—0.52	1.67	0.05	1.000	1.000
0.06	2121.6	—0.52	1.67	0.06	1.000	1.000
0.07	2121.6	—0.52	1.67	0.07	1.000	1.000
0.08	2121.7	—0.52	1.67	0.08	1.000	1.000
0.09	2121.7	—0.52	1.67	0.09	1.000	1.000
0.10	2121.7	—0.52	1.67	0.10	1.000	1.000
0.2	2121.9	—0.52	1.67	0.20	1.000	1.000
0.3	2122.1	—0.52	1.67	0.30	1.000	1.000
0.4	2122.3	—0.52	1.67	0.40	1.000	1.000
0.5	2122.5	—0.52	1.67	0.50	0.999	1.000
0.6	2122.7	—0.52	1.67	0.60	0.999	1.000
0.7	2122.9	—0.52	1.67	0.70	0.999	1.000
0.8	2123.1	—0.52	1.67	0.80	0.999	1.000
0.9	2123.3	—0.52	1.77	0.90	0.999	1.000
1.0	2123.6	—0.52	1.67	1.00	0.999	1.000
1.5	2124.6	—0.52	1.67	1.50	0.998	1.000
2.0	2125.6	—0.52	1.67	2.00	0.998	1.000
2.5	2126.6	—0.52	1.67	2.51	0.997	1.000
3.0	2127.7	—0.52	1.67	3.01	0.997	1.000
3.5	2128.7	—0.52	1.67	3.51	0.996	1.000
4.0	2129.7	—0.52	1.67	4.02	0.996	1.000
4.5	2130.8	—0.52	1.67	4.52	0.995	1.000
5.0	2131.8	—0.52	1.67	5.02	0.995	1.000
5.5	2132.8	—0.52	1.68	5.53	0.995	1.000
6.0	2133.9	—0.52	1.68	6.04	0.994	1.000
6.5	2134.9	—0.52	1.68	6.54	0.994	1.000
7.0	2135.9	—0.52	1.68	7.05	0.993	1.000
7.5	2136.9	—0.52	1.68	7.56	0.993	1.000
8.0	2138.0	—0.52	1.68	8.06	0.992	1.000
8.5	2139.0	—0.52	1.68	8.57	0.992	1.000
9.0	2140.0	—0.52	1.68	9.08	0.991	1.000
9.5	2141.1	—0.52	1.68	9.59	0.991	1.000
10.0	2142.1	—0.52	1.68	10.10	0.990	1.000
15.0	2152.4	—0.52	1.69	15.22	0.985	1.000
20.0	2162.7	—0.52	1.70	20.40	0.980	1.000
25.0	2173.0	—0.52	1.71	25.62	0.975	0.999

Table II.18 (*Continued*)

p	w	μ	k	f	α/α_0	γ/γ_0
30	2183.4	−0.53	1.71	30.90	0.971	0.999
35	2193.8	−0.53	1.72	36.22	0.966	0.999
40	2204.2	−0.53	1.73	41.60	0.961	0.999
45	2214.7	−0.53	1.74	47.03	0.956	0.999
50	2225.2	−0.53	1.75	52.52	0.951	0.999
55	2235.8	−0.53	1.76	58.05	0.947	0.999
60	2246.4	−0.53	1.76	63.64	0.942	0.999
65	2257.0	−0.54	1.77	69.29	0.937	0.998
70	2267.6	−0.54	1.78	74.99	0.933	0.998
75	2278.3	−0.54	1.79	80.75	0.928	0.998
80	2288.9	−0.54	1.80	86.56	0.923	0.998
85	2299.5	−0.54	1.81	92.42	0.919	0.998
90	2310.2	−0.54	1.81	98.35	0.915	0.998
95	2320.8	−0.54	1.82	104.33	0.910	0.998
100	2331.4	−0.54	1.83	110.37	0.906	0.998

$T = 1400$ K

p	w	μ	k	f	α/α_0	γ/γ_0
0.01	2201.6	−0.52	1.67	0.01	1.000	1.000
0.02	2201.6	−0.52	1.67	0.02	1.000	1.000
0.03	2201.6	−0.52	1.67	0.03	1.000	1.000
0.04	2201.6	−0.52	1.67	0.04	1.000	1.000
0.05	2201.7	−0.52	1.67	0.05	1.000	1.000
0.06	2201.7	−0.52	1.67	0.06	1.000	1.000
0.07	2201.7	−0.52	1.67	0.07	1.000	1.000
0.08	2201.7	−0.52	1.67	0.08	1.000	1.000
0.09	2201.7	−0.52	1.67	0.09	1.000	1.000
0.10	2201.8	−0.52	1.67	0.10	1.000	1.000
0.2	2202.0	−0.52	1.67	0.20	1.000	1.000
0.3	2202.2	−0.52	1.67	0.30	1.000	1.000
0.4	2202.4	−0.52	1.67	0.40	1.000	1.000
0.5	2202.6	−0.52	1.67	0.50	1.000	1.000
0.6	2202.8	−0.52	1.67	0.60	0.999	1.000
0.7	2203.0	−0.52	1.67	0.70	0.999	1.000
0.8	2203.2	−0.52	1.67	0.80	0.999	1.000
0.9	2203.4	−0.52	1.67	0.90	0.999	1.000
1.0	2203.6	−0.52	1.67	1.00	0.999	1.000
1.5	2204.5	−0.52	1.67	1.50	0.999	1.000
2.0	2205.5	−0.52	1.67	2.00	0.998	1.000
2.5	2206.5	−0.52	1.67	2.51	0.998	1.000
3.0	2207.5	−0.52	1.67	3.01	0.997	1.000
3.5	2208.5	−0.52	1.67	3.51	0.997	1.000
4.0	2209.5	−0.52	1.67	4.01	0.996	1.000
4.5	2210.5	−0.52	1.67	4.52	0.996	1.000
5.0	2211.5	−0.52	1.67	5.02	0.995	1.000
5.5	2212.5	−0.52	1.67	5.53	0.995	1.000
6.0	2213.5	−0.52	1.68	6.03	0.994	1.000
6.5	2214.5	−0.52	1.68	6.54	0.994	1.000
7.0	2215.5	−0.52	1.68	7.04	0.994	1.000
7.5	2216.4	−0.52	1.68	7.55	0.993	1.000
8.0	2217.4	−0.52	1.68	8.06	0.993	1.000
8.5	2218.4	−0.52	1.68	8.57	0.992	1.000
9.0	2219.4	−0.52	1.68	9.07	0.992	1.000
9.5	2220.4	−0.52	1.68	9.58	0.991	1.000
10.0	2221.4	−0.52	1.68	10.09	0.991	1.000
15.0	2231.3	−0.52	1.69	15.21	0.986	1.000

Table II.18 (*Continued*)

p	w	μ	k	f	α/α_0	γ/γ_0
20.0	2241.2	—0.52	1.70	20.37	0.982	1.000
25.0	2251.1	—0.52	1.70	25.58	0.977	0.999
30	2261.1	—0.52	1.71	30.83	0.973	0.999
35	2271.1	—0.52	1.72	36.13	0.968	0.999
40	2281.1	—0.53	1.73	41.48	0.964	0.999
45	2291.2	—0.53	1.73	46.88	0.959	0.999
50	2301.3	—0.53	1.74	52.33	0.955	0.999
55	2311.5	—0.53	1.75	57.82	0.950	0.999
60	2321.6	—0.53	1.76	63.37	0.946	0.999
65	2331.8	—0.53	1.76	68.97	0.942	0.999
70	2342.1	—0.53	1.77	74.61	0.937	0.999
75	2352.3	—0.53	1.78	80.31	0.933	0.998
80	2362.5	—0.54	1.79	86.06	0.929	0.998
85	2372.8	—0.54	1.80	91.86	0.925	0.998
90	2383.0	—0.54	1.80	97.71	0.920	0.998
95	2393.2	—0.54	1.81	103.61	0.916	0.998
100	2403.5	—0.54	1.82	109.57	0.912	0.998
			$T=1500$ K			
0.01	2278.9	—0.52	1.67	0.01	1.000	1.000
0.02	2278.9	—0.52	1.67	0.02	1.000	1.000
0.03	2278.9	—0.52	1.67	0.03	1.000	1.000
0.04	2278.9	—0.52	1.67	0.04	1.000	1.000
0.05	2278.9	—0.52	1.67	0.05	1.000	1.000
0.06	2279.0	—0.52	1.67	0.06	1.000	1.000
0.07	2279.0	—0.52	1.67	0.07	1.000	1.000
0.08	2279.0	—0.52	1.67	0.08	1.000	1.000
0.09	2279.0	—0.52	1.67	0.09	1.000	1.000
0.10	2279.0	—0.52	1.67	0.10	1.000	1.000
0.2	2279.2	—0.52	1.67	0.20	1.000	1.000
0.3	2279.4	—0.52	1.67	0.30	1.000	1.000
0.4	2279.6	—0.52	1.67	0.40	1.000	1.000
0.5	2279.8	—0.52	1.67	0.50	1.000	1.000
0.6	2280.0	—0.52	1.67	0.60	0.999	1.000
0.7	2280.2	—0.52	1.67	0.70	0.999	1.000
0.8	2280.4	—0.52	1.67	0.80	0.999	1.000
0.9	2280.6	—0.52	1.67	0.90	0.999	1.000
1.0	2280.8	—0.52	1.67	1.00	0.999	1.000
1.5	2281.7	—0.52	1.67	1.50	0.999	1.000
2.0	2282.7	—0.52	1.67	2.00	0.998	1.000
2.5	2283.6	—0.52	1.67	2.51	0.998	1.000
3.0	2284.6	—0.52	1.67	3.01	0.997	1.000
3.5	2285.6	—0.52	1.67	3.51	0.997	1.000
4.0	2286.5	—0.52	1.67	4.01	0.997	1.000
4.5	2287.5	—0.52	1.67	4.52	0.996	1.000
5.0	2288.4	—0.52	1.67	5.02	0.996	1.000
5.5	2289.4	—0.52	1.67	5.53	0.995	1.000
6.0	2290.3	—0.52	1.67	6.03	0.995	1.000
6.5	2291.3	—0.52	1.68	6.54	0.994	1.000
7.0	2292.2	—0.52	1.68	7.04	0.994	1.000
7.5	2293.2	—0.52	1.68	7.55	0.994	1.000
8.0	2294.2	—0.52	1.68	8.05	0.993	1.000
8.5	2295.1	—0.52	1.68	8.56	0.993	1.000
9.0	2296.1	—0.52	1.68	9.07	0.992	1.000
9.5	2297.0	—0.52	1.68	9.58	0.992	1.000

Table II.18 (*Continued*)

p	w	μ	k	f	α/α_0	γ/γ_0
10.0	2298.0	−0.52	1.68	10.09	0.991	1.000
15.0	2307.5	−0.52	1.69	15.19	0.987	1.000
20.0	2317.1	−0.52	1.69	20.34	0.983	1.000
25.0	2326.7	−0.52	1.70	25.54	0.979	1.000
30	2336.3	−0.52	1.71	30.77	0.975	0.999
35	2345.9	−0.52	1.72	36.05	0.970	0.999
40	2355.6	−0.52	1.72	41.38	0.966	0.999
45	2365.3	−0.52	1.73	46.75	0.962	0.999
50	2375.0	−0.53	1.74	52.17	0.958	0.999
55	2384.8	−0.53	1.74	57.63	0.954	0.999
60	2394.6	−0.53	1.75	63.13	0.950	0.999
65	2404.4	−0.53	1.76	68.69	0.946	0.999
70	2414.3	−0.53	1.77	74.29	0.942	0.999
75	2424.1	−0.53	1.77	79.93	0.938	0.999
80	2434.0	−0.53	1.78	85.63	0.934	0.998
85	2443.9	−0.53	1.79	91.37	0.930	0.998
90	2453.8	−0.53	1.79	97.16	0.926	0.998
95	2463.6	−0.54	1.80	103.00	0.922	0.998
100	2473.5	−0.54	1.81	108.88	0.918	0.998

REFERENCES

1. D. N. Astrov, "Development of low-temperature thermometry and thermophysics," Izmeritel'naya Tekh. No. 2, pp. 23–25 (1980).
2. I. V. Bogoyavlenskiĭ, N. G. Bereznyak, and B. N. Esel'son, "Measurement of the liquid-crystal equilibrium diagram for solutions of helium isotopes," Zh. Eksp. Teor. Fiz., *47*(2), 480–483 (1964).
3. I. V. Bogoyavlenskiĭ, L. V. Karnatsevich, and V. G. Konareva, "Experimental study of the equation of state of helium (He^4 and He^3) in the temperature range between 3.3 and 14 K," Fiz. Nizk. Temp., *4*(5), 549–561 (1978).
4. I. V. Bogoyavlenskiĭ and S. I. Yurchenko, "Measurement of the molar volumes of liquid solutions of He^3 and He^4 under pressures up to 100 atm. in the temperature range 1.5–4.2 K," Fiz. Nizk. Temp., *2*(11), 1379–1387 (1976).
5. A. A. Vasserman, "Estimating the weights of experimental data for comparison between the equations of state of a real gas," Teplofiz. Vys. Temp., *19*(5), 1103–1105 (1981).
6. A. A. Vasserman and A. Ya. Kreizerova, "Optimization of the number of coefficients in the equation of state," Teplofiz. Vys. Temp., *16*(6), 1185–1188 (1978).
7. O. A. Dobrovol'skiĭ and I. F. Golubev, "Measurement of the density of helium," Gozovaya Promyshlennost', No. 7, 53–54 (1965).
8. B. N. Esel'son et al., Svoistva zhidkogo i tverdogo geliya (Properties of Liquid and Solid Helium), Izd-vo Standartov, 1978, 127 p.
9. A. B. Kalenkov, Eksperimental'noe issledovanie szhimaemosti geliya (Experimental Study of the Compressibility of Helium), Candidate Thesis, MEI, Moscow, 1976, 24 p.
10. A. I. Karnus and N. S. Rudenko, "Denisty of helium-4 in the temperature range 14–54 K at pressures up to 110 atm," in: Teplofizicheskie svoistva veshchestv pri nizkikh temperaturakh. Materialy 1-go Veseoyuznogo soveshchaniya, 16–19 Fevralya 1971 (Thermophysical Properties of Fluids at Low Temperatures. Proc. First All-Union Conf., 16–19 February 1971), Moscow, 1972.
11. W. H. Keesom, Helium [Russian translation, IIL, Moscow, 1949, 542 p.].

310 REFERENCES

12. G. A. Korn and T. M. Korn, Mathematical Handbook for Scientists and Engineers, McGraw-Hill, 1968 [Russian translation, Nauka, Moscow, 1978, 474 p.].
13. V. F. Kukarin, V. G. Martynets, E. V. Matizen, and A. G. Sartakov, "Experimental study of the p-ρ-T dependence near the vaporization critical point," Fiz. Nizk. Temp., 6(5), 549–559 (1980).
14. V. A. Medvedev and V. K. Yaruntsev, "Isobaric thermal capacity of helium-4 in the region of hydrogen temperatures," in: Thermophysical Properties of Materials, Collection, GSSSD No. 7, Izd. Standartov, Moscow, 1973, pp. 17–20.
15. V. R. Petrov, Eksperimental'noe issledovanie szhimaemosti geliya i smeseĭ gelii-argon (Experimental Study of the Compressibility of Helium and Helium-Argon Mixtures), Candidate Thesis, MEI, Moscow, 1972, 24 p.
16. P. V. Popov, Eksperimental'noe issledovanie p, v, T-zavisimosti geliya-4 pri nizkikh temperaturakh (Experimental Study of the p, v, T-Dependence of Helium-4 at Low Temperatures), Candidate Thesis, MEI, Moscow, 1981, 20 p.
17. P. V. Popov, V. A. Rabinovich, and V. I. Chernyshov, "Experimental study of the compressibility of helium-4 in the subcritical region," in: Thermophysical Properties of Materials, Collection, GSSSD No. 13, Izd. vo Standartov, Moscow, 1979, pp. 137–140.
18. V. G. Skripka, "Virial coefficients of helium below 30 K," Tr. VNIIKIMASh, No. 10, Mashinostroenie, Moscow, 1965, pp. 153–161.
19. V. G. Skripka, "Second virial coefficients of oxygen, nitrogen, argon, helium, neon, hydrogen, deuterium, and krypton at temperatures below 150 K," Tr. VNIIKIMASh, No. 10, Mashinstroenie, Moscow, 1965, pp. 163–183.
20. V. N. Taran, Issledovanie termodinamicheskikh svoĭstva geliya kak nizkotemperaturnogo khladoagenta (Investigation of the Thermodynamic Properties of Helium as a Low-Temperature Coolant), Candidate Thesis, OTIKhP, Odessa, 1970, 28 p.
21. V. V. Sychev, A. A. Vasserman, A. D. Kozlov, et al., Thermodynamic Properties of Nitrogen, Hemisphere, Washington, 1987 (originally published in Russian by Standards Publishing House, Moscow, 1977).
22. V.V. Sychev, A. A. Vasserman, V. A. Zagoruchenko, et al., Thermodynamic Properties of Methane, Hemisphere, Washington, 1987 (originally published in Russian by Standards Publishing House, Moscow, 1979).
23. D. Hudson, Statistics for Physicists [Russian translation, Mir, Moscow, 1970, 296 p.].
24. N. V. Tsederberg, V. N. Popov, and N. A. Morozova, Teplofizicheskie svoĭstva gelia (Thermophysical Properties of Helium), Gosenergoizdat, Moscow-Leningrad, 1961, 120 p.
25. N. V. Tsederberg, V. N. Popov, and N. A. Morozova, Termodinamicheskie i teplofizicheskie svoĭstva geliya (Thermodynamic and Thermophysical Properties of Helium), Atomizdat, Moscow, 1969, 276 p.
26. N. V. Tsederberg, V. N. Popov, and V. R. Petrov, "Experimental study of the compressibility of helium at temperatures between 20 and −150°C and pressures between 20 and 400 bar," Teploenergetika, No. 6, 87–89 (1972).
27. F. P. van Agt and H. Kamerlingh Onnes, "Isotherms of mon-atomic substances and their binary mixtures. XXV. The same of diatomic substances. XXXI. The compressibility of hydrogen and helium gas between 90 and 14 K," Commun. Phys. Lab. Univ. Leiden, No. 176b, pp. 15–29 (1925).
28. G. Ahlers, "Properties of He^4 near the γ-phase," Phys. Rev., 135(1A), 10–16 (1964).
29. E. Ambler and R. P. Hudson, "An examination of the helium vapor-pressure scale of temperature using a magnetic thermometer," J. Res. NBS, 56(2), 99–104 (1956).
30. E. Ambler and R. P. Hudson, "An examination of the 1955 helium vapor-pressure scale of temperature," J. Res. NBS, 57(1/2), 23–25 (1956).
31. M. S. Anderson and C. A. Swenson, "Characteristics of germanium resistance thermometer from 1 to 35 K and the magnetic temperature scale," Rev. Sci. Instrum., 49(8), 1027–1033 (1978).
32. S. Angus, K. M. de Reuck, and R. D. McCarthy, Helium, International Thermodynamic Tables of the Fluid State, vol. 4, Pergamon Press, 1977, 265 p.

33. Argon, Helium and Rare Gases, vol. 1, ed. by Cook, N., Interscience, New York, 1961, 394 p.
34. K. R. Atkins, Liquid Helium, Cambridge, Univ. Press, 1959, 312 p.
35. K. R. Atkins and C. E. Chase, "The velocity of first sound in liquid helium," Proc. Phys. Soc. London, A *64* (Pt. 9-No. 381), 826–833 (1951).
36. K. R. Atkins and R. A. Stasior, "First sound in liquid helium at high pressures," Can. J. Phys., *31*(7), 1156–1164 (1953).
37. M. Barmas and I. Rudnick, "Velocity and attenuation of first sound near the λ-point of helium," Phys. Rev., *170*(1), 224–238 (1968).
38. E. P. Bartlett, "The compressibility isotherms of hydrogen, nitrogen and mixtures of these gases at 0 °C and pressures to 1000 atmospheres," J. Am. Chem. Soc., *49*, 687–701 (1927).
39. J. J. M. Beenakker, F. H. Varekamp, and A. van Itterbeek, "The isotherms of the hydrogen isotopes and their mixtures with helium at the boiling point of hydrogen," Physica, *25*(1), 9–24 (1959).
40. R. Berman and C. F. Mate, "Some thermal properties of helium and their relation to the temperature scale," Phil. Mag., *3*(29), 461–469 (1958).
41. R. Berman and J. Poulter, "On the latent heat and vapor density of helium," Phil. Mag., *43*(345), 1047–1054 (1952).
42. R. Berman and C. A. Swenson, "Absolute temperature scale between 4.2 and 5.2 K," Phys. Rev., *95*(21), 311–314 (1954).
43. K. H. Berry, "P-v isotherms of He^4 at low temperature," Metrologia, *8*(3), 125 (1972).
44. K. H. Berry, "Measurements of the second and third virial coefficients of He^4 in the range 2.6–27.1 K," Mol. Phys., *37*(1), 317–318 (1979).
45. A. L. Blancett, K. R. Hall, and F. B. Canfield, "Isotherms for the He-Ar system at 0 °C, and −50 °C up to 700 atm.," Physica, *47*(1), 75–91 (1970).
46. B. Bleaney and F. Simon, "The vapor pressure curve of liquid helium below the λ-point," Trans. Faraday Soc., *35*(9), 1205–1214 (1939).
47. J. D. A. Boks and H. Kamerlingh Onnes, "Isotherms of monatomic substances and their binary mixtures. XXIII. Isotherms of helium from 20° to −259 °C," Commun. Phys. Lab. Univ. Leiden, No. 170a, 3–9 (1924).
48. M. E. Boyd, S. Y. Larsen, and H. Plumb, "Second virial coefficient of He^4 in the temperature range from 2 to 20 K," J. Res. NBS, *72A*(2), 155–156 (1968).
49. P. W. Bridgeman, "The compressibility of five gases to high pressures," Proc. Am. Acad. Arts Sci., *59*, 173–211 (1924).
50. T. C. Briggs, "Compressibility data for helium over the temperature range −5 °C to 80 °C and at pressures to 800 atmospheres," U.S. Dept. Interior, Bureau of Mines, R.I. 7352, pp. 1–39 (1970).
51. T. C. Briggs, B. J. Dalton, and R. E. Barieau, "Compressibility data for helium at 0 °C and pressures to 800 atmospheres," U.S. Dept. Interior, Bureau of Mines, R.I. 7287, pp. 1–54 (1969).
52. E. S. Burnett, "Compressibility determinations without volume measurements," J. Appl. Mech., Trans. ASME, *3A*, 136–140 (1936).
53. F. Canfield, T. W. Leland, and R. Kobayashi, "Compressibility factors for helium-nitrogen mixtures," J. Chem. Eng. Data, *10*(2), 92–96 (1965).
54. T. C. Cetas and C. A. Swenson, "A paramagnetic salt temperature scale 0.9 to 18 K," Metrologia *8*(2), 46–64 (1972).
55. C. E. Chase, E. Maxwell, and W. E. Millet, "The dielectric constant of liquid helium," Physica, *27*(12), 1129–1145 (1961).
56. J. R. Clement, "Some apparent anomalies in low-temperature heat capacities," Phys. Rev., *93*(6), 1420–1421 (1954).
57. J. R. Clement, J. K. Logan, and J. Gaffney, "Liquid helium vapor pressure equation," Phys. Rev., *100*(2), 743–744 (1955).

312 REFERENCES

58. R. K. Crawford and W. B. Daniels, "Experimental determination of the p-T melting curves of Kr, Ne and He," J. Chem. Phys., *55*(12), 5651–5656 (1971).
59. L. I. Dana and H. Kamerlingh Onnes, "Further experiments with liquid helium. Preliminary determinations of the latent heat of vaporization of liquid helium," Commun. Phys. Lab. Univ. Leiden, No. 179c (1926).
60. H. van Dijk, Temperature, Its Measurement and Control, vol. 2, Reinhold Publ. Co., New York, p. 199, 1955.
61. H. van Dijk and M. Durieux, "Thermodynamic temperature scale (T_{L55}) in the liquid helium region," Physica, *24*(1), 1–19 (1958).
62. H. van Dijk, M. Durieux, J. R. Clement, and J. K. Logan, "The 1958 He^4 scale of temperatures. Part 2. Tables for the 1958 temperature scale," J. Res. NBS, *64A*(1), 4–17 (1960).
63. D. D. Dillard, M. Waxman, and R. L. Robinson Jr., "Volumetric data and virial coefficients for helium, krypton and helium-krypton mixtures," J. Chem. Eng. Data, *23*(4), 269–274 (1978).
64. J. S. Dugdale and J. P. Franck, "The thermodynamic properties of solid and fluid helium-3 and helium-4 above 3 K at high densities," Phil. Trans. Roy. Soc. London, *257A*, 1–29 (1964).
65. J. S. Dugdale and F. E. Simon, "Thermodynamic properties and melting of solid helium," Proc. Roy. Soc. London, *218A*(113), 291–310 (1953).
66. J. H. Dymond and E. B. Smith, The Virial Coefficient of Gases. A Critical Compilation, Clarendon Press, Oxford, pp. 172–181, 1969.
67. F. J. Edeskuty and R. H. Sherman, "P-v-T relations of liquid He^3 and He^4," Proc. Fifth Intern. Conf. on Low Temp. Phys. Chem., Washington, 1957, Univ. Wisconsin Press, 1958, pp. 102–106.
68. M. H. Edwards, "Refractive index of He^4 liquid," Can. J. Phys., *36*(7), 884–898 (1958).
69. M. H. Edwards, "Equation of state of helium four," Proc. Eleventh Intern. Conf. on Low Temp. Phys., vol. 1, St. Andrews, pp. 231–235, 1968.
70. D. O. Edwards and R. C. Pandorf, "Heat capacity and other properties of hexagonal close-packed helium-4," Phys. Rev., *140*(3A), 816–825 (1965).
71. D. O. Edwards and R. C. Pandorf, "Heat capacity and other properties of body-centered cubic He^4," Phys. Rev., *144*(1), 143–151 (1966).
72. M. H. Edwards and W. C. Woodbury, "Compressibility of liquid He^4," Can. J. Phys., *39*, 1833–1841 (1961).
73. M. H. Edwards and W. C. Woodbury, "Saturated He^4 near its critical temperature," Phys. Rev., *129*(5), 1911–1918 (1963).
74. Z. E. H. A. El Hadi and M. Durieux, "The density of the saturated vapor of He^4," Physica, *41*(2), 305–319 (1969).
75. Z. E. H. A. El Hadi, M. Durieux, and H. van Dijk, "The density of liquid He^4 under its saturated vapor pressure," Physica, *41*(2), 289–304 (1969).
76. D. L. Elwell and H. Meyer, "Coefficient of thermal expansion and related properties of liquid He^4 under pressure," Phys. Rev., *164*(1), 245–255 (1967).
77. R. A. Erickson and L. D. Robert, "The measurement and the calculation of the liquid vapor pressure—temperature scale from 1 to 4.2 K," Phys. Rev., *93*(5), 957–962 (1954).
78. J. C. Findlay, A. Pitt, H. Grayson-Smith, and J. O. Wilhelm, "The velocity of sound in liquid helium," Phys. Rev., *54*(7), 506–509 (1938).
79. J. C. Findlay, A. Pitt, H. Grayson-Smith, and J. O. Wilhelm, "The velocity of sound in liquid helium under pressure," Phys. Rev., *56*(1), 122 (1939).
80. T. W. Gibby, T. C. Tanner, and I. Mason, "The pressures of gaseous mixtures. II. Helium and hydrogen and their intermolecular forces," Proc. Roy. Soc. London, *122A*, 283–304 (1929).
81. A. R. M. Glassford and J. L. Smith Jr., "Pressure-volume-temperature and internal energy data for helium from 4.2 to 20 K between 100 and 1300 atm." Cryogenics, *6*(4), 193–206 (1966).

82. C. J. Grebenkemper and J. P. Hagen, "Dielectric constant of liquid helium," Phys. Rev., 80(1), 89 (1950).
83. E. R. Grilly, "Compressibility of liquid He^4 as a function of pressure," Phys. Rev., 149(1), 97–101 (1966).
84. E. R. Grilly, "Pressure-volume-temperature relations in liquid and solid He^4," J. Low Temp. Phys., 11(1/2), 33–36 (1973).
85. E. R. Grilly and R. L. Mills, "Melting properties of He^3 and He^4 up to 3500 kg/cm^2," Ann. Phys. (NY), 8, 1–23 (1959).
86. E. R. Grilly and R. L. Mills, "PvT relations in He^4 near the melting curve and the λ-line," Ann. Phys. (NY), 18, 250–263 (1962).
87. D. T. Grimsrud and J. H. Werntz, "Measurements of the velocity of sound in He^3 and He^4 gas at low temperatures with implications for the temperature scale," Phys. Rev., 157, 181–190 (1967).
88. D. Gugan and G. W. Michel, "Measurements of the polarizability and of the second and third virial coefficients of He^4 in the range 4.2–27.1 K," Mol. Phys., 39(3), 783–785 (1980).
89. K. R. Hall and F. B. Canfield, "Isotherms for the $He-N_2$ system at $-190\,°C$, $-170\,°C$ and $-160\,°C$ up to 700 atm.," Physica, 47(2), 219–226 (1970).
90. R. W. Hill and O. V. Lounasmaa, "The specific heat of liquid helium," Phil. Mag., 2, 143–148 (1957).
91. R. W. Hill and O. V. Lounasmaa, "The thermodynamic properties of liquid helium," Phil. Trans. Roy. Soc. London, 252A, 357–395 (1960).
92. L. Holborn and J. Otto, "Uber die Isothermen von Stickstoff, Sauerstoff und Helium," Z. Physik, 10, 367–376 (1922).
93. L. Holborn and J. Otto, "Uber die Isothermen von Helium, Stockstoff und Argon unterhalb 0°," Z. Physik, 30, 320–328 (1924).
94. L. Holborn and J. Otto, "Uber die Isothermen einiger Gase bis 400° and ihre Bedeutung für das Gasthermometer," Z. Physik, 23, 77–94 (1924).
95. L. Holborn and J. Otto, "Uber die Isothermen einiger Gase zwischen +400° und −183°," Z. Physik, 33, 1–11 (1925).
96. L. Holborn and H. Schultze, "Uber die Druckwage und die Isothermen von Luft, Argon und Helium zwischen 0 and 200°C," Ann. Phys. (Leipzig), 47, 1089–1111 (1915).
97. F. A. Holland, J. A. M. Huggill, and G. O. Jones, "The solid-fluid equilibrium of helium above 5000 atm. pressure," Proc. Roy. Soc. London, 207A(1089), 268–277 (1951).
98. A. van Itterbeek, "Mesures sur les properties cinetiques et caloriques de l'helium gaseux aux temperatures de helium liquide," J. Phys. Radium, Ser. VIII 8, No. 8 (1938).
99. A. van Itterbeek and W. de Laet, "Measurements on the velocity of sound in helium gas at liquid helium temperatures," Physica, 24(1), 59–67 (1958).
100. H. Kamerlingh Onnes, "Further experiments with liquid helium. F. Isotherms of monatomic gases. XII. Thermal properties of helium," Commun. Phys. Lab. Univ. Leiden, No. 124b (1911).
101. H. Kamerlingh Onnes and J. D. A. Boks, "Further experiments with liquid helium. U. Isotherms of helium at 4.2 K and lower. V. The variation of density of liquid helium below the boiling point," Commun. Phys. Lab. Univ. Leiden, No. 170b, 13–23 (1924).
102. H. Kamerlingh Onnes and S. Weber, "Further experiments with liquid helium. O. On the measurement of very low temperatures. XXV. The determination of the temperatures which are obtained with liquid helium, especially in connection with measurements of the vapor-pressure of helium," Commun. Phys. Lab. Univ. Leiden, No. 147b, 17–35 (1915).
103. W. H. Keesom, "Sur l'echelle de temperature pour le domain de l'helium liquid," Commun. Phys. Lab. Univ. Leiden, Suppl. 71d, 42–46 (1932).
104. W. H. Keesom and K. Clusius, "Die Unwandlung flüssiges Helium I—flüssiges Helium II under Druck," Commun. Phys. Lab. Univ. Leiden, No. 216b, 9–14 (1932).
105. W. H. Keesom and K. Clusius, "Uber die spezifische Wärme des flüssigen Heliums," Commun. Phys. Lab. Univ. Leiden, No. 219e, 42–58.

106. W. H. Keesom and A. P. Keesom, "Isopycnals of liquid helium I." Commun. Phys. Lab. Univ. Leiden, No. 224d, 14–20.
107. W. H. Keesom and A. P. Keesom, Isopycnals of liquid helium II," Commun. Phys. Lab. Univ. Leiden, No. 224e, 21–24.
108. W. H. Keesom and A. P. Keesom, "Measurements concerning the specific heats of solid helium and the melting heat of helium," Physica, *3*(2), 105–117 (1936).
109. W. H. Keesom and H. H. Kraak, "The compressibility of helium gas between 2.6 and 4.2 K," Physica's Grav., *2*(1), 37–44 (1935).
110. W. H. Keesom and W. K. Walstra, "The second virial coefficient of helium at temperatures of liquid and solid hydrogen," Physica, *13*(4/5), 225–230 (1947).
111. W. H. Keesom, S. Weber, and D. Schmidt, "New measurements on the vapor pressure curve of liquid helium II—Commun. Phys. Lab. Univ. Leiden, *18* (No. 202C), 24–37 (1929).
112. W. E. Keller, Helium-3 and Helium-4, Plenum Press, New York, 1969, 417 p.
113. W. E. Keller, "Pressure-volume isotherms of He^4 below 4.2 K," Phys. Rev., *100*(6), 1790 (1955).
114. W. E. Keller, "Pressure-volume isotherms of He^3 between 1.5 and 3.8 K," Phys. Rev., *98*(6), 1571–1575 (1955).
115. E. C. Kerr, "Density of liquid He^4," J. Chem. Phys., *26*(3), 511–514 (1957).
116. E. C. Kerr and R. H. Sherman, "The molar polarizability of He^3 at low temperatures and its density dependence," J. Low Temp. Phys., *3*, 451–461 (1970).
117. E. C. Kerr and R. D. Taylor, "The molar volume and expansion coefficient of liquid He^4," Ann. Phys. (NY), *26*(2), 292–306 (1964).
118. H. A. Kierstead, "Pressure coefficient and phase diagram of He^4 near the upper lambda point," Phys. Rev., *138A*(6), 1594–1599 (1965).
119. H. A. Kierstead, "Melting pressure and α-γ solid-phase transformation of He^4 near the upper lambda point," Phys. Rev., 114(1), 166–169 (1966).
120. H. A. Kierstead, "Lambda curve of liquid He^4," Phys. Rev., *162*(1), 153–161 (1967).
121. H. A. Kierstead, "Pressure on the critical isochore of He^4," Phys. Rev., *3A*(1), 329–339 (1971).
122. H. A. Kierstead, "PVT surface of He^4 near its critical point," Phys. Rev. *7A*(1), 242–251 (1973).
123. J. E. Kilpatrick, W. E. Keller, E. F. Hammel, and N. Metropolis, "Second virial coefficients of He^3 and He^4," Phys. Rev., *94*(5), 1103–1110 (1954).
124. J. Kistemaker, "The vaporization heat of liquid helium from 0 to 5.2 K and the value of h calculated from the thermodynamical behavior of helium," Physica, *12*, 281–288 (1946).
125. J. Kistemaker and W. H. Keesom, "Isotherms of helium gas from 2.7 to 1.7 K," Physica, *12*(4), 227–240 (1946).
126. P. S. Ku and B. F. Dodge, "Compressibility of binary systems helium-nitrogen and carbon dioxide-ethylene," J. Chem. Eng. Data, *12*(2), 158–164 (1967).
127. H. Landolt and R. Börnstein, Zahlenwerte und Funktionen aus Physik, Chemie, Astronomie, Geophysik und Technik, 6 Aufl., B2, Teil 4, 1961.
128. O. V. Lounasmaa and L. Kaunisto, "Direct measurement of $(\partial p/\partial T)_v$ of liquid helium near the λ-curve," Ann. Acad. Sci. Fennicae Series A.6 (Physica), No. 59, 1–16 (1960).
129. O. V. Lounasmaa and E. Kojo, "The specific heat c_v of liquid helium near the λ-curve at various densities," Ann. Acad. Sci. Fennicae, Serie A.6, No. 36, 1–26 (1959).
130. D. T. Mage and D. L. Katz, "Enthalpy determinations on the helium-nitrogen system," AIChE J., *12*(1), 137–144 (1966).
131. E. Mathias, C. A. Crommelin, H. Kamerlingh Onnes, and J. C. Swallow, "Further experiments with liquid helium. X. The rectilinear diameter of helium," Commun. Phys. Lab. Univ. Leiden, No. 172b, 13–22 (1925).

132. R. D. McCarty, "Thermodynamic properties of helium-4 from 2 to 1500 K at pressures up to 10^8 Pa," J. Phys. Chem. Ref. Data, 2(4), 923–1042 (1973).†
133. A. Michels and H. Wouters, "Isotherms of helium between 0° and 150°C up to 200 Amagat," Physica, 8(8), 923–931 (1941).
134. R. L. Mills and E. R. Grilly, "Melting curves of He^3, He^4, H_2, D_2, Ne, N_2, and O_2 up to 3500 kg/cm^2," Phys. Rev., 99(2), 480–486 (1955).
135. R. L. Mills and E. R. Grilly, "The volume change on melting of He^3 and He^4 up to 3500 kg/cm^2," Proc. Fifth Intern. Conf. on Low Temp. Phys., Wisconsin, 1957, Univ. of Wisconsin Press, pp. 106–108, 1958.
136. R. L. Mills, D. H. Liebenberg, and J. C. Bronson, "Equation of state and melting properties of He^4 from measurements to 20 kbar," Phys. Rev., 21B(11), 5137–5147 (1980).
137. M. R. Moldover, "Scaling of the specific-heat singularity of He^4 near its critical point," Phys. Rev., 182(1), 342–352 (1969).
138. G. P. Nijhoff and W. H. Keesom, "Isotherms of monatomic substances and their binary mixtures. XXVI. Isotherms of helium at −183° and −201.5°C and pressures of 3 to 8 atmospheres," Commun. Phys. Lab. Univ. Leiden, No. 188b, 19–23 (1927).
139. G. P. Nijhoff, W. H. Keesom, and B. Iliin, "Isotherms of monatomic substances and their binary mixtures. XXVII. Isotherms of helium between −103.6° and −259.9°C and at pressures of 1.5 to 14 atmospheres," Commun. Phys. Lab. Univ. Leiden, No. 188c, 27–28 (1927).
140. J. Palacios Martinez and H. Kamerlingh Onnes, "Isotherms de substances monoatomiques et de leurs melanges binaires. XXI. Déterminations d'isotherms de l'hydrogene et de l'hélium à basse temperature, faites en vue d'examiner si la compressibilité de ces gazs est influencée par les quanta," Commun. Phys. Lab. Univ. Leiden, No. 164, 3–26 (1922).
141. J. R. Pellam and C. F. Square, "Ultrasonic velocity and absorption in liquid helium," Phys. Rev., 72(12), 1245–1252 (1947).
142. F. M. Penning and H. Kamerlingh Onnes, "Isothermes de substances monoatomiques et de leurs mélanges binaires. XXII. Isothermes de l'hélium entre −205°C et −258°C," Commun. Phys. Lab. Univ. Leiden, No. 165c (1923).
143. H. Plumb and G. Cataland, "Acoustical thermometer and the National Bureau of Standards Provisional Temperature Scale 2-20 K (1965)," Metrologia, 2, 127–139 (1966).
144. J. A. Provine and F. B. Canfield, "Isotherms for the He-Ar system at −130, −115, and −90°C up to 700 atm.," Physica, 52(1), 79–91 (1971).
145. P. R. Roach, "Pressure-density-temperature surface of He^4 near the critical point," Phys. Rev., 170(1), 213–223 (1968).
146. P. R. Roach and D. H. Douglas Jr., "Coexistence curve of He^4 near the critical point," Phys.Rev. Letters, 17(21), 1083–1086 (1966).
147. P. R. Roach and D. H. Douglas Jr., "Denisty of He^4 near the critical point," Phys. Rev. Letters, 19(6), 287–290 (1967).
148. J. R. Roebuck and H. Osterberg, "The Joule-Thomson effect in helium," Phys. Rev., 43, 60–69 (1933).
149. J. S. Rogers, R. J. Tainsh, M. S. Anderson, and C. A. Swenson, "Comparison between gas thermometer, acoustic and platinum resistance temperature scale between 2 and 20 K," Metrologia, 4(2), 47–59 (1968).
150. R. L. Rusby and C. A. Swenson, "A new determination of the helium vapor pressure scale using CMN magnetic thermometer and the NPL-75 gas thermometer scale," Metrologia, 16(2), 73–87 (1980).
151. D. Schmidt and W. H. Keesom, "New measurements of liquid helium temperatures. I. The boiling point of helium. II. The vapor pressure curve of liquid helium," Physica, 4(10), 963–972 (1937).

†Changed in proof.—T.B.S.

152. W. G. Schneider and J. A. H. Duffie, "Compressibility of gases at high temperatures. II. Second virial coefficient of helium in the temperature range 0°C to 600°C," J. Chem. Phys., *17*(9), 751-754 (1949).
153. R. H. Sherman, S. G. Sydoriak, and T. R. Roberts, "The 1962 He^3 scale of temperatures. 4. Tables," J. Res. NBS, *68A*, 579-588 (1964).
154. F. Simon, M. Ruhemann, and W. A. M. Edwards, "Untersuchungen über die Schmelzkurve des Heliums. I.," Z. Phys. Chem., *B2*, 340-344 (1929).
155. F. Simon, M. Ruhemann, and W. A. M. Edwards, "Untersuchung über die Schmelzkurve des Heliums. II.," Z. Phys. Chem., *B6*, 62-77 (1929).
156. G. C. Straty and E. D. Adams, "He^4 melting curve below 1 K," Phys. Rev. Letters, *17*(6), 290-292 (1966).
157. L. Stroud, J. E. Miller, and L. W. Brandt, "Compressibility of helium at $-10°$ to 130°F and pressures to 4000 psia," J. Chem. Eng. Data, *5*(1), 51-52 (1960).
158. J. A. Sullivan and R. E. Sonntag, "Compressibilities for helium at temperatures from 70 to 120 K and pressures to 690 atm.," Cryogenics, *7*(1), 13-17 (1967).
159. C. A. Swenson, "The liquid-solid transformation in helium near absolute zero," Phys. Rev., *79*(4), 626-631 (1950).
160. C. A. Swenson, "The liquid-solid transformation in helium from 1.6 to 4 K," Phys. Rev., *86*(6), 870-876 (1952).
161. C. S. Swenson, "The blocked capillary method for determining melting pressures: The melting curve of helium from 1.5 to 4 K," Phys. Rev., *89*(3), 538-544 (1953).
162. C. A. Swenson, "Relationship from 1 to 34 K between a Paramagnetic Salt Temperature Scale and other scales: an addendum," Metrologia, *9*, 99-101 (1973).
163. H. ter Harmsel, H. van Dijk, and M. Durieux, "The heat of vaporization of helium," Physica, *36*(4), 620-636 (1967).
164. F. H. Varekamp and I. I. Beenakker, "The equation of state of hydrogen isotopes and their mixtures with helium at the boiling point of hydrogen," Physica, *25*(1), 899-904 (1959).
165. J. H. Vignos and H. A. Fairbank, "Sound measurements in liquid and solid He^3, He^4, and He^3-He^4 mixtures," Phys. Rev., *147*(1), 185-197 (1966).
166. G. W. Weems and N. L. Miller, "Compressibility factor for helium," U.S. Dept. Interior, Bureau of Mines, R.I. 7223, Washington, D.C., 1969, 41 p.
167. D. White, T. Rubin, P. Camky, and H. L. Johnston, "The virial coefficients of helium from 2 to 300 K," J. Phys. Chem., *64*(11) 1607-1612 (1960).
168. R. Wiebe, V. L. Gaddy, and C. Heins, Jr., "The compressibility isotherms of helium at temperatures from -70 to 200°C and at pressures to 100 atmospheres," J. Am. Chem. Soc., *53*(72), 1721-1724 (1931).
169. J. Wilks, The Properties of Liquid and Solid Helium, Clarendon Press, Oxford, 1967, 703 p.
170. R. J. Witonsky and J. G. Miller, "Compressibility of gases. IV. The Burnett method applied to gas mixtures at higher temperatures. The second virial coefficients of the helium-nitrogen system from 175° to 475°C," J. Am. Chem. Soc., *85*, 282-286 (1963).
171. R. D. Worley, M. W. Zemansky, and H. A. Boorse, "The vapor-pressure curve of helium between 4.2 and 4.8 K," Phys. Rev., *93*(1), 45-46 (1954).
172. J. L. Yntena and W. G. Schneider, "Compressibility of gases at high temperatures, III. The second virial coefficient of helium in the temperature range 600°C to 1200°C," J. Chem. Phys., *18*(5), 641-646 (1950).
173. J. Zelmanov, "Joule-Thomson effect in helium at low temperatures," J. Phys. USSR, *3*(1), 43-52 (1940).
174. J. Zelmanov, "Specific heat and enthalpy of helium at low temperatures," J. Phys. USSR, *8*(3), 129-134 (1944).
175. H. W. Woolley, R. B. Scott, and F. G. Brickwedde, "Compilation of Thermal Properties of Hydrogen in Its Various Isotopic and Ortho-Para Modifications," J. Res. NBS, *41*, 379 (1948).[†]

[†]Changed in proof.—T.B.S.

RAYMOND H. FOGLER LIBRARY
DATE DUE

BOOKS ARE SUBJECT TO